AF576050

Soil–Water Conservation, Erosion, and Landslide

Soil–Water Conservation, Erosion, and Landslide

Editor

Su-Chin Chen

MDPI • Basel • Beijing • Wuhan • Barcelona • Belgrade • Manchester • Tokyo • Cluj • Tianjin

Editor
Su-Chin Chen
Department of Soil and water conservation
National Chung-Hsing University
Taichung
Taiwan

Editorial Office
MDPI
St. Alban-Anlage 66
4052 Basel, Switzerland

This is a reprint of articles from the Special Issue published online in the open access journal *Water* (ISSN 2073-4441) (available at: www.mdpi.com/journal/water/special_issues/Soil_Conservation).

For citation purposes, cite each article independently as indicated on the article page online and as indicated below:

LastName, A.A.; LastName, B.B.; LastName, C.C. Article Title. *Journal Name* **Year**, *Volume Number*, Page Range.

ISBN 978-3-0365-3432-9 (Hbk)
ISBN 978-3-0365-3431-2 (PDF)

Contents

Editorial

Soil–Water Conservation, Erosion and Landslide

Su-Chin Chen [1,2]

[1] Department of Soil and Water Conservation, National Chung Hsing University, 145 Xingda Road, Taichung 40227, Taiwan; scchen@nchu.edu.tw
[2] Innovation and Development Centre of Sustainable Agriculture (IDCSA), National Chung Hsing University, 145 Xingda Road, Taichung 40227, Taiwan

Citation: Chen, S.-C. Soil–Water Conservation, Erosion and Landslide. *Water* **2022**, *14*, 665. https://doi.org/10.3390/w14040665

Received: 25 November 2021
Accepted: 16 February 2022
Published: 21 February 2022

Publisher's Note: MDPI stays neutral with regard to jurisdictional claims in published maps and institutional affiliations.

In the wake of climate change, extreme storm events, catastrophic disasters (including soil erosion, debris and landslide formation, loss of life, etc.) have surged. These disasters are more common in mountainous regions, and could be a result of tectonic, climatic, and/or human activities [1–3]. Over the past two decades, more than 300 natural disasters occur annually around the globe, affecting over four billion and cost around USD 2.97 trillion [4,5]. The 2021 state of the environment notes that disasters are continuing to take a heavy toll on life and assets, severely affecting and rolling back the development gains of countries [6]. In addition, Mohammed et al. [7] and the sixth Intergovernmental Panel on Climate Change report [8] note with confidence that human-induced climate change is the dominant driver in sediment related natural disasters. In assessing the influence of climate change on soil erosion and sediment yield, Chen et al. [9] illustrate an increase in these events under the A1B-climate change scenario. This study highlights the importance of incorporating climate change in sediment-related disaster models. One of the most important transboundary rivers in China, the Lancang-Mekong River has been shown to cause major sediment loads in the last decade in Asian Rivers with a mean annual loss of 5350 t ha^{-1} $year^{-1}$ [10].

In sight of this, this Special Issue aimed to contribute towards improving our knowledge and understanding on the processes and mechanics of soil erosion and landslides, as these are among the main natural disasters affecting the globe. This is crucial in developing the right tools and models for soil and water conservation, disaster mitigation, and early warning systems. Several novel tools and methodologies are presented in this Special Issue.

Several novel tools and methodologies are presented in this Special Issue, which consists of 19 articles, covering a wide range of topics, including landslide prediction models, soil erosion and sediment yield estimation, flood simulations, dam breach, and rainfall-runoff models.

Wu and Yeh [11] developed an improved landslide probability model from existing models by including long-term landslide inventory and rainfall factors, which can further be used to predict landslides based on future changes in rainfall patterns. In addition to the aforementioned landslide probability model, Wu et al. [12] argue that landslide susceptibility assessments after extreme rainfall events is equally critical. Four methods are evaluated based on 12 landslide related factors which formed the basis for the landslide susceptibility assessment. The methods include Landslide ratio-based logistic regression (LRBLR), Frequency Ration Method (FRM), Instability Index Method (IIM), and Weight of Evidence Method (WEM). Among these, the LRBLR method is shown to be the best in landslide susceptibility assessment. The article by Wu and Lin [13] presents a method applying rainfall analysis, spatiotemporal landslide analysis, and comparison analysis of rainfall induced landslide and earth quakes to evaluate how landslides would be active after specific rainfall events. In classifying landslides, Wan et al. [14] presented a methodology that applied hyperspectral data instead of solely relying on digital elevation models. The model can differentiate bare land to landslides, which has often been a complex task.

In addition to the application of complex models, some authors [15,16] argued that the ability of vegetation in mitigating soil erosion and subsequent large sediment yield under extreme rainfall events and runoff is not well investigated. Different vegetation communities enacted at 1, 11, 15, 25, and 40 years were evaluated. Findings showed root biomass can reach up to ~11 mg/cm^3, which reduces slope runoff velocity by 48%, while increasing runoff resistance by 35 times. The results suggest the importance of indigenous knowledge in reducing the impacts of sediment related disasters. Chen, Guo, and Wang [16] compared farmland with revegetated gullies and demonstrated that revegetating gullies could lower soil erodibility by 31–78% and could further improve critical shear stress by up to four times, and stable conditions were possible in approximately 18 years. The importance of riparian vegetation in stabilizing river banks is illustrated by Zhu et al. [17]. The authors show that healthy native alpine swamp can enhance riverbank stability and could further delay the development of tensile cracks.

The article by Liu et al. [18] describes a novel shallow water equation based methodology that could simulate flood routing in complex terrains. Hung et al. [19] used flume tests to study dam breach and the resulting seismic signals induced. The authors conclude overtopping discharge and lateral sliding masses are significant in influencing the evolution of dam breach. The resulting dam breach model from the study is important for dam breach warning, which could save lives in the event of a catastrophic dam breach associated with floods.

Mosavi et al. [20] presents a novel machine learning model (Weighted sub-space random forest) to map susceptibility of water erosion of the soil. In applying the model, 19 factors are applied (some of these are aspect, curvature, slope length, flow accumulation, normalized vegetative index, soil texture, lithology, etc.). Such a tool is crucial in watershed conservation, especially in a world generating enormous data that requires super-fast models. In another instance, Lee et al. [21] evaluates seven machine learning models for time-saving to estimate the rainfall-erosivity factor (R-factor) used in the Universal Soil Loss Equation (USLE). Their findings show that deep neural networks are very efficient estimating the R-factor given monthly precipitation, maximum daily precipitation, and maximum hourly precipitation.

Finally, the article by Lee, Lu, and Huang [21] investigate the interaction between overfall types and scour at bridges. When a bridge pier is in the maximum scour location, it induces more scour due to disturbances caused by the water jet and the pier; hence, more attention is required to protect the pier. This is especially important in areas prone to earthquakes, which may cause the riverbed to uplift.

In conclusion, the Special Issue presents several articles. They are broadly categorized into five themes: (i) those with emphasis on soil erosion and how climate change has worsened natural disasters; (ii) those investigating and developing landslide related models; (iii) articles regarding flood models and dam breach processes; (iv) articles focusing on the application of recent technologies, such as machine learning in addressing sediment related disasters; and (v) studies dedicated at the protection of riverine structures. These articles have indeed contributed immensely towards understanding soil erosion and landslide processes, and the corresponding necessary tools to foster resilience. This further aligns with some of the post 2015 global frameworks, such as the Sendai Framework which articulates the need for improved understanding of disasters in all its dimensions.

Funding: This research received no external funding.

Institutional Review Board Statement: Not applicable.

Informed Consent Statement: Not applicable.

Conflicts of Interest: The authors declare no conflict of interest.

References

1. Noguchi, K.; Tsou, C.-Y.; Ishikawa, Y.; Higaki, D.; Wu, C.-Y. Tree-Ring Based Chronology of Landslides in the Shirakami Mountains, Japan. *Water* **2021**, *13*, 1185. [CrossRef]
2. Yan, R.-X.; Peng, J.-B.; Zhang, J.-Y.; Wang, S.-k. Static Liquefaction Capacity of Saturated Undisturbed Loess in South Jingyang Platform. *Water* **2020**, *12*, 2298. [CrossRef]
3. Wen, Y.; Gao, P.; Mu, X.; Li, M.; Su, Y.; Wang, H. Experimental Study on Landslides in Terraced Fields in the Chinese Loessial Region under Extreme Rainfall. *Water* **2021**, *13*, 270. [CrossRef]
4. Prasad, A.S.; Francescutti, L.H. Natural Disasters. In *International Encyclopedia of Public Health*, 2nd ed.; Quah, S.R., Ed.; Academic Press: Oxford, UK, 2017; pp. 215–222.
5. United Nations Disaster Risk Reduction. *Human Cost of Disasters—An Overview of the Last 20 Years 2000–2019*; United Nations Disaster Risk Reduction: Geneva, Switzerland , 2021.
6. World Meteorological Organization. *State of Climate in 2021: Extreme Events and Major Impacts*; WMO: Geneva, Switzerland, 2021.
7. Mohammed, S.; Abdo, H.G.; Szabo, S.; Pham, Q.B.; Holb, I.J.; Linh, N.T.T.; Anh, D.T.; Alsafadi, K.; Mokhtar, A.; Kbibo, I.; et al. Estimating Human Impacts on Soil Erosion Considering Different Hillslope Inclinations and Land Uses in the Coastal Region of Syria. *Water* **2020**, *12*, 2786. [CrossRef]
8. IPCC. Summary for Policymakers. In *Climate Change 2021: The Physical Science Basis. Contribution of Working Group I to the Sixthssessment Report of the Intergovernmental Panel on Climate Change*; IPCC: Geneva, Switzerland, 2021.
9. Chen, C.-N.; Tfwala, S.S.; Tsai, C.-H. Climate Change Impacts on Soil Erosion and Sediment Yield in a Watershed. *Water* **2020**, *12*, 2247. [CrossRef]
10. Chuenchum, P.; Xu, M.; Tang, W. Estimation of Soil Erosion and Sediment Yield in the Lancang–Mekong River Using the Modified Revised Universal Soil Loss Equation and GIS Techniques. *Water* **2020**, *12*, 135. [CrossRef]
11. Wu, C.-Y.; Yeh, Y.-C. A Landslide Probability Model Based on a Long-Term Landslide Inventory and Rainfall Factors. *Water* **2020**, *12*, 937. [CrossRef]
12. Wu, C. Landslide Susceptibility Based on Extreme Rainfall-Induced Landslide Inventories and the Following Landslide Evolution. *Water* **2019**, *11*, 2609. [CrossRef]
13. Wu, C.; Lin, C. Spatiotemporal Hotspots and Decadal Evolution of Extreme Rainfall-Induced Landslides: Case Studies in Southern Taiwan. *Water* **2021**, *13*, 2090. [CrossRef]
14. Wan, S.; Lei, T.C.; Ma, H.L.; Cheng, R.W. The Analysis on Similarity of Spectrum Analysis of Landslide and Bareland through Hyper-Spectrum Image Bands. *Water* **2019**, *11*, 2414. [CrossRef]
15. Chang, E.; Li, P.; Li, Z.; Su, Y.; Zhang, Y.; Zhang, J.; Liu, Z.; Li, Z. The Impact of Vegetation Successional Status on Slope Runoff Erosion in the Loess Plateau of China. *Water* **2019**, *11*, 2614. [CrossRef]
16. Chen, Z.; Guo, M.; Wang, W. Variations in Soil Erosion Resistance of Gully Head along a 25-Year Revegetation Age on the Loess Plateau. *Water* **2020**, *12*, 3301. [CrossRef]
17. Zhu, H.; Gao, P.; Li, Z.; Fu, J.; Li, G.; Liu, Y.; Li, X.; Hu, X. Impacts of the Degraded Alpine Swamp Meadow on Tensile Strength of Riverbank: A Case Study of the Upper Yellow River. *Water* **2020**, *12*, 2348. [CrossRef]
18. Liu, D.; Tang, J.; Wang, H.; Cao, Y.; Bazai, N.A.; Chen, H.; Liu, D. A New Method for Wet-Dry Front Treatment in Outburst Flood Simulation. *Water* **2021**, *13*, 221. [CrossRef]
19. Hung, C.-Y.; Tseng, I.-F.; Chen, S.-C.; Feng, Z.-Y. On Dam Failure Induced Seismic Signals Using Laboratory Tests and on Breach Morphology Due to Overtopping by Modeling. *Water* **2021**, *13*, 2757. [CrossRef]
20. Mosavi, A.; Sajedi-Hosseini, F.; Choubin, B.; Taromideh, F.; Rahi, G.; Dineva, A.A. Susceptibility Mapping of Soil Water Erosion Using Machine Learning Models. *Water* **2020**, *12*, 1995. [CrossRef]
21. Lee, W.-L.; Lu, C.-W.; Huang, C.-K. A Study on Interaction between Overfall Types and Scour at Bridge Piers with a Moving-Bed Experiment. *Water* **2021**, *13*, 152. [CrossRef]

Article

Susceptibility Mapping of Soil Water Erosion Using Machine Learning Models

Amirhosein Mosavi [1,2,*], Farzaneh Sajedi-Hosseini [3,*], Bahram Choubin [4], Fereshteh Taromideh [5], Gholamreza Rahi [6] and Adrienn A. Dineva [7]

1 Environmental Quality, Atmospheric Science and Climate Change Research Group, Ton Duc Thang University, Ho Chi Minh City 758307, Vietnam
2 Faculty of Environment and Labour Safety, Ton Duc Thang University, Ho Chi Minh City 758307, Vietnam
3 Department of Reclamation of Arid and Mountainous Regions, Faculty of Natural Resources, University of Tehran, Karaj 31585-77871, Iran
4 Soil Conservation and Watershed Management Research Department, West Azarbaijan Agricultural and Natural Resources Research and Education Center, AREEO, Urmia 57169-63963, Iran; b.choubin@areeo.ac.ir
5 Department of Irrigation, Sari Agricultural Sciences and Natural Resources University, Sari 48181-68984, Iran; ftaromide@stu.sanru.ac.ir
6 Natural Resources and Watershed Management Research Department, Bushehr Agricultural and Natural Resources Research and Education Center, Bushehr 75156-43373, Iran; gh.raghi@areeo.ac.ir
7 Kalman Kando Faculty of Electrical Engineering, Obuda University, 1034 Budapest, Hungary; dineva.adrienn@kvk.uni-obuda.hu
* Correspondence: amirhosein.mosavi@tdtu.edu.vn (A.M.); farzsajedi@ut.ac.ir (F.S.-H.)

Received: 3 May 2020; Accepted: 10 July 2020; Published: 14 July 2020

Abstract: Soil erosion is a serious threat to sustainable agriculture, food production, and environmental security. The advancement of accurate models for soil erosion susceptibility and hazard assessment is of utmost importance for enhancing mitigation policies and laws. This paper proposes novel machine learning (ML) models for the susceptibility mapping of the water erosion of soil. The weighted subspace random forest (WSRF), Gaussian process with a radial basis function kernel (Gaussprradial), and naive Bayes (NB) ML methods were used in the prediction of the soil erosion susceptibility. Data included 227 samples of erosion and non-erosion locations through field surveys to advance models of the spatial distribution using predictive factors. In this study, 19 effective factors of soil erosion were considered. The critical factors were selected using simulated annealing feature selection (SAFS). The critical factors included aspect, curvature, slope length, flow accumulation, rainfall erosivity factor, distance from the stream, drainage density, fault density, normalized difference vegetation index (NDVI), hydrologic soil group, soil texture, and lithology. The dataset cells of samples (70% for training and 30% for testing) were randomly prepared to assess the robustness of the different models. The functional relevance between soil erosion and effective factors was computed using the ML models. The ML models were evaluated using different metrics, including accuracy, the kappa coefficient, and the probability of detection (POD). The accuracies of the WSRF, Gaussprradial, and NB methods were 0.91, 0.88, and 0.85, respectively, for the testing data; 0.82, 0.76, and 0.71, respectively, for the kappa coefficient; and 0.94, 0.94, and 0.94, respectively, for POD. However, the ML models, especially the WSRF, had an acceptable performance regarding producing soil erosion susceptibility maps. Maps produced with the most robust models can be a useful tool for sustainable management, watershed conservation, and the reduction of soil and water loss.

Keywords: water erosion; susceptibility; Gaussian process; climate change; radial basis function kernel; weighted subspace random forest; extreme events; extreme weather; naive Bayes; feature selection; machine learning; hydrologic model; simulated annealing; earth system science

1. Introduction

Soil conservation is of utmost importance for sustainable development, food security, and environmental protection [1]. Understanding soil erosion is considered to be an essential practice for soil conservation programs around the world [2]. Currently, soil erosion has increasingly become known as a severe concern for sustainable agriculture, water resource management, and modern civilization [3]. Soil erosion is a significant menace for soil, ecology, and for humanity since the long-term production of soil productive capacity is profoundly affected by the destruction and leaching of soil's organic and topsoil matters [4]. Soil erosion is an intricate process that depends on the plant cover and land use, watershed topography, soil properties, climate, and land management practices. In the last century, soil erosion has intensified due to human activity and is an environmental problem [5]. Primary soil segregates when the rainfall or water flow power is greater than the soil's resistance to corrosion [6]. Generally, there are different types of water erosion, such as sheet, gully, landslide, debris flow, streambank, etc. [7].

In semiarid regions, such as Iran, soil erosion is a significant crisis [8] and can be considered to be one of the critical problems concerning agricultural development, natural resources, and the environment [9]. In such regions, water is limited, and there are many sources of sediment [10]. The high input of sediment in upstream rivers increases the water turbidity, reduces the lifespan of dams owing to reservoir siltation, and negatively affects water quality and biological activity [8]. According to scholars, the mean annual rate of soil erosion in Iran is about 25 tons/ha/year, which is four times more than the mean yearly rate around the world [11,12]. Therefore, the susceptibility mapping of soil erosion is necessary for controlling this critical problem.

Rather than using traditional and experimental models, such as the universal soil loss equation (USLE) [13] and multi-criteria decision-making methods [14], that have been used in water erosion assessments, machine learning (ML) models are known to be successful methods [15,16]. Different ML methods, such as support vector machine (SVM), boosted regression trees (BRT), random forest (RF), naive Bayes (NB), and artificial neural network (ANN), have been used for landslides [17–23], debris flows [24–26], and gully erosion [27–30]. For instance, Angileri et al. [15] used the stochastic gradient tree boost (SGT) for water erosion susceptibility mapping in central-northern Sicily, Italy. The results indicated that the applied model had excellent reliability (accuracy from 0.87 to 0.92). Recently, Garosi et al. [31] applied the RF, SVM, and NB models, along with the generalized additive model (GAM), to predict the gully erosion susceptibility in the Ekbatan Dam drainage basin, Iran. The results indicated that the RF model had the highest performance (accuracy = 92.4%) among the models tested. Svoray et al. [16] used different ML models, namely, SVM, ANN, and decision trees (DT), for predicting the gully erosion in a watershed scale in Israel and compared them with the results from topographic threshold (TT) and analytic hierarchy process (AHP) methods. The results indicated that the ML models produced better performances than the AHP and TT methods. Mao et al. [32] evaluated the soil erosion in the Shiqiaopu catchment, Hubei province, China, using SVM and ANN models. They optimized the parameters of the SVM using the particle swarm optimization (PSO) algorithm. The results indicated that the SVM had higher accuracy in comparison with the ANN model. Rahmati et al. [28] compared the ML models of SVM, ANN, RF, and BRT when predicting the gully erosion susceptibility in the Kashkan watershed, Iran. The results indicated that the performance of the RF and SVM models for predicting the gully occurrences in the watershed were better than the other models.

Due to the advancement of ML models, applying and evaluating novel methods in water erosion studies can help to accurately predict hazardous areas, especially in developing countries where soil erosion data are incomplete. The current study tried to predict water erosion susceptibility using two novel ML models, namely, a weighted subspace random forest (WSRF) and a Gaussian process with a radial basis function kernel (Gaussprradial), for the first time and compared their results with the NB model. Therefore, the primary purposes of this study were: (i) to identify the more significant factors regarding soil erosion through feature selection, (ii) to compare the performance of the novel predictive models (i.e., WSRF and Gaussprradial) with a model previously used for this application (i.e., NB), and (iii) the prediction of the spatial susceptibility of soil erosion induced by water.

2. Materials and Methods

2.1. Study Area

The Nur-Rood watershed is located in the southwest of the Haraz watershed, in the north of Iran. The watershed lies within 51°26′–52°19′ E and 36°01′–36°16′ N (Figure 1). The elevation of the watershed ranges from 732 to 4333 m. There are six rain gauge stations in the region provided the long-term mean annual data from 1976 to 2016 that were were used in this study. The study area is about 1297 km^2 and is located upstream of the Haraz dam. The main application of this dam is to provide drinking and agriculture water for five cities (i.e., Amol, Babol, Babolsar, Nur, and Mahmoodabad) in the Mazandaran province. According to the literature, the watershed generates water with a high sediment load such that it causes a reduction in the dam's capacity [8,33]. Therefore, identifying the hazardous areas can help to control the upstream erosion and aid with providing sustainable watershed management in the Nur-Rood watershed.

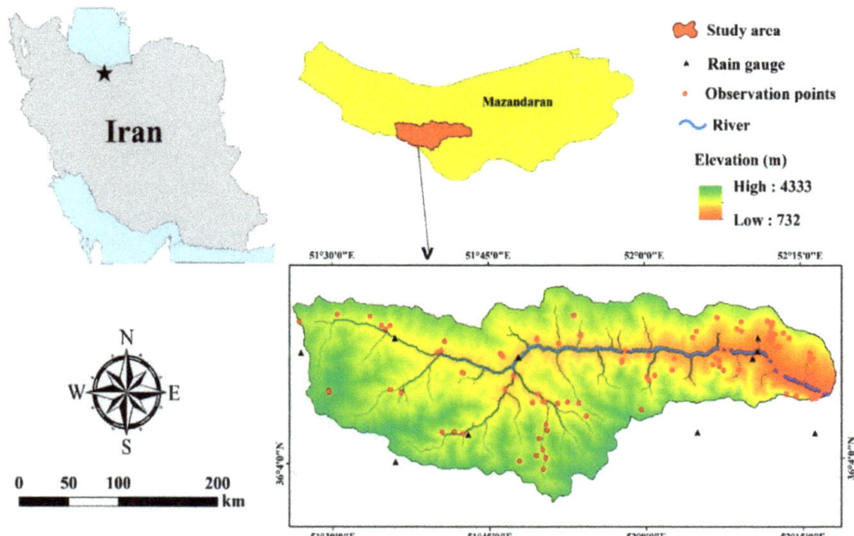

Figure 1. Location of the Nur-Rood watershed, Mazandaran province, Iran.

2.2. Methodology

The methodology consisted of several fundamental building blocks to ensure the accuracy of the susceptibility prediction. Figure 2 presents the schematic of the methodology workflow from data sampling to the susceptibility prediction. The method consisted of five sections: (i) preparation and collection of the relevant factors for soil erosion modeling; (ii) extraction of the erosion and non-erosion locations by the field observations; (iii) selection of the essential factors using the simulated annealing feature selection (SAFS) algorithm; (iv) water erosion modeling using the Gaussprradial, NB, and WSRF models in the Nur-Rood watershed; and (v) evaluating the models' performance.

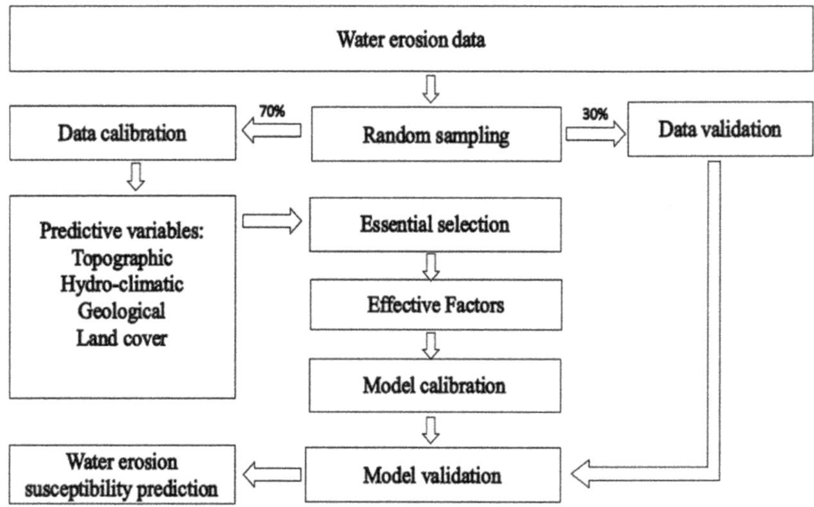

Figure 2. Schematic representation of the proposed method.

2.2.1. Field Data

It was necessary to know the locations of eroded and non-eroded areas for susceptibility mapping of the Nur-Rood watershed. Therefore, the locations (i.e., x and y coordinates) of 227 area (116 erosion locations and 111 non-erosion locations) were sampled through field surveys to model the water erosion susceptibility based on a binary scale (occurrence/non-occurrence). According to Sajedi-Hosseini et al. [8], the recorded soil erosion areas include different kinds of water erosions (such as sheet, rill, gully, and mass movements).

2.2.2. Predictive Variables

In this study, according to the literature review, 19 relevant factors regarding soil erosion were collected and prepared, including the topographic, hydro-climatic, geological, and land cover factors. The attributes of the factors are presented in Table 1. A brief description of each of the predictive factors is presented afterward.

Table 1. Characteristics of the considered factors for susceptibility mapping of water erosion.

Factors	Range/Class
Topographic factors:	
Elevation	732 to 4333 (m)
Slope	0 to 473.8 (%)
Aspect	Flat, north, northeast, east, southeast, south, southwest, west, northwest
Slope length (SL)	0 to 5613 (m)
Curvature	−35 to 25
Hydro-climate factors:	
Drainage density (DD)	0 to 3 (km/km^2)
Distance from stream (DFS)	0 to 5135 (m)
Topographic wetness index (TWI)	6 to 21.3
Stream power index (SPI)	0 to 150,768
Flow accumulation (FA)	0 to 1,630,717 (pixel)
Precipitation (PCP)	0 to 768 (mm)
Rainfall erosivity factor (R)	272 to 2078
Hydrologic soil group (HSG)	B [1], C [2], D [3]

Table 1. *Cont.*

Factors	Range/Class
	Geological factors [4]:
Fault density (FD)	0 to 2.4 (km/km^2)
Lithology	TRJs, Pr, Mm.s.l, Pd, Odi, Tre, PZ2bvt, Tre1, Qs.D, Ebv, Tra.bv, Jl, Ek, K1bvt, Ktzl, Pldv, Jk, K2l2, Eksh
Soil texture	Sandy loam, loamy sand, loam, clay loam, sandy clay loam, clay
	Land-cover factors:
Normalized difference vegetation index (NDVI)	−0.07 to 0.63
Land use	Rangeland, residential, forest, agriculture, rock
Distance from road (DFR)	0 to 18,978 (m)

[1] Silt loam types of soils with a moderate infiltration rate. [2] Sandy clay loam types of soils with low infiltration rates. [3] Clay loam, silty clay loam, sandy clay, silty clay, or clay with the highest runoff potential. [4] Definition of the geological factors include; TRJs: Dark grey shale and sandstone; Pr: Dark grey medium - bedded to massive limestone; Mm.s.l: Marl, calcareous sandstone, sandy limestone and minor conglomerate; Pd: Red sandstone and shale with subordinate sandy limestone: Odi: Diorite; Tre: Thick bedded grey o'olitic limestone; PZ2bvt: Basaltic volcanic tuff; Tre1: Thin bedded, yellow to pinkish argillaceous limestone with worm tracks; Qs.D: Unconsolidated wind-blown sand deposit including sand dunes; Ebv: Basaltic volcanic rocks; Tra.bv: Triassic, andesitic and basaltic volcanics; Jl: Light grey, thin-bedded to massive limestone; Ek: Well bedded green tuff and tuffaceous shale; K1bvt: Basaltic volcanic tuff; Ktzl: Thick bedded to massive, white to pinkish orbitolina bearing limestone; Pldv: Rhyolitic to rhyodacitic volcanics; Jk: Conglomerate, sandstone and shale with plantremains and coal seams; K2l2: Thick-bedded to massive limestone; Eksh: Greenish-black shale, partly tuffaceous with intercalations of tuff.

Topographic Parameters

The topographic parameters included the elevation, slope, aspect, slope length (SL), and curvature (Figure 3). These factors are influential regarding soil erosion velocity [34]. The different elevations (Figure 3a), aspects (Figure 3c), and curvature (Figure 3e) cause different conditions of evaporation, soil temperature, soil moisture, and solar radiation, which have different effects on the soil erosion. Furthermore, slope (Figure 3b) and SL (Figure 3d) affect the runoff velocity and volume, where a steeper slope or a longer SL can increase the soil erosion by water [8].

The topographic factors were produced using a digital elevation model (DEM) with a cell size of 30 m in the ArcGIS 10 software (Environmental Systems Research Institute, Redlands, CA, USA).

Hydro-Climate Factors

The hydro-climate factors included the drainage density (DD), distance from the stream (DFS), topographic wetness index (TWI), stream power index (SPI), flow accumulation (FA), precipitation (PCP), rainfall erosivity factor (R), and hydrologic soil group (HSG) (Figure 4). The DD (Figure 4a) is calculated from the sum of the length of all streams in the watershed area. The DD values depend on the permeability and resistance of the surface and deeper soil layers that affect water erosion [8]. Regarding the DFS (Figure 4b), the regions near streams are more susceptible to soil erosion [35]. The DD and DFS layers were created using line density and Euclidian distance tools, respectively, in geographic information system (GIS). TWI (Figure 4c) shows the soil moisture and water-saturated area of the watershed. SPI (Figure 4d) indicates the potential for erosion due to the water flow, in which higher values indicate a higher potential. TWI and SPI were produced using the SAGA GIS 2.0.7 software (SAGA User Group Association, Hamburg, Germany). The flow accumulation (FA) function (Figure 4e) computes the sum of the weight of all accumulated pixels upstream [36], which is most important for showing the water-accumulated pixels that affect the water erosion. The PCP (Figure 4f) and R (Figure 4g) were the climate factors considered to affect soil erosion. Their effects depend on soil attributes such as the soil texture, soil organic matter, and soil structure. The PCP map is produced by the mean annual precipitation of the gauge stations in the study area. The R factor is directly related to the soil erodibility. The best method for calculating it is a direct measurement of soil erosion in plots [37]. However, in this study, according to Takal et al. [38], an empirical equation was used to calculate this factor, as follows.

$$R = 0.0483P^{1.61}, \quad (1)$$

where R is the precipitation erosivity index (MJ·mm·ha^{-1}·hr^{-1}) and P is the mean annual precipitation (mm).

The HSG (Figure 4h) indicates the infiltration and runoff generation rates that affect soil erosion. This layer is extracted from the digital soil map of the world [39] and it includes three groups: B, C, and D. Group B has moderately low runoff potential when completely humid. Soils in this group have 50 to 90% sand, 10 to 20% clay, and have sandy loam or loamy sand textures. Water transition across the soil is unrestricted. Group C soils have moderately high runoff potential when completely humid. They have less than 50% sand, 20 to 40% clay, and include sandy clay loam, silty clay loam, loam, silt loam, and clay loam textures. Group D soils have high runoff potential and the infiltration across the soil is very limited [40].

Geological Factors

Geological factors include the fault density (FD), lithology, and soil texture (Figure 5). The FD affects infiltration and runoff, which can affect soil erosion. Furthermore, the existence of a fault can accelerate the mass movements [41]. The layer of FD (Figure 5a) was produced in the ArcGIS environment by using the line density tool on the fault layer. The lithology has the greatest effect on erosion control. Erosion depends on the exposed material weathering attributes or the lithology [42,43]. The lithology map (Figure 5b) was taken from a geological survey done by the Iranian department of environment and had a scale of 1:100,000. The other important factor is soil texture (Figure 5c). Porosity and soil texture, along with the soil profile and surface, are the dominant soil attributes that influence soil erosion. An increase in the clay value of the soil causes a decrease in soil erosion [44]. The soil textures of the study area were clay, clay loam, loam, loamy sand, sandy clay loam, and sandy loam (Figure 5c).

Land Cover Factors

The land cover factors considered were the normalized difference vegetation index (NDVI), land-use, and distance from road (DFR) (Figure 6). The NDVI (Figure 6a) was extracted from Landsat satellite images for June 2018. The NDVI values range from −1 to 1 [45]. A watershed with a higher NDVI provides higher resistance against soil erosion [9,46]. The land uses of the study area included rangeland, residential, forest, agriculture, and rock (Figure 6b). The land-use map was received from the Iranian Water Resources Management Company (IWRMC). Roads are one of the man-made features that increase the availability of materials for transformation and increase the sediment yield in the watershed. Moreover, roads increase the runoff speed through collecting and concentrating the surface runoff in the given areas (such as near bridges); therefore, faster flows increase the erosion. The DFR layer (Figure 6c) was calculated using the line density tool within the ArcGIS environment.

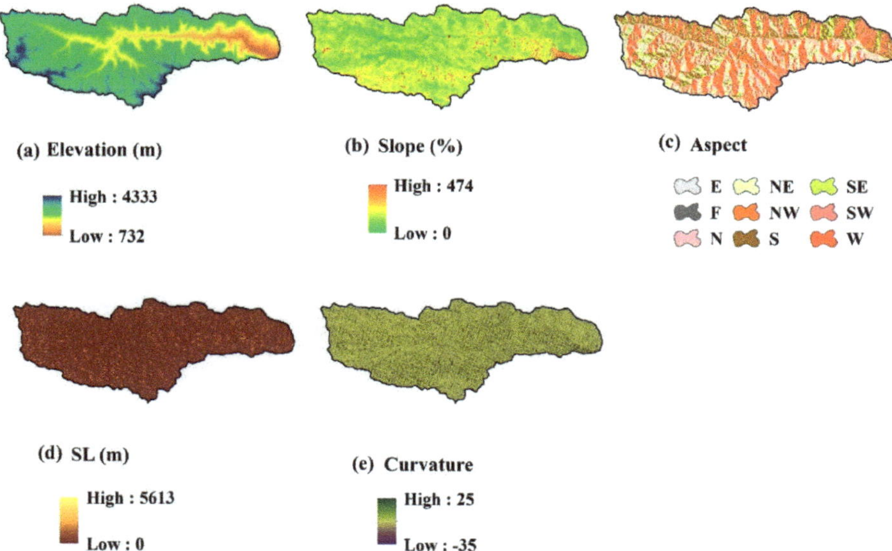

Figure 3. Topographic factors: (**a**) elevation, (**b**) slope, (**c**) aspect, (**d**) slope length (SL), and (**e**) curvature.

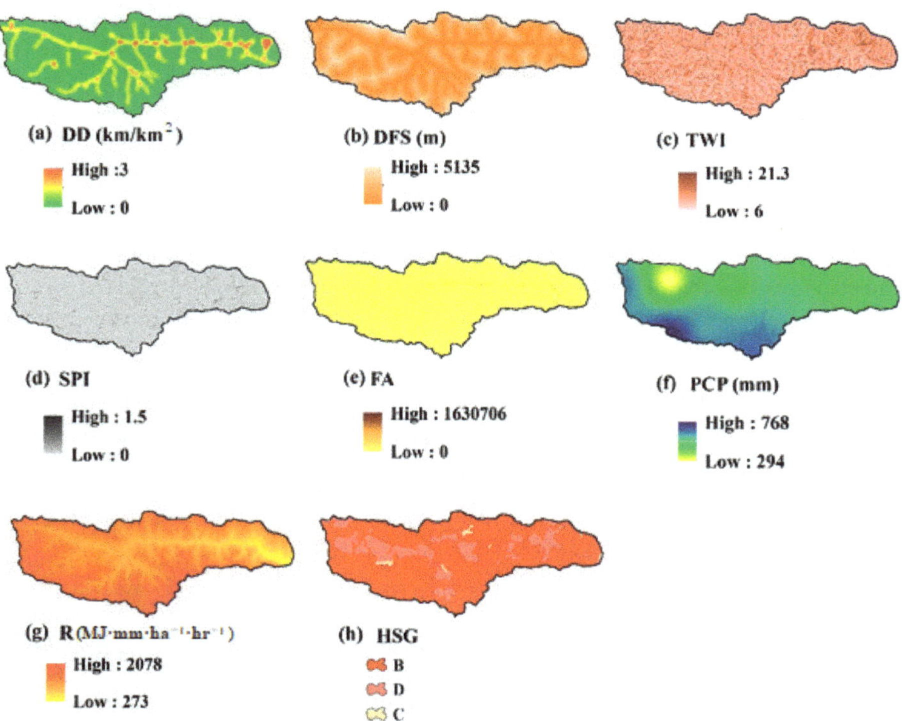

Figure 4. Hydro-climate factors: (**a**) drainage density (DD), (**b**) distance from the stream (DFS), (**c**) topographic wetness index (TWI), (**d**) stream power index (SPI), (**e**) flow accumulation (FA), (**f**) precipitation (PCP), (**g**) rainfall erosivity factor (R), and (**h**) hydrologic soil group (HSG).

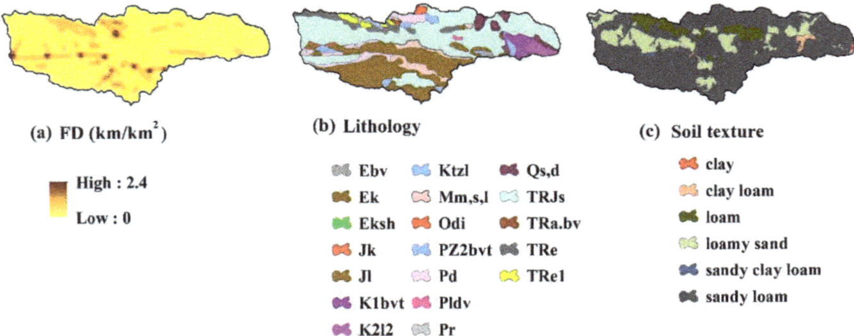

Figure 5. Geological factors: (**a**) fault density (FD), (**b**) lithology, and (**c**) soil texture.

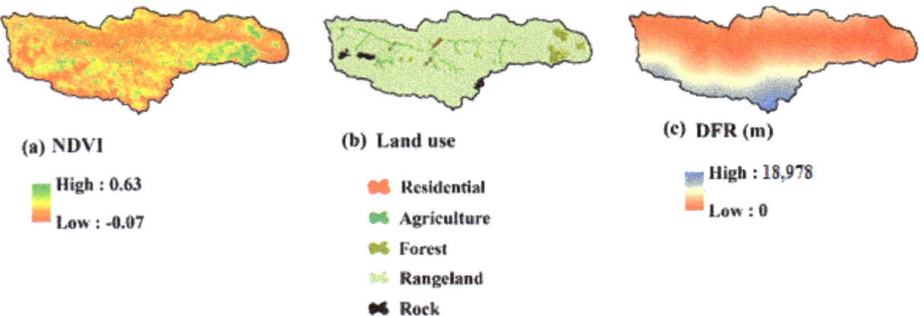

Figure 6. Land cover factors: (**a**) normalized difference vegetation index (NDVI), (**b**) land use, and (**c**) distance from road (DFR).

2.2.3. Feature Selection

To select the most important factors in the water erosion of soil based on parsimonious objectives from the large number of factors considered, the simulated annealing feature selection (SAFS) model was used. The SAFS method is based on the minimum energy configuration theory, whereby a solid is gradually cooled such that its structure is frozen [47]. Many studies have used this method for feature selection in environmental fields, such as flash-flood hazard assessment [48], dust and air quality evaluation [49], and earth fissure hazard prediction [50]; see Bertsimas and Tsitsiklis [47] for more details of the SAFS method.

In the current research, the SAFS was conducted using the k-fold (k = 10) cross-validation methodology and it was implemented in the Caret package [51] of the R software (4.0.2, R Core Team, Vienna, Austria).

2.2.4. Weighted Subspace Random Forest (WSRF)

Xu et al. [52] suggested a new random forest, namely, the WSRF model, which involves weighting the input variables and afterward opting for the variables that ensure each subspace always includes informative attributes. The WSRF model is implemented as multi-thread processes. This algorithm categorizes very high-dimensional data and sparse data with random forests made using small subspaces. A new variable weighting manner is applied for the variable subspace choice rather than the traditional random variable sampling in the random forest model [53]. More details of the WSRF model are presented in Xu et al. [52] and Zhao et al. [53]. The WSRF model was implemented using the "wsrf" package [53] in the R software using the k-fold (k = 10) cross-validation procedure.

2.2.5. Naive Bayes (NB)

The NB classifiers are a set of assortment algorithms that use Bayes' Theorem. This is a family of algorithms, where every pair of features being categorized is independent of each other; on the other hand, all of them share a common principle. The dataset is categorized into two sections:

- A response vector, which includes the value of the class variable.
- The feature matrix, which includes all the rows of the dataset and each row contains all the dependent features.

According to the primary naive Bayes hypothesis, each element must be independent and equal [54,55]; see Webb et al. [56] for more details of the NB model. The NB model was done using the k-fold (k = 10) cross-validation method in the "klar" package [57,58] within the R software.

2.2.6. The Gaussian Process with a Radial Basis Function Kernel (Gaussprradial)

Gaussian process regression is a vigorous, non-parametric Bayesian method used for solving regression problems and modeling unknown functions [59,60]. It can capture the different relationships between inputs and output variables by applying a hypothetically infinite number of parameters and allowing the dataset to determine the level of complexity via Bayesian inference [61]. The Gaussian process is parametrized using a kernel. One of the benefits of Gaussian process regression is the flexibility in choosing the kernel; furthermore, the different kernels can be combined to perform the regression [59]. In this study, the radial basis function network (RBF) was used to perform the Gaussian process. The Gaussprradial was performed in the R software using the "kernlab" package [62] using the k-fold (k = 10) cross-validation approach.

2.2.7. Model Calibration and Validation

The database, including the predictand and predictors, was randomly divided into the training (70%) and testing (30%) sets. A k-fold (k = 10) cross-validation methodology was used to calibrate the models. The models were assessed using testing datasets after the calibration using the features selected by the SAFS. Here, for the assessment of the models' performances, three classification evaluation metrics were used: accuracy, kappa, and the probability of detection (POD). The models' performances were represented as accuracy percentages. Kappa indicates the probability of agreement by chance using the likelihood of the model classification [63]. The metrics are computed as follows:

$$\text{Accuracy} = \frac{H + CN}{H + FA + M + CN}, \tag{2}$$

where H (the number of hits), FA (the number of false alarms), M (the number of misses), and CN (the number of correct negatives) were computed from a contingency table.

$$\text{Kappa} = \frac{\text{Accuracy} - Pe}{1 - Pe}, \tag{3}$$

where Pe is the expected probability of chance agreement [64] that is computed using Equation (4):

$$Pe = \frac{(H + FA)(H + M) + (M + CN)(FA + CN)}{(H + FA + M + CN)^2} \tag{4}$$

The POD is a metric used to quantify the possibility of finding a specific detect. The POD is significantly linked to the subject of risk evaluation and probabilistic analyses of the components' integrity. The POD is the ratio of the correct predicted data to the total number observed occurrences. It ranges from 0 to 1, where 1 indicates a perfect score [49,50]. The metric is calculated using Equation (5):

$$\text{POD} = \frac{H}{H + M}. \tag{5}$$

3. Results and Discussion

3.1. Feature Selection Results

A relatively large number of factors, such as elevation, slope, aspect, SL, curvature, DD, DFS, TWI, SPI, FA, PCP, R, HSG, FD, lithology, soil texture, NDVI, land use, and DFR, were used in the current study to predict water erosion. The results of the feature selection using the SAFS algorithm are shown in Table 2. As can be seen, the minimum and maximum selected features were 8 and 14 variables, respectively, in the folds number of 8 (accuracy = 0.84, Kappa = 0.67) and 3 (accuracy = 0.92, Kappa = 0.83). The fold number 6 provided the worst performance (accuracy = 0.74, Kappa = 0.48), whereas the fold number 10 provided the best performance (accuracy = 0.93, Kappa = 0.87).

Table 2. Selected factors in each fold using the simulated annealing feature selection (SAFS) method.

Fold	Number of Selected Features	Selected Features	Accuracy	Kappa
1	10	Aspect, elevation, DFR, FA, lithology, HSG, NDVI, R, SL, soil texture	0.85	0.69
2	9	Aspect, DF, DFS, FA, lithology, NDVI, PCP, slope, TWI	0.75	0.49
3	14	DD, DF, DFR, DFS, FA, lithology, HSG, NDVI, R, PCP, slope, TWI, soil texture, SL	0.92	0.83
4	10	aspect, curvature, DD, DF, DFS, lithology, NDVI, SL, SPI, soil texture	0.91	0.82
5	9	Curvature, elevation, DF, lithology, HSG, NDVI, R, SL, SPI	0.86	0.72
6	9	Curvature, aspect, DD, DFR, FA, HSG, land use, NDVI, R, SL, slope, soil texture	0.74	0.48
7	9	DFR, DFS, FA, lithology, HSG, land use, NDVI, R, soil texture	0.90	0.81
8	8	DF, DFS, FA, NDVI, R, PCP, SL, TWI	0.84	0.67
9	13	aspect, curvature, DD, DF, DFS, lithology, land use, NDVI, R, SL, slope, soil texture, TWI	0.89	0.78
10	11	Aspect, DD, DF, FA, lithology, land use, NDVI, R, SL, SPI, soil texture	0.93	0.87
Average	10.2	-	0.86	0.72

According to the 10-fold results, the selected factors should be between the minimum and maximum selected features and should be mostly equal to the mean selected factors across all folds. However, the percentage of selected factors in all folds can be a good criterion for selecting the final variables [48–50]. Figure 7 shows the percentage of selected factors in all folds. Twelve variables had a frequency of at least 50% across all folds. As can be seen, the NDVI with a 100% frequency (F) was selected in all folds. However, the important role of the vegetation and NDVI is obvious and shown in previous studies [8,9,46]. Followed the NDVI (F = 100%), the variables of lithology (F = 80%), R (F = 80%), SL (F = 80%), FD (F = 70%), FA (F = 70%), soil texture (F = 60%), DFS (F = 60%), aspect (F = 50%), curvature (F = 50%), HSG (F = 50%), and DD (F = 50%) were selected.

Although the feature selection has largely not been done in studies on the water erosion of soil, the importance of these selected variables in the water erosion of soil is demonstrated by previous studies, such as those of De Baets et al. [46], Md. Rejaur et al. [9], Sajedi-Hosseini et al. [8], Di Stefano et al. [42], Arabameri et al. [43], Lin et al. [37], Choubin et al. [41], Auzet et al. [44], and Nekhay et al. [35].

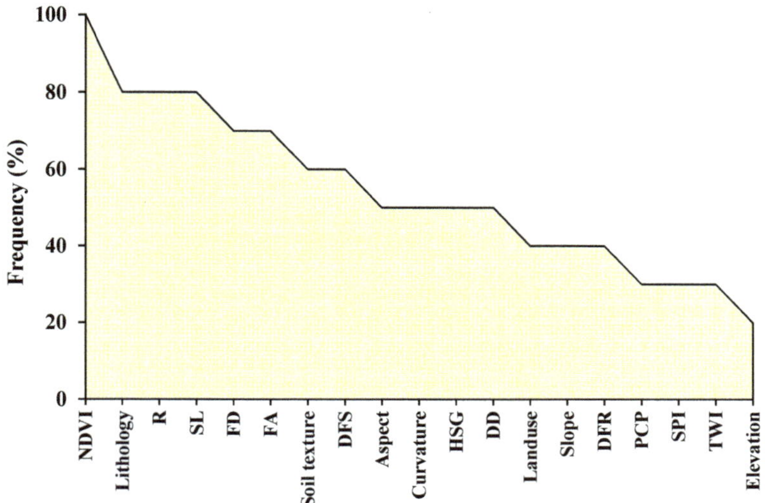

Figure 7. The percentage of selected factors in all folds.

3.2. *Results of Water Erosion Modeling*

The calibration of models was conducted using the "tunelength" function in the Caret R package [51]. The performance results of the three models (WSRF, Gaussprradial, and NB) were evaluated using the three statistics of accuracy, kappa, and the probability of detection (POD), which are presented in Table 3.

Table 3. The performances of the models using the testing dataset.

Statistic	WSRF	Gaussprradial	NB
Accuracy	0.91	0.88	0.85
Kappa	0.82	0.76	0.71
POD	0.94	0.94	0.94

WSRF: Weighted Subspace Random Forest, NB: Naive Bayes.

As can be seen from Table 3, the evaluation of the models' performance indicated that the WSRF model had a higher accuracy (accuracy = 0.91), followed by the Gaussprradial (accuracy = 0.88) and NB (accuracy = 0.85) models. According to Monserud and Leemans [65], the kappa values indicated that all three models were in the "very good" degree of agreement (i.e., 0.70 < Kappa < 0.85) (Table 3). However, like the accuracy, the kappa statistic for the WSRF model (Kappa = 0.82) was more than the Gaussprradial (Kappa = 0.76) and NB (Kappa = 0.71) models. Regarding the POD, the WSRF, Gaussprradial, and NB models showed an equal performance (POD = 0.94) (Table 3).

Generally, the evaluation of the applied machine learning (ML) models in this study indicated an acceptable performance for all the ML models. However, regarding the accuracy and kappa values, the models' performances were ranked as follows: WSRF > Gaussprradial > NB. A direct comparison between the results of this study and previous ones is not possible because the application of the WSRF and Gaussprradial models was undertaken for the water erosion of soil for the first time. However, two novel ML models (WSRF and Gaussprradial) applied in this study indicated a better performance than the NB model that has previously been used in this field. Previous studies have indicated the accurate performance of the NB model in the assessment of soil erosion, such as Weihua et al. [66] and Nhu et al. [67].

3.3. Spatial Prediction of Water Erosion Susceptibility

After the calibration and validation of the models, the maps of the soil erosion probability were predicted using the values of the pixels throughout the study area. Then, the probability maps were classified into five susceptibility classes of very low, low, medium, high, and very high based on the classification method of natural breaks through the ArcGIS software (Figure 8).

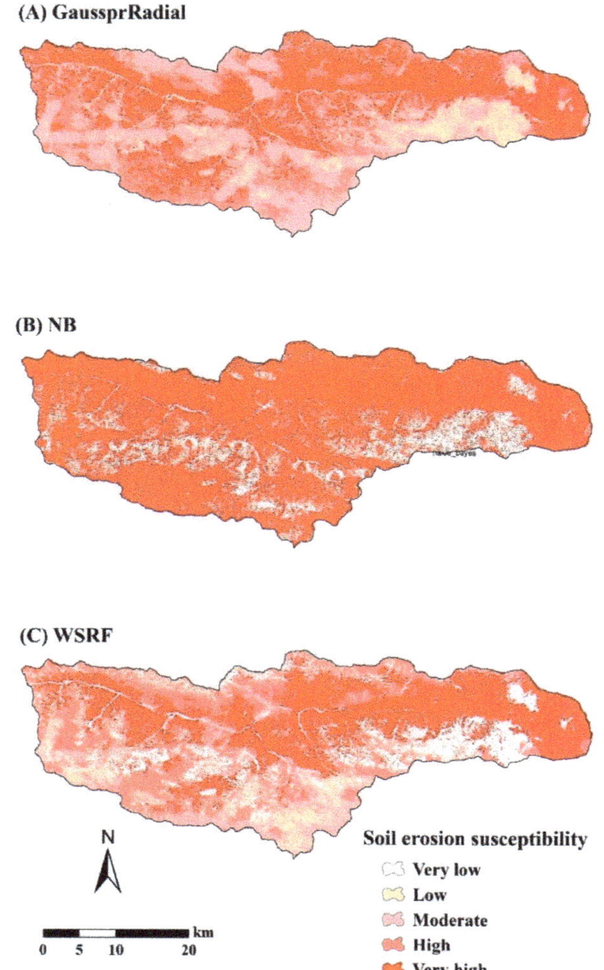

Figure 8. Spatial prediction of water erosion using various methods: (**A**) Gaussprradial, (**B**) NB, and (**C**) WSRF.

The area of the susceptibility classes found using each model is presented in Table 4. As can be seen, the Gaussprradial model predicted most of the area in the moderate class (about 450 km^2, 34.69% of the study area). The sum of the areas for low and very low classes was less than 7% (about 85 km^2) (Figure 8A). According to the NB model, more than 65% of the study area (about 850 km^2) was located in very high susceptibility zones (Figure 8B). Results of the WSRF model indicated that the classes of the very low, low, moderate, high, and very high susceptibilities covered 11.91% (154.52 km^2),

9.76% (126.6 km^2), 20.25% (262.64 km^2), 28.66% (371.87 km^2), and 29.42% (381.7 km^2) of the study area, respectively (Figure 8C).

Table 4. Area of the susceptibility classes found using each model.

Susceptibility Class	Gaussprradial		NB		WSRF	
	Area (km^2)	Area (%)	Area (km^2)	Area (%)	Area (km^2)	Area (%)
Very low	0.08	0.01	115.73	8.92	154.52	11.91
Low	84.87	6.54	75.49	5.82	126.60	9.76
Moderate	450.02	34.69	93.00	7.17	262.64	20.25
High	386.76	29.81	163.39	12.59	371.87	28.66
Very high	375.60	28.95	849.72	65.50	381.70	29.42

Although the predicted models indicated different areas for each class, there was something in common for all predicted maps. By comparing the predicated maps (Figure 8) with the NDVI map (Figure 5a), it was clear that the susceptibility maps were approximately matched with the NDVI and lithology maps. For example, the green areas in the east of the region on the NDVI map (Figure 6a) had higher values of NDVI, which correspond to lower values on the water erosion susceptibility maps (Figure 7). Furthermore, the higher susceptibly values (Figure 8) corresponded to the TRJ lithology (Figure 5b). TRJs include dark grey shale, claystone, siltstone, and sandstone of the Shemshak formation. In this formation, various kinds of water erosion, such as rill, riverbank, gully, and badland erosions can be seen. This agrees with the SAFS results, which indicated that the NDVI and lithology were the most important variables during the feature selection.

4. Conclusions

This study focused on the probability of water erosion occurring in the Nur-Rood watershed. Using the SAFS model, the most important factors were selected among nineteen parameters, namely, NDVI, lithology, R, SL, FD, FA, soil texture, DFS, aspect, curvature, HSG, and DD. Based on the performance analysis of the machine learning (ML) models, the two novel applied ML models of WSRF (accuracy = 0.91, Kappa = 0.82) and Gaussprradial (accuracy = 0.88, Kappa = 0.76) displayed better performances than the NB (accuracy = 0.85, Kappa= 0.71) model that has previously been used in this field. The predicted maps created using the ML models indicated the different areas for each susceptibility class but it was obvious that the susceptibility maps were approximately matched with the NDVI and lithology maps (which were identified as the most important variables). One of the main limitations in this study that also occurs in other spatial modeling studies is that different scales are used for the input variables all over the world. Although all of the input variables were resampled into the same spatial resolution, the data collection and sampling of them were not on the same scale; this is an inevitable limitation for the time being. It may be the case that the data availability of the NDVI (30 m resolution) helped this variable to be the most important variable during the soil erosion modeling compared with the other variables (such as the soil dataset and lithology with scales of 1:100,000 or more). Despite these limitations, producing the water erosion susceptibility maps in developing countries can be a useful tool for sustainable management, the conservation of watersheds, the reduction of soil degradation, and alleviating water quality decline.

Author Contributions: Conceptualization, A.M.; data curation, F.S.-H. and B.C.; formal analysis, F.S.-H., B.C., and G.R.; investigation, F.S.-H., B.C., F.T., and G.R.; methodology, F.S.-H. and B.C.; project administration, F.S.-H. and F.T.; resources, F.T. and G.R.; software, B.C. and G.R.; supervision, A.M. and A.A.D.; validation, F.T.; visualization, G.R. and A.A.D.; writing—original draft, A.M.; writing—review and editing, F.T. All authors have read and agreed to the published version of the manuscript.

Funding: This research received no external funding.

Conflicts of Interest: The authors decorate no conflict of interest.

References

1. Morgan, R.P.C. *Soil Erosion and Conservation*; John Wiley & Sons: Hoboken, NJ, USA, 2005; ISBN 9788578110796.
2. Vanmaercke, M.; Poesen, J.; Verstraeten, G.; de Vente, J.; Ocakoglu, F. Sediment yield in Europe: Spatial patterns and scale dependency. *Geomorphology* **2011**, *130*, 142–161. [CrossRef]
3. Panagos, P.; Borrelli, P.; Meusburger, K.; Alewell, C.; Lugato, E.; Montanarella, L. Estimating the soil erosion cover-management factor at the European scale. *Land Use Policy* **2015**, *48*, 38–50. [CrossRef]
4. García-Ruiz, J.M.; Beguería, S.; Lana-Renault, N.; Nadal-Romero, E.; Cerdà, A. Ongoing and Emerging Questions in Water Erosion Studies. *Land Degrad. Dev.* **2017**, *28*, 5–21. [CrossRef]
5. Scott, A. *Water Erosion in the Murray-Darling Basin: Learning from the Past*; Elsevier: New York, NY, USA, 2001.
6. Sharma, A.; Tiwari, K.N.; Bhadoria, P.B.S. Effect of land use land cover change on soil erosion potential in an agricultural watershed. *Environ. Monit. Assess.* **2011**, *173*, 789–801. [CrossRef]
7. Alkharabsheh, M.M.; Alexandridis, T.K.; Bilas, G.; Misopolinos, N.; Silleos, N. Impact of Land Cover Change on Soil Erosion Hazard in Northern Jordan Using Remote Sensing and GIS. *Procedia Environ. Sci.* **2013**, *19*, 912–921. [CrossRef]
8. Sajedi-Hosseini, F.; Choubin, B.; Solaimani, K.; Cerdà, A.; Kavian, A. Spatial prediction of soil erosion susceptibility using a fuzzy analytical network process: Application of the fuzzy decision making trial and evaluation laboratory approach. *Land Degrad. Dev.* **2018**, *29*, 3092–3103. [CrossRef]
9. Md. Rejaur, R.; Shi, Z.H.; Chongfa, C. Land use/land cover change analysis using geo- information technology: Two case studies in Bangladesh and China. *Int. J. Geoinformatics* **2009**, *5*, 25–37.
10. Vaezi, A.R.; Abbasi, M.; Bussi, G.; Keesstra, S. Modeling Sediment Yield in Semi-Arid Pasture Micro-Catchments, NW Iran. *Land Degrad. Dev.* **2017**, *28*, 1274–1286. [CrossRef]
11. Afshar, F.A.; Ayoubi, S.; Jalalian, A. Soil redistribution rate and its relationship with soil organic carbon and total nitrogen using 137Cs technique in a cultivated complex hillslope in western Iran. *J. Environ. Radioact.* **2010**, *101*, 606–614. [CrossRef]
12. Khalili Moghadam, B.; Jabarifar, M.; Bagheri, M.; Shahbazi, E. Effects of land use change on soil splash erosion in the semi-arid region of Iran. *Geoderma* **2015**, *241*, 210–220. [CrossRef]
13. Pradeep, G.S.; Krishnan, M.V.N.; Vijith, H. Identification of critical soil erosion prone areas and annual average soil loss in an upland agricultural watershed of Western Ghats, using analytical hierarchy process (AHP) and RUSLE techniques. *Arab. J. Geosci.* **2015**, *8*, 3697–3711. [CrossRef]
14. Choubin, B.; Rahmati, O.; Tahmasebipour, N.; Feizizadeh, B.; Pourghasemi, H.R. Application of fuzzy analytical network process model for analyzing the gully erosion susceptibility. *Adv. Nat. Technol. Hazards Res.* **2019**, *48*, 105–125. [CrossRef]
15. Angileri, S.E.; Conoscenti, C.; Hochschild, V.; Märker, M.; Rotigliano, E.; Agnesi, V. Water erosion susceptibility mapping by applying Stochastic Gradient Treeboost to the Imera Meridionale River Basin (Sicily, Italy). *Geomorphology* **2016**, *262*, 61–76. [CrossRef]
16. Svoray, T.; Michailov, E.; Cohen, A.; Rokah, L.; Sturm, A. Predicting gully initiation: Comparing data mining techniques, analytical hierarchy processes and the topographic threshold. *Earth Surf. Process. Landf.* **2012**, *37*, 607–619. [CrossRef]
17. Lee, S.; Ryu, J.H.; Lee, M.J.; Won, J.S. Use of an artificial neural network for analysis of the susceptibility to landslides at Boun, Korea. *Environ. Geol.* **2003**, *44*, 820–833. [CrossRef]
18. Pourghasemi, H.R.; Rossi, M. Landslide susceptibility modeling in a landslide prone area in Mazandarn Province, north of Iran: A comparison between GLM, GAM, MARS, and M-AHP methods. *Theor. Appl. Climatol.* **2017**, *130*, 609–633. [CrossRef]
19. Pradhan, B. A comparative study on the predictive ability of the decision tree, support vector machine and neuro-fuzzy models in landslide susceptibility mapping using GIS. *Comput. Geosci.* **2013**, *51*, 350–365. [CrossRef]
20. Taner San, B. An evaluation of SVM using polygon-based random sampling inlandslide susceptibility mapping: The Candir catchment area(western Antalya, Turkey). *Int. J. Appl. Earth Obs. Geoinf.* **2014**, *26*, 399–412. [CrossRef]
21. Dou, J.; Yamagishi, H.; Pourghasemi, H.R.; Yunus, A.P.; Song, X.; Xu, Y.; Zhu, Z. An integrated artificial neural network model for the landslide susceptibility assessment of Osado Island, Japan. *Nat. Hazards* **2015**, *78*, 1749–1776. [CrossRef]

22. Gorsevski, P.V.; Brown, M.K.; Panter, K.; Onasch, C.M.; Simic, A.; Snyder, J. Landslide detection and susceptibility mapping using LiDAR and an artificial neural network approach: A case study in the Cuyahoga Valley National Park, Ohio. *Landslides* **2016**, *13*, 467–484. [CrossRef]
23. Hong, H.; Pourghasemi, H.R.; Pourtaghi, Z.S. Landslide susceptibility assessment in Lianhua County (China): A comparison between a random forest data mining technique and bivariate and multivariate statistical models. *Geomorphology* **2016**, *259*, 105–118. [CrossRef]
24. Yuan, L.; Zhang, Q.; Li, W.; Zou, L. Debris Flow Hazard Assessment Based on Support Vector Machine. In *2006 IEEE International Symposium on Geoscience and Remote Sensing*; IEEE: Piscataway, NJ, USA, 2006; Volume 11, pp. 4221–4224.
25. Chang, T.C. Risk degree of debris flow applying neural networks. *Nat. Hazards* **2007**, *42*, 209–224. [CrossRef]
26. Chang, T.C.; Chao, R.J. Application of back-propagation networks in debris flow prediction. *Eng. Geol.* **2006**, *85*, 270–280. [CrossRef]
27. Eustace, A.H.; Pringle, M.J.; Denham, R.J. A risk map for gully locations in central Queensland, Australia. *Eur. J. Soil Sci.* **2011**, *62*, 431–441. [CrossRef]
28. Rahmati, O.; Tahmasebipour, N.; Haghizadeh, A.; Pourghasemi, H.R.; Feizizadeh, B. Evaluation of different machine learning models for predicting and mapping the susceptibility of gully erosion. *Geomorphology* **2017**, *298*, 118–137. [CrossRef]
29. Arabameri, A.; Rezaei, K.; Cerda, A.; Lombardo, L.; Rodrigo-Comino, J. GIS-based groundwater potential mapping in Shahroud plain, Iran. A comparison among statistical (bivariate and multivariate), data mining and MCDM approaches. *Sci. Total Environ.* **2019**, *658*, 160–177. [CrossRef] [PubMed]
30. Pourghasemi, H.R.; Yousefi, S.; Kornejady, A.; Cerdà, A. Performance assessment of individual and ensemble data-mining techniques for gully erosion modeling. *Sci. Total Environ.* **2017**, *609*, 764–775. [CrossRef]
31. Garosi, Y.; Sheklabadi, M.; Conoscenti, C.; Pourghasemi, H.R.; Van Oost, K. Assessing the performance of GIS-based machine learning models with different accuracy measures for determining susceptibility to gully erosion. *Sci. Total Environ.* **2019**, *664*, 1117–1132. [CrossRef]
32. Mao, D.; Zeng, Z.; Wang, C.; Lin, W. Support Vector Machines with PSO Algorithm for Soil Erosion Evaluation and Prediction. In Proceedings of the Third International Conference on Natural Computation (ICNC 2007), Haikou, China, 24–27 August 2007; IEEE: Piscataway, NJ, USA, 2007; Volume 1, pp. 656–660.
33. Azareh, A.; Rafiei Sardooi, E.; Choubin, B.; Barkhori, S.; Shahdadi, A.; Adamowski, J.; Shamshirband, S. Incorporating multi-criteria decision-making and fuzzy-value functions for flood susceptibility assessment. *Geocarto Int.* **2019**. [CrossRef]
34. Salcedo-Sanz, S.; Ghamisi, P.; Piles, M.; Werner, M.; Cuadra, L.; Moreno-Martínez, A.; Izquierdo-Verdiguier, E.; Muñoz-Marí, J.; Mosavi, A.; Camps-Valls, G. Machine Learning Information Fusion in Earth Observation: A Comprehensive Review of Methods, Applications and Data Sources. *Inf. Fusion.* **2020**, *22*, 480–545. [CrossRef]
35. Nekhay, O.; Arriaza, M.; Boerboom, L. Evaluation of soil erosion risk using Analytic Network Process and GIS: A case study from Spanish mountain olive plantations. *J. Environ. Manag.* **2009**, *90*, 3091–3104. [CrossRef]
36. Emadi, M.; Taghizadeh-Mehrjardi, R.; Cherati, A.; Danesh, M.; Mosavi, A.; Scholten, T. Predicting and Mapping of Soil Organic Carbon Using Machine Learning Algorithms in Northern Iran. *Remote Sens.* **2020**, *12*, 2234.
37. Lin, B.S.; Thomas, K.; Chen, C.K.; Ho, H.C. Evaluation of soil erosion risk for watershed management in Shenmu watershed, central Taiwan using USLE model parameters. *Paddy Water Environ.* **2016**, *14*, 19–43. [CrossRef]
38. Yu, B.; Rosewell, C.J. A Robust estimator of the R-factor for the universal soil loss equation. *Trans. Am. Soc. Agric. Eng.* **1996**, *39*, 559–561. [CrossRef]
39. Food and Agriculture Organization of the United Nations. *Digital Soil Map of the World and Derived Soil Properties*; Food and Agriculture Organization of the United Nations: Bangkok, Thailand, 2003.
40. Kadam, A.K.; Kale, S.S.; Pande, N.N.; Pawar, N.J.; Sankhua, R.N. Identifying Potential Rainwater Harvesting Sites of a Semi-arid, Basaltic Region of Western India, Using SCS-CN Method. *Water Resour. Manag.* **2012**, *26*, 2537–2554. [CrossRef]
41. Choubin, B.; Borji, M.; Mosavi, A.; Sajedi-Hosseini, F.; Singh, V.P.; Shamshirband, S. Snow avalanche hazard prediction using machine learning methods. *J. Hydrol.* **2019**, *577*, 123929. [CrossRef]

42. Di Stefano, C.; Ferro, V.; Porto, P.; Tusa, G. Slope curvature influence on soil erosion and deposition processes. *Water Resour. Res.* **2000**, *36*, 607–617. [CrossRef]
43. Arabameri, A.; Pourghasemi, H.R. Spatial Modeling of Gully Erosion Using Linear and Quadratic Discriminant Analyses in GIS and R. In *Spatial Modeling in GIS and R for Earth and Environmental Sciences*; Elsevier: Amsterdam, The Netherlands, 2019; pp. 299–321. [CrossRef]
44. Auzet, A.V.; Poesen, J.; Valentin, C. Soil patterns as a key controlling factor of soil erosion by water. *Catena* **2002**, *46*, 85–87. [CrossRef]
45. Crippen, R.E. Calculating the vegetation index faster. *Remote Sens. Environ.* **1990**, *34*, 71–73. [CrossRef]
46. De Baets, S.; Poesen, J.; Reubens, B.; Wemans, K.; De Baerdemaeker, J.; Muys, B. Root tensile strength and root distribution of typical Mediterranean plant species and their contribution to soil shear strength. *Plant Soil* **2008**, *305*, 207–226. [CrossRef]
47. Bertsimas, D.; Tsitsiklis, J. Simulated annealing. *Stat. Sci.* **1993**. [CrossRef]
48. Hosseini, F.S.; Choubin, B.; Mosavi, A.; Nabipour, N.; Shamshirband, S.; Darabi, H.; Haghighi, A.T. Flash-flood hazard assessment using ensembles and Bayesian-based machine learning models: Application of the simulated annealing feature selection method. *Sci. Total Environ.* **2020**, *711*. [CrossRef] [PubMed]
49. Choubin, B.; Abdolshahnejad, M.; Moradi, E.; Querol, X.; Mosavi, A.; Shamshirband, S.; Ghamisi, P. Spatial hazard assessment of the PM10 using machine learning models in Barcelona, Spain. *Sci. Total Environ.* **2020**, *701*. [CrossRef]
50. Choubin, B.; Mosavi, A.; Alamdarloo, E.H.; Hosseini, F.S.; Shamshirband, S.; Dashtekian, K.; Ghamisi, P. Earth fissure hazard prediction using machine learning models. *Environ. Res.* **2019**, *179*. [CrossRef] [PubMed]
51. Max, K.; Weston, S.; Wing, J.; Williams, A.; Keefer, C.; Engelhardt, A. Package 'caret'. Classification and Regression Training. *Adsabs Harv. Edu* **2011**, *138*, 454–466.
52. Xu, B.; Huang, J.Z.; Williams, G.; Wang, Q.; Ye, Y. Classifying very high-dimensional data with random forests built from small subspaces. *Int. J. Data Warehous. Min.* **2012**, *8*, 44–63. [CrossRef]
53. Zhao, H.; Williams, G.J.; Huang, J.Z. Wsrf: An R package for classification with scalable weighted subspace random forests. *J. Stat. Softw.* **2017**, *77*. [CrossRef]
54. Rennie, J.D.M.; Shih, L.; Teevan, J.; Karger, D. Tackling the Poor Assumptions of Naive Bayes Text Classifiers. In Proceedings of the 20th International Conference on Machine Learning (ICML-03), Washington, DC, USA, 21–24 August 2003; Volume 2, pp. 616–623.
55. Schneider, K.-M. A Comparison of Event Models for Naive Bayes Anti-Spam E-Mail Filtering. In Proceedings of the 10th Conference of the European Chapter of the Association for Computational Linguistics, Budapest, Hungary, 12–17 April 2003.
56. Webb, G.I.; Boughton, J.R.; Wang, Z. Not so naive Bayes: Aggregating one-dependence estimators. *Mach. Learn.* **2005**, *58*, 5–24. [CrossRef]
57. Roever, C.; Raabe, N.; Luebke, K.; Ligges, U.; Szepannek, G. *Package "klaR"*; Springer: Berlin, Germany, 2020.
58. Weihs, C.; Ligges, U.; Luebke, K.; Raabe, N. klaR Analyzing German Business Cycles. In *Data Analysis and Decision Support*; Springer: Berlin/Heidelberg, Germany, 2005; pp. 335–343.
59. Schulz, E.; Speekenbrink, M.; Krause, A. A tutorial on Gaussian process regression: Modelling, exploring, and exploiting functions. *J. Math. Psychol.* **2018**, *85*, 1–16. [CrossRef]
60. Gershman, S.J.; Blei, D.M. A tutorial on Bayesian nonparametric models. *J. Math. Psychol.* **2012**, *56*, 1–12. [CrossRef]
61. Williams, C.K.I. Prediction with Gaussian Processes: From Linear Regression to Linear Prediction and Beyond. In *Learning in Graphical Models*; Springer: Dodlek, The Netherlands, 1998; pp. 599–621.
62. Alexandros, A.; Smola, A.; Hornik, K. *Kernel-Based Machine Learning Lab-Package 'kernlab'*; Springer: Berlin, Germany, 2019.
63. Darabi, H.; Choubin, B.; Rahmati, O.; Torabi Haghighi, A.; Pradhan, B.; Kløve, B. Urban flood risk mapping using the GARP and QUEST models: A comparative study of machine learning techniques. *J. Hydrol.* **2019**, *569*, 142–154. [CrossRef]
64. Beucher, A.; Møller, A.B.; Greve, M.H. Artificial neural networks and decision tree classification for predicting soil drainage classes in Denmark. *Geoderma* **2019**, *352*, 351–359. [CrossRef]
65. Monserud, R.A.; Leemans, R. Comparing global vegetation maps with the Kappa statistic. *Ecol. Modell.* **1992**, *62*, 275–293. [CrossRef]

66. Weihua, L.; Yonggang, W.; Dianhui, M.; Yan, Y. Region Assessment of Soil Erosion Based on Naive Bayes. In Proceedings of the 2007 International Conference on Computational Intelligence and Security, CIS 2007, Harbin, China, 15–19 December 2007.
67. Nhu, V.-H.; Shirzadi, A.; Shahabi, H.; Singh, S.K.; Al-Ansari, N.; Clague, J.J.; Jaafari, A.; Chen, W.; Miraki, S.; Dou, J.; et al. Shallow Landslide Susceptibility Mapping: A Comparison between Logistic Model Tree, Logistic Regression, Naïve Bayes Tree, Artificial Neural Network, and Support Vector Machine Algorithms. *Int. J. Environ. Res. Public Health* **2020**, *17*, 2749. [CrossRef] [PubMed]

Article

Climate Change Impacts on Soil Erosion and Sediment Yield in a Watershed

Ching-Nuo Chen [1,*], Samkele S. Tfwala [2] and Chih-Heng Tsai [3]

1 Department of Tropical Agriculture and International Cooperation, National Pingtung University of Science and Technology, Pingtung 91201, Taiwan
2 Department of Soil and Water Conservation, National Chung Hsing University, Taichung 40227, Taiwan; samkelet@email.nchu.edu.tw
3 Research and Development Foundation, National Cheng-Kung University, Tainan 70101, Taiwan; jht581212@gmail.com
* Correspondence: ginrochen@mail.npust.edu.tw

Received: 2 July 2020; Accepted: 7 August 2020; Published: 10 August 2020

Abstract: This study analyzed the influence of climate change on sediment yield variation, sediment transport and erosion deposition distribution at the watershed scale. The study was based on Gaoping River basin, which is among the largest basins in southern Taiwan. To carry out this analysis, the Physiographic Soil Erosion Deposition (PSED) model was utilized. Model results showed a general increase in soil erosion and deposition volume under the A1B-S climate change scenario. The situation is even worsened with increasing return periods. Total erosion volume and total sediment yield in the watershed were increased by 4–25% and 8–65%, respectively, and deposition volumes increased by 2–23%. The study showed how climate change variability would influence the watershed through increased sediment yields, which might even worsen the impacts of natural disasters. It has further illustrated the importance of incorporating climate change into river management projects.

Keywords: climate change; soil erosion; sediment yield; PSED Model

---

1. Introduction

Due to the increasing severity of global warming and climate change effect in recent years, extreme hydrographic phenomena have frequently been observed. Climate change has increased precipitation concentration, volume and intensity, which has significantly impacted runoff and soil erosion in many watersheds [1–3]. The sediments generated from watershed erosion are transported to rivers via surface runoff, and they are the main composition of river sediments and a major source of reservoir or river dam sediment deposition [4]. The degree of soil erosion has a significant impact on the evolution of river channels, influencing river stability, flood prevention safety and river remediation planning. Hence, the control of sediment yield is crucial in watershed management, especially since it usually involves high costs. In a review of sediment management strategies in Taiwan and the barriers to their implementation, Wang et al. [5] highlighted how technical barriers are driven primarily by engineering and costs. This was in reference to methods such as construction of upstream sediment structures and hydraulic and mechanical dredging. Moreover, in some regions, climate change is projected to decrease the overall soil erosion potential due to decrease in rainfall [6].

Several studies have focused on the impacts of climate change on precipitation volume, runoff volume, erosion volume and sediment yield. General Circulation Models (GCMs) have been applied to analyze the impacts of climate change on precipitation characteristic [7–9] and river runoff [10,11]. The Soil and Water Assessment Tool (SWAT) model seems to be the most favored by researchers when evaluating the impacts of climate change on flow rate, soil erosion and sediment yield in a watershed area [12,13]. Thodsen et al. [14] applied the High Resolution Limited Area Model (HIRHAM) regional climate model to investigate the impact of climate change on suspended load transport rate of Danish rivers. In modeling flow rate and sediment yield for high flow-rate rivers under the A2 scenario in the rain season of 2050, Phan et al. [15] showed an increase of 11.4% and 15.3%, respectively. Cousino et al. [16] utilized SWAT to provide hydrological insights for the Maumee River watershed, showing a reduction by 10% in flow, while sediment yield increased by 11%. Most recently, Zhou et al. [17] used the SWAT model to evaluate the impacts of climate change on flow and sediment yield in northeast China. In northern Iran, Azari et al. [18] reported an annual increase of 5% in annual streamflow and more than 35% in sediment yield. Zhang et al. [19] generated climatic conditions for future periods 2020–2039, 2050–2069 and 2080–2099, and their results demonstrated an increase of 13% in sediment yield.

The abovementioned methods do not couple the computations of slope erosion and river sediment transport. Instead, watershed erosion is calculated first based on slope information, followed by watershed sediment yield based on sediment delivery ratio, a river sediment transport model or runoff volume with a flow rate-sediment transport rating curve. The estimation of slope erosion, regardless of applying the Universal Soil Loss Equation (USLE), Revised Universal Soil Loss Equation (RUSLE) [20] or Modified Universal Soil Loss Equation (MUSLE), is carried out by empirical model. Empirical models are established based on erosion data of the initial location via inductive analysis. Therefore, they have application and location limitations. Moreover, the erosion volume estimated by empirical models is the total erosion volume rather than the time-dependent or spatial-dependent erosion volume. Furthermore, watersheds and river systems are complex and sediment boundary conditions for carrying out river sediment transport simulations by the abovementioned models cannot be obtained directly. Typically, sediment transport volume is estimated by the flow rate-sediment transport rating curves established through hydrological observation stations. Nonetheless, sediment transport rates estimated by rating curves have large uncertainties, especially with high discharge values [21].

To understand the impacts of climate change on erosion volume, sediment yield and erosion distribution for a watershed, this study utilized a Physiographic Soil Erosion-Deposition (PSED) model proposed by Chen, Tsai and Tsai [4]. Unlike empirical models, the PSED model has physical mechanisms that enable simultaneous computation of slope erosion and river sediment transport. Additionally, this model does not require the flow rate-sediment rating curve as a boundary condition, thus eliminating uncertainties that come with their adoption. Findings from this study will serve as a useful reference for decision-making authorities in planning appropriate strategies and corresponding measures to prevent water- and soil-related disasters.

2. Materials and Methods

2.1. Study Area

Gaoping River, which is the largest river in Taiwan, was used for the analysis. It is located in the southwest of Taiwan and has a total length of 171 km, covering 3257 km^2. It originates from west of the Central Mountain Range near Yushan. The upstream end is connected to Laonong River, and its main branches include Qishan, Ailiao and Zhuokou (Figure 1). The topographic height varies significantly, descending along the direction from northeast to southwest, with a maximum difference of nearly 4000 m. The area above 1000 m accounts for 47.45% of the total drainage basin, that between 100 and 1000 m accounts for 32.38% and finally, the area below 100 m accounts for 20.17%. The average slope of the river bed is approximately 1/150, with 1/15, 1/100 and 1/1000 for the upstream, midstream and downstream sections, respectively. Additionally, the sections' lengths are 37, 68 and 66 km for the upstream, midstream and downstream sections, respectively.

Time- and location-based precipitation distribution in the basin varies widely. Near the Central Mountain Range it is large (~3400 mm), whereas in the plain and coastal areas it is significantly smaller

(~2000 mm). Precipitation is concentrated from May to October, which accounts for 90% of the annual precipitation. Average annual runoff volume is ~8.45 billion m^3, of which 7.69 billion m^3 (91%) occur in the wet season. The average annual sediment transport volume is 35.61 million tons, with 10,934 tons of sediment transport per km^2 of drainage basin area.

Figure 1. The drainage network in the Gaoping River basin.

The Gaoping River basin is among the worst basins in Taiwan in terms of sediment yield and is highly vulnerable to sediment deposition. Statistical data from the Water Resources Agency of Taiwan, MOEA, estimate a total dredging sediment volume of 94,780,000 m^3 for Gaoping River basin between 2009 to 2014 [22]. The major contributions to such high rates include the high-slope landform and concentrated precipitation. The geological map in Figure 2 shows that the watershed mainly constitutes sand gravel and sandstone, rendering it vulnerable to erosion. Stefanidis and Stathis [6], in their assessment of soil erosion in a catchment, showed that vulnerable geological subsoil and the steep slopes favor the development of erosion phenomena. At Gaoping, about 80% of the annual precipitation is from an average of 3–4 typhoons per year [21], which fall between June and September each year. Precipitation in these events is characterized by high intensity and short duration, leading to enormous volumes of sediment yield. The situation is even worsened by additional factors, such as climate change, which has caused drastic changes in precipitation [23].

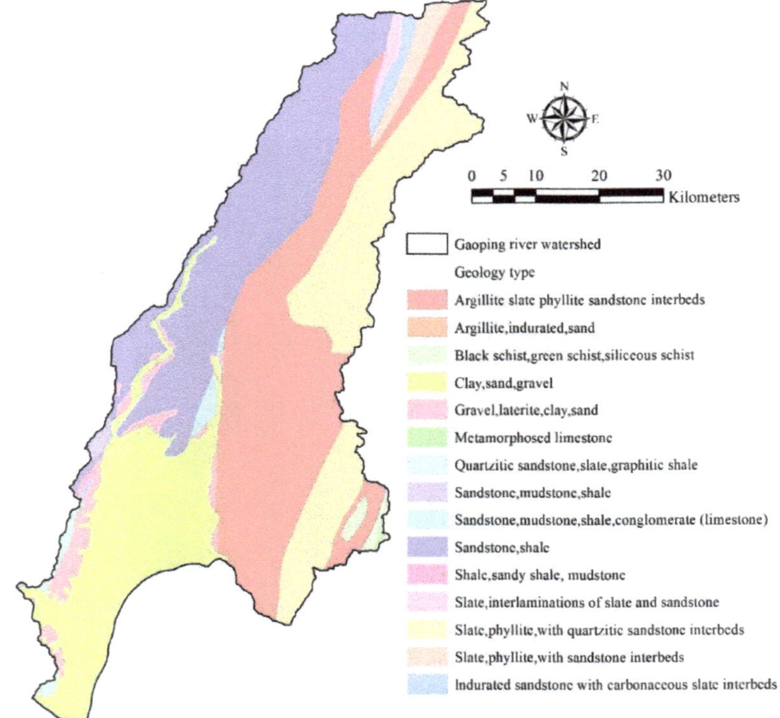

Figure 2. Geological map of Gaoping watershed [24].

2.2. Long Term Climate and Hydrological Changes

According to Taiwan's hydrological data from the past 55 years, there has been a gradual increase in precipitation and typhoon intensity. Both the frequencies of occurrence and the severity of flooding are showing an increasing trend. Figures 3 and 4 show the long-term climate and hydrological data from 1962 to 2016 recorded by the Kaohsiung weather station in the study area. From Figure 3, the annual average temperature in the study area exhibited an increasing trend. Beginning in 1997, the annual average temperature was higher than the average temperature in the past 55 years (24.8 °C) and after 2000, the increase of annual average temperature was more significant. Precipitation after 1996 was more significant and most of the annual precipitation was higher than the long-term annual average precipitation of 1770 mm. The deviation between high and low precipitation became larger and the duration of high and low precipitation became shorter (Figure 4).

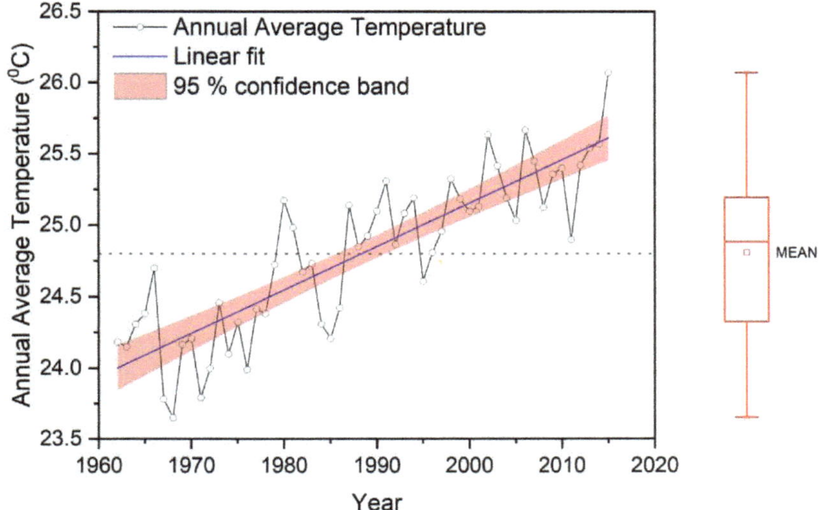

Figure 3. Changes in annual average temperature recorded by the Kaohsiung weather station, 1960–2016.

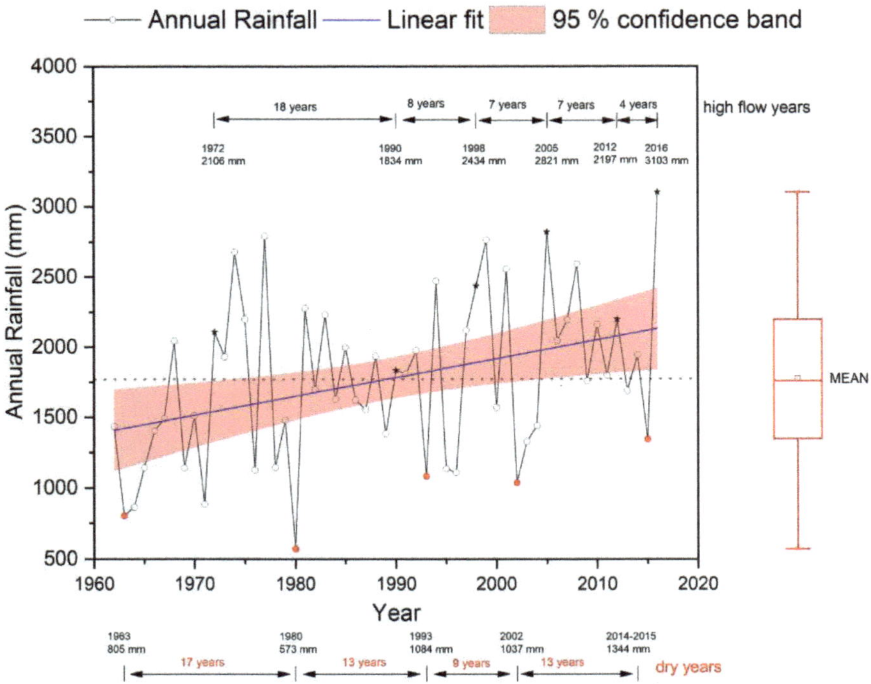

Figure 4. Changes in annual precipitation recorded by the Kaohsiung Weather Station, 1960–2016.

2.3. The Physiographic Soil Erosion-Deposition Model (PSED Model)

The PSED Model [4,25] is a physical mechanism-based model that was developed by integrating the Geographic Information System (GIS) with a Physiographic Precipitation-Runoff model. It incorporates the effects of slope and river channel erosion, entrainment and deposition on river bed erosion-deposition for a watershed. Based on topography, landform, river system, land use and soil characteristics of the watershed, the PSED Model utilizes GIS to partition the watershed area into non-structural computational cells. The computed cells are then classified into slope cells, river cells and special cells. Esri ArcMap 10.7 was utilized to obtain hydrological and physiographical data within each cell. In addition, the extension modules of ArcMap (spatial analysis, hydrologic model, 3D Analyst and Network Analyst) and their object-oriented programming language were used.

The model consists mainly of two parts; a water flow simulation and a soil erosion-deposition simulation. Water flow simulation calculates the transport of precipitation runoff in the watershed area. The continuity equation of water flow is as follows:

$$A_i \frac{\partial h_i}{\partial t} = \sum_k Q_{i,k}(h_i, h_k) + P_{ei}(t) \tag{1}$$

where: t is time; A_i is area of the i cell; h_i and h_k represent the water stage of the i and k cell, respectively; $Q_{i,k}$ denotes the flow rate (discharge) from the k cell into its neighboring i cell; and P_{ei} expresses the effective rainfall volume per second in the i cell, which is equal to the effective rainfall per second in the i cell multiplied by its area. Depending on the topography and landform information, the watershed area can be divided into several computed cells and the water level change of each cell should then satisfy the continuity equation of water flow and flow rate simulation, as expressed in Equation (1).

In the soil erosion-deposition simulation for the watershed, simulations for slope cells and river cells were calculated separately. The sediment transport rate and river bed erosion profile of each cell were simulated by using the suspended load equation (Equation (2)), the river bed variation continuity equations (Equation (3)) [26], and the river bed load transport equation.

$$\frac{\partial V_{si}}{\partial t} = \sum_k Q_{SC_{i,k}} + Q_{sei} - Q_{sdi} + R_{DTi} \tag{2}$$

$$(1-\lambda)\frac{\partial V_{di}}{\partial t} = \sum_k Q_{SB_{i,k}} - Q_{sei} + Q_{sdi} - R_{DTi} \tag{3}$$

where: V_{si} is the soil volume of water body in i cell (= $A_i \times D_i \times C_i$); D is the cell water depth; C is the suspended load volume concentration; λ is the porosity; V_{di} is the alluvium volume of cell i; $Q_{SC_{i,k}}$ and $Q_{SB_{i,k}}$ denote the suspended and the river bed load flow rates, respectively, from the k cell into its neighboring i cell; Q_{sei} represents the entrainment rate of ground surface soil or river bed sediment of the i-th cell; Q_{sdi} expresses the deposition rate of river bed sediment for i cell; and R_{DTi} is the precipitation separation rate of the i cell.

2.4. Computational Cells

Figures 5–7 show the Digital Elevation Model (DEM), soil map and land use maps of Gaoping River Basin, respectively. The basin ranges from 0 m to more than 3000 m above sea level and is dominated mostly by forest land at elevations higher than 500 m. Based on the DEM, land use, soil maps, road system maps, and slope maps (Figure 8) the basin was divided by Esri ArcMap 10.7 into 17,635 computational cells for further analysis.

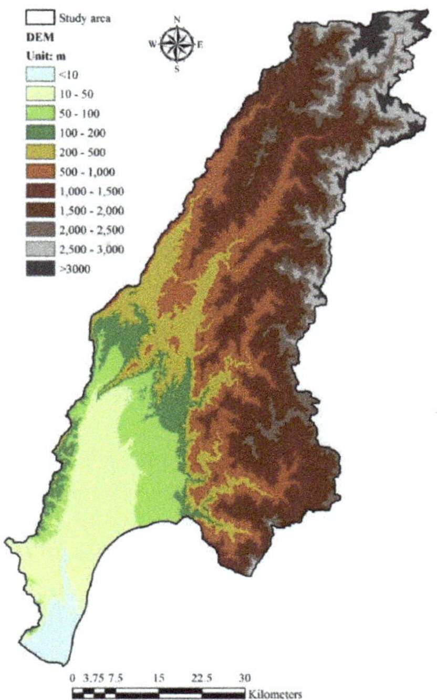

Figure 5. Digital elevation model (DEM) of the Gaoping River watershed.

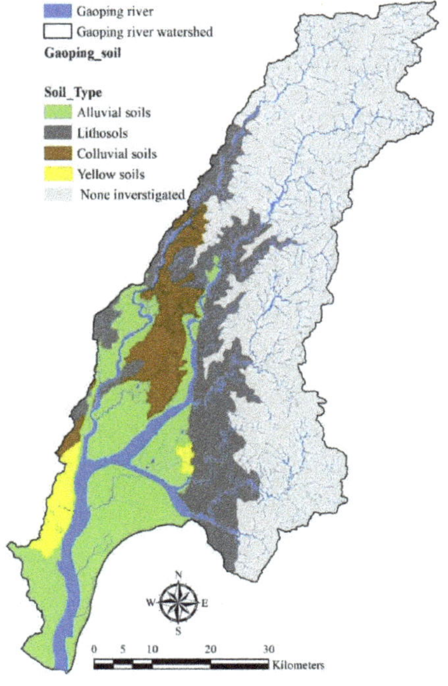

Figure 6. Soil map of the Gaoping River watershed [27].

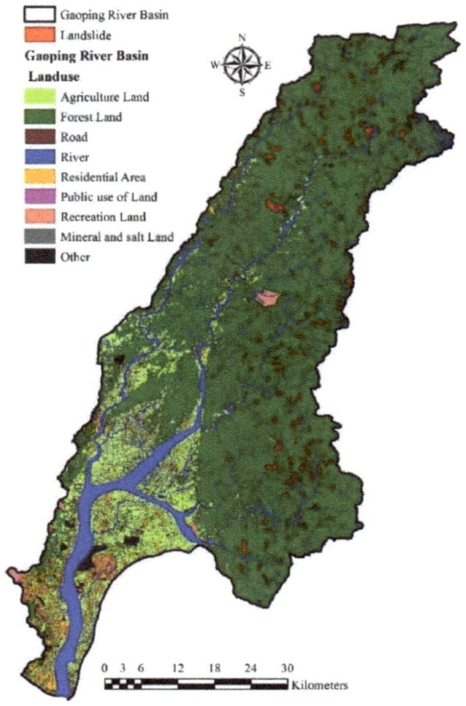

Figure 7. Digital land use map of the Gaoping River watershed [28].

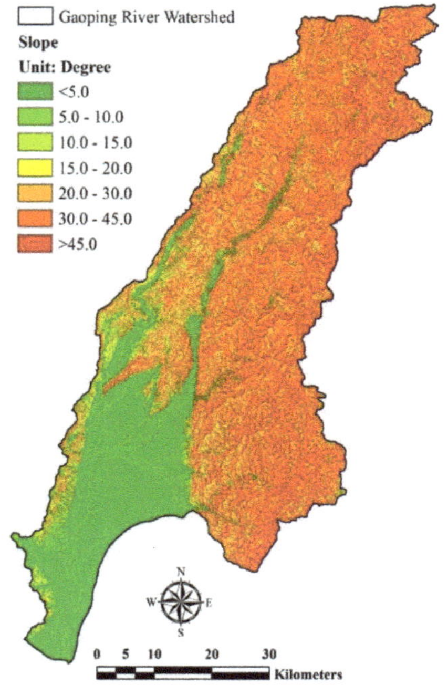

Figure 8. Slope map of Gaoping River watershed.

2.5. Input Data and Model Setup

In the basin, there are 28 precipitation monitoring stations established by the Central Weather Bureau; however, they are not evenly distributed. In order to minimize computational errors, we applied the Thiessen polygons method to determine the controlling area of each precipitation monitoring station. The reader should note that the Thiessen polygons were not used to derive the weighted precipitation. Instead, precipitation data for each station were used as the precipitation volume of computed cells in the control area for the same precipitation monitoring station (Figure 9).

Figure 9. Effective area for precipitation gauging stations in the Gaoping River watershed.

PSED simulated sediment transport was validated by comparing the flow rate and sediment transport data collected by sediment monitoring stations of the 7th River Management Office of Water Resources Agency, MOEA, which are located at the downstream of the Gaoping River and its tributaries. They include the Shanlin Bridge of Qishan River, Tachin Bridge of Laonong River, Sandimen Bridge of Ailiao River, and Lilin Bridge of Gaoping River (Figure 10).

The future scenario in this study was set from 2020 to 2039 and the corresponding baseline was set from 1980 to 1999. The Taiwan Climate Change Projection and Information Platform Project (TCCIP) provided downscaled precipitation data at 5 km^2 resolution. To process the data, TCCIP relies on 24 GCMs as described in the Intergovernmental Panel on Climate Change (IPCC), Fourth Assessment Report (AR4) [29]. Additionally, IPCC identifies A2, A1B and B1 as the most probable scenarios; hence, this study adopted the A1B-S for analysis, which is regarded as a worse scenario and is similar to the A1B scenario. The worst-case scenario is primarily obtained through subtracting or adding one standard deviation between the estimated values of GCMs from the multi-model ensemble of all GCMs [30]. Monthly precipitation scenario information was further combined with a weather generator to evaluate the impact of climate change on daily precipitation volume.

The baseline was defined by precipitation data from 1980 to 1999. Historical daily precipitation data from the monitoring stations were used as input files for the climate derived models. These were then applied to generate the daily precipitation data representing the future climate change scenario. In addition, the daily precipitation data of the baseline scenario were combined with the precipitation distribution in the watershed area to translate into precipitation profiles to be used by the PSED model.

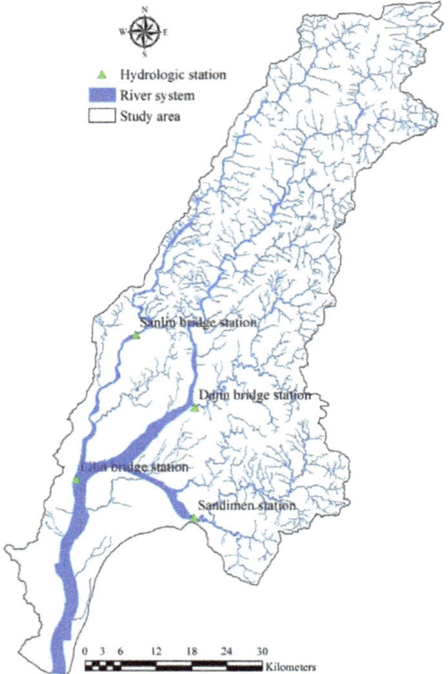

Figure 10. Sediment monitoring stations on the Gaoping River and its tributaries.

2.6. Model Verification

Typhoon Morakot (2009), the most disastrous storm to have hit Taiwan in the last century, was used to validate runoff and suspended load hydrographs from the PSED model. We further compared actual discharge and sediment transport from each hydrological station to simulated data. Simulated and observed flow hydrograph from Lilin Bridge is shown in Figure 11. The peak of the simulated hydrograph coincided with that from the observed data, and the hydrograph shapes are similar. This suggests that the model can be successfully applied.

Since the hydrological monitoring stations along the Gaoping River system do not have suspended load concentration data hydrographs, the historical discharge and suspended load data from sediment monitoring stations downstream of the main branch of the Gaoping River were used to establish the correlation between discharge and sediment transport volume of each hydrological monitoring station. These served as the basis for validating the suspended load concentration hydrograph obtained from the numerical model. The simulated discharge and suspended load transport rate under Typhoon Morakot in 2009 were plotted onto the correlation diagram between the observed discharge and sediment transport rate for the Sanlin, Dajin, Sandimen and Lilin bridge Stations (Figures 12–15). In these figures, points are actual historical measured data. The solid line represents the regression relation between discharge and sediment transport rate for each hydrological monitoring station. It is noted from the figures that the simulated correlation between discharge and sediment transport rate was consistent with the correlation between discharge and sediment transport rate of the actually observed data for all monitoring stations except Lilin Bridge station, particularly under high discharge. The model slightly overestimated data at this station (Figure 15). Nonetheless, the model also indicated reasonable estimates on suspended load and suspended load transport.

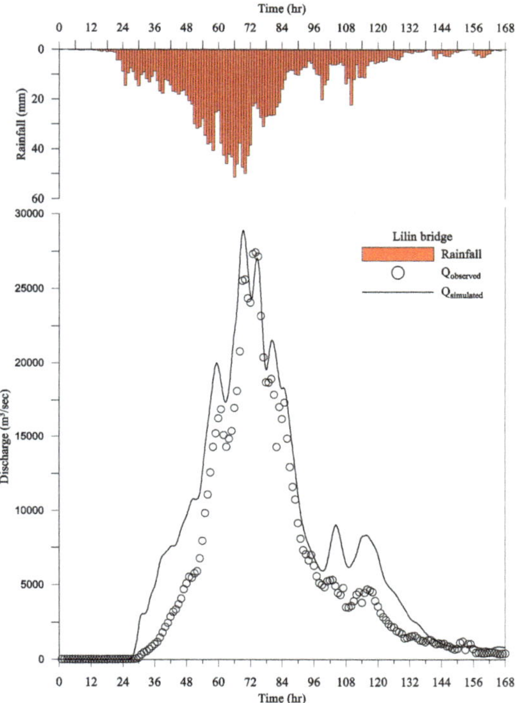

Figure 11. Simulated and observed discharge during Typhoon Morakot in 2009 at Lilin Bridge station.

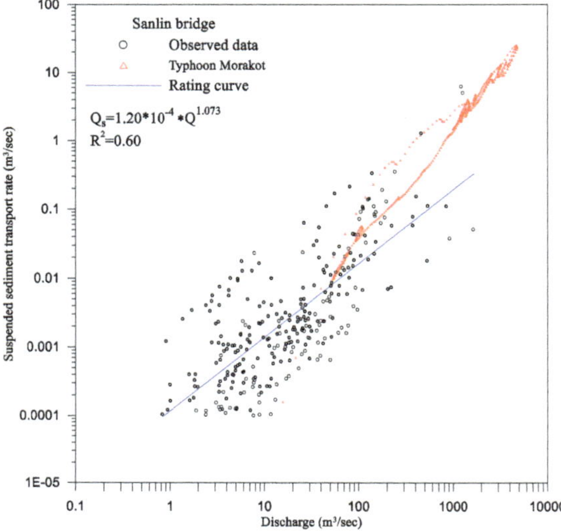

Figure 12. Simulated and observed correlation between flow discharge and sediment transport rate at Sanlin Bridge station.

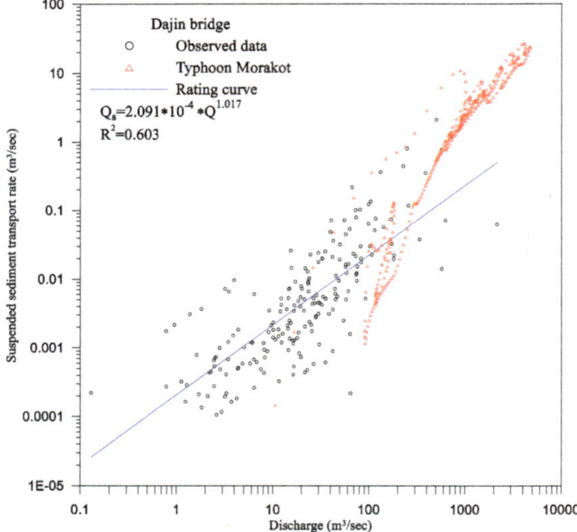

Figure 13. Simulated and observed correlation between flow discharge and sediment transport rate at Dajin Bridge station.

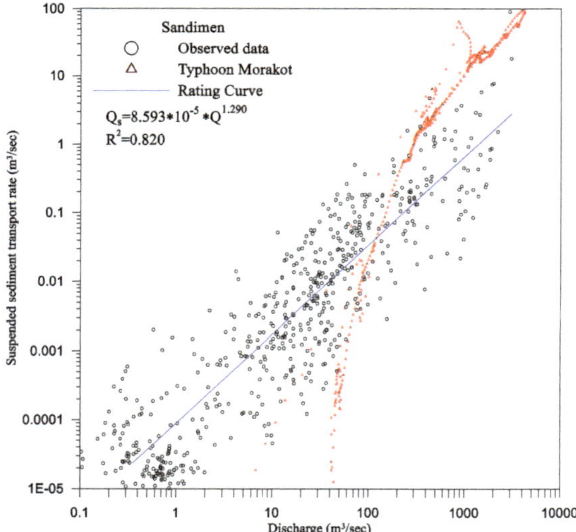

Figure 14. Simulated and observed relationship between flow discharge and sediment transport rate at Sandimen Bridge station.

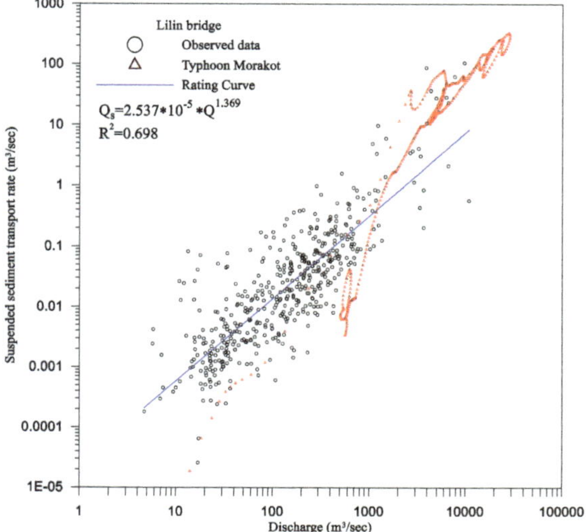

Figure 15. Simulated and observed relationship between flow discharge and sediment transport rate at Lilin Bridge station.

3. Results

3.1. Impacts of Climate Change on Erosion Volume and Sediment Yield

Different precipitation types, rainfall distribution, rainfall intensities and precipitation volume will result in different runoff processes, leading to different erosion volumes and sediment yields in a watershed. The precipitation volumes collected by each precipitation monitoring station in the watershed area for each return period (2, 5, 10, 25, 50, 100 and 200 year return period) under the baseline and A1B-S scenarios were used to calculate the average maximum rainfall intensity and average rainfall for each return period via the controlled area weighted method. The above return periods were selected as they are the standards used for most engineering designs in Taiwan. The precipitation results for the baseline were then compared with the results of the A1B-S scenario. Table 1 shows the comparison of average maximum rainfall intensity and average rainfall between the baseline and selected scenarios of each return period. Average maximum rainfall intensity increased more than average rainfall, suggesting that under the influence of climate change, not only did the precipitation volume increase but also the precipitation intensity.

Table 1. Average maximum rainfall intensity and average rainfall increase rates under baseline and A1B-S scenarios for various return periods.

Return Period	Average Maximum Rainfall Intensity (mm/hr)			Average Annual Rainfall (mm)		
	Baseline (1980–1999)	A1B-S (2020–2039)	Increase Rate (%)	Baseline (1980–1999)	A1B-S (2020–2039)	Increase Rate (%)
2	27.52	29.14	5.89	411.18	434.90	5.77
5	39.00	41.90	7.43	584.82	627.39	7.28
10	46.53	51.63	10.97	701.12	776.94	10.81
25	55.10	65.92	19.65	843.42	1004.26	19.07
50	61.12	78.24	27.99	957.97	1212.17	26.53
100	66.96	91.98	37.36	1094.72	1466.32	33.95
200	72.75	107.35	47.56	1280.60	1794.54	40.13

Simulated erosion volume and sediment yield under the baseline and A1B-S scenarios for various return periods are shown in Table 2. Total erosion volume and sediment yield under the A1B-S scenario for various return periods are greater than under the baseline. The increase in the total sediment yield rate was higher than that of the total erosion volume. The total erosion volume and total sediment yield increases by 4–25% and 8–65%, respectively, when compared to the baseline. This implies that climate change contributed to 15% and 36% increases in soil erosion volume and sediment yield, respectively, when compared to the baseline average.

Table 2. Soil erosion and sediment yield increase rates under baseline and A1B-S scenarios for various return periods.

Return Period	Total Erosion (m^3)			Total Sediment Yield (m^3)		
	Baseline (1980–1999)	A1B-S (2020–2039)	Increase Rate (%)	Baseline (1980–1999)	A1B-S (2020–2039)	Increase Rate (%)
2	25,025,101	26,158,002	4.53	3,610,538	3,919,368	8.55
5	33,683,874	35,566,538	5.59	6,312,775	7,083,252	12.21
10	38,589,952	41,576,563	7.74	8,280,890	9,711,170	17.27
25	43,557,305	49,077,034	12.67	10,631,451	13,811,304	29.91
50	46,687,318	54,601,150	16.95	12,325,453	17,496,549	41.95
100	49,482,826	59,940,637	21.13	14,010,777	21,564,822	53.92
200	52,023,499	65,223,943	25.37	15,708,576	25,927,062	65.05

3.2. Climate Change Effect on Erosion and Erosion Distribution

Soil erosion simulation results for the studied area were compiled and summarized in Table 3. The least return period indicated a 0.54% increase in area, while for under 200 year return period there was a 2.3% increase. The impacts of climate change on erosion were found to be lower when compared to other areas within the Asian region. Pal and Chakrabortty [31] simulated the impacts of climate change on soil erosion in a sub-tropical monsoon dominated watershed based on a RUSLE model, and found erosion to increase by 33% under a 15 year return period.

Table 3. Comparison of the increase rate of soil erosion area increase under the baseline and A1B-S scenarios for various return periods.

Return Period	Area Increase (m^2)	Increase Rate (%)
2	11,888,256	0.54
5	16,749,128	0.7
10	29,890,573	1.22
25	42,155,931	1.67
50	47,859,021	1.92
100	50,239,407	1.97
200	58,921,624	2.3

Climate change was also shown to cause larger erosion depths (Figure 16). Except in a few selected areas, a greater percentage indicates an increasing trend, and erosion depth increased with increasing return period. This is in line with observations made by Giang et al. [32], who predicted that Asian countries would be among the hardest hit regions globally.

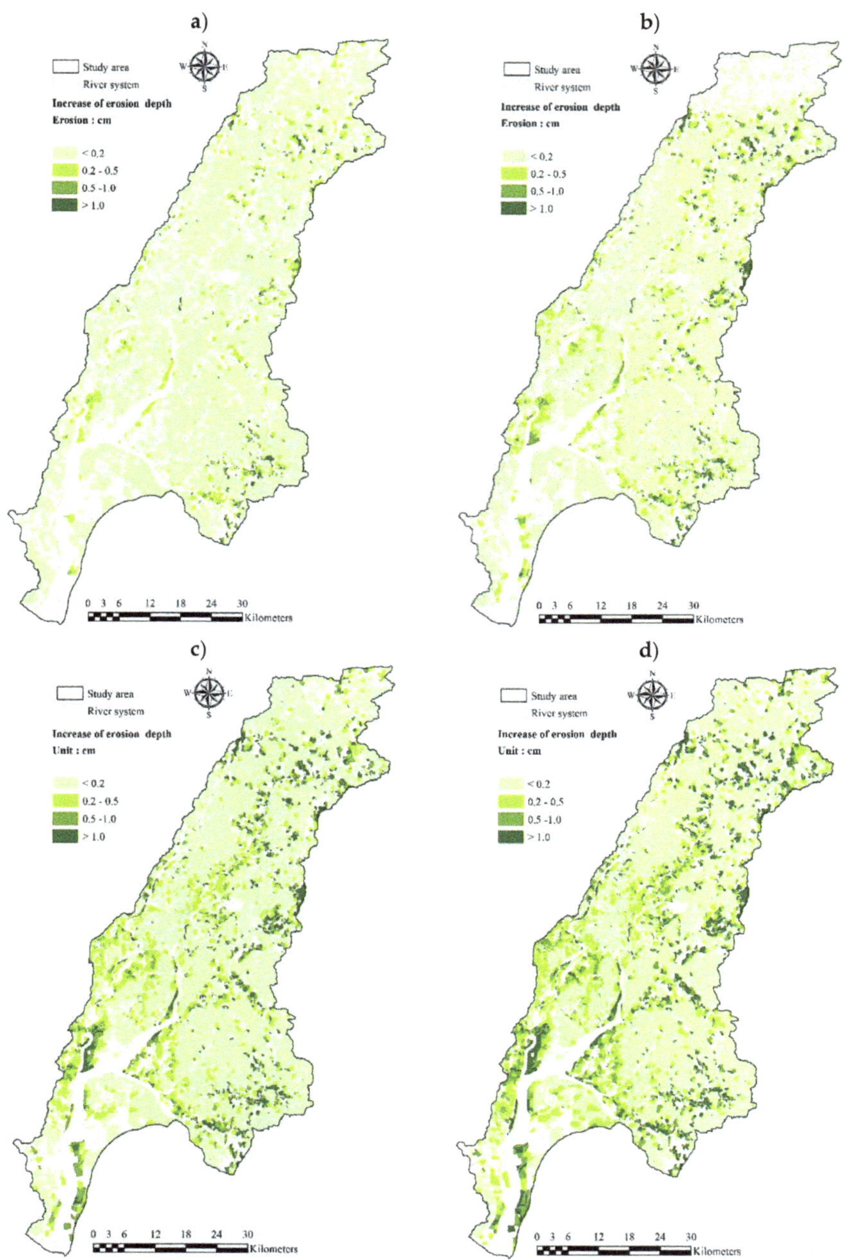

Figure 16. Spatial distribution of erosion depth increase under the influence of climate change under (**a**) 10 year, (**b**) 25 year, (**c**) 100 year and (**d**) 200 year return periods.

3.3. Climate Change Effects on Deposition Volume and the Deposition Distribution

Deposition distributions under baseline and climate change scenarios for 10 and 100 year return periods are shown in Figures 17 and 18, respectively. High deposition was observed mainly at the confluences; between Gaoping and Qishan River (zone A), Gaoping River and Laonong River (zone B), Ailiao River Gaoshu Bridge and Ailiao weir (zone C) and in the middle and downstream of Gaoping River (zone D). Zone areas are shown in Figure 19. Large deposition at these areas is attributed to widening of the cross sections, low river bed slopes and low flow rates. Deposition in each computational cell was calculated by multiplying the deposition height of a cell by its area, and the total deposition of all cells was simply the summation of the volumes in each cell. A summary of the deposition under the different return periods is shown in Table 4.

Figure 20 shows the spatial distribution of deposition depth increase under the baseline and climate change scenarios. Similar increase patterns were observed between the simulated cases, with a larger deposition depth increase located in the middle and downstream of river channels. Deposition depth increase was highest in the main channel as expected, and increased with increasing return periods indicated by dark green areas in Figure 20.

Table 4. Increase in volume of estimated sediment deposition under baseline (1980–1999) and A1B-S (2020–2039) scenarios for various return periods.

Return Period	Baseline (m^3) (1980–1999)	A1B-S (m^3) (2020–2039)	Deposition Volume Increase for Baseline and A1B-S Scenarios (m^3)	Increase Rate (%)
2	9,499,196	9,720,588	221,392	2.33
5	11,984,107	12,425,532	441,425	3.68
10	13,292,933	14,141,714	848,781	6.39
25	14,618,375	16,252,862	1,634,487	11.18
50	15,424,874	17,836,223	2,411,349	15.63
100	16,180,722	19,313,894	3,133,172	19.36
200	16,855,567	20,857,945	4,002,378	23.75

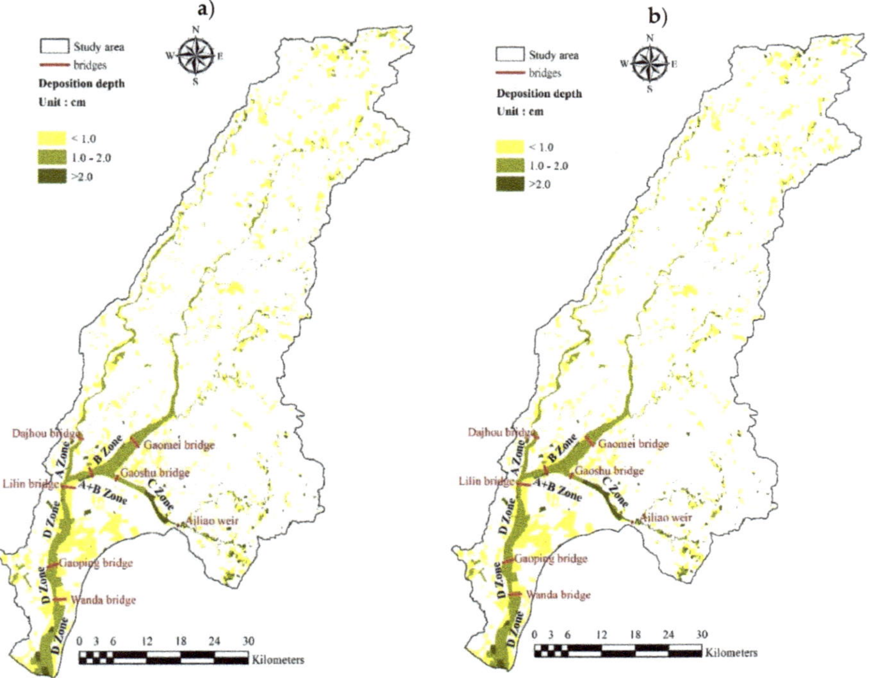

Figure 17. Deposition depth distribution under 10 year return period for (**a**) baseline and (**b**) climate change scenarios.

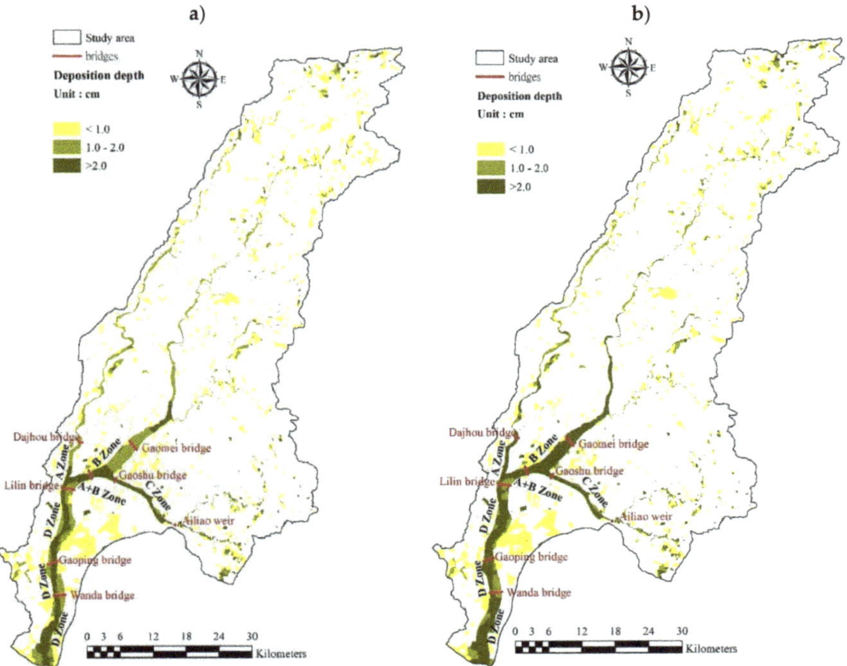

Figure 18. Deposition depth distribution under 100 year return period for (**a**) baseline and (**b**) climate change scenarios.

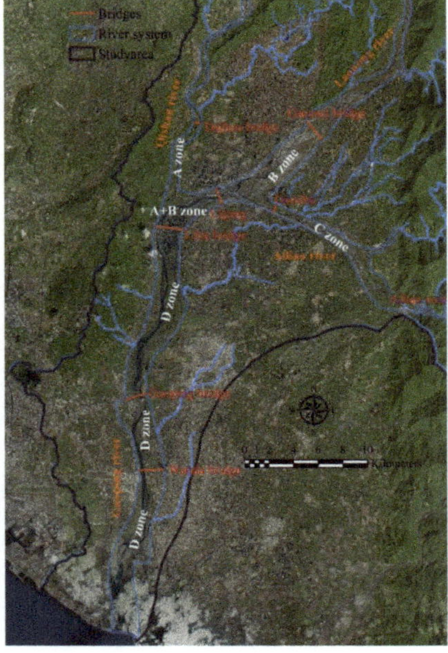

Figure 19. Location of various river sections in the Gaoping River basin that are vulnerable to soil deposition.

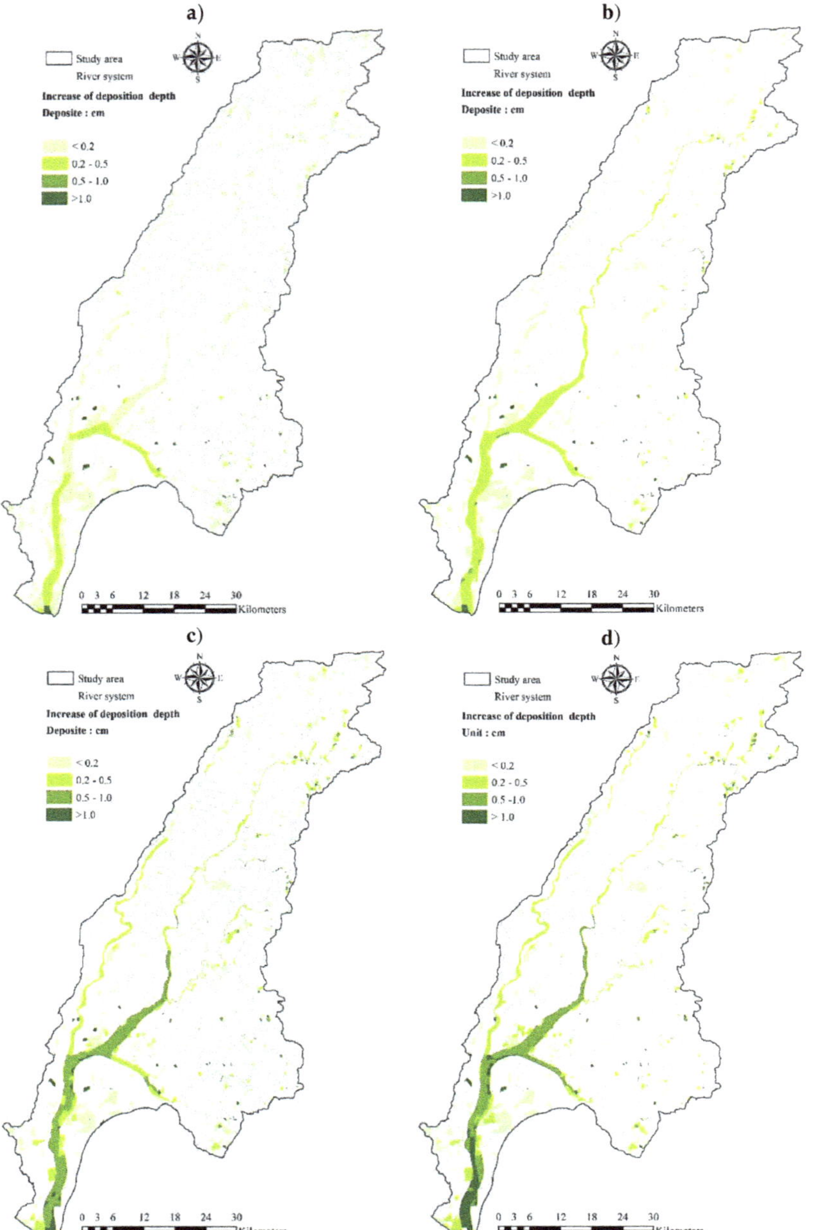

Figure 20. Spatial distribution of deposition depth increase under the influence of climate change for (**a**) 10 year, (**b**) 25 year, (**c**) 100 year and (**d**) 200 year return periods.

Gaoping River basin exhibited high erosion volume and sediment yield. According to statistical data reported by the Water Resources Agency [22], 76,870,000 m^3 have been dredged between 2010 and 2013. Table 5 shows the dredged volume of each year, while Figure 21 indicates the location of the dredged site. Dredged locations coincide with high deposition areas computed by the PSED model, as illustrated by Figure 21.

Table 5. The actual dredged sediment deposition amounts from 2010 to 2013 in each river [22].

River	Actual Dredged Amount (10^4 m^3)			
	2010	2011	2012	2013
Laonong	947.26	1516.37	855.74	623.36
Zhuokou	27.68	75.00	6.49	44.60
Qishan	423.07	241.70	187.77	244.46
Ailiao	555.21	689.79	184.91	139.99
Gaoping	534.61	203.68	104.33	81.54
Total	2487.83	2726.54	1339.24	1133.95

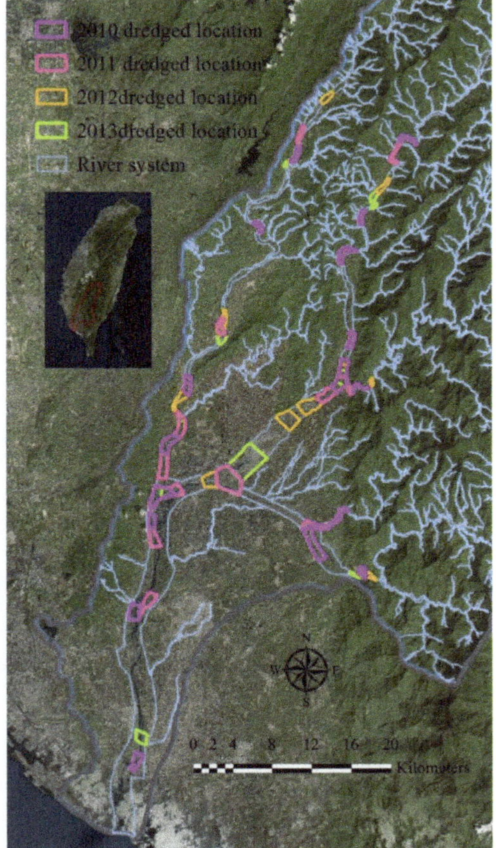

Figure 21. Dredged locations in Gaoping River basin from 2010 to 2013.

In each year, a huge amount of money is needed to carry out the dredging works at Gaoping River. With climate change increasing the deposition rates, not only will there be more pressure on financial resources, but flooding risk is also expected to increase. Hence, appropriate structures and policies should be put in place in order to redress climate change impacts. Proposed strategies include identifying and mapping areas more prone to soil erosion and implementing river management and stability measures. Planning the overall river basin operations and management strategies can effectively control the sediment yield in a watershed, hence reducing sediment deposition in river channels due to soil erosion and eliminating flooding disasters due to limited water passage.

4. Conclusions

This study applied a numerical model to investigate the impacts of climate change on erosion volume, sediment yield and erosion deposition in a watershed. The results showed that precipitation under the A1B-S climate change scenario would significantly increase soil erosion volume, sediment yield and sediment transport rate. Total erosion volume and total sediment yield in the watershed under the A1B-S scenario for various return periods increased by 4–25% and 8–65%, respectively, from 2 year to 200 year return periods. Climate change further increased deposition volume by 2–23% relative to the baseline and by 13% relative to the baseline average. Deposition was found to mostly occur at the river confluences, river middle and at the downstream end of Gaoping River. The study clearly revealed the adverse impacts climate change is likely to bring to this basin; hence, appropriate conservation measures are suggested.

Author Contributions: Conceptualization, methodology, formal analysis, C.-N.C.; data curation, C.-H.T.; validation, writing—review and editing, S.S.T. All authors have read and agreed to the published version of the manuscript.

Funding: This research was funded by the Ministry of Science and Technology (MOST) of Taiwan, grant number MOST-107-2221E-020-006.

Acknowledgments: The authors extend their gratitude toward the anonymous reviewers for their valuable comments to improve the quality of the manuscript.

Conflicts of Interest: The authors declare no conflict of interest.

References

1. Diodato, N.; Filizola, N.; Borrelli, P.; Panagos, P.; Bellocchi, G. The Rise of Climate-Driven Sediment Discharge in the Amazonian River Basin. *Atmosphere* **2020**, *11*, 208. [CrossRef]
2. Babur, M.; Shrestha, S.; Bhatta, B.; Datta, A.; Ullah, H. Integrated Assessment of Extreme Climate and Landuse Change Impact on Sediment Yield in a Mountainous Transboundary Watershed of India and Pakistan. *J. Mt. Sci.* **2020**, *17*, 624–640. [CrossRef]
3. Gupta, S.; Kumar, S. Simulating Climate Change Impact on Soil Erosion Using Rusle Model—A Case Study in a Watershed of Mid-Himalayan Landscape. *J. Earth Syst. Sci.* **2017**, *126*, 43. [CrossRef]
4. Chen, C.-N.; Tsai, C.-H.; Tsai, C.-T. Simulation of Sediment Yield from Watershed by Physiographic Soil Erosion–Deposition Model. *J. Hydrol.* **2006**, *327*, 293–303. [CrossRef]
5. Wang, H.-W.; Kondolf, M.; Tullos, D.; Kuo, W.-C. Sediment Management in Taiwan's Reservoirs and Barriers to Implementation. *Water* **2018**, *10*, 1034. [CrossRef]
6. Stefanidis, S.; Stathis, D. Effect of Climate Change on Soil Erosion in a Mountainous Mediterranean Catchment (Central Pindus, Greece). *Water* **2018**, *10*, 1469. [CrossRef]
7. Jha, M. Impacts of Climate Change on Streamflow in the Upper Mississippi River Basin: A Regional Climate Model Perspective. *J. Geophys. Res.* **2004**, *109*, 1–12. [CrossRef]
8. Abbaspour, K.C.; Faramarzi, M.; Ghasemi, S.S.; Yang, H. Assessing the Impact of Climate Change on Water Resources in Iran. *Water Resour. Res.* **2009**, *45*, 1–16. [CrossRef]
9. Zhang, R.; Corte-Real, J.; Moreira, M.; Kilsby, C.; Birkinshaw, S.; Burton, A.; Fowler, H.J.; Forsythe, N.; Nunes, J.P.; Sampaio, E.; et al. Downscaling Climate Change of Water Availability, Sediment Yield and Extreme Events: Application to a Mediterranean Climate Basin. *Int. J. Climatol.* **2019**, *39*, 2947–2963. [CrossRef]
10. Chang, T.J.; Hsu, M.H.; Lin, G.F.; Lai, J.S.; Pan, T.Y. *Investigation on Analysis Method of Flood Vulnerability and Risk*; Water Resources Agency: Taipei, Taiwan, 2010.
11. Zarghami, M.; Abdi, A.; Babaeian, I.; Hassanzadeh, Y.; Kanani, R. Impacts of Climate Change on Runoffs in East Azerbaijan, Iran. *Glob. Planet. Chang.* **2011**, *78*, 137–146. [CrossRef]
12. Kumar, N.; Tischbein, B.; Kusche, J.; Laux, P.; Beg, M.K.; Bogardi, J.J. Impact of Climate Change on Water Resources of Upper Kharun Catchment in Chhattisgarh, India. *J. Hydrol. Reg. Stud.* **2017**, *13*, 189–207. [CrossRef]

13. Leta, O.T.; El-Kadi, A.I.; Dulai, H.; Ghazal, K.A. Assessment of Climate Change Impacts on Water Balance Components of Heeia Watershed in Hawaii. *J. Hydrol. Reg. Stud.* **2016**, *8*, 182–197. [CrossRef]
14. Thodsen, H.; Hasholt, B.; Kjærsgaard, J.H. The Influence of Climate Change on Suspended Sediment Transport in Danish Rivers. *Hydrol. Process.* **2008**, *22*, 764–774. [CrossRef]
15. Phan, D.B.; Wu, C.C.; Hsieh, S.C. Impact of Climate Change on Stream Discharge and Sediment Yield in Northern Viet Nam. *Water Resour.* **2011**, *38*, 827–836. [CrossRef]
16. Cousino, L.K.; Becker, R.H.; Zmijewski, K.A. Modeling the Effects of Climate Change on Water, Sediment, and Nutrient Yields from the Maumee River Watershed. *J. Hydrol. Reg. Stud.* **2015**, *4*, 762–775. [CrossRef]
17. Zhou, Y.; Xu, Y.; Xiao, W.; Wang, J.; Huang, Y.; Yang, H. Climate Change Impacts on Flow and Suspended Sediment Yield in Headwaters of High-Latitude Regions—A Case Study in China's Far Northeast. *Water* **2017**, *9*, 966. [CrossRef]
18. Azari, M.; Moradi, H.R.; Saghafian, B.; Faramarzi, M. Climate Change Impacts on Streamflow and Sediment Yield in the North of Iran. *Hydrol. Sci. J.* **2016**, *61*, 123–133. [CrossRef]
19. Zhang, S.; Li, Z.; Lin, X.; Zhang, C. Assessment of Climate Change and Associated Vegetation Cover Change on Watershed-Scale Runoff and Sediment Yield. *Water* **2019**, *11*, 1373. [CrossRef]
20. Renard, K.G.; Foster, G.R.; Weesies, G.A.; Porter, J.P. Rusle: Revised Universal Soil Loss Equation. *J. Soil Water Consv.* **1991**, *46*, 30–33.
21. Tfwala, S.; Wang, Y.-M. Estimating Sediment Discharge Using Sediment Rating Curves and Artificial Neural Networks in the Shiwen River, Taiwan. *Water* **2016**, *8*, 53. [CrossRef]
22. Water Resources Agency. *Assessment of the Efficiency of Dredging Engineering in the Kaoping River Watershed*; Ministry of Economic Affairs: Taichung, Taiwan, 2014.
23. Yu, S.-W.; Tsai, L.L.; Talling, P.J.; Lin, A.T.; Mii, H.-S.; Chung, S.-H.; Horng, C.-S. Sea Level and Climatic Controls on Turbidite Occurrence for the Past 26kyr on the Flank of the Gaoping Canyon Off Sw Taiwan. *Mar. Geol.* **2017**, *392*, 140–150. [CrossRef]
24. Central Geological Survey. *Geological Information Service*; Affairs, M.O.E., Ed.; Central Geological Survey: Taipei, Taiwan, 2000.
25. Chen, C.N.; Tsai, C.H.; Tsai, C.T. Simulation of Runoff and Suspended Sediment Transport Rate in a Basin with Multiple Watersheds. *Water Resour. Manage.* **2011**, *25*, 793–816. [CrossRef]
26. Lin, C.-P.; Chen, C.-N.; Wang, Y.-M.; Tsai, C.-H.; Tsai, C.-T. Spatial Distribution of Soil Erosion and Suspended Sediment Transport Rate for Chou-Shui River Basin. *J. Earth Syst. Sci.* **2014**, *123*, 1517–1539. [CrossRef]
27. Taiwan Agricultural Research Institute. *Soil Maps of Taiwan*; Council of Agriculture: Taichung, Taiwan, 2004.
28. National Land Surveying and Mapping Center. *Land Use Maps of Taiwan*; Ministry of the Interior: Taichung, Taiwan, 2006.
29. Intergovernmental Panel on Climate Change. *Climate Change 2014: Synthesis Report*; IPCC: Geneva, Switzerland, 2014; p. 151.
30. Chen, C.-N.; Tfwala, S. Impacts of Climate Change and Land Subsidence on Inundation Risk. *Water* **2018**, *10*, 157. [CrossRef]
31. Pal, S.C.; Chakrabortty, R. Simulating the Impact of Climate Change on Soil Erosion in Sub-Tropical Monsoon Dominated Watershed Based on Rusle, Scs Runoff and Miroc5 Climatic Model. *Adv. Space Res.* **2019**, *64*, 352–377. [CrossRef]
32. Giang, P.Q.; Giang, L.T.; Toshiki, K. Spatial and Temporal Responses of Soil Erosion to Climate Change Impacts in a Transnational Watershed in Southeast Asia. *Climate* **2017**, *5*, 22. [CrossRef]

Article

Variations in Soil Erosion Resistance of Gully Head Along a 25-Year Revegetation Age on the Loess Plateau

Zhuoxin Chen [1], Mingming Guo [1,2] and Wenlong Wang [1,*]

[1] State Key Laboratory of Soil Erosion and Dryland Farming on the Loess Plateau, Institute of Water and Soil Conservation, Northwest A&F University, Xianyang 712100, China; xiyu.zxchen@foxmail.com (Z.C.); guomingming@iga.ac.cn (M.G.)

[2] Key Laboratory of Mollisols Agroecology, Northeast Institute of Geography and Agroecology, Chinese Academy of Sciences, Harbin 150081, China

* Correspondence: wlwang@nwafu.edu.cn

Received: 10 October 2020; Accepted: 19 November 2020; Published: 24 November 2020

Abstract: The effects of vegetation restoration on soil erosion resistance of gully head, along a revegetation age gradient, remain poorly understood. Hence, we collected undisturbed soil samples from a slope farmland and four grasslands with different revegetation ages (3, 10, 18, 25 years) along gully heads. Then, these samples were used to obtain soil detachment rate of gully heads by the hydraulic flume experiment under five unit width flow discharges (2–6 m^3 h). The results revealed that soil properties were significantly ameliorated and root density obviously increased in response to restoration age. Compared with farmland, soil detachment rate of revegetated gully heads decreased 35.5% to 66.5%, and the sensitivity of soil erosion of the gully heads to concentrated flow decreased with revegetation age. The soil detachment rate of gully heads was significantly related to the soil bulk density, soil disintegration rate, capillary porosity, saturated soil hydraulic conductivity, organic matter content and water stable aggregate. The roots of 0–0.5 and 0.5–1.0 mm had the highest benefit in reducing soil loss of gully head. After revegetation, soil erodibility of gully heads decreased 31.0% to 78.6%, and critical shear stress was improved by 1.2 to 4.0 times. The soil erodibility and critical shear stress would reach a stable state after an 18-years revegetation age. These results allow us to better evaluate soil vulnerability of gully heads to concentrated flow erosion and the efficiency of revegetation.

Keywords: soil erosion; gully erosion; vegetation restoration; soil erodibility; land use

---

1. Introduction

Soil erosion is recognized as a global environmental problem, which severely damages infrastructure, causes land degradation and water pollution, and threatens the safety of human production and life [1–3]. In the past few decades, many scholars have made many efforts to study the process and mechanism of soil erosion, establish many soil erosion prediction models and try to control soil erosion [4–7]. At present, a set of soil erosion control measures system integrating engineering measures, agricultural measures, and biological measures has been formed [8–10], especially vegetation measures play an extremely important role in soil erosion control [11,12].

Previous studies have shown that revegetation can effectively reduce soil erosion. For example, Wang et al. [13] found that soil detachment capacity of abandoned farmland was 1.02 to 2.29 times greater than four restored lands. Li et al. [14] reported that the ratios of the soil detachment capacity of cropland to those of orchard, shrubland, woodland, grassland, and wasteland were 7.14, 12.29, 25.78, 28.45, and 46.43, respectively. The improvement of soil erosion resistance by revegetation is mainly controlled by the combination of soil properties and root traits [15–18]. In terms of soil properties, many studies have verified that the revegetation significantly affects soil erosion by changing the soil bulk density, organic matter content, and water-stable aggregate [19–21]. Furthermore, the vegetation root zone is the dynamic interface of soil–plant–atmosphere continuum in partitioning rain and irrigation water into evaporation, transpiration, runoff, and deep drainage [22,23], but is also the home of "green water" which is the source of plant nutrition [24]. Especially, the vegetation root systems also play a great role in protecting soil against flow scouring by affecting soil water movement [25,26]. Root-permeated soils exhibited lower erosion rates primarily through increasing the required shear stress before detachment [27]. Moreover, root growth can bind and bond soil particles and aggregate, thus, enhances soil resistance to erosion [28]. Some root parameters, for example root biomass, length density, and surface area density, were used to estimate the effect of root on soil detachment [21,28–31]. De Baets and Poesen [25] found that soil detachment rate reduced exponentially with increasing root biomass. Some studies also showed that soil detachment was related to root architecture and fibrous root was more effective than tap root in reducing soil loss [24,31]. However, the most of previous studies only focus on the impact of revegetation on soil erosion resistance of hillslopes. In the watershed dominated by gully erosion, the gully head is the main source of soil erosion, but the effect of revegetation on soil erosion resistance of gully heads remains unclear. Therefore, there is a strong need to understand the effect of revegetation on soil erosion of gully head by concentrated flow to develop a more reasonable vegetation model.

Notably, in the gully region of the Loess Plateau, about 63% of total runoff is generated from the loess tableland with a gentle slope of 1–5°, which can initiate gully headcut erosion and contribute 86.3% of total sediment [32]. The gully headcut erosion by concentrated flow became the main sediment resource. At present, the gully headcut erosion was controlled effectively due to the implementation of a series of control measures (e.g., the "Three Protection Belts" and the "Green for Grain" project), which, to some extent, was attributed to the fact that the revegetation improves the soil resistance of gully heads to concentrated flow [21,33]. Since some ecological restoration projects were conducted, land use has changed dramatically in the Loess Plateau [34]. Hence, the land use has changed, and the natural succession of vegetation was promoted [35]. With progression in natural restoration of grassland, soil physical and chemical properties and vegetation characteristics (e.g., coverage, community structure, species composition and diversity, and root diameter, density, and diameter distribution) varied greatly [36,37]. These changes would result in dynamic variations in soil erosion resistance. However, the response of soil erosion resistance to vegetation succession process mainly focused on the hillslope in the hilly-gully region of Loess Plateau [15,16], and few studies were conducted to explore the response of soil resistance of gully heads by concentrated flow to vegetation succession process.

Therefore, to evaluate the effect of revegetation process on soil erosion resistance of gully heads and optimize revegetation measures for controlling gully headcut erosion in the gully region of the Loess Plateau, we selected four grasslands with different revegetation ages (3, 10, 18, 25 years) along gully heads with the slope farmland as the control. This study aimed to (1) quantify the effect of revegetation age on soil detachment by concentrated flow, (2) clear the relationships between soil detachment rate and soil and root properties, and (3) confirm the dynamic variation in soil erosion resistance of gully head with revegetation age.

2. Material and Method

2.1. Study Area

The study was conducted in the Nanxiaohegou watershed in the Xifeng Research Station of Soil and Water Conservation (35°41′–35°44′ N, 107°30′–107°37′ E). The watershed has an area of 36.3 km^2 and altitudes ranging from 1050 to 1423 m above mean sea level (Figure 1) in the typical gully region of the Loess Plateau. The climate is temperate continental semiarid. The mean temperature is 10 °C, and the frost-free period is 160–180 days. Annual precipitation is approximately 523 mm, which has the characteristic of annual variation and uneven distribution during the year. In the form of short heavy storms, 58.8% of the rainfall occurs from July to September. The soil type is yellow loamy soil. The original vegetation has disappeared due to human activities. Gully headcut erosion is the main resource of sediment yield in the watershed. Since the 1970s, some soil and water conservation projects, for example the "three protection belts" project and the "Green for Grain" project and so on, were implemented to control soil and water loss, and the vegetation cover of the Loess Plateau increased to 59.6% in 2013. Additionally, the land use has undergone tremendous changes [38]. These efforts also effectively stabilized the gully heads and thus contained the gully headcut erosion [33]. At present, the annual soil erosion module is effectively controlled at the level of 2440 t km^{-2} a^{-1} in the study area, and the vegetation communities comprise mainly planted forests and shrubs and native secondary herbaceous plants [21].

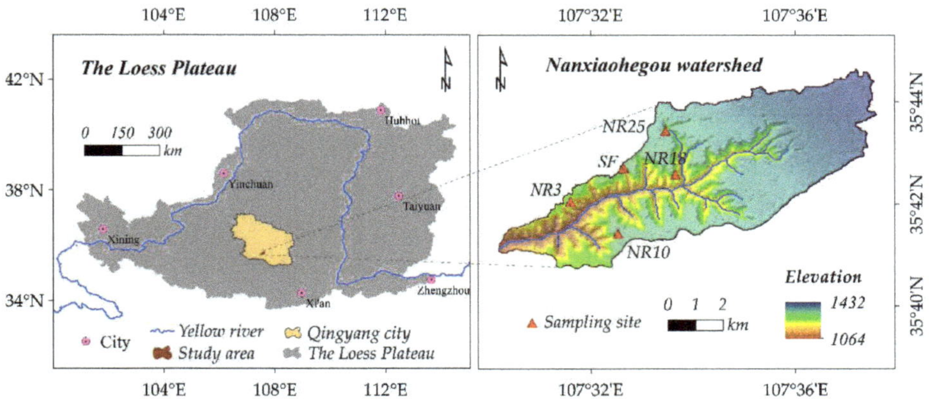

Figure 1. Location of the study area on the Loess Plateau and the location of sampling sites in the Nanxiaohegou watershed.

2.2. Sampling Sites Selection

During our investigation of gully heads, some cracks developed near gully heads. Kompani-Zare et al. [39] and Guo et al. [21] stated that soil samples from 0 to 30 cm depth near the gully heads (the distance was less than 5 m) can represent soil properties of the gully heads. Therefore, in consideration of collapsibility and vertical joints development of loess, the sampling plots were established about 1.0 m meters from gully heads to ensure safety. As a result, four natural restoration grasslands with different ages (3, 10, 18, 25 years) were selected (Figure 1). The natural restoration age was confirmed by consulting the village elders and scientists at the scientific experimental station. The slope aspect and gradient, elevation, soil type, and previous farming practices of the selected sites were similar to minimize the effects of these factors on the experimental results. For comparison, one corn-planted farmland site, with a topography similar to that of the grasslands, was selected as a control. The basic information of the five selected sites is listed in Table 1.

Table 1. Basic information of the selected five sampling sites.

Site Code	Restoration Age (yr)	Slope (°)	Coverage (%)	Altitude (m)	Dominant Communities	Main Companion Species
SF	0	2.4	—	1420	*Zea mays*	*Setaria viridis*
NR3	3	2.9	72.3	1405	*Artemisia capillaris*	*Artemisia sacrorum*
NR10	10	2.4	80.8	1401	*Artemisia sacrorum*	*Artemisia capillaris*
NR18	18	3.2	93.4	1390	*Artemisia sacrorum*	*Artemisia capillaris*
NR25	25	3.1	91.2	1380	*Bothriochloa ischaemum*	*Artemisia sacrorum* + *Lespedeza daurica*

2.3. Sampling and Measurement of Soil and Root

In this study, seven soil and root property parameters including soil bulk density, capillary porosity, soil disintegration rate, soil water-stable aggregate, soil saturated hydraulic conductivity, organic matter content, and root mass density were measured. Firstly, three repeated sampling plots (5 m × 5 m) were established in each of gully head sampling sites with the litter layer removed, and topsoil samples (0–30 cm) were collected. Then, three cutting rings (200 cm^3) were used to randomly collect soil samples in each plot, and a total of nine samples were oven-dried at 105 °C for 24 h to determine the soil bulk density of each gully head site. Similarly, the other 9 soil samples were also collected by cutting rings of 200 cm^3 to determine the soil saturated hydraulic conductivity by applying the constant water head test method. Three cutting rings (100 cm^3) were used to collect soil samples for the measurement of soil capillary porosity [33]. Three man-made steel cubical boxes (5 cm in length) were used to collect soil samples for measuring soil disintegration rate by using a disintegration box [14,40]. Lastly, the other three samples were randomly collected in each plot to form a mixed sample. A total of 45 mixed samples were obtained and used for laboratory analyses of organic matter content and water-stable aggregate and its stability. These mixed soil samples were air-dried at room temperature, with large roots and organic residues manually removed. Sieves with apertures (0.25, 0.5, 1.0, 2.5, and 5.0 mm) were used to test the water-stable aggregate. The potassium dichromate external heating method was used to measure the soil organic matter content.

2.4. Hydraulic Flume Experiments

A hydraulic flume experiment was conducted to determine the soil resistance to concentrated flow upstream gully heads (Figure 2). The size of the flume was 2.0 m long and 0.15 m wide similar to the one used by De Baets et al. [28,31], which was enough to make water flow along the slope soil. An opening (0.5 m length and 0.1 m wide) was set at the bottom of the flume, and a metal sample box with the same size was used to collect undisturbed soil samples so that the surface of the soil sample was at the same level of the flume surface. The space between sampling box and flume edge sealed with painter's mastic to prevent boundary effects. According to the study of Guo et al. [40], the flume experiment was carried out under five different unit width flow discharges of 2, 3, 4, 5, 6 $m^3\ s^{-1}$, and thus, a total of 100 samples (5 sites × 5 flow discharges × 4 replications) were collected to measure soil resistance of gully heads. To simulate real flow generation conditions, the soil should be saturated by using a watering pot before experiment. During the experiment, a portable flow meter instrument (LS300-A) with 1.5% accuracy was used to measure flow velocity which was regarded as the flow velocity scoring soil area. Runoff and sediment samples were collected with sampling tanks, and the sampling time was recorded. The measured flow velocity was modified according to flow regime [41]. Sampled sediment was oven-dried at 105 °C for 24 h to determine the soil loss amount (*SLA*, kg). Thus, the soil detachment rate (D_r, kg $m^{-2}\ s^{-1}$) could be calculated as follows: please check font size, please check all reference citation

$$D_r = \frac{SLA}{AT} \tag{1}$$

where *SLA* is the oven-dry mass of every sediment sample (kg), *T* is the experimental period (s) and *A* is the soil sample area (m^2). In addition, the relative soil detachment rate (RD_r) was calculated as the ratio between D_r for the root-permeated soil samples and that for the farmland topsoil samples, tested at the same condition [28].

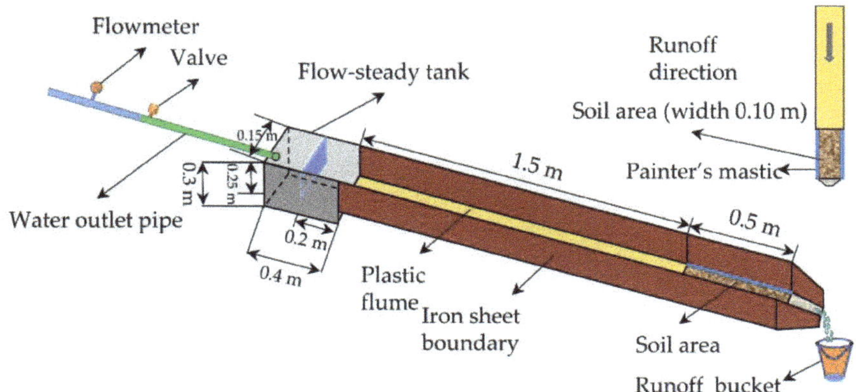

Figure 2. Skitch of scouring flume for determining soil erosion resistance of gully heads.

In addition, the flow depth (h, m) and shear stress (τ, Pa) were also calculated as follows:

$$h = \frac{q}{vw} \quad (2)$$

$$\tau = \rho g h S \quad (3)$$

where h is flow depth (m), q is the flow discharge (m^3 s^{-1}), w is the width of the flume (m), v is the mean flow velocity (m s^{-1}). ρ is the water mass density (kg m^{-3}), g is the gravity constant (m s^{-2}), and S is the slope steepness (m m^{-1}).

After each scouring test, a steel cubical box (10 cm in length) was used to take soil sample in the center of soil area of scouring flume, and the sample was soaked in tap water for about one hour to increase the dispersion of soil and then were placed on a 0.25 mm sieve and washed with tap water using low-pressure head. The living roots, plant debris and some pebbles were left on the sieve. Only the living roots were picked out carefully using tweezers one by one [15]. The washed roots were classified into 4 levels (0–0.5, 0.5–1.0, 1.0–2.0, and >2.0 mm) by vernier caliper and then were oven-dried for 24 h at 65 °C and weighed to calculate root mass density (RBD, kg m^{-3}).

2.5. Parameter Calculation after the Experiments

Soil particle is detached when flow shear stress exceeds the critical shear stress [6]. Soil erodibility parameter (K_r) and critical shear stress (τ_c) were estimated for every natural restoration stage as the slope coefficient and intercept on the abscissa axis of the regression line between soil detachment rate and shear stress as described in the WEPP model as follow:

$$D_r = K_r(\tau - \tau_c) \quad (4)$$

Generally, soil detachment rate can be considered as zero when root reached the infinity. To quantify the relationship between detachment rate and root mass density, the Hill curve was selected to simulate the relationship between them [21,31,42]. The Hill curve is expressed as follows [43]:

$$RD_r = \frac{KX_r{}^a}{X_r{}^a + b}, \ (K > 0,\ a < 0,\ b > 0) \quad (5)$$

where RD_r is relative soil detachment rate; X_r is root mass density; K, a and b are constants. The parameter a determines the shape of the curve, b determines the steepness of the curve and K is the asymptote of D_r for infinitesimal X_r values. Additionally, the Hill curve can be used to evaluate the ability of roots to increase soil resistance against concentrated flow erosion. According to Li et al. [44], b^(1/a) is plant specific and can be used as an index to compare the effectiveness of different plant roots in reducing soil erosion rates: the lower $b^{\wedge}(1/a)$, the more effective the plant root. When the value of X_r is $b^{\wedge}(1/a)$, the soil detachment rates is reduced by 50%.

2.6. Statistical Analysis and Plotting

The analysis of one-way ANOVA followed by multiple comparisons with LSD was applied to assess the differences of soil properties (Soil bulk density, soil capillary porosity, soil disintegration rate, saturated hydraulic conductivity, organic matter content, and water-stable aggregate) and root mass density among the five revegetation ages. All soil and root variables of each revegetation ages were tested whether the data exhibited a normal distribution and variance homogeneity by Shapiro-Wilk test and Levene test, respectively. If the data failed to meet the two conditions, the Kruskal–Wallis test was performed for the above analysis. The interaction effect of flow discharge and revegetation age was detected using a two-way ANOVA. Pearson's correlation analysis was used to determine linkages among soil properties, root mass density, and soil detachment rate. Relationships among soil detachment rate, soil properties, flow shear stress and restoration age were analyzed by the regression method. The data analyses were conducted in SPSS v. 16.0 statistical software (IBM Corp., Armonk, NY, USA). The figure plotting was conducted by Origin v. 2020 (OriginLab Corp., Northampton, MA, USA).

3. Results and Discussion

3.1. Effect of Revegetation on Soil and Root Properties of Gully Heads

Figure 3 illustrates that the six soil properties of gully heads exhibited a significant increase or decrease with revegetation age. Compared with slope farmland, the soil bulk density (SBD) and soil disintegration rate (SDR) of revegetated gully heads significantly decreased by 5.7–18.6% and 28.8–80.5%, respectively ($p < 0.05$, Figure 3a,c), while the soil capillary porosity (SCP), saturated soil hydraulic conductivity (SHC), organic matter content (OMC) and water-stable aggregate (WSA) significantly increased by 3.9–13.8%, 17.4–236.2%, 34.2–221.8%, and 27.7–64.4%, respectively ($p < 0.05$, Figure 3b,d–f). Figure 4 illustrates that the roots of 1–2 mm in slope farmland had the relatively higher root mass density (RMD, 0.20 kg m^{-3}) and accounted for 39% of total RMD. Notably, after revegetation, the RMD of >2.0 mm was significantly greater than those of the other three root diameters, and it can account for 40–61% of total RMD. When revegetation age was greater than 3 years, there was a significant difference in RMD among four root diameter levels ($p < 0.05$). In addition, we found that the RMD of four root diameters (except for >2.0 mm) showed a non-significant increase in the first three-years and then significantly increased.

These results were similar to previous findings regarding the effects of revegetation on soil properties [45–47]. In fact, the improvement of soil properties of gully heads with revegetation age can be attributed to the accumulation of fresh plant residues in surface soil as well as roots and decomposed root residues in subsurface soil [48]. These materials can be directly transformed into soil organic matter and thus provide energy/carbon sources and nutrients for soil microorganisms [49,50], further promoting the development of soil aggregation and enhancing the cohesion of soil particles [51]. Hence, vegetation restoration would induce the formation of macroaggregates and increase the water stability of aggregates [19,52].

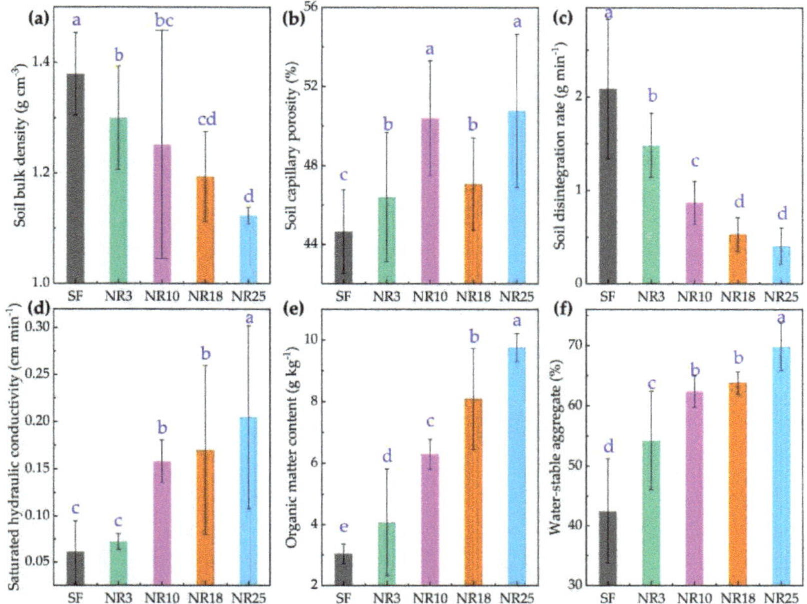

Figure 3. Variation in soil properties with revegetation age. Note: Bar means the 95% confidence interval (95% CI). Different lowercase letters indicate significant difference among different revegetation ages ($p < 0.05$). (**a**) Soil bulk density (SBD); (**b**) Soil capillary porosity (SCP); (**c**) Soil disintegration rate (SDR); (**d**) Saturated soil hydraulic conductivity (SHC); (**e**) Organic matter content (OMC); (**f**) Water-stable aggregate (WSA).

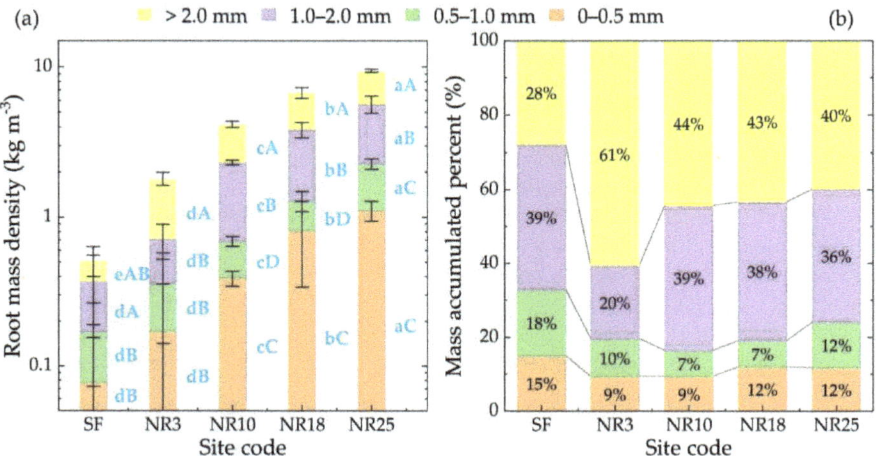

Figure 4. Changes in root mass density (**a**) and its proportion (**b**) of different root diameters with revegetation age. Note: Bar means the 95% confidence interval (95% CI). SF refers to the slope farmland. NR3, NR10, NR18, NR25 represents the 3, 10, 18, and 25 years of natural restoration time, respectively. Different capital letters for the same restoration age indicate a significant difference among different root diameters ($p < 0.05$), and different lowercase letters for the same root diameter level indicate a significant difference among different revegetation ages ($p < 0.05$). (**a**) Root mass density; (**b**) Mass accumulated percent.

3.2. Effect of Revegetation Age on Soil Detachment of Gully Heads

As illustrated in Figure 5a, the D_r of gully heads showed a significant decrease during the 25-year revegetation. This result was not agreed with the conclusion of Wang et al. [16] who stated that soil detachment capacity of sloped lands fluctuated with abandonment time, and the soil detachment capacity of the slope farmland was significantly greater than those of the abandoned farmlands. The difference was mainly attributed to the great difference in erosion environment (e.g., plant type, geomorphological feature, climate) significantly affecting the succession process [36,47]. The mean D_r of slope farmland was 1.6 to 3.0 times greater than those of revegetated gully heads, which indicated that the revegetation played a role in enhancing the soil resistance of gully heads to concentrated flow erosion.

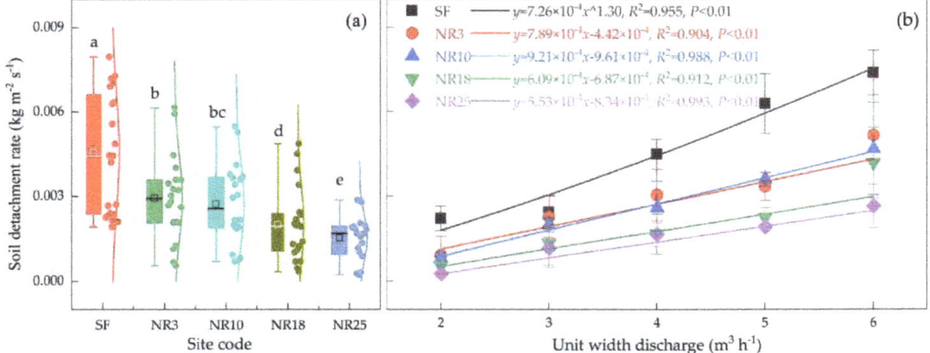

Figure 5. Change in soil detachment rate of gully heads with restoration age (**a**), and its relationships with flow discharge (**b**). Note: Bar means the 95% confidence interval (95% CI). SF refers to the slope farmland. NR3, NR10, NR18, NR25 represents the 3, 10, 18 and 25 years of revegetation age, respectively. The different lowercase letters indicate a significant difference among different revegetation ages ($p < 0.05$). (**a**) Site code; (**b**) Unit width discharge.

Figure 5b shows the D_r of gully heads of slope farmland and four restored grasslands varied with flow discharge. The optimal relationships between D_r and flow discharge were fitted, which can reflect the response of D_r of gully heads to concentrated flow induced by rainstorms of different recurrence intervals. It was found that the response of D_r of slope farmland to flow discharge could be expressed as a power function ($y = m \times x^n$), and the n-value was greater than 1, indicating the soil loss of gully heads increases at an increased speed with increasing flow. However, for the restored gully heads, the optimal relationships between D_r and flow discharge could be described by a series of linear functions ($y = p \times x + q$), and the p-value decreased with revegetation age, indicating that the sensitivity of D_r of the gully heads to concentrated flow erosion gradually decreased with increasing restoration age. Besides, the interacted effect of revegetation age and unit width discharge significantly affected D_r ($p < 0.001$) (Table 2).

Table 2. Summary of two-way ANOVAs tests.

Source	SS	Df	MS	F	p-Value
Revegetation age	1.08×10^{-4}	4	2.69×10^{-5}	151.44	<0.001
Unit width discharge	1.75×10^{-4}	4	4.38×10^{-5}	246.48	<0.001
Revegetation age × Unit width discharge	2.29×10^{-5}	16	1.43×10^{-6}	8.07	<0.001
Error	1.33×10^{-5}	75	1.78×10^{-7}		
Total	0.0011	100			

3.3. Response of Soil Detachment to Soil Properties

Figure 6 showed that D_r was positively correlated with soil bulk density and soil disintegration rate ($p < 0.01$), but negatively correlated with capillary porosity, saturated hydraulic conductivity, organic matter and water-stable aggregate of >0.25 mm ($p < 0.01$). Regression analysis showed that D_r increased with soil bulk density as a power function (Figure 7a), which showed an opposite trend with the Wang et al. [13] and Yu et al. [15]. Lower soil bulk density was caused by greater root physical and soil organisms' activities, and thus a soil with lower bulk density was harder to be detached. Additionally, D_r decreased with capillary porosity as a logarithmic function (Figure 7b), which was caused by physically binding and chemically bonding effect of root improving soil structure and porosity and hence increasing soil resistance to erosion [28,53]. Soil disintegration rate referred to the dispersion speed of soil contacting with water, which is an important factor determining soil resistance to erosion [14]. In this study, the soil disintegration rate decreased with the revegetation time (Figure 3c) and D_r increased linearly with an increase in soil disintegration rate (Figure 7c). This is attributed to root wedging mechanism preventing soil from detaching that roots can bind soil and tie surface soil layer into strong and stable subsurface soil layer [14,54]. Saturated soil hydraulic conductivity is an integrating parameter for several physical characteristics such as bulk density, porosity, and mechanical composition. The conclusion that the D_r decreased with increasing soil hydraulic conductivity by a power function is reasonable (Figure 7d) because this study and previous research findings have also indicated that changes in soil bulk density and porosity influenced soil detachment and also were affected by revegetation (e.g., Neves et al. [55]; Zhang et al. [56]). A negative power function was found between D_r and soil organic matter content (Figure 7e). The accumulation of soil organic matter in soil could promote the formation of aggregate and enhance the cohesion of soil particles [57]. Hence, water-stable aggregate also was an indicator determining soil resistance to flow erosion [19]. The D_r decreased as a linear function of water-stable aggregate of >0.25 mm (Figure 7f). The results were in agreement with the findings of Li et al. [14]. However, in Wang et al. [13,16] studies, no significant relationships were found between D_r and organic matter and water-stable aggregate of >0.25 mm, probably caused by small variations of the two factors in their studies and difference in land use between their studies and this study (Podwojewski et al. [58]).

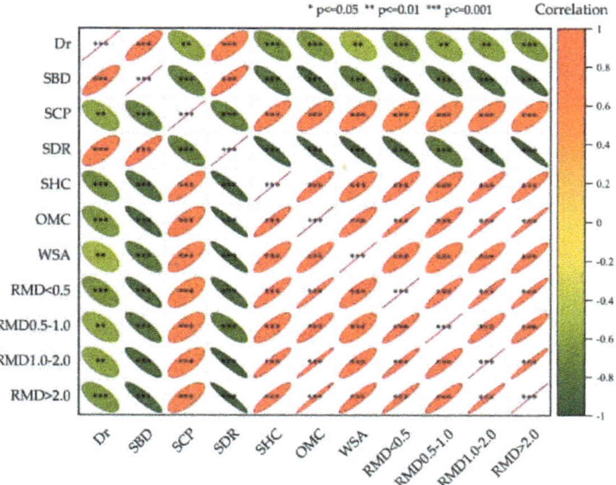

Figure 6. Correlation matrix among soil detachment rate, soil properties, and root mass density. Note: Dr, SBD, SCP, SDR, SHC, OMC, WAS, RMD < 0.5, RMD 0.5–1.0, RMD 1.0–2.0, and RMD > 2.0 refers to the soil detachment rate, soil bulk density, soil capillary porosity, soil disintegration rate, saturated soil hydraulic conductivity, organic matter content, water-stable aggregate, root mass density of <0.5 mm, root mass density of 0.5–1.0 mm, root mass density of 1.0–2.0 mm, and root mass density of >2.0 mm, respectively.

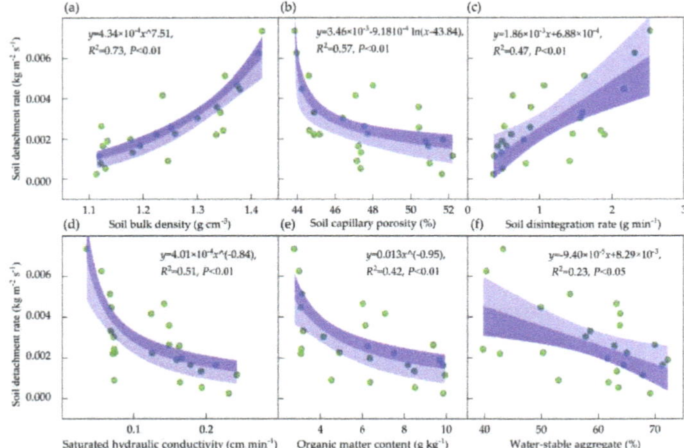

Figure 7. Relationships between soil detachment rate and soil properties. (**a**) Soil bulk density (SBD); (**b**) Soil capillary porosity (SCP); (**c**) Soil disintegration rate (SDR); (**d**) Saturated soil hydraulic conductivity (SHC); (**e**) Organic matter content (OMC); (**f**) Water-stable aggregate (WSA).

3.4. Response of Soil Loss of Gully-Head to Root Traits

Significant correlation was found between D_r and RMD of 0–0.5 mm, 0.5–1.0 mm, 1.0–2.0 mm and >2.0 mm ($p < 0.01$, Figure 6), of which the RMD of 0–0.5 mm had the highest correlation with D_r, indicating that roots of each diameter level had the significant impact on soil erosion of gully heads, especially the fibrous root of 0–0.5 mm. Furthermore, the Hill curve could well simulate the relationships between D_r and RMD of different root diameters with R^2 varying from 0.42 to 0.57 (Figure 8). As illustrated in Figure 8, the D_r showed a rapid decrease when RMD of 0–0.5, 0.5–1.0, 1.0–2.0 and >2.0 mm ranged from 0 to 0.25 kg m^{-3}, 0 to 0.3 kg m^{-3}, 0 to 0.5 kg m^{-3}, and 0 to 1.0 kg m^{-3}, respectively, implying that soil erosion of gully heads could be controlled once vegetation restoration or root growth in soil. Although the roots were limited in density and flexible in early revegetation stage, whereas, roots can contribute to soil cohesion and additional strength, and be crucial in reduction of soil erosion [28,59]. Additionally, root system can bind soil and tie surface soil layer into strong and stable subsurface soil layer [54]. Well-developed root system had great physical binding and chemical bonding effect that could well bind soil particles and soil aggregates together and enhance soil resistance to erosion [16,28,42].

In addition, judged by fitted efficiency (R^2), the optimal results were found in RMD of 0–0.5 mm (Figure 8a), indicating that fibrous root of 0–0.5 mm is the optimal root system reducing soil loss of gully heads. However, Li et al. [44] reported that the ability of plant roots to decrease soil erosion mainly depended on the number of fibrous roots <1.0 mm. Shangguan et al. [53] also found a similar result but recommended root surface area density as the root variable. The reason may be that plant species with contrasting root architectures have a different erosion reduction effect [25]. Additionally, Amezketa [60] and Gyssels et al. [61] reported that monocotyledonous plants are superior to dicotyledonous plants and grasses are better than cereals in stabilizing soil aggregates.

According to Li et al. [44], b^(1/a) can be used as an indicator to compare the effectiveness of different diameter roots in reducing soil erosion. The lower b^(1/a), the more effective the diameter root. The relatively lower b^(1/a) values (0.132 and 0.131 kg m^{-3}) were found in the roots of 0–0.5 mm and 0.5–1.0 mm than 1.0–2.0 mm and >2.0 mm (Figure 8), indicating that the 0–0.5 mm and 0.5–1.0 mm are the most effective roots in reducing soil erosion of gully heads. However, De Baets et al. [25] study the effect of the mixed community of four grasses [*Lolium perenne* (variety: tove), L. *perenne* (variety: starlet), *Festuca rubra* (variety: echo), and *F. arundinacea* (variety: starlet)] on SDR and found the b^(1/a)

value is 0.79 kg m^{-3} that is greater than that of our study. The result fully indicated that the different plant communities had the markedly different influences on reducing soil erosion. The result also suggested us reasonably choosing plant species with different root architectures and root diameters for revegetation at gully heads.

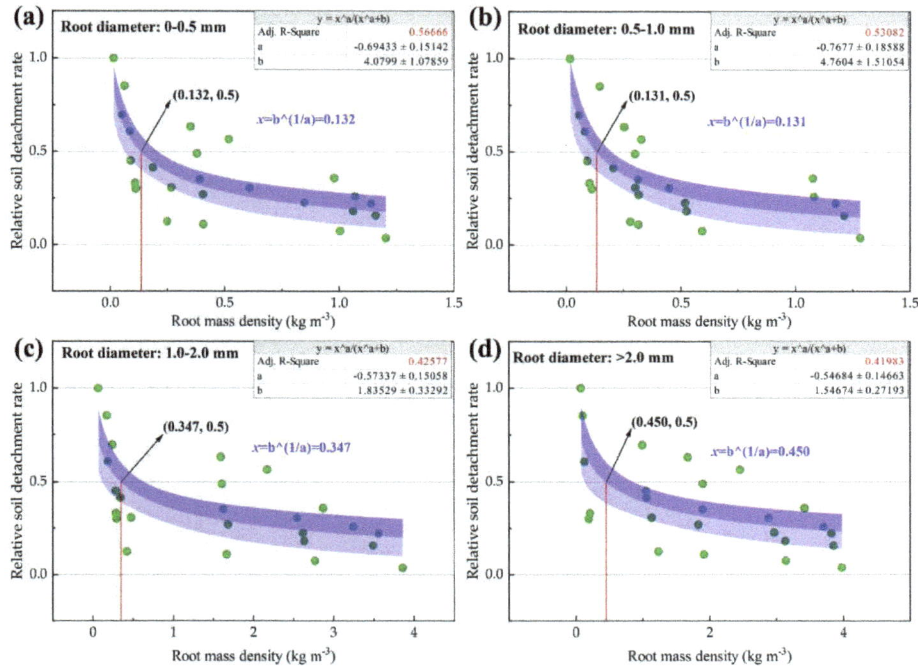

Figure 8. Relationships between relative soil detachment rate and root mass density of different root diameters. (**a**) Root of 0–0.5 mm; (**b**) Root of 0.5–1.0 mm; (**c**) Root of 1.0–2.0 mm; (**d**) Root > 2.0 mm.

3.5. *Effect of Revegetation Age on Soil Erosion Resistance of Gully Head*

Rill soil erodibility parameter (K_r) and critical shear stress (τ_c) were employed to characterize the soil erosion resistance of gully heads [13,21], and were determined by the WEPP model (Equation (4)). The fitted linear function between D_r and shear stress was illustrated in Figure 9. The slope of the fitted line is equal to the K_r, and the K_r of the restored grasslands were 31% to 78.6% less than that of slope farmland. In addition, we found that the soil erodibility of 3-year restored grassland rapidly declined by 31% compared with the slope farmland, indicating the short-term revegetation can rapidly reduce soil erodibility of gully heads. The K_r of grasslands in this study ranged from 0.0009 to 0.0029 s m^{-1}, which were less than those reported by Li et al. [14]. Wang et al. [13] found that averaged K_r of restored lands of abandoned farmland was 0.0024 s m^{-1} that was close to those of this study. The difference was mainly caused by differences in land use, plant species and restoration time. The soil samples were taken from different vegetation restoration models (korshinsk peashrub, black locust, Chinese pine and mixed forest of amorpha and Chinese pine) in the study of Wang et al. [13], and the restoration age (37 years) was greater than that of this study (3 to 25 years). Regression analysis found that the K_r decreased with restoration time in an exponential function and showed a slight change when restored age was greater than 18 years (Figure 10).

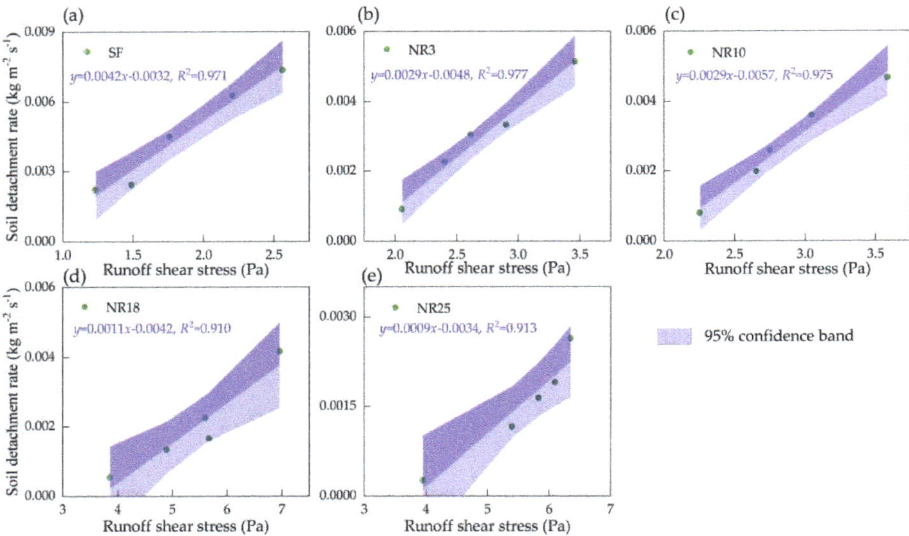

Figure 9. Relationship between soil detachment rate and shear stress under different revegetation ages. (**a**) SF; (**b**) NR3; (**c**) NR10; (**d**) NR18; (**e**) NR25.

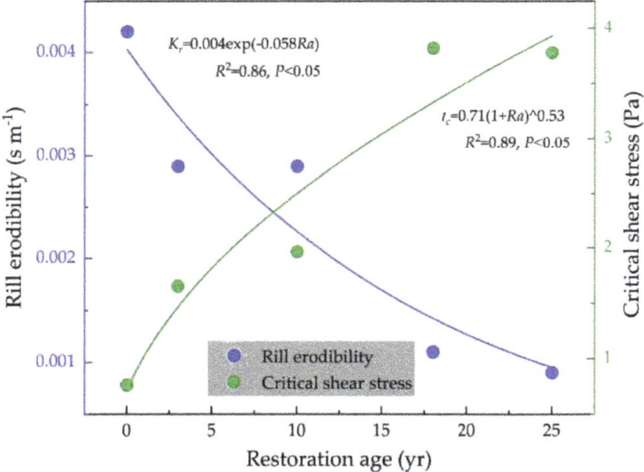

Figure 10. Relationships between soil erodibility and critical shear stress and revegetation age.

In addition, the τ_c increased with restoration age by a power function (Figure 10). However, the result was inconsistent with the finding of Wang et al. [16] that critical shear stress varied with restoration age in a nonlinear pattern, reaching the minimum at the restored age of 18. The difference in the temporal change of critical shear stress between Wang et al. [13] and this study was caused probably by differences in soil properties and vegetation characteristics of the sampling sites. Compared with slope farmland, the τ_c of the grasslands was improved by 1.2 to 4.0 times, while τ_c of restored land had a little decrease when restored time was more than 18 years (Figure 10). The result further indicated that revegetation can effectively improve the soil erosion resistance of gully head to concentrated flow, and the critical shear stress would reach a stable state after a 18-year revegetation.

4. Conclusions

This study was carried out to explore the effect of revegetation age on soil erosion resistance of gully heads in the gully region of the Loess Plateau. The results showed that revegetation significantly improved soil properties and promoted root accumulation of gully heads. The mean D_r of slope farmland was 1.6 to 3.0 times greater than those of revegetated gully heads. The revegetation can effectively weaken the sensitivity of soil erosion of the gully heads to concentrated flow. The D_r of gully heads was positively related to bulk density and disintegration rate and negatively related to soil capillary porosity, saturated soil hydraulic conductivity, organic matter content, and water-stable aggregate. Roots of 0–0.5 mm and 0.5–1.0 mm were the most effective roots in reducing soil erosion of gully head, and the native plant species with rich root of 0.5–1.0 mm and 0–0.5 mm were recommended as the first choice for revegetation to restrain gully headcut erosion. Revegetation can reduce soil erodibility of gully heads by 31% to 78.6% and improve the critical shear stress by 1.2 to 4.0 times. This study allows us to better evaluate soil vulnerability of gully head to concentrated flow erosion during revegetation. Further studies are needed to quantify the effect of the different combinations of vegetation types with different root architecture types on soil erosion resistance of gully heads.

Author Contributions: Conceptualization, Z.C. and M.G.; data curation, Z.C.; formal analysis, Z.C. and M.G.; funding acquisition, W.W.; investigation Z.C. and M.G.; methodology, Z.C.; project administration, W.W.; resources, W.W.; software, Z.C.; supervision, W.W.; validation, W.W.; writing—original draft, Z.C.; writing—review and editing, M.G. and W.W. All authors have read and agreed to the published version of the manuscript.

Funding: This work was supported by the National Natural Science Foundation of China (42077079; 41571275).

Acknowledgments: All authors thank anonymous reviewers for insightful comments on the original manuscript.

Conflicts of Interest: The authors declare no conflict of interest.

References

1. Mohammed, S.; Al-Ebraheem, A.; Holb, I.J.; Alsafadi, K.; Dikkeh, M.; Pham, Q.B.; Linh, N.T.T.; Szabo, S. Soil management effects on soil water erosion and runoff in central Syria—A comparative evaluation of general linear model and random forest regression. *Water* **2020**, *12*, 2529. [CrossRef]
2. Hao, H.X.; Cheng, L.; Guo, Z.L.; Wang, L.; Shi, Z.H. Plant community characteristics and functional traits as drivers of soil erodibility mitigation along a land degradation gradient. *Land Degrad. Dev.* **2020**, *31*, 1851–1863. [CrossRef]
3. Zarris, D.; Lykoudi, E.; Panagoulia, D. Assessing the impacts of sediment yield on the sustainability of major hydraulic systems. In Proceedings of the International Conference "Protection and Restoration of the Environment VIII", Chania, Greece, 3–7 July 2006.
4. Brazier, R.E.; Beven, K.J.; Freer, J.; Rowan, J.S. Equifinality and uncertainty in physically based soil erosion models: Application of the GLUE methodology to WEPP–the water erosion prediction project–for sites in the UK and USA. *Earth Surf. Process. Landf.* **2000**, *25*, 825–845. [CrossRef]
5. Kinnell, P.; Risse, L.M. USLE-M: Empirical modeling rainfall erosion through runoff and sediment concentration. *Soil Sci. Soc. Am. J.* **1998**, *62*, 1667–1672. [CrossRef]
6. Nearing, M.A.; Foster, G.R.; Lane, L.J.; Finkner, S.C. A process-based soil erosion model for USDA-water erosion prediction project technology. *Trans. ASAE* **1989**, *32*, 1587–1593. [CrossRef]
7. Efthimiou, N.; Lykoudi, E.; Panagoulia, D.; Karavitis, C. Assessment of soil susceptibility to erosion using the EPM and RUSLE models: The case of Venetikos river catchment. *Glob. Nest. J.* **2016**, *18*, 164–179.
8. Guo, W.Z.; Kang, H.L.; Wang, W.L.; Guo, M.M.; Chen, Z.X. Erosion-reducing effects of revegetation and fish-scale pits on steep spoil heaps under concentrated runoff on the Chinese Loess Plateau. *Land Degrad Dev.* **2020**. [CrossRef]
9. Gramlich, A.; Stoll, S.; Stamm, C.; Walter, T.; Prasuhn, V. Effects of artificial land drainage on hydrology, nutrient and pesticide fluxes from agricultural fields—A review. *Agric. Ecosyst. Environ.* **2018**, *266*, 84–99. [CrossRef]
10. Kervroëdan, L.; Armand, R.; Saunier, M.; Ouvry, J.; Faucon, M. Plant functional trait effects on runoff to design herbaceous hedges for soil erosion control. *Ecol. Eng.* **2018**, *118*, 143–151. [CrossRef]

11. Sun, L.; Zhang, G.H.; Luan, L.L.; Liu, F. Temporal variation in soil resistance to flowing water erosion for soil incorporated with plant litters in the Loess Plateau of China. *Catena* **2016**, *145*, 239–245. [CrossRef]
12. Chang, E.; Li, P.; Li, Z.; Su, Y.; Zhang, Y.; Zhang, J.; Liu, Z.; Li, Z. The impact of vegetation successional status on slope runoff erosion in the Loess Plateau of China. *Water* **2019**, *11*, 2614. [CrossRef]
13. Wang, B.; Zhang, G.H.; Shi, Y.Y.; Zhang, X.C. Soil detachment by overland flow under different vegetation restoration models in the Loess Plateau of China. *Catena* **2014**, *116*, 51–59. [CrossRef]
14. Li, Z.W.; Zhang, G.H.; Geng, R.; Wang, H.; Zhang, X.C. Land use impacts on soil detachment capacity by overland flow in the Loess Plateau, China. *Catena* **2015**, *124*, 9–17. [CrossRef]
15. Yu, Y.C.; Zhang, G.H.; Geng, R.; Sun, L. Temporal variation in soil detachment capacity by overland flow under four typical crops in the Loess Plateau of China. *Biosyst. Eng.* **2014**, *122*, 139–148. [CrossRef]
16. Wang, B.; Zhang, G.H.; Shi, Y.Y.; Zhang, X.C.; Ren, Z.P.; Zhu, L.J. Effect of natural restoration time of abandoned farmland on soil detachment by overland flow in the Loess Plateau of China. *Earth Surf. Process. Landf.* **2013**, *38*, 1725–1734. [CrossRef]
17. Scherer, U.; Zehe, E.; Tr Bing, K.; Kai, G. Prediction of soil detachment in agricultural loess catchments: Model development and parameterisation. *Catena* **2012**, *90*, 63–75. [CrossRef]
18. Zhang, G.H.; Tang, M.K.; Zhang, X.C. Temporal variation in soil detachment under different land uses in the Loess Plateau of China. *Earth Surf. Process. Landf.* **2010**, *34*, 1302–1309. [CrossRef]
19. Bernard, B.; Roose, E. Aggregate stability as an indicator of soil susceptibility to runoff and erosion; validation at several levels. *Catena* **2002**, *47*, 133–149.
20. Knapen, A.; Poesen, J.; Govers, G.; Baets, S.D. The effect of conservation tillage on runoff erosivity and soil erodibility during concentrated flow. *Hydrol. Process.* **2008**, *22*, 1497–1508. [CrossRef]
21. Guo, M.M.; Wang, W.L.; Wang, T.C.; Wang, W.X.; Kang, H.L. Impacts of different vegetation restoration options on gully head soil resistance and soil erosion in loess tablelands. *Earth Surf. Process. Landf.* **2020**, *45*, 1038–1050. [CrossRef]
22. Lazarovitch, N.; Vanderborght, J.; Jin, Y.; van Genuchten, M.T. The root zone: Soil physics and beyond. *Vadose Zone J.* **2018**, *17*, 1–6. [CrossRef]
23. Panagoulia, D. Hydrological modelling of a medium-size mountainous catchment from incomplete meteorological data. *J. Hydrol.* **1992**, *137*, 279–310. [CrossRef]
24. Falkenmark, M. Land and water integration and river basin management. *FAO Land Water Bull.* **1995**, *1*, 15–16.
25. Baets, S.D.; Poesen, J. Empirical models for predicting the erosion-reducing effects of plant roots during concentrated flow erosion. *Geomorphology* **2010**, *118*, 425–432. [CrossRef]
26. Guo, M.M.; Wang, W.L.; Shi, Q.H.; Chen, T.D.; Kang, H.L.; Li, J.M. An experimental study on the effects of grass root density on gully headcut erosion in the gully region of China's Loess Plateau. *Land Degrad. Dev.* **2019**, *30*, 2107–2125. [CrossRef]
27. Khanal, A.; Fox, G.A. Detachment characteristics of root-permeated soils from laboratory jet erosion tests. *Ecol. Eng.* **2017**, *100*, 335–343. [CrossRef]
28. De Baets, S.; Poesen, J.; Gyssels, G.; Knapen, A.; Adili, A.A. Effects of grass roots on the erodibility of topsoils during concentrated flow. *Geomorphology* **2006**, *76*, 54–67. [CrossRef]
29. Mamo, M.; Bubenzer, G.D. Detachment rate, soil erodibility, and soil strength as influenced by living plant roots part II: Field study. *Trans. ASAE* **2001**, *5*, 1175–1181. [CrossRef]
30. Mamo, M.; Bubenzer, G.D. Detachment rate, soil erodibility, and soil strength as influenced by living plant roots part I: Laboratory study. *Trans. ASAE* **2001**, *5*, 1167–1174. [CrossRef]
31. Baets, S.D.; Poesen, J.; Knapen, A.; Galindo, P. Impact of root architecture on the erosion-reducing potential of roots during concentrated flow. *Earth Surf. Process. Landf.* **2007**, *32*, 1323–1345. [CrossRef]
32. Shi, Q.; Wang, W.; Guo, M.; Chen, Z.; Feng, L.; Zhao, M.; Xiao, H. The impact of flow discharge on the hydraulic characteristics of headcut erosion processes in the gully region of the Loess Plateau. *Hydrol. Process.* **2019**, *34*, 718–729. [CrossRef]
33. Guo, M.M.; Wang, W.L.; Kang, H.L.; Yang, B. Changes in soil properties and erodibility of gully heads induced by vegetation restoration on the Loess Plateau, China. *J. Arid Land* **2018**, *10*, 712–725. [CrossRef]
34. Fu, B.; Chen, L.; Ma, K.; Zhou, H.; Wang, J. The relationships between land use and soil conditions in the hilly area of the loess plateau in northern Shaanxi, China. *Catena* **2000**, *39*, 69–78. [CrossRef]

35. Xiong, Y.; Zhou, J.Z.; Chen, L.; Jia, B.J.; Sun, N.; Tian, M.Q.; Hu, G.H. Land use pattern and vegetation cover dynamics in the three Gorges Reservoir (TGR) intervening Basin. *Water* **2020**, *12*, 2036. [CrossRef]
36. Jiao, J.Y.; Tzanopoulos, J.; Xofis, P.; Bai, W.J.; Ma, X.H.; Mitchley, J. Can the study of natural vegetation succession assist in the control of soil erosion on abandoned croplands on the Loess Plateau, China? *Restor. Ecol.* **2007**, *15*, 391–399. [CrossRef]
37. Wang, B.; Liu, G.B.; Xue, S.; Zhu, B.B. Changes in soil physico-chemical and microbiological properties during natural succession on abandoned farmland in the Loess Plateau. *Environ. Earth Sci.* **2011**, *62*, 915–925. [CrossRef]
38. Chen, Y.P.; Wang, K.B.; Lin, Y.S.; Shi, W.Y.; Song, Y.; He, X.H. Balancing green and grain trade. *Nat. Geosci.* **2015**, *8*, 739–741. [CrossRef]
39. Kompani-Zare, M.; Soufi, M.; Hamzehzarghani, H.; Dehghani, M. The effect of some watershed, soil characteristics and morphometric factors on the relationship between the gully volume and length in Fars Province, Iran. *Catena* **2011**, *86*, 150–159. [CrossRef]
40. Guo, M.M.; Wang, W.L.; Kang, H.L.; Yang, B.; Li, J.M. Changes in soil properties and resistance to concentrated flow across a 25-year passive restoration chronosequence of grasslands on the Chinese Loess Plateau. *Restor. Ecol.* **2020**, *28*, 104–114. [CrossRef]
41. Luk, S.H.; Merz, W. Use of the salt tracing technique to determine the velocity of overland flow. *Sci. Total Environ.* **1992**, *5*, 289–301.
42. Li, Y.; Xu, X.Q.; Zhu, X.M.; Tian, J.Y. Effectiveness of plant roots on increasing the soil permeability on the Loess Plateau. *Chin. Sci. Bull.* **1992**, *37*, 1735–1738.
43. Hill, A.V. The possible effects of the aggregation of the molecules of haemoglobin on its dissociation curves. *J. Physiol.* **1910**, *40*, 4–7.
44. Li, Y.; Zhu, X.M.; Tian, J.Y.; Al-Ebraheem, A. Effectiveness of plant roots to increase the anti-scouribility of soil on the Loess Plateau. *Chin. Sci. Bull.* **1991**, *36*, 2077–2082.
45. Gros, R.; Jocteur Monrozier, L.; Bartoli, F.; Chotte, J.L.; Faivre, P. Relationships between soil physico-chemical properties and microbial activity along a restoration chronosequence of alpine grasslands following ski run construction. *Appl. Soil Ecol.* **2004**, *27*, 7–22. [CrossRef]
46. Li, Y.Y.; Shao, M.A. Change of soil physical properties under long-term natural vegetation restoration in the Loess Plateau of China. *J. Arid. Environ.* **2006**, *64*, 77–96. [CrossRef]
47. Jiao, F.; Wen, Z.M.; An, S.S. Changes in soil properties across a chronosequence of vegetation restoration on the Loess Plateau of China. *Catena* **2011**, *86*, 110–116. [CrossRef]
48. Zhao, Y.G.; Wu, P.T.; Zhao, S.W.; Feng, H. Variation of soil infiltrability across a 79-year chronosequence of naturally restored grassland on the Loess Plateau, China. *J. Hydrol.* **2013**, *504*, 94–103. [CrossRef]
49. Six, J.; Elliott, E.T. Aggregation and soil organic matter accumulation in cultivated and native grassland soils. *Soil Sci. Soc. Am. J.* **1998**, *62*, 1367–1377. [CrossRef]
50. Six, J.; Bossuyt, H.; Degryze, S.; Denef, K. A history of research on the link between (micro)aggregates, soil biota, and soil organic matter dynamics. *Soil Tillage Res.* **2004**, *79*, 7–31. [CrossRef]
51. Six, J.; Paustian, K.; Elliott, E.T.; Combrink, C. Soil structure and organic matter I. distribution of aggregate-size classes and aggregate-associated carbon. *Soil Sci. Soc. Am. J.* **2000**, *64*, 681–689. [CrossRef]
52. Kong, A.Y.Y.; Six, J.; Bryant, D.C.; Denison, R.F.; Kessel, C.V. The relationship between carbon input, aggregation, and soil organic carbon stabilization in sustainable cropping systems. *Soil Sci. Soc. Am. J.* **2005**, *69*. [CrossRef]
53. Shangguan, Z.P.; Zheng, C.Z. Soil anti-scouribility enhanced by plant roots. *J. Integr. Plant Biol.* **2005**, *47*, 676–682.
54. Adili, A.A.; Azzam, R.; GiovanniSpagnoli, S.J. Strength of soil reinforced with fiber materials (Papyrus). *Soil Mech. Found. Eng.* **2012**, *48*, 241–247. [CrossRef]
55. Neves, C.S.V.J.; Feller, C.; Guimarães, M.F.; Medina, C.C.; Tavares Filho, J.; Fortier, M. Soil bulk density and porosity of homogeneous morphological units identified by the cropping profile method in clayey oxisols in Brazil. *Soil Tillage Res.* **2003**, *71*, 109–119. [CrossRef]
56. Zhang, B.J.; Zhang, G.H.; Yang, H.Y.; Wang, H. Soil resistance to flowing water erosion of seven typical plant communities on steep gully slopes on the Loess Plateau of China. *Catena* **2019**, *173*, 375–383. [CrossRef]

57. Fattet, M.; Fu, Y.; Ghestem, M.; Ma, W.; Foulonneau, M.; Nespoulous, J.; Le Bissonnais, Y.; Stokes, A. Effects of vegetation type on soil resistance to erosion: Relationship between aggregate stability and shear strength. *Catena* **2011**, *87*, 60–69. [CrossRef]
58. Podwojewski, P.; Orange, D.; Jouquet, P.; Valentin, C.; Nguyen, V.T.; Janeau, J.L.; Tran, D.T. Land-use impacts on surface runoff and soil detachment within agricultural sloping lands in Northern Vietnam. *Catena* **2008**, *74*, 109–118. [CrossRef]
59. Gyssels, G.; Poesen, J. The importance of plant root characteristics in controlling concentrated flow erosion rates. *Earth Surf. Process. Landf.* **2003**, *28*, 371–384. [CrossRef]
60. Amézketa, E. Soil aggregate stability: A review. *J. Integr. Plant Biol.* **1999**, *14*, 83–151. [CrossRef]
61. Gyssels, G.; Poesen, J.; Bochet, E.; Li, Y. Impact of plant roots on the resistance of soils to erosion by water: A review. *Prog. Phys. Geogr.* **2005**, *29*, 189–217. [CrossRef]

Publisher's Note: MDPI stays neutral with regard to jurisdictional claims in published maps and institutional affiliations.

Article

Estimation of Soil Erosion and Sediment Yield in the Lancang–Mekong River Using the Modified Revised Universal Soil Loss Equation and GIS Techniques

Pavisorn Chuenchum [1], Mengzhen Xu [2] and Wenzhe Tang [1,*]

1 Institute of Hydraulic Structures Engineering and Construction Management, and State Key Laboratory of Hydroscience and Engineering, Tsinghua University, Beijing 100084, China; huw18@mails.tsinghua.edu.cn
2 River Research Institute, and State Key Laboratory of Hydroscience and Engineering, Tsinghua University, Beijing 100084, China; mzxu@mail.tsinghua.edu.cn
* Correspondence: twz@mail.tsinghua.edu.cn; Tel.: +86-10-6279-4324

Received: 9 December 2019; Accepted: 29 December 2019; Published: 31 December 2019

Abstract: The Lancang–Mekong River basin, as an important transboundary river in Southeast Asia, is challenged by rapid socio-economic development, especially the construction of hydropower dams. Furthermore, substantial factors, such as terrain, rainfall, soil properties and agricultural activity, affect and are highly susceptible to soil erosion and sediment yield. This study aimed to estimate average annual soil erosion in terms of spatial distribution and sediment deposition by using the revised universal soil loss equation (RUSLE) and GIS techniques. This study also applied remote sensing and available data sources for soil erosion analysis. Annual soil erosion in most parts of the study area range from 700 to 10,000 t/km^2/y with a mean value of 5350 t/km^2/y. Approximately 45% of the total area undergoes moderate erosion. Moreover, the assessments of sediment deposition and erosion using the modified RUSLE and the GIS techniques indicate high sediment erosion along the flow direction of the mainstream, from the upper Mekong River to the Mekong Delta. The northern part of the upper Mekong River and the central and southern parts of the lower Mekong River are the most vulnerable to the increase in soil erosion rates, indicating sediment deposition.

Keywords: soil erosion; sediment yield; RUSLE; Lancang–Mekong River basin

---

1. Introduction

Soil erosion affects and challenges the world's environment and natural resources [1–7], and economic and environmental dimensions with negative impacts can affect soil erosion, further resulting in low agricultural productivity, ecological collapse and high sedimentation [6–10]. Approximately 84% of the degraded lands around the world are associated with the most relevant issues about the environment with water and wind as the main agents of erosion [7,11–13]. The average soil erosion by water is estimated to exceed 2000 t/km^2/y with this type of erosion mainly occurring on croplands in tropical areas [14,15]. Human activities and climate change can also be triggered at a much higher rate thus simulating erosion [8,16–22]. Soil erosion by human activities is reportedly 10–15 times faster than any natural process [23]. For instance, approximately 80% of agricultural areas around the world face high to extreme erosion, and the amount of generated sediments can worsen the turbidity of rivers and increase further the concentration of pollutants [24–26]. Moreover, soil erosion and sediment yield can affect humans and the environment severely if sediment quantity exceeds the standard measurement value of aquatic organisms.

Soil erosion is the main part of the initial process of sediment delivery to rivers; in this initial process, displaced soil particles are transformed into sediments due to the influence of an agent of erosion. The amount of sediments can decrease the potential storage capacity of reservoirs and the performance of hydraulic structures [10,27–30]. According to Reference [31], approximately 0.5% to 1% of sediment depositions affect the annual loss of storage capacity of reservoirs around the world, indicating that most dams will likely be left with only 50% of their corresponding volumes by the 2050s. Reference [32] affirms that sediments currently occupy 40% of the reservoir storages in Asia, indicating high loss of storage capacity. These circumstances affect the long-term sustainability of water sources for hydropower dams. The supposedly low sediment yield from the trapping of dams may also cause shoreline erosion, bank erosion and loss of riparian vegetation [33–36].

Lancang–Mekong River basin, as an important transboundary river in Southeast Asia, is one of the largest rivers causing high sediment loads in Asian rivers. According to Reference [37], the average annual sediment load and the specific sediment yield in the Lancang–Mekong River basin is approximately 160 Mt/y and 200 t/km^2/y, respectively. The upper Mekong basin contributes approximately 50% of the amount of sediments in the Lancang–Mekong River basin [37–39]. Moreover, the Lancang–Mekong River basin is beset by soil erosion and sediment problems because of rapid socio-economic development, population growth, land deterioration and deforestation in the last 50 years, and the problem is most especially caused by the development of hydropower dams in the region [38,40–42]. Many areas are easily vulnerable to soil erosion due to the influence of rainfall, runoff and human activities. In the last few years, the Lancang–Mekong River basin has eroded at an average rate of 5000 t/km^2/y [33] which is a moderate erosion level, and it tends to increase in intensity continuously from climate change and land-use change. Conversely, sediment yield in the river basin is decreasing from 250 t/km^2/y to 209 t/km^2/y, because the sediment quantity is trapped by hydropower dams. Historical sediment load (1960–2013) from China to the lower Mekong River indicates clearly that the amount of sediment loads heavily decreased from 84.7 Mt/y to 10.8 Mt/y and 147 Mt/y to 66 Mt/y at Chiang Saen and Pakse stations, respectively [43].

Previous research attempted to study emphatically the sediment issue in the Lancang–Mekong River basin and some parts of the basin as a means to accumulate knowledge and information for policymakers. The study of sediments in this river can be divided into two main groups. The first group of previous research focused on the changes in sediment load from the construction and operation of dams in the upper Mekong Basin. Reference [44] considered the changing sediment load in the lower Mekong basin because of the possible effects of the cascade dams in the Lancang. Reference [45] considered the effect of sediments from the Manwan Dam in both pre-dam and post-dam stages. Reference [45] estimated the sediment load of the lower Mekong River basin by classifying the rating curve of suspended sediment concentrations obtained from adjacent stations. Reference [46] investigated the nature and magnitude of changing sediment load and their trends in the Lancang–Mekong River basin using available sediment data from 1965 to 2003. Reference [34] analysed the suspended sediment flux and the sediment supply in the lower Mekong River basin using high-frequency measurements obtained from specific stations in Vietnam. Most research in the first group reveals that the construction and operation of dams in the upper Mekong River basin affect negatively the sediment load in this river due to the trapped sediments in the reservoirs. Sediment load also appears with constantly decreasing trends. Meanwhile, the second group of previous research focused on the sediment trapping efficiency of dams in the Lancang–Mekong River basin. Reference [28] analysed and predicted the sediment trapping efficiency of reservoirs in the mainstream of Lancang River. Reference [47] developed an estimation technique for the sediment trapping efficiency of existing and planned reservoirs in the Mekong River using Brune's method. Reference [48] estimated sediment yield based on geomorphic characteristics, tectonic history and available sediment data and, subsequently, considered the cumulative sediment trapping of dams. Most research in the second group indicates that the majority of sediment loads are trapped in existing dams in the upper Mekong River basin, and they will be further trapped if planned dams are operated officially in the near future. However, most of the above studies concentrated only on sediment load data and used the trapped sediment load data of dams obtained from observation stations. Conversely, studies on soil erosion in the Lancang–Mekong River basin requiring both field surveys and other techniques are rare.

Some studies on soil erosion apply the universal soil loss equation (USLE) in combination with GIS and remote sensing techniques to analyse the spatial distributions and patterns of soil erosion in the Lancang–Mekong River basin. The method is convenient for soil erosion analysis, because it can estimate long-term soil erosion. References [49,50] estimated soil erosion in the upper Mekong River basin in Yunnan Province using USLE and analysed spatial patterns with environmental factors. Reference [51] assessed the conserved water and soil ecosystems in Yunnan Province using remote sensing techniques. Reference [52] analysed the spatial distribution of soil erosion in north-western Yunnan (Lancang River) based on the revised universal soil loss equation (RULSE) and GIS techniques. Reference [10] estimated the impact of soil erosion on the reservoirs in Yunnan Province using USLE. Reference [53] conducted a soil loss vulnerability analysis of the Mekong River basin by applying USLE. Nonetheless, the above studies identified the limitations of the USLE model, including the development of input data for new areas to satisfy the long-term data requirements, difficulties in assessing gully erosion and large-scale areas, estimation of soil loss only and insufficient computation of sediment deposition. The RUSLE model was developed accordingly to improve the estimation of potential soil erosion. The input factors in RULSE can be used by using values from the literature or adapted for empirical and statistical data in combination with GIS software. In addition, the RUSLE results are valid in terms of estimating the risks of water erosion.

Previous studies mostly investigated the changing sediment load and the sediment trapping caused by dam construction and operation. Nonetheless, the understanding of soil erosion and soil deposition is also highly important. Soil erosion, as the main part of the sediment process, can be used to plan countermeasures for the Lancang–Mekong River basin. Previous studies also emphasized that soil erosion research should focus on the simulation of sediment erosion, but they did not consider sediment deposition. Hence, this research aimed to develop methods to calculate sediment deposition and erosion based on the RUSLE model and GIS techniques and, subsequently, evaluate the impact of soil erosion on hydropower dams in the Lancang–Mekong River basin. This study only considered suspended sediment despite the limitation of the model. In addition, the factors that can influence potential and actual soil erosion in the Lancang–Mekong River basin were also determined. The simulation period of the study covered from 2000 to 2015 depending on the available data in the analysis.

2. Study Area

The Lancang–Mekong River basin is a transboundary river in Southeast Asia (Figure 1). Originating from China's Qinghai–Tibet Plateau, the source of the river is located in Yushu of Qinghai Province. By its name, Lancang River represents the upper Mekong River basin in China, while the downstream part is located in Yunnan Province. Along with the river portions in Myanmar, Laos PDR, Thailand, Cambodia, and Vietnam, the lower Mekong River basin has a length of 4909 km and a coverage area of 795,000 km^2 [38,54,55]. The average annual water discharge is approximately 475 km^3 [38,44,54,56]. Thus, approximately 24% of the total area comprises the upper Mekong River basin, with contribution rates of 15% to 20% of the water flow to the Lancang–Mekong River basin. Most areas comprise complex mountains and hills and deep valleys [10,38,53]. In addition, approximately 76% of the total area is developed by major tributary systems from the lower Mekong Basin, especially Lao People's Democratic Republic (Laos PDR) [38,53]. The elevation of the basin varies from 0 to 6549 m above sea level. The different elevations have varying distributed agriculture depending on climatic zones and temperature. Moreover, various elevations of the river have water development projects, such as cascade hydropower dams, in both the mainstream and the sub-basins. Furthermore, soil erosion in the Lancang–Mekong River basin results in sediment deposition. The sediments affect the dams in all statuses (i.e., operation, under construction, and planned). The dam system of the Lancang–Mekong River basin comprises 133 dams [38,57,58] including those in the mainstream and the sub-basins.

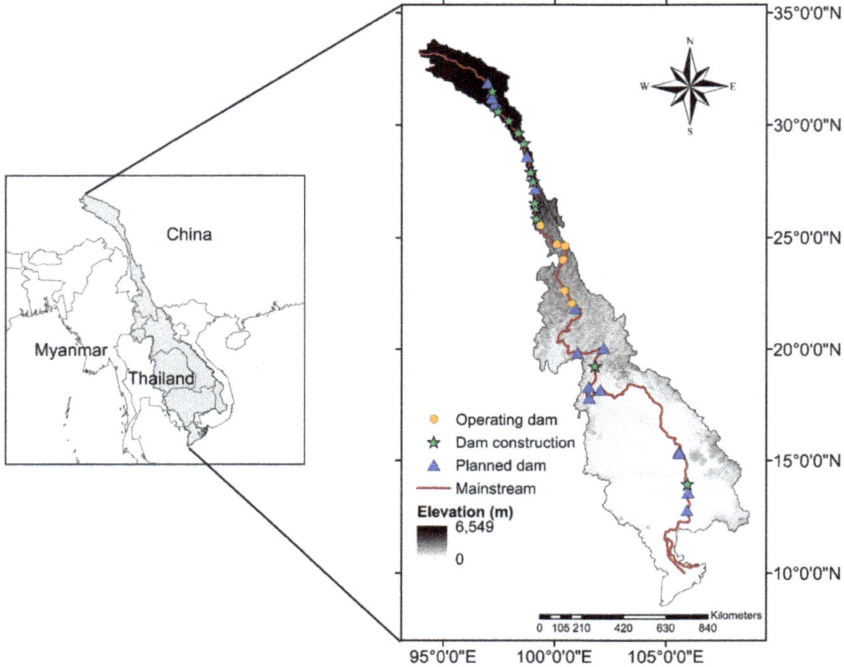

Figure 1. Location and elevation of the study area and location of dams in the mainstream river.

3. Materials and Methods

3.1. RUSLE

The RUSLE model, based on the USLE model, was developed by the US Department of Agriculture. The RUSLE is an empirical soil erosion model and has been recognised as a standard method to calculate the risk of average soil erosion on land. The RUSLE is also the most popular model for estimating average soil erosion in water [59], and it is simple to integrate with GIS and remote sensing [10,60–62]. Furthermore, RUSLE can provide international applicability and comparability for the results and methods, as the model can be adapted and applied in many regions globally. The RUSLE model can be expressed as follows:

$$A = R \times K \times LS \times C \times P \tag{1}$$

where:

A is the mean annual soil loss (t/ha·y);
R is the rainfall erosivity factor (MJ·mm/ha·hr·y);
K is the soil erodibility factor (t·hr/MJ·mm);
LS is the topographic factor (dimensionless);
C is the cropping management factor (dimensionless); and
P is the support practice factor (dimensionless).

The assessment of soil erosion in the Lancang–Mekong River basin can be classified into five levels according to the Soil Erosion Standard Document–Technological Standard of Soil and Water Conservation (SD238-87) of Reference [63].

3.1.1. Rainfall Erosivity Factor

Rainfall plays an important role in the process of soil erosion and sedimentation and leads to water erosion, such as splash erosion, sheet erosion, rill erosion and gully erosion, caused by water flow. Soil particles, which are transported away from a site by the flow, are those detached by rainfall impact [64]. Therefore, high-potential erosion can be determined by rainfall intensity and storm duration. Normally, the relationship between total storm energy (E) and maximum 30 min intensity (I30) can be regarded as the *R* factor, as reported by Reference [65]. Given the limitation of precipitation data about the river, the *R* factor is derived from the Asian Precipitation Highly Resolved Observational Data Integration Towards Evaluation of the Extreme Events (APHRODITE) for the period from 2000 to 2015 which also correspond to the daily gridded precipitation data for Monsoon Asia [66]. This project is developed from the daily rain gauge data for the Asia region and cover nearly 12,000 stations. This study has selected the highest fine-gridded resolution (spatial resolution of 5 km) of available precipitation data. For the conditions in the Lancang–Mekong River basin, this study chose the formula of the *R* factor from References [41,67] which applied the assessment of the *R* factor in Southern China. Equation (2) is appropriate, because the climate and area conditions in Southern China are almost uniform to those in the Lancang–Mekong River basin.

$$R = \sum_{i=1}^{12} (-1.15527 + 1.792P_i) \tag{2}$$

where R is the rainfall erosivity factor (MJ·mm/ha·hr·y), and P_i is the monthly rainfall (mm).

3.1.2. Soil Erodibility Factor

The effect of soil characteristics and soil properties on soil erosion can be represented by the soil erodibility factor (K), because this factor shows the physical and chemical properties of the soil through the equations related to soil texture, soil organic matter and percentages of sand, silt, and clay. Furthermore, the K factor is based on soil permeability and particle size distribution. The K factor is strongly related with the R factor through the soil erosion rate per kinematic energy of rainfall erosivity index. The observed data of the local soil properties in the Lancang–Mekong River basin are extremely difficult to access. Thus, the soil data in this study were derived from the SoilGrids map which is developed and maintained by ISRIC–World Soil Information. This study used the available soil data grid with a spatial resolution of 1 km. The data on soil properties were analysed using the methods in References [68,69], in which the percentages of silt, clay, sand and organic carbon fraction were calculated by Equations (3)–(6). Soil erodibility was computed according to the method in Reference [70] as shown in Equation (7). Then, the unit of the K factor was transferred to the International System of Units (SI) [70]. This method is widely used for the analysis of the K factor for soil properties such as soil structure and particle-size distribution [10,53,68,69,71,72].

$$f_{csand} = \left\{0.2 + 0.3\exp\left[-0.256m_s\left(1 - \frac{m_{silt}}{100}\right)\right]\right\} \tag{3}$$

$$f_{cl-si} = \left(\frac{m_{silt}}{m_c + m_{silt}}\right)^{0.3} \tag{4}$$

$$f_{orgC} = \left\{1 - \frac{0.25orgC}{orgC + \exp[3.72 - 2.95orgC]}\right\} \tag{5}$$

$$f_{hisand} = \left\{1 - \frac{0.7\left(1 - \frac{m_s}{100}\right)}{\left(1 - \frac{m_s}{100}\right) + \exp\left[-5.51 + 22.9\left(1 - \frac{m_s}{100}\right)\right]}\right\} \tag{6}$$

$$K = f_{csand} \times f_{cl-si} \times f_{orgC} \times f_{hisand} \tag{7}$$

where K is the soil erodibility factor, f_{csand} is the function of high-coarse sand content in soil, f_{cl-si} is the function of clay and silt in soil, f_{orgC} is the function of organic carbon content in soil, f_{hisand} is the function of high sand content in soil, m_s is the percentage of sand fraction content (particles with diameters from 0.05 to 2 mm) (%), m_{silt} is the percentage of silt fraction content (particles with diameters from 0.002 to 0.05 mm) (%), m_c is the percentage of clay fraction content (particles with diameters of <0.002) (%), and $orgC$ is the percentage of organic carbon content of the layer (%).

3.1.3. Topographic Factor

The topographic factor (*LS*) includes slope length (*L*) and slope steepness (*S*), which are the two important influencing parameters of soil erosion. Both GIS and remote sensing techniques were applied to access the *LS* factor in the RUSLE equation using the digital elevation model (DEM) [73]. For a large area, grid resolution is important for soil erosion estimation [74]. Changes in grid size affect steepness values, both directly and indirectly. The *L* factor depends on grid size and steepness, while the *S* factor affects steepness only. Hence, if the DEM data have a high resolution, then the model output can increase the accuracy of the *LS* factor in the RUSLE model [75,76]. Digital elevation model images with a 1 km resolution were downloaded from the US Geological Survey (https://earthexplorer.usgs.gov). Past researchers applied high-resolution DEM images for soil erosion determination because of these images' good accuracy and reliability [6,10,16,20,49–51,53,60,62,73,76–79]. The calculation of the *LS* factor can be based on the RUSLE principle by using the GIS software as explained in References [20,73,78,80–82]. The *S* factor was calculated in two conditions (Equations (8) and (9)), and the *L* factor was computed with Equation (10). Then, the *LS* factor in each grid cell was coupled in Equation (11).

$$S_{factor} = 10.8\sin\theta + 0.03; \text{ slope gradients} < 9\% \tag{8}$$

$$S_{factor} = 16.8\sin\theta + 0.50; \text{ slope gradients} \geq 9\% \tag{9}$$

$$L_{factor} = \left(\frac{\lambda}{22.12}\right) \times \left(\frac{\frac{\left(\frac{\sin\theta}{0.0896}\right)}{(3\sin\theta \times 0.8 + 0.56)}}{1 + \frac{\left(\frac{\sin\theta}{0.0896}\right)}{(3\sin\theta \times 0.8 + 0.56)}}\right) \tag{10}$$

$$LS = L_{factor} \times S_{factor} \tag{11}$$

where λ is the length of the slope, L_{factor} denotes the slope length factors, and S_{factor} is the slope steepness factor.

3.1.4. Cropping Management Factor

Vegetation cover is one of the most important factors affecting the erosion process and the development of rivers [64]. Moreover, vegetation cover can shield the soil surface from the impact of falling rain and slow down the velocity and scouring power of runoffs. Normally, vegetation cover can be depicted by the cropping and management practices in an area through the C factor. The range of the C factor is between 1 and 0. If the C factor is equal to 1, then no vegetation cover (i.e., bare land) exists in that area. If the C factor is close to 0, then strong vegetation cover exists, indicating protection against soil erosion.

The product of remote sensing data from the Moderate Resolution Imaging Spectroradiometer (MODIS), with a cell size of 250 m in spatial resolution, was applied. The MODIS is a good choice for large-area coverages. The normalized difference vegetation index (*NDVI*) was used in this study to estimate the C factor following the method of [83]. The detailed equations were given by Equations (12) and (13). The MODIS' remote sensing can investigate all months, from the historical period to the present (2000–2015), to investigate the study area.

$$C = \frac{(-NDVI + 1)}{2} \tag{12}$$

$$NVDI = \frac{(NIR - RED)}{(RED + NIR)} \tag{13}$$

where C is the cropping management factor, *NDVI* is the average of the normalized difference vegetation index, *NIR* is surface spectral reflectance in the near-infrared band, and *RED* is the surface spectral reflectance in the red band. Both *NIR* and *RED* were extracted from the MODIS images. In reflecting the vegetation cover and the agricultural activities in the Lancang–Mekong River basin, the five months of January, April, July, October and December [38] were selected from 2000 to 2015. The average *NDVI* was calculated from these data covering 16 years.

3.1.5. Support Practice Factor

The support practice factor was used to express the effect of land use and land cover on soil erosion. The *P* factor describes the change in potential erosion by flowing water through the effect of supporting conservation practices such as contouring, buffer strips and terraced contour farming [6,53,65,77,84]. The maximum value of the *P* factor is usually set to 1.0 to mean no erosion control solution. A decreasing value of the *P* factor means that flowing water is reduced in terms of both volume and velocity. Moreover, a decreasing *P* also means reduced intensity of sediment deposition on the surface [85]. Given the many limitations, the *P* factor was determined on the basis of the land cover type from the *C* factor (Table 1) as suggested by [86]. Land-use type was obtained from the product of the MODIS' remote sensing with a cell size of 250 m for the spatial resolution.

Table 1. Land cover classification and the *C* and *P* factors [86].

Land Cover of the RUSLE	*C* Factor	*P* Factor
Urban area	0.1	1.0
Bare land	0.35	1.0
Dense forest	0.001	1.0
Sparse forest	0.01	1.0
Mixed forest and cropland	0.1	0.8
Cropland	0.5	0.5
Paddy field	0.1	0.5
Dense grassland	0.08	1.0
Sparse grassland	0.2	1.0
Mixed grassland and cropland	0.25	0.8
Wetland	0.05	1.0
Water body	0.01	1.0
Permanent ice and snow	0.001	1.0

3.1.6. Application of GIS Tools

The input data, such as rainfall, types of land use, and land cover, terrain and soil properties, in the RUSLE model were imported and calculated using the functions in ArcGIS 10.5. The five factors were analysed according to the spatial resolution and the coordinate system of their original data. The final results of the quantitative output of soil erosion were generated as the maximum grid with 5 km of spatial resolution depending on the original data. Soil erosion in the Lancang–Mekong River basin was analysed using the results of two types of erosion (i.e., potential soil erosion and actual soil erosion), as shown in Equation (1), in the spatial distribution. The *R*, *K*, *L*, and *S* factors were considered as potential soil erosion, whereas the *R*, *K*, *LS*, *C*, and *P* factors were examined as actual soil erosion.

3.2. Descriptive Statistics in the RUSLE Model

Soil erosion can be identified in each factor of the RUSLE model, indicating the influence of soil erosion on a specific area [6]. The RUSLE model is transformed into logarithmic form in Equation (15), and multiple linear regression must be applied to examine the relationships among all factors, as shown in Equation (16), and the effects on the soil erosion rate.

$$\ln(A) = \ln(R \times K \times LS \times C \times P) \quad (14)$$

$$\ln(A) = \ln(R) + \ln(K) + \ln(LS) + \ln(C) + \ln(P) \quad (15)$$

$$\ln(A) = \beta_0 + \beta_i(\ln R) + \beta_j(\ln K) + \beta_k(\ln LS) + \beta_l(\ln C) + \beta_h(\ln P) \quad (16)$$

where $\ln(A)$ is the logarithm of soil erosion rate, $\ln(R, K, LS, C, \text{and } P)$ denotes the logarithmic value of the input factors in the RUSLE model, β_0 is the intercept of soil erosion rate (constant term), and $\beta_{i\text{–}h}$ is the estimated regression coefficient of each explanatory variable. Different units of the input factors are reflected through the standard coefficient (β) in Equation (16). The factors of multiple linear regression in logarithmic form can be explained as follows: if one of the factors in the RUSLE model increases by 1% in standard deviation, then $\beta_{i\text{–}k}$ percent of the standard deviation leads to an increased value of soil erosion rate (A). This study sets the statistical significance level at 95% confidence in SPSS. Nonetheless, given the differences in the spatial resolutions of the input factors, some factors (K, LS, C, and P) were estimated as 5 km (A and R factors) in spatial resolution using the spatially averaged values assigned in the function of ArcGIS.

3.3. Technique of Sediment Yield Estimation

References [20,87] proposed a new technique to estimate sediment yield or sediment deposition in each sub-basin of Thailand by modifying the original RUSLE model. They regarded the suspended sediment flow from one grid cell to the other grids as dependent on the sediment yield of the original grid cell (S_y) and the average sediment yield capacity of sub-basin (S_c). If S_y is greater than S_c, then the sediment moves to the next site. By contrast, if S_c is more than S_y, then the sediment is deposited. S_y is calculated using the individual parameters in each grid cell (Equation (17)). In the same way, S_c is calculated using the original RUSLE model with the area-averaged parameters (Equation (18)). This technique was only developed for the assessment of suspended sediment. It is not appropriate for analysing the total sediment form (i.e., bed load and suspended sediment).

$$S_y = f(I_1, I_2, \ldots, I_5) \quad (17)$$

$$S_C = f\left(\frac{\sum_{i=1}^{n} I_1}{A_{ba\sin}}, \frac{\sum_{i=1}^{n} I_2}{A_{ba\sin}}, \ldots, \frac{\sum_{i=1}^{n} I_5}{A_{ba\sin}}\right) \quad (18)$$

$$D_i \quad \text{if } S_y < S_c \quad (19)$$

$$T_i \quad \text{if } S_y > S_c \quad (20)$$

where S_y is sediment yield, S_c is sediment capacity, I_i represents the parameters in the RUSLE model (R, K, LS, C, and P), A_{basin} is an area of the sub-basin, n is the number of data in each sub-basin, D_i is the sediment deposition in a cell i, and T_i is the sediment transportation in cell i. S_y is the result of actual soil erosion by computing from the RUSLE input factors. S_c is calculated from the summation of each parameter in the RUSLE model dividing an area of the sub-basin. The five outcomes then are multiplied as S_c.

The above technique can show the spatial distribution of sediment yield and sediment deposition in the Lancang–Mekong River basin, indicating an integrated consideration of the sediment issue which is the main problem for water development projects in this river. Furthermore, the technique is extremely useful in studying the influence of dam construction on sediment budget, because the loss of storage capacity of dams and the reduced transport of sediments downstream are caused by sedimentation which, in turn, is the result of soil erosion [32]. Dam design and sediment management in operations planning can be arranged properly if the sediment budget of the river is primarily determined in dam construction.

3.4. Observed Sediment Data

The results from net sediment mapping or sediment deposition and erosion mapping are estimated and compared with the observed sediment data from relevant organizations, such as MRC, and the literature for verification [32,44,56,88,89]. The present study collected, from 15 stations, the average sediment load and specific sediment yield (SSY) data for each sub-basin (Figure 2) from the years 1952 to 2011 (60 years) to cover the whole basin (see Supplementary Materials) which is the time period of the data collection. Sediment loads were estimated from the suspended sediment concentration (SSC) and water discharge using the sediment rating curve, and the SSY data in the Lancang–Mekong River basin were estimated based on historical geological and geomorphological characteristics of each sub-basin [48] and historical sediment load. The results of this study will be verified with SSY in each sub-basin only. Each observational station is a representative of a sub-basin in the Lancang–Mekong River basin for verification between observed SSY (1952–2011) and estimated SSY from the modified RUSLE model (2000–2015) (see Table 4).

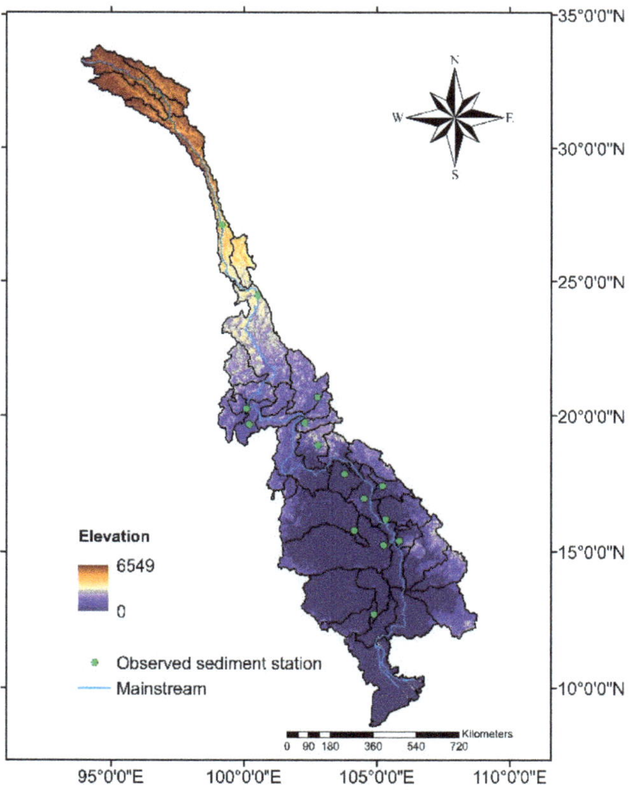

Figure 2. Location of sediment observational stations in the Lancang–Mekong River basin.

4. Results

4.1. Soil Erosion Factors

4.1.1. Rainfall Erosivity Factor

The values of the R factor were analysed using Equation (2). Figure 3a shows the spatial distribution of the R factor for the Lancang–Mekong River basin. The range of the R factor was 65.6–524.3 MJ·mm/(ha·hr·y) with a mean of 294.9 MJ·mm/(ha·hr·y). The standard deviation was 80.3. The lowest values for the R factor were distributed mostly in the upper Mekong River basin or Lancang River in China. Meanwhile, the highest values for the R factor were distributed primarily in the sub-basins of Laos PDR and Cambodia and the Mekong Delta, because those areas are located along the direction of monsoon storms from the South China Sea in seasonal cycles. According to the results, the R factor increased from the lower basin to the upper basin, a scenario explaining the influence of climate and temperature on the river.

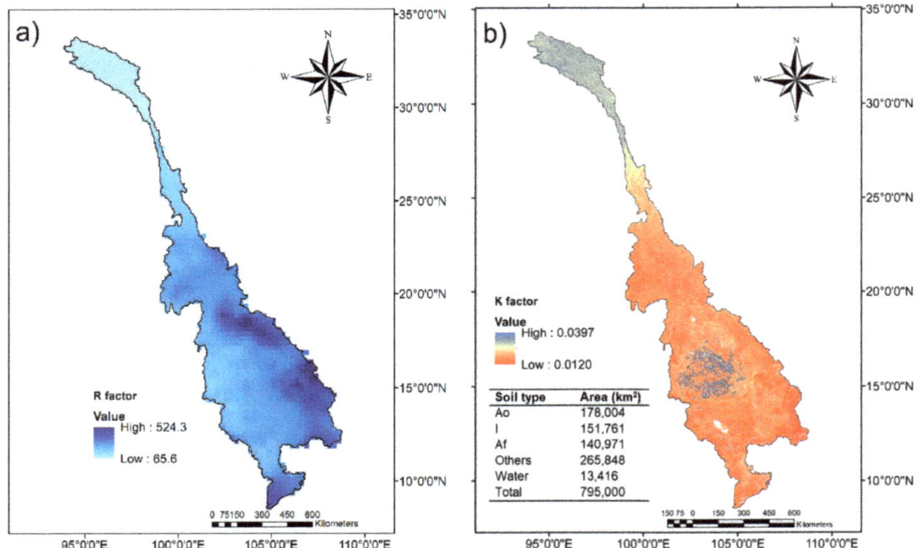

Figure 3. (**a**) R factor and (**b**) K factor.

4.1.2. Soil Erodibility Factor

Major soil groups in the Lancang–Mekong River basin (Figure 3b) were determined using the SoilGrids database of ISRIC–World Soil Information [90,91]. The K factor was calculated with Equations (3)–(7). The range of the K factor was 0.012–0.0397 t/(hr·MJ·mm), with an average of 0.0258 t/(hr·MJ·mm). The standard deviation was 0.0012. The spatial distribution in Figure 3b indicated that the K factor decreased from the upper basin to the lower basin, but some areas of the Mun and Chi River basins in Thailand had high K values. In the Lancang–Mekong River basin, the highest elevation areas were identified by the highest K values, whereas the lowest elevation areas were identified by the lowest K values. This result corresponded with the findings in Reference [10], in which the K values correlated with the variation of the terrain; moreover, highly significant K values were found for high elevation areas such as mountains. Orthic Acrisols (Ao), Lithosoils (I) and Ferric Acrisols (Af) are the largest areas in the Lancang–Mekong River basin, and they accounted for approximately 59% of the total basin, while the other soil groups accounted for 39%.

4.1.3. Topographic Factor

Topographic factor was the most influential factor of soil erosion due to the flowing water from rainfall and runoff. The *LS* factor was considered from the elevation map of the Lancang–Mekong River basin (Figure 1) and the calculations of Equations (8)–(10). The range of elevation in the study area is from 0 to 6549 m above sea level, and the elevation mean was 3274 m. The basin with high elevation is mainly located in the upper Mekong River basin, and the elevation gradually decreases in the central part of the basin. More than 65% of the natural area has a slope gradient of >9%, and this area is mainly situated in the upper Mekong River basin. Slopes from 10° to 70° account for approximately 59%. Thus, the results of the *LS* factor were in the range of 0–336 (Figure 4a), and its mean value was 168. In addition, the areas represented by the *LS* values were below 60. The slope is steep, and the slope length is short. The areas with relatively high *LS* values were located in the upper part of the river, while the those with relatively low *LS* values were situated in the central part of the Mekong Delta.

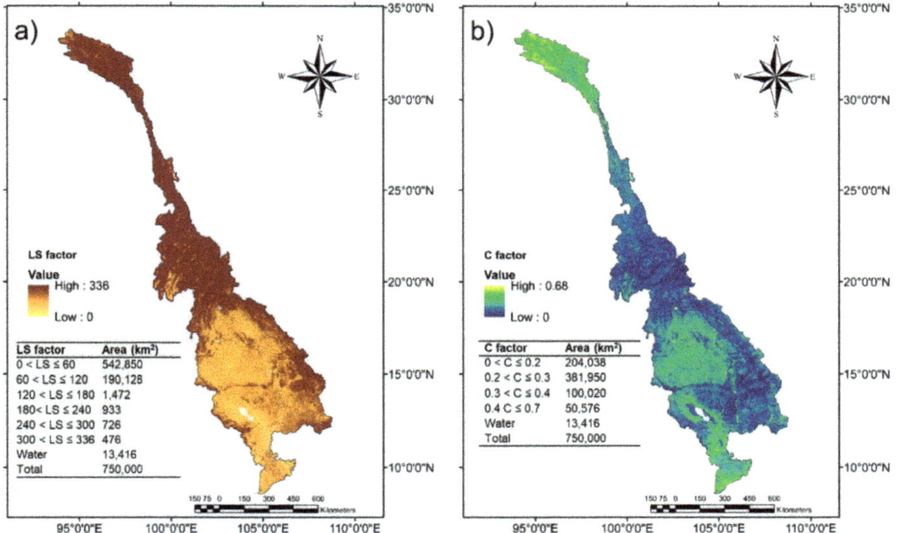

Figure 4. (**a**) *LS* factor and (**b**) *C* factor.

4.1.4. Cropping Management Factor

The *C* factor was applied using the *NDVI* analysis from the MODIS satellite images and the calculation in Equation (11). The *C* factor varied from 0 to 0.7 (Figure 4b). The *C* mean and the standard deviation were 0.34 and 0.076, respectively. Most lands in the study area are forests in parts of China, Laos PDR and Cambodia, and they were represented by relatively low values of the *C* factor. Conversely, relatively high values for the *C* factor were shown in the upper Mekong River basin in China, Thailand, and the Mekong Delta.

4.1.5. Support Practice Factor

The values for the *P* factor were determined following the suggestion in Reference [86] (Table 1). The change in *C* values to *P* values was applied with the functions in ArcGIS. The *P* values were 0.5, 0.8, and 1 (Figure 5). Nearly 52% of the *P* values were between 0.8 and 1, and they represent the largest portion. Thus, most areas in the basin are forests and lands with vegetation cover, indicating that soil is protected from agents of erosion. The areas with relatively high and low *P* values corresponded to similar areas for the *C* values (see Section 4.1.4).

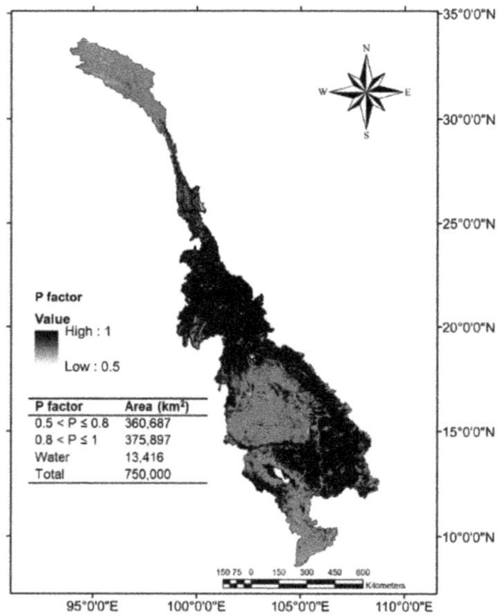

Figure 5. *P* factor.

4.2. Potential and Actual Soil Erosion

Soil erosion was divided into two types: potential and actual soil erosion. Potential erosion (*R*, *K*, *L*, and *S*) was defined as a natural erosion process without cropping management (*C*) and support practice (*P*) factors. If potential soil erosion is combined with the *C* and *P* factors, then it can be considered actual soil erosion (*R*, *K*, *LS*, *C*, and *P*). Potential soil erosion was calculated on the basis of four factors by using ArcGIS and GIS techniques. The range of potential soil erosion rate was 5000–25,000 t/km^2/y (Figure 6a). The average potential soil erosion was 15,000 t/km^2/y, and the standard deviation was 4623. The findings on spatial distribution demonstrates high-potential soil erosion in most areas in the basin. Thus, all the factors were computed for actual soil erosion (Figure 6b) which is the real-world soil erosion in the Lancang–Mekong River basin. Actual soil erosion was in the range of 700–10,000 t/km^2/y. The mean actual soil erosion was 5350 t/km^2/y, and the standard deviation was 1470. Most of the relatively high soil erosion rates were located in the north part of upper Mekong River basin and Mekong Delta. Some parts of Thailand had values close to the mean actual soil erosion. The results of the potential erosion and actual soil erosion manifested notable differences. The potential soil erosion rate was differentiated by the *C* and *P* factors because of the forest and agricultural areas. Hence, the *C* and the *P* factors play important roles in decreasing soil erosion, and they can reduce the effect by 2.5–7 times in the basin. The *C* factor indicates the capability to absorb the impact of raindrops, reduce the velocity and scouring power of runoff and reduce the runoff volume by increasing percolation into soil. Meanwhile, the *P* factor indicates the capability to decrease the amount and rate of water runoff and soil erosion with supporting cropland practices such as cross-slope cultivation, contouring farming and strip cropping.

4.3. Soil Erosion Risk Mapping

The results of actual soil erosion can be classified into five categories (Figure 7) according to the Soil Erosion Standard Document–Technological Standard of Soil and Water Conservation (SD238-87) [63]. Table 2 shows the soil erosion in the study area ranging from minimal erosion to extreme erosion. Most of the soil erosion in the Lancang–Mekong River basin (45% of the total area) is moderate erosion. However, the soil erosion rate is higher than 5000 t/km^2/y hence their classification as high erosion and extreme erosion; the corresponding areas comprise 37% of the total area, including the high-elevation areas in China, the plateau in Thailand, Tonle Sap, and the Mekong Delta. By contrast, a low erosion rate was found mostly in Laos PDR and some parts of Cambodia because of their forest areas.

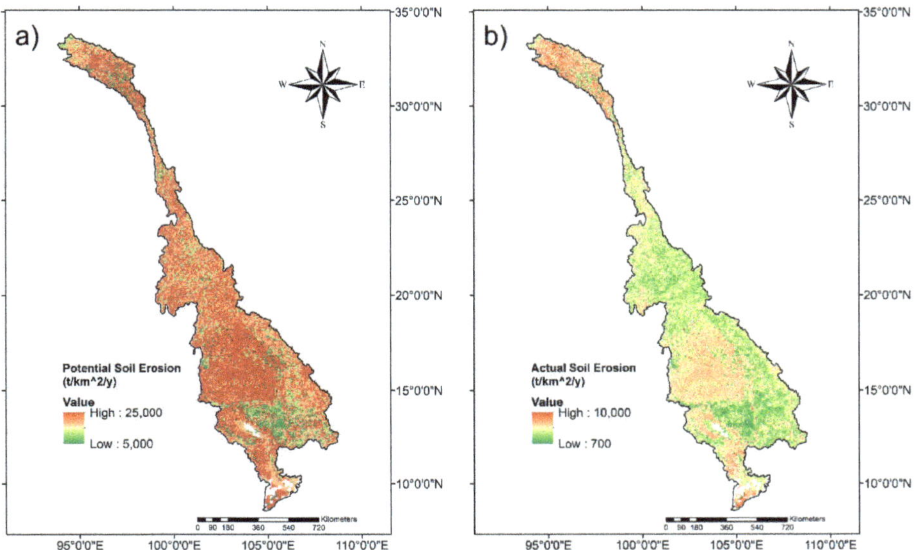

Figure 6. (**a**) Potential soil erosion; (**b**) Actual soil erosion.

Table 2. Soil erosion in the Lancang–Mekong River basin.

Level	Soil Loss (t/km^2/y)	Area (km^2)	Percentage of Total Area
Minimal erosion	<500	-	-
Low erosion	500–2500	125,450	16
Moderate erosion	2500–5000	335,942	45
High erosion	5000–8000	253,342	34
Extreme erosion	>8000	21,850	3
Water		13,416	2
Total		750,000	100

The analytical results on the correlation between soil erosion rate and all input factors in the RUSLE model using SPSS are shown in Table 3. The hypotheses of all factors were determined on the basis of a 95% confidence (i.e., level of statistical significance). The results were then used to build the multiple linear regression in logarithmic form for the soil erosion rate and all the RUSLE factors of the Lancang–Mekong River basin.

$$\ln(A) = 0.168 \times \ln(R) + 0.364 \times \ln(K) + 0.898 \times \ln(LS) + 0.184 \times \ln(C) + 0.246 \times \ln(P) \quad (21)$$

Equation (21) is given by the values of standardized coefficients that are strongly related with all the RUSLE factors of the soil erosion rate. The strongest influencing factor for soil erosion in the study area was the *LS* factor (β = 0.898). Therefore, slope length and slope steepness directly affect soil erosion. In other words, soil erosion likely occurs because of gravity erosion and water erosion in an area.

Table 3. Standardized coefficients of factors in the RUSLE model.

Independent Variable	Standardized Coefficient (β)	Significance
ln(*R*)	0.168	0.000
ln(*K*)	0.364	0.000
ln(*LS*)	0.898	0.000
ln(*C*)	0.184	0.000
ln(*P*)	0.246	0.000

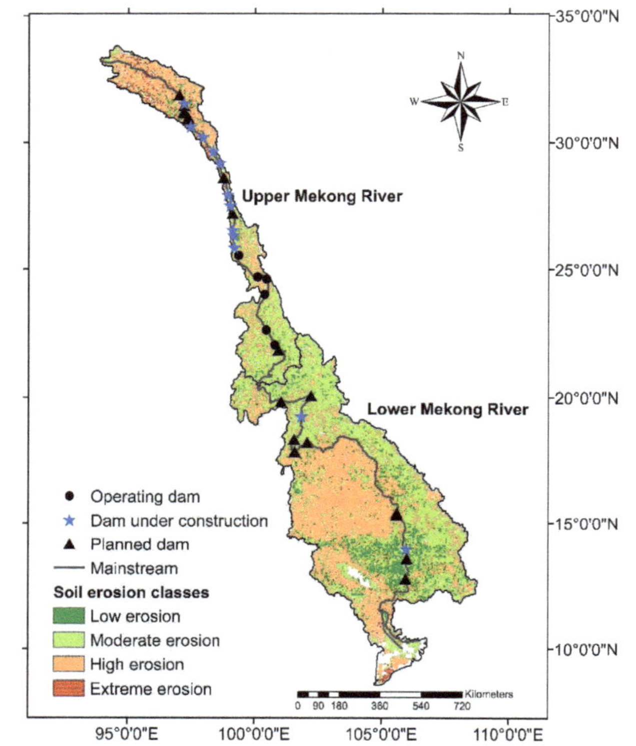

Figure 7. Soil erosion classification and location of dams in the Lancang–Mekong River basin.

4.4. Estimation of Sediment Deposition Areas

The assessment of sediment yield or sediment deposition areas in the Lancang–Mekong River basin was computed by modifying the RUSLE model according to Equations (17) and (18). The RUSLE model was determined using the spatially average parameters for the estimation of sediment yield capacity in each sub-basin. The results of the sediment yield capacity for the study area are presented in Figure 8a. Most of the sub-basins have high sediment yield capacities. Some sub-basins have low sediment yield capacity in the central part and the north part of upper Mekong River basin. The size of the sub-basin and the elevations directly result in sediment yield capacity. The average sediment yield capacity (S_c) was compared with the estimation of sediment yield (S_y) to assess the sediment deposition and sediment erosion in each grid. If the result of S_y was higher than S_c, then sediment erosion appeared in each grid cell. Conversely, if S_y was lower than S_c, then sediment deposition appeared in each grid cell. The results of sediment deposition and sediment erosion capacities in each grid cell are shown in Figure 8b. Sediment erosion is presented as positive values, whereas sediment deposition is presented in negative values. Potential sediment deposition and erosion are in the range from less than −3000 to more than 6200 t/km^2/y. The mean was 2105 t/km^2/y, and the standard deviation was 2033.

Relatively high sediment erosion occurs along the flow direction of the mainstream, including the north part of upper Mekong River basin, Laos PDR, Tonle Sap and Mekong Delta. Meanwhile, most areas in Yunnan Province, Thailand and Cambodia have high sediment depositions. The sediment deposition and erosion results can be verified using the observed sediment data from the 15 stations along the Lancang–Mekong River basin. The scatter plot of the whole basin, which was based on the observed sediment data and the assessment data on sediment yield from the RUSLE model, shows good results with a correlation higher than 0.9 (Figure 9). The RUSLE model and the technique used to assess sediment deposition and erosion can be applied in the research and prediction of soil erosion and SSY.

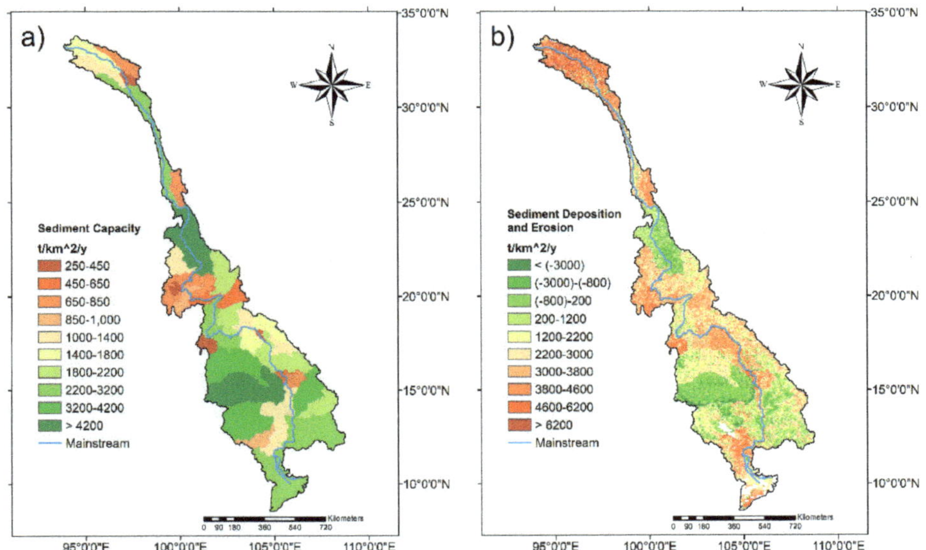

Figure 8. (**a**) Sediment capacity in each sub-basin, (**b**) Deposited and eroded sediments in each sub-basin.

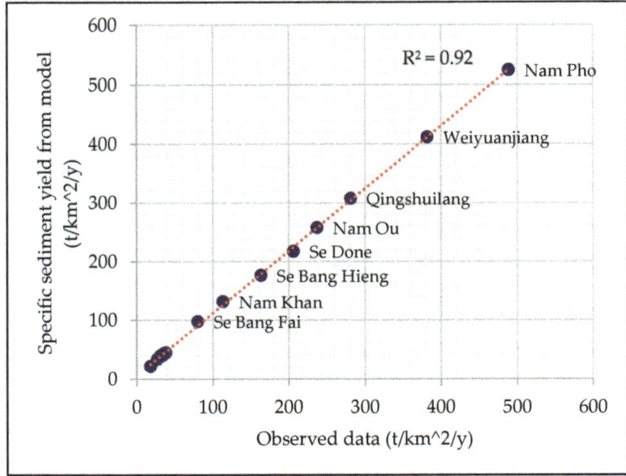

Figure 9. Comparative result of the whole basin based on observed data and the data of the modified RUSLE model by using the sediment estimation technique.

5. Discussion

5.1. Soil Erosion Rate in the Lancang–Mekong River Basin

This study focused on the assessment of soil erosion rate and sedimentation in the Lancang–Mekong River basin using the RUSLE model and GIS techniques with related available data. The river as the research object had many data limitations, and accessing input data to develop research on soil erosion, sediment yield capacity and sediment transport was difficult. This study attempted to utilize previous research on soil erosion in the Lancang–Mekong River basin [10,49,50,53,79], as no other evidence and information exist on how much the soil erosion rate has changed in this basin. The average soil erosion in the previous research ranges from 1400 to 8500 t/km^2/y. Our results are in the range near the mean values of the previous research. The spatial pattern of soil erosion occurrence in the north part of upper Mekong River basin is generally consistent with the findings of [10,49,50,53,79], but some spatial soil erosion results differ in other areas, especially in the lower part of Mekong River. Furthermore, we presumed that the different results can be attributed to the computation of the *R* factor (the main factor in soil erosion) which is influenced by differences in rainfall data. Each of the available rainfall data were previously developed for the purpose of individual projects. Nonetheless, if the *R* factor was developed from local rainfall stations in the six riparian countries, then the soil erosion rates can be regarded accurate and be further improved. Meanwhile, the results of descriptive statics in the RULSE model clearly showed that the *LS* factor is the most influential factor for soil erosion and sediment yield in the Lancang–Mekong River basin, especially in the upper Mekong River. Most studies on soil erosion and sedimentation claim that the geographical features of the Lancang–Mekong River basin, such as its steep slopes and the slope length of its hills and mountains, are affected directly by the occurrence of soil erosion in specific areas, and these sediments are transformed when transported along the river [10,32,35,41,47,49–53,79]. Consequently, the mitigation measures currently used to reduce soil erosion need to further consider solutions related with the *LS* factor such as the implementation of check dams and the application of vegetation cover. In order to consider the analytical results on the correlation between soil erosion rate and all input factors in the RUSLE model using SPSS, the *LS* factor is the strongest influencing factor on soil erosion in the study area. Nonetheless, the analytical results may not be quite effective, because the *LS* factor varies greatly in the river basin against other factors. Therefore, this section should be considered by regarding the different geological and geomorphological characteristics of the river basin such as mountains, piedmont and lowland. Moreover, different altitudinal conditions are also important conditions that directly affect the RUSLE input factors. This issue needs to be improved correctly for understanding the influencing factor on soil erosion in each feature of the river basin. In addition, the case study on potential and actual soil erosion verifies the ability of the *C* and the *P* factors to protect and reduce soil erosion. Natural vegetation covers, such as the forests in Laos PDR and Cambodia, can decrease soil erosion at rates greater than those of agricultural areas in Thailand. Hence, if forested areas are transformed into agricultural activities, then the soil erosion rate will increase remarkably, especially in upstream areas [20].

The Soil Erosion Standard Document–Technological Standard of Soil and Water Conservation (SD238-87) [63] was applied in this study to classify the soil erosion rate in the Lancang–Mekong River basin. One of the reasons is that the river has not been evaluated using the standard on soil erosion classification. Previous research used the soil erosion classification in References [63,92]. However, the number of classifications in Reference [92] is lower than that in Reference [63] and, thus, does not correspond with the results of our study. The highest value of severe erosion according to Reference [92] is greater than 3300 t/km^2/y, while extreme erosion according to Reference [63] is greater than 8000 t/km^2/y. The values differ considerably in terms of soil erosion classification. We suppose that the soil erosion classification should depend on the researcher's discretion and the suitability of research results until a set of criteria is developed by relevant credible agencies such as the Lancang–Mekong Cooperation (LMC) or the Mekong River Commission (MRC).

5.2. Estimation of Sediment Yield Using the Modified RUSLE Model

Soil erosion is the initial process of the sedimentation process of a river channel. The Lancang–Mekong River basin faces the challenge of sediment starvation due to the implementation of water development projects, especially hydropower dams. Most studies confirm that sediments have started to decrease continuously because of sediment trapping by hydropower dams [28,44,47,48,93]. Therefore, sediment yield capacity and sediment deposition should be analysed by relevant organizations and the six riparian country governments when drafting the needed solutions. However, the field measurement of sediment aspects is very difficult due to the limitations of equipment and nations' borders in the Lancang–Mekong River basin. Hence, the modified RUSLE model was used for the estimation of sediment yield. This method was also clearly applied to understand the sediment deposition and erosion.

The developed RUSLE model and new technique used to assess sediment yield capacity and sediment deposition areas were appropriate, and the observed sediment data and the sediment yield results from the RUSLE simulation were well correlated despite the limitation of investigating a large field survey area. The consistency between observed sediment data and the RUSLE results can also improve the accuracy of soil erosion prediction and the analyses of sediment yield capacity and sediment deposition. Nevertheless, the observed sediment data from the 15 stations were insufficient for validation, especially for the upper part of the Mekong River and all the sub-basins. This study could only access two stations (i.e., Jiuzhou and Gajiu) in the upper Mekong River basin. If other sediment data regarding the upper Mekong River can be acquired for analysis, then the effectiveness of the RUSLE model with the abovementioned technique can be effectively assessed. Besides, the results of sediment yield in some of the sub-basins may have been overestimated. Problems may have also been caused by the analysis of the RUSLE input factors which are also likely overestimated values. Additionally, the assessment of sediment deposition and erosion using the modified RUSLE model may have led to overestimated results for the sub-basins. A previous study [20] also showed the same trend for Thailand after applying the modified RUSLE model. Therefore, in the application of the method, the abovementioned limitations should be considered for model enhancement. The method proposed in this study is useful in furthering the research and analysis of sediment load at reservoirs and sediment transport in the Lancang–Mekong River basin. Furthermore, the results can be used as basis to understand the physical process of sedimentation in each sub-basin.

The result in Table 4 shows a quite good comparability of the observed and estimated SSY from the RUSLE model. Almost half of the sub-basins were in approximately 5–10% of the percentage error, while the remaining sub-basins were estimated at more than 10% from the observed values. The sub-basins have a high sediment quantity. The modified RUSLE model can be a well-known simulation. Conversely, if sub-basins have a low sediment quantity, the model shows low performance for sediment yield estimation. These causes may occur from two important factors including the spatial resolution of the RUSLE input factors and the features of the river basins. For the spatial resolution in the analysis, this study chose a rather coarse grid (5 km) resolution despite the limitations of the input data sources. The model can be well-captured in some specific areas from the influence of grid resolution. If this study can be applied to a spatial resolution of 1 km, the sediment yield estimation may be improved efficiently [20]. Meanwhile, the features of the river basins directly affect the sediment yield estimation, especially rainfall from changing climate and land-use change from human activity. Most land in the sub-basins, which have greater values of percentage error (10–29%), have changed from forest areas to agricultural areas (among other types), especially Nam Chi, Nam Mun, Nam Songkhram, and Nam Ngum. This issue created inaccuracies in the analysis of the C and P factors, because the C factor was considered from the MODIS satellite image using the remote sensing techniques, and the P factor was also estimated from the C factor [86]. Furthermore, sub-basins, which are overestimated values, have features without high slopes when comparing with other sub-basins. Hence, the modified RUSLE model may be able to consider areas with better slopes which is quite consistent with Reference [20]. Totally, these factors may be the causes of the problem in the study of sediment yield estimation in the Lancang–Mekong River basin.

Table 4. Comparative results between observed SSY and estimated SSY from model.

Sub-Basin	Area (km^2)	Observed SSY (t/km^2/y)	Estimated SSY from Model (t/km^2/y)	Percentage Error (%)
Qingshuilang	87,205	281	308	10%
Weiyuanjiang	120,000	382	412	8%
Nam Pho	184,845	489	525	7%
Nam Chi	43,100	18	22	22%
Nam Kam	2360	35	42	20%
Nam Khan	5800	113	122	8%
Nam Mae Ing	5700	38	45	18%
Nam Mun	116,000	27	34	26%
Nam Ngum	5220	36	44	22%
Nam Ou	19,700	237	258	9%
Nam Songkhram	4650	31	40	29%
Se Bang Fai	4520	80	98	23%
Se Bang Hieng	19,400	163	177	9%
Se Done	5760	206	218	6%
St. Sen	14,000	33	40	21%

5.3. Soil Erosion Impact on Dams

Soil erosion can negatively affect hydropower dams in the Lancang–Mekong River basin. Sediments resulting from soil erosion can decrease the storage capacity of dams used to generate electricity and for other purposes. The upper Mekong basin, especially in the northern area, is classified as having extreme and high soil erosion, indicating increased vulnerability to soil erosion rate. Dams under construction and planned dams may also face the risk of increased sedimentation once they become operational. The dams located in the central and south parts of upper Mekong River basin are relatively less risky than those in the north part, because soil erosion in those areas have low and moderate erosion levels. Previous studies [10,53,79] obtained results that coincide with our research for the analysis of soil erosion impact on dams in the upper Mekong River basin. Meanwhile, soil erosion in the lower Mekong River basin, especially from the Khorat Plateau (Thailand) to the Mekong Delta, can also generate sedimentation problems due to the high occurrences of soil loss. The agricultural activities in these areas mainly cause the increase in the soil erosion rate. A dam under construction (Don Sahong) and four planned dams (Ban Koum, Phu Ngoy, Stung Treng and Sambor) may be threatened by soil erosion due to the impact of sub-basins in the Khorat Plateau in Thailand, particularly the confluence with the Lancang–Mekong River's mainstream. In addition, extreme soil erosion occurs in the Mekong Delta. Many studies affirm that the Mekong Delta is the most vulnerable area in terms of risk of soil loss [33–36].

The impact of soil erosion on dams in the Lancang–Mekong River's mainstream can be analysed in two parts based on the water sources of the river, namely, the upper Mekong River (with three river areas from Lancang basin) and the lower Mekong River (composed of the northern highlands, Khorat Plateau, Tonle Sap and Mekong Delta). The upper Mekong River basin covers 180,000 km^2 or approximately 24% of the study area, while the lower Mekong River basin covers 570,000 km^2 or approximately 76%. The soil erosion modulus of the upper Mekong River basin is 235.7 t/km^2/y. Its percentage relative to total soil erosion is approximately 36%, even if this area is smaller than the lower Mekong River basin. The soil erosion modulus of the lower Mekong River basin is 198.2 t/km^2/y which represents approximately 64% of the total occurrence of soil erosion. The results of the soil erosion modulus can be explained as that the reservoirs located in the upper Mekong River basin are more vulnerable from soil erosion and increased sediment. Consequently, dams are likely to be at risk of decreasing storage capacity continually. Our results are consistent with the findings of past studies on the impact of soil erosion on dams and sediment trapping. For instance, Reference [47] reported that the sediment trapping rates of dams under construction and the planned dams in the Lancang–Mekong River basin will increase from 51% to 69% due to the high heterogeneity of specific sediment yield in the different parts of the basin, and much higher trapped sediment load is predicted because of soil erosion resulting from socio-economic development. More than 50% of the sediment load (approximately 140 Mt) in the Lancang–Mekong River basin is expected to be trapped annually. Furthermore, more than 60% of the sediment load originates from China's end of the Lancang–Mekong River's mainstream. Existing dams, dams under construction and planned dams are expected to have the highest impact on storage capacity due to the fact of sediment load. Reference [28] reported that the main dams in the Lancang River, such as Manwan, Gongguoqiao, Dachaoshan and Jinghong, have sediment trapping rates between 30% and 70% because of the high sediment yield in the Lancang–Mekong River's mainstream and sub-basins. The storage capacity of reservoirs will continuously decrease from the sediment load due to the soil erosion in the reservoir upstream.

5.4. Delineating Sediment Form

This study endeavoured to estimate the sediment yield by considering factors such as soil erosion, gully erosion and rill erosion. These erosions are not the only sources of sediment into the river channel, because sediment yield is fundamentally controlled by climatic conditions, geomorphologic characteristics and anthropogenic forcing [22,48]. Some sediments are formed by erosion in the river channel. Our analysis did not take other factors into account in this study. This study could not consider erosion in the river channel due to the limitation of the modified RUSLE model which solely analyses erosion on land. For the study of channel deformations and changing river morphology, a hydrodynamic model is needed. Besides, sediment data (suspended and bed load sediment) for the Lancang–Mekong River basin are insufficient, because a number of measuring stations continue to be unavailable. This is the main limitation for further study in the basin. The results in this research can be considered together with erosion in the river channel using a hydrodynamic model; it would be able to show the sediment process on both land and in the river.

6. Conclusions

The RUSLE model was integrated with GIS techniques in this study to assess soil erosion and sediment yield in the Lancang–Mekong River basin. The impact of soil erosion on hydropower dams was also considered. The findings indicate that soil erosion occurs in all areas of the Lancang–Mekong River basin, accounting for 5350 t/km^2/y of its average soil erosion rate or approximately 45% of the basin. The north part of the upper Mekong River basin and some parts of Thailand have higher terrains than the other areas, and they have good vegetation cover and support practice. Furthermore, the *LS* factor showed that this factor was the strongest influencing factor for soil erosion in the study area. The spatial distribution of soil erosion also indicated that the norther part of the upper Mekong River basin and the central and southern parts of the lower Mekong River basin are the most vulnerable areas in terms of increased soil erosion rates due to the movement of sediments to the river. Hence, the dams in this river are highly threatened by sediment problems.

The value of pursuing research on the sediment capacity of each sub-basin of the Lancang–Mekong River basin are summarized as follows. The size of the sub-basins and their elevation directly affect the sediment capacity of the river. Moreover, the spatial distribution of sediment deposition and erosion indicates that relatively high sediment erosion occurs along the flow direction of the mainstream, from the northern part of upper Mekong River basin to the Mekong Delta. The findings on sediment yield estimation from the modified RUSLE model and the observed sediment data were in good agreement and had high correlation. The proposed technique can be applied in the assessment of sediment yield capacity and sediment deposition in the Lancang–Mekong River basin.

The modified RUSLE method was successfully applied to the assessment of the amount and spatial distribution of soil erosion and sediment deposition in the Lancang–Mekong River basin. The method can be applied not only to this river but also to other important areas. This study can help policymakers and relevant organizations improve their decision making based on the provided valuable information on soil erosion and sedimentation in this region.

Supplementary Materials: The following are available online at http://www.mdpi.com/2073-4441/12/1/135/s1, Table S1. Lists and location of hydropower dams and reservoirs in Lancang-Mekong River's mainstream, Table S2. Average observed SL and SSY from 1962 to 2010 based on catchment area of the station, Table S3. Data sources for the analysis of the RUSLE factors in this study, Figure S1. Locations of the sediment load observed stations.

Author Contributions: Conceptualization, P.C., M.X. and W.T.; Methodology, P.C., M.X. and W.T. Software, M.X. and W.T.; Formal analysis, P.C.; Investigation, M.X. and W.T.; Resources, P.C. and W.T.; Data curation, P.C. and W.T.; Writing—original draft preparation, P.C.; Writing—review and editing, M.X. and W.T.; Visualization, P.C.; Supervision, M.X. and W.T.; Funding acquisition, M.X. and W.T. All authors have read and agreed to the published version of the manuscript.

Funding: This research was funded by the National Natural Science Foundation of China (Grant Nos. 51579135, 51379104 and 51079070), the State Key Laboratory of Hydroscience and Engineering (Grant Nos. 2013-KY-5 and 2015-KY-5), the Chinese Academy of Sciences (XDA23090401) and the National Key Research and Development Program of China (2016YFC0402407).

Acknowledgments: Our sincerest appreciation to the Lancang–Mekong Cooperation and the Mekong River Commission for providing us the observed sediment data. We would also like to thank the APHRODITE Project for allowing us the use of the precipitation product. We are also thankful for the technical recommendations of Prem Rangsiwanichpong and the encouragement from my beloved wife, Usa Chuenchum.

Conflicts of Interest: The authors declare no conflict of interest.

References

1. Bakker, M.M.; Govers, G.; Kosmas, C.; Vanacker, V.; Van Oost, K.; Rounsevell, M. Soil erosion as a driver of land-use change. *Agric. Ecosyst. Environ.* **2005**, *105*, 467–481. [CrossRef]
2. Borrelli, P.; Robinson, D.A.; Fleischer, L.R.; Lugato, E.; Ballabio, C.; Alewell, C.; Meusburger, K.; Modugno, S.; Schutt, B.; Ferro, V.; et al. An assessment of the global impact of 21st century land use change on soil erosion. *Nat. Commun.* **2017**, *8*, 2013. [CrossRef]
3. Ighodaro, I.D.; Lategan, F.S.; Yusuf, S.F. The impact of soil erosion on agricultural potential and performance of Sheshegu community farmers in the Eastern Cape of South Africa. *J. Agric. Sci.* **2013**, *5*, 140–147. [CrossRef]
4. Littleboy, M.; Freebairn, D.M.; Hammer, G.L.; Silburn, D.M. Impact of soil-erosion on production in cropping systems. II. Simulation of production and erosion risks for a wheat cropping system. *Soil Res.* **1992**, *30*, 775–788. [CrossRef]
5. Parveen, R.; Kumar, U. Integrated approach of universal soil loss equation (USLE) and geographical information system (GIS) for soil loss risk assessment in Upper South Koel Basin, Jharkhand. *J. Geogr. Inf. Syst.* **2012**, *4*, 588–596. [CrossRef]
6. Pham, T.G.; Degener, J.; Kappas, M. Integrated universal soil loss equation (USLE) and Geographical Information System (GIS) for soil erosion estimation in A Sap basin: Central Vietnam. *Int. Soil Water Conserv. Res.* **2018**, *6*, 99–110. [CrossRef]
7. Pimentel, D. Soil erosion: A food and environmental threat. *Environ. Dev. Sustain.* **2006**, *8*, 119–137. [CrossRef]
8. Julien, P.Y. *Erosion and Sedimentation*, 2nd ed.; Cambridge University Press: Cambridge, UK, 2010.
9. Morris, G.L.; Fan, J. *Reservoir Sedimentation Handbook*; McGraw-Hill Book Co.: New York, NY, USA, 1998; p. 805.
10. Zhou, Q.W.; Yang, S.T.; Zhao, C.S.; Cai, M.Y.; Ya, L. A soil erosion assessment of the Upper Mekong River in Yunnan Province, China. *Mt. Res. Dev.* **2014**, *34*, 36–47. [CrossRef]
11. Blanco-Canqui, H.; Lal, R. *Principles of Soil Conservation and Management*, 1st ed.; Springer: Haarlem, The Netherlands, 2008; p. 617.
12. Food and Agriculture Organization of the United Nations. *Status of the World's Soil Resources (SWSR)—Main Report*; Food and Agriculture Organization of the United Nations: Rome, Italy, 2015.

13. Toy, T.J.; Foster, G.R.; Renard, K.G. *Soil Erosion: Processes, Prediction, Measurement, and Control*; John Wiley & Sons, Inc.: Hoboken, NJ, USA, 2002.
14. Food and Agriculture Organization of the United Nations. *Soil Change: Impacts and Responses*; FAO: Rome, Italy, 2015.
15. Van Oost, K.; Quine, T.A.; Govers, G.; De Gryze, S.; Six, J.; Harden, J.W.; Ritchie, J.C.; McCarty, G.W.; Heckrath, G.; Kosmas, C.; et al. The impact of agricultural soil erosion on the global carbon cycle. *Science* **2007**, *318*, 626–629. [CrossRef]
16. Asselman, N.E.M.; Middelkoop, H.; Van Dijk, P.M. The impact of changes in climate and land use on soil erosion, transport and deposition of suspended sediment in the River Rhine. *Hydrol. Process.* **2003**, *17*, 3225–3244. [CrossRef]
17. Bouraoui, F.; Grizzetti, B.; Granlund, K.; Rekolainen, S.; Bidoglio, G. Impact of climate change on the water cycle and nutrient losses in a Finnish catchment. *Clim. Chang.* **2004**, *66*, 109–126. [CrossRef]
18. Komori, D.; Rangsiwanichpong, P.; Inoue, N.; Ono, K.; Watanabe, S.; Kazama, S. Distributed probability of slope failure in Thailand under climate change. *Clim. Risk Manag.* **2018**, *20*, 126–137. [CrossRef]
19. Mukundan, R.; Pradhanang, S.M.; Schneiderman, E.M.; Pierson, D.C.; Anandhi, A.; Zion, M.S.; Matonse, A.H.; Lounsbury, D.G.; Steenhuis, T.S. Suspended sediment source areas and future climate impact on soil erosion and sediment yield in a New York City water supply watershed, USA. *Geomorphology* **2013**, *183*, 110–119. [CrossRef]
20. Rangsiwanichpong, P.; Kazama, S.; Gunawardhana, L. Assessment of sediment yield in Thailand using revised universal soil loss equation and geographic information system techniques. *River Res. Appl.* **2018**, *34*, 1113–1122. [CrossRef]
21. Shrestha, B.; Babel, M.S.; Maskey, S.; Van Griensven, A.; Uhlenbrook, S.; Green, A.; Akkharath, I. Impact of climate change on sediment yield in the Mekong River basin: A case study of the Nam Ou basin, Lao PDR. *Hydrol. Earth Syst. Sci.* **2013**, *17*, 1–20. [CrossRef]
22. Syvitski, J.P.M.; Peckham, S.D.; Hilberman, R.; Mulder, T. Predicting the terrestrial flux of sediment to the global ocean: A planetary perspective. *Sediment. Geol.* **2003**, *162*, 5–24. [CrossRef]
23. Wilkinson, B.H.; McElroy, B.J. The impact of humans on continental erosion and sedimentation. *Geol. Soc. Am. Bull.* **2007**, *119*, 140–156. [CrossRef]
24. Li, J.P.; Dong, S.K.; Liu, S.L.; Yang, Z.F.; Peng, M.C.; Zhao, C. Effects of cascading hydropower dams on the composition, biomass and biological integrity of phytoplankton assemblages in the middle Lancang-Mekong River. *Ecol. Eng.* **2013**, *60*, 316–324. [CrossRef]
25. Pimentel, D.; Harvey, C.; Resosudarmo, P.; Sinclair, K.; Kurz, D.; Mcnair, M.; Crist, S.; Shpritz, L.; Fitton, L.; Saffouri, R.; et al. Environmental and economic costs of soil erosion and conservation benefits. *Science* **1995**, *267*, 1117–1123. [CrossRef]
26. Tang, W.Z.; Shan, B.Q.; Zhang, H.; Zhang, W.Q.; Zhao, Y.; Ding, Y.K.; Rong, N.; Zhu, X.L. Heavy metal contamination in the surface sediments of representative limnetic ecosystems in Eastern China. *Sci. Rep.* **2014**, *4*, 7152. [CrossRef]
27. Eroglu, H.; Cakir, G.; Sivrikaya, F.; Akay, A.E. Using high resolution images and elevation data in classifying erosion risks of bare soil areas in the Hatila Valley Natural Protected Area, Turkey. *Stoch. Environ. Res. Risk Assess.* **2010**, *24*, 699–704. [CrossRef]
28. Fu, K.D.; He, D.M. Analysis and prediction of sediment trapping efficiencies of the reservoirs in the mainstream of the Lancang River. *Chin. Sci. Bull.* **2007**, *52*, 134–140. [CrossRef]
29. Vaezi, A.R.; Abbasi, M.; Keesstra, S.; Cerda, A. Assessment of soil particle erodibility and sediment trapping using check dams in small semi-arid catchments. *Catena* **2017**, *157*, 227–240. [CrossRef]
30. Wang, Z.Y.; Wu, B.S.; Wang, G.Q. Fluvial processes and morphological response in the Yellow and Weihe Rivers to closure and operation of Sanmenxia Dam. *Geomorphology* **2007**, *91*, 65–79. [CrossRef]
31. World Commission on Dams [WCD]. *Dams and Development: A New Framework for Decision-Making*; Earth Scan: London, UK, 2000.
32. Walling, D.E. Human impact on the sediment loads of Asian rivers. In *Sediment Problems and Sediment Management in Asian River Basins*; IAHS: Wallingford, UK, 2011; Volume 349, pp. 37–51.
33. Anthony, E.J.; Brunier, G.; Besset, M.; Goichot, M.; Dussouillez, P.; Nguyen, V.L. Linking rapid erosion of the Mekong River delta to human activities. *Sci. Rep.* **2015**, *5*, 14745. [CrossRef]

34. Ha, D.T.; Ouillon, S.; Vinh, G.V. Water and suspended sediment budgets in the lower Mekong from high-frequency measurements (2009–2016). *Water* **2018**, *10*, 846.
35. Manh, N.V.; Dung, N.V.; Hung, N.N.; Kummu, M.; Merz, B.; Apel, H. Future sediment dynamics in the Mekong Delta floodplains: Impacts of hydropower development, climate change and sea level rise. *Glob. Planet Chang.* **2015**, *127*, 22–33. [CrossRef]
36. Walling, D.E. The changing sediment load of the Mekong River. *AMBIO* **2008**, *37*, 150–157. [CrossRef]
37. Milliman, J.D.; Syvitski, J.P.M. Geomorphic tectonic control of sediment discharge to the ocean—The importance of small mountainous rivers. *J. Geol.* **1992**, *100*, 525–544. [CrossRef]
38. Mekong River Commission. *State of the Basin Report 2010*; Mekong River Commission: Phnom Penh, Cambodia, 2010.
39. Mekong River Commission. *The Study on the Sustainable Management and Development of the Mekong River Basin, Including Impacts of Mainstream Hydropower Projects*; Mekong River Commission: Phnom Penh, Cambodia, 2019.
40. Kondolf, G.M.; Schmitt, R.J.P.; Carling, P.; Darby, S.; Arias, M.; Bizzi, S.; Castelletti, A.; Cochrane, T.A.; Gibson, S.; Kummu, M.; et al. Changing sediment budget of the Mekong: Cumulative threats and management strategies for a large river basin. *Sci. Total Environ.* **2018**, *625*, 114–134. [CrossRef]
41. Li, L.; Wang, Y.; Liu, C. Effects of land use changes on soil erosion in a fast developing area. *Int. J. Environ. Sci. Technol.* **2014**, *11*, 1549–1562. [CrossRef]
42. Ta, T.K.O.; Nguyen, V.L.; Tateishi, M.; Kobayashi, I.; Tanabe, S.; Saito, Y. Holocene delta evolution and sediment discharge of the Mekong River southern Vietnam. *Quat. Sci. Rev.* **2002**, *21*, 1807–1819. [CrossRef]
43. Koehnken, L. *Discharge Sediment Monitoring Project (DSMP) 2009–2013: Summary and Analysis of Results Analysis of Results*; Mekong River Commission: Phnom Penh, Cambodia, 2014; pp. 5–14.
44. Kummu, M.; Varis, O. Sediment-related impacts due to upstream reservoir trapping, the Lower Mekong River. *Geomorphology* **2007**, *85*, 275–293. [CrossRef]
45. Fu, K.D.; He, D.M.; Lu, X.X. Sedimentation in the Manwan reservoir in the Upper Mekong and its downstream impacts. *Quat. Int.* **2008**, *186*, 91–99. [CrossRef]
46. Liu, C.; He, Y.; Des Walling, E.; Wang, J.J. Changes in the sediment load of the Lancang-Mekong River over the period 1965–2003. *Sci. China Technol. Sci.* **2013**, *56*, 843–852. [CrossRef]
47. Kummu, M.; Lu, X.X.; Wang, J.J.; Varis, O. Basin-wide sediment trapping efficiency of emerging reservoirs along the Mekong. *Geomorphology* **2010**, *119*, 181–197. [CrossRef]
48. Kondolf, G.M.; Rubin, Z.K.; Minear, J.T. Dams on the Mekong: Cumulative sediment starvation. *Water Resour. Res.* **2014**, *50*, 5158–5169. [CrossRef]
49. Yao, H.R.; Yang, Z.F.; Cui, B.S. Soil erosion and its environmental background at Lancang Basin of Yunnan Province. *Bull. Soil Water Conserv.* **2005**, *25*, 5–14. (In Chinese)
50. Yao, H.R.; Yang, Z.F.; Cui, B.S. Spatial analysis on soil erosion of Lancang River Watershed in Yunnan Province under the support of GIS. *Geogr. Res.* **2006**, *25*, 421–429. (In Chinese)
51. Yu, D.Y.; Pan, Y.Z.; Long, Z.H.; Wang, Y.Y.; Liu, X. Value evaluation of conserving water and soil for ecosystem supported by remote sensed technique in Yunnan Province. *J. Soil Water Conserv.* **2006**, *20*, 174–178. (In Chinese)
52. Peng, J.; Li, D.D.; Zhang, Y.Q. Analysis of spatial characteristics of soil erosion in mountain areas of northwestern Yunnan based on GIS and RUSLE. *J. Mt. Sci.* **2007**, *25*, 548–556. (In Chinese)
53. Thuy, H.T.; Lee, G. Soil loss vulnerability assessment in the Mekong River Basin. *J. Korean Geo Environ. Soc.* **2017**, *18*, 37–47. [CrossRef]
54. Mekong River Commission. *Overview of the Hydrology of the Mekong Basin*; Mekong River Commission: Phnom Penh, Cambodia, 2005; p. 73.
55. Mekong River Commission. *Assessment of Basin-Wide Development Scenarios—Main Report*; Mekong River Commission: Phnom Penh, Cambodia, 2011; p. 228.
56. Kummu, M.; Lu, X.X.; Rasphone, A.; Sarkkula, J.; Koponen, J. Riverbank changes along the Mekong River: Remote sensing detection in the Vientiane-Nong Khai area. *Quat. Int.* **2008**, *186*, 100–112. [CrossRef]
57. International Centre for Environmental Management. *MRC Strategic Environmental Assessment (SEA) of Hydropower on the Mekong Mainstream: Summary of the Final Report*; Mekong River Commission: Hanoi, Vietnam, 2010; p. 23.

58. WLE Graeter Mekong. *Dams in the Mekong River Basin: Commissioned, Under Construction, and Planned Dams*; WLE Graeter Mekong: Colombo, Sri Lanka, 2016.
59. Ozcan, A.U.; Erpul, G.; Basaran, M.; Erdogan, H.E. Use of USLE/GIS technology integrated with geostatistics to assess soil erosion risk in different land uses of Indagi Mountain Pass-Cankiri, Turkey. *Environ. Geol.* **2008**, *53*, 1731–1741. [CrossRef]
60. Fu, B.J.; Zhao, W.W.; Chen, L.D.; Zhang, Q.J.; Lu, Y.H.; Gulinck, H.; Poesen, J. Assessment of soil erosion at large watershed scale using RUSLE and GIS: A case study in the Loess Plateau of China. *Land Degrad. Dev.* **2005**, *16*, 73–85. [CrossRef]
61. Karaburun, A. Estimating potential erosion risks in Corlu using the GIS-based RUSLE method. *Fresenius Environ. Bull.* **2009**, *18*, 1692–1700.
62. Lufafa, A.; Tenywa, M.M.; Isabirye, M.; Majaliwa, M.J.G.; Woomer, P.L. Prediction of soil erosion in a Lake Victoria basin catchment using a GIS-based Universal Soil Loss model. *Agric. Syst.* **2003**, *76*, 883–894. [CrossRef]
63. Ministry of Water Resources of China. *Soil Erosion Rate Standard, Technological Standard of Soil and Water Conservation SD238-87*; Water Resources and Electric Power Press: Beijing, China, 1988. (In Chinese)
64. Wang, Z.Y.; Lee, J.H.W.; Melching, C.S. *River Dynamics and Integrated River Management*; Tsinghua University Press: Beijing, China; Springer: Berlin/Heidelberg, Germany, 2015.
65. Wischmeier, W.H. *Predicting Rainfall Erosion Losses: A Guide to Conservation Planning*; Agricultural Handbook No. 537; U.S. Department of Agriculture: Washington, DC, USA, 1978; pp. 285–291.
66. Yatagai, A.; Kamiguchi, K.; Arakawa, O.; Hamada, A.; Yasutomi, N.; Kitoh, A. APHRODITE constructing a long-term daily gridded precipitation dataset for Asia based on a dense network of rain gauges. *Bull. Am. Meteorol. Soc.* **2012**, *93*, 1401–1415. [CrossRef]
67. Zhou, F.J.; Chen, M.H.; Lin, F.X.; Huang, Y.H.; Lu, C.L. The rainfall erosivity index in Fujian Province. *J. Soil Water Conserv.* **1995**, *9*, 13–18.
68. Williams, J.R. Chapter 25: The epic mode. In *Computer Models of Watershed Hydrology*; Singh, V.P., Ed.; Water Resource Publications: Woodbridge, VA, USA, 1995; pp. 909–1000.
69. Williams, J.R.; Renard, K.G.; Dyke, P.T. Epic—A new method for assessing erosions effect on soil productivity. *J. Soil Water Conserv.* **1983**, *38*, 381–383.
70. Foster, G.R.; Mccool, D.K.; Renard, K.G.; Moldenhauer, W.C. Conversion of the universal soil loss equation to SI metric units. *J. Soil Water Conserv.* **1981**, *36*, 355–359.
71. Rammahi, A.A.; Khassaf, S.I. Estimation of soil erodibility factor in RUSLE equation for Euphrates river watershed using GIS. *Int. J. Geomate* **2018**, *14*, 164–169. [CrossRef]
72. Wawer, R.; Nowocień, E.; Podolski, B. Real and calculated K-USLE erodibility factor for selected polish soils. *Pol. J. Environ. Stud.* **2005**, *14*, 655–658.
73. Panagos, P.; Borrelli, P.; Meusburger, K. A new European slope length and steepness factor (LS-Factor) for modeling soil erosion by water. *Geosciences* **2015**, *5*, 117–126. [CrossRef]
74. Wang, G.Q.; Jiang, H.; Xu, Z.X.; Wang, L.J.; Yue, W.F. Evaluating the effect of land use changes on soil erosion and sediment yield using a grid-based distributed modelling approach. *Hydrol. Process.* **2012**, *26*, 3579–3592. [CrossRef]
75. Desmet, P.J.J.; Govers, G. A GIS procedure for automatically calculating the USLE LS factor on topographically complex landscape units. *J. Soil Water Conserv.* **1996**, *51*, 427–433.
76. Liu, H.H.; Fohrer, N.; Hormann, G.; Kiesel, J. Suitability of S factor algorithms for soil loss estimation at gently sloped landscapes. *Catena* **2009**, *77*, 248–255. [CrossRef]
77. Kim, S.H.; Julien, P.Y. Soil erosion modeling using RUSLE and GIS on the MIHA watershed. *Water Eng. J.* **2006**, *7*, 29–41.
78. Mccool, D.K.; Brown, L.C.; Foster, G.R.; Mutchler, C.K.; Meyer, L.D. Revised slope steepness factor for the universal soil loss equation. *Trans. ASAE* **1987**, *30*, 1387–1396. [CrossRef]
79. Suif, Z.; Yoshimura, C.; Valeriano, S.; Seingheng, H. Spatially distributed model for soil erosion and sediment transport in the Mekong River Basin. *Int. Water Technol. J.* **2013**, *3*, 197–206.
80. Renard, K.G.; Foster, G.R.; Weesies, G.A.; McCool, D.K.; Yoder, D.C. *Predicting Soil Erosion by Water: A Guide to Conservation Planning with the Revised Universal Soil Loss Equation (RUSLE)*; Agriculture Handbook Number 703; USDA: Washington, DC, USA; Agricultural Research Service: Washington, DC, USA, 1997.

81. Van Remortel, R.D.; Hamilton, M.E.; Hickey, R. Estimating the LS factor for RUSLE through iterative slope length processing of digital elevation data within Arclnfo grid. *Cartography* **2001**, *30*, 27–35. [CrossRef]
82. Zhang, H.; Wei, J.; Yang, Q.; Baartman, J.E.M.; Gai, L.; Yang, X.; Li, S.; Yu, J.; Ritsema, C.J.; Geissen, V. An improved method for calculating slope length (λ) and the LS parameters of the Revised Universal Soil Loss Equation for large watersheds. *Geoderma* **2017**, *308*, 36–45. [CrossRef]
83. Durigon, V.L.; Carvalho, D.F.; Antunes, M.A.H.; Oliveira, P.T.S.; Fernandes, M.M. NDVI time series for monitoring RUSLE cover management factor in a tropical watershed. *Int. J. Remote Sens.* **2014**, *35*, 441–453. [CrossRef]
84. Kuok, K.; Mah, D.; Chiu, P. Evaluation of C and P factors in universal soil loss equation on trapping sediment: Case study of Santubong River. *J. Water Resour. Prot.* **2013**, *5*, 1149–1154. [CrossRef]
85. Toy, T.J.; Foster, G.R. Mined lands, construction sites, and reclaimed lands. In *Guidelines for the Use of Revised Universal Soil Loss Equation (RUSLE)*; Joe, R.G., Ed.; Version 1.06; Western Regional Coordinating Centre Office of Surface Mining: Denver, CO, USA; The Office of Technology Transfer: Denver, CO, USA, 1998.
86. Yang, D.W.; Kanae, S.; Oki, T.; Koike, T.; Musiake, K. Global potential soil erosion with reference to land use and climate changes. *Hydrol. Process.* **2003**, *17*, 2913–2928. [CrossRef]
87. Kaffas, K.; Hrissanthou, V.; Sevastas, S. Modeling hydromorphological processes in a mountainous basin using a composite mathematical model and ArcSWAT. *Catena* **2018**, *162*, 108–129. [CrossRef]
88. He, D.M.; Hsiang, T.K. Facilitating regional sustainable development through integrated multi-objective utilization, management of water resources in the Lancang-Mekong River Basin. *J. Chin. Geogr.* **1997**, *7*, 9–21.
89. Mekong River Commission. *Databases of Mekong River Commission*; Mekong River Commission (MRC): Phnom Penh, Cambodia, 2010.
90. Hengl, T.; De Jesus, J.M.; MacMillan, R.A.; Batjes, N.H.; Heuvelink, G.B.M.; Ribeiro, E.; Samuel-Rosa, A.; Kempen, B.; Leenaars, J.G.B.; Walsh, M.G.; et al. SoilGrids1km—Global soil information based on automated mapping. *PLoS ONE* **2014**, *9*, e105992. [CrossRef]
91. Wei, S.G.; Hengl, T.; De Jesus, J.M.; Yuan, H.; Dai, Y.J. Mapping the global depth to bedrock for land surface modeling. *J. Adv. Model. Earth Syst.* **2017**, *9*, 65–88.
92. Organization for Economic Co-operation and Development. *Environmental Performance of Agriculture in OECD Countries Since 1990*; OECD: Paris, France, 2008; pp. 179–386.
93. Kondolf, G.M.; Gao, Y.X.; Annandale, G.W.; Morris, G.L.; Jiang, E.H.; Zhang, J.H.; Cao, Y.T.; Carling, P.; Fu, K.D.; Guo, Q.C.; et al. Sustainable sediment management in reservoirs and regulated rivers: Experiences from five continents. *Earths Future* **2014**, *2*, 256–280. [CrossRef]

Article

Estimating Human Impacts on Soil Erosion Considering Different Hillslope Inclinations and Land Uses in the Coastal Region of Syria

Safwan Mohammed [1], Hazem G. Abdo [2], Szilard Szabo [3], Quoc Bao Pham [4,5], Imre J. Holb [6,7], Nguyen Thi Thuy Linh [8,9,*], Duong Tran Anh [10], Karam Alsafadi [11], Ali Mokhtar [12,13,14], Issa Kbibo [15], Jihad Ibrahim [15] and Jesus Rodrigo-Comino [16,17]

1 Institution of Land Utilization, Technology and Regional Planning, University of Debrecen, 4032 Debrecen, Hungary; safwan@agr.unideb.hu
2 Geography Department, Faculty of Arts and Humanities, University of Tartous, 51003 Tartous, Syria; hazemabdo1990@gmail.com
3 Department of Physical Geography and Geoinformatics, Faculty of Science and Technology, University of Debrecen, 4032 Debrecen, Hungary; szabo.szilard@science.unideb.hu
4 Environmental Quality, Atmospheric Science and Climate Change Research Group, Ton Duc Thang University, Ho Chi Minh City 700000, Vietnam; phambaoquoc@tdtu.edu.vn
5 Faculty of Environment and Labour Safety, Ton Duc Thang University, Ho Chi Minh City 700000, Vietnam
6 Institute of Horticulture, University of Debrecen, Böszörményi út 138, 4032 Debrecen, Hungary; holbimre@gmail.com
7 Hungarian Academy of Sciences, Plant Protection Institute, Herman Ottó út 15, 1022 Budapest, Hungary
8 Institute of Research and Development, Duy Tan University, Danang 550000, Vietnam
9 Faculty of Environmental and Chemical Engineering, Duy Tan University, Danang 550000, Vietnam
10 Ho Chi Minh City University of Technology (HUTECH) 475A, Dien Bien Phu, Ward 25, Binh Thanh District, Ho Chi Minh City 700000, Vietnam; ta.duong@hutech.edu.vn
11 Department of Geography and GIS, Faculty of Arts, Alexandria University, Alexandria 25435, Egypt; karam.alsafadi@alexu.edu.eg
12 State of Key Laboratory of Soil Erosion and Dryland Farming on Loess Plateau, Institute of Soil and Water Conservation, Northwest Agriculture and Forestry University, Yangling 712100, China; ali.mokhtar@agr.cu.edu.eg
13 Chinese Academy of Sciences and Ministry of Water Resources, Yangling 712100, China
14 Department of Agricultural Engineering, Faculty of Agriculture, Cairo University, Giza 12613, Egypt
15 Department of Soil and Water Science, Faculty of Agriculture, Tishreen University, Lattakia, Syria; dr.e.harsanyi.de@gmail.com (I.K.); samerkiw1990@gmail.com (J.I.)
16 Department of Physical Geography, University of Trier, 54296 Trier, Germany; jesus.rodrigo@uv.es
17 Soil Erosion and Degradation Research Group, Department of Geography, Valencia University, Blasco Ibàñez, 28, 46010 Valencia, Spain
* Correspondence: nguyentthuylinh58@duytan.edu.vn

Received: 30 August 2020; Accepted: 4 October 2020; Published: 7 October 2020

Abstract: Soils in the coastal region of Syria (CRoS) are one of the most fragile components of natural ecosystems. However, they are adversely affected by water erosion processes after extreme land cover modifications such as wildfires or intensive agricultural activities. The main goal of this research was to clarify the dynamic interaction between erosion processes and different ecosystem components (inclination, land cover/land use, and rainy storms) along with the vulnerable territory of the CRoS. Experiments were carried out in five different locations using a total of 15 erosion plots. Soil loss and runoff were quantified in each experimental plot, considering different inclinations and land uses (agricultural land (AG), burnt forest (BF), forest/control plot (F)). Observed runoff and soil loss varied greatly according to both inclination and land cover after 750 mm of rainfall (26 events). In the cultivated areas, the average soil water erosion ranged between 0.14 ± 0.07 and 0.74 ± 0.33 kg/m^2; in the BF plots, mean soil erosion ranged between 0.03 ± 0.01 and 0.24 ± 0.10 kg/m^2. The lowest amount of erosion was recorded in the F plots where the erosion ranged between 0.1 ± 0.001 and 0.07 ± 0.03 kg/m^2. Interestingly, the General Linear Model revealed that all factors (i.e., inclination, rainfall and land use) had a significant ($p < 0.001$) effect on the soil loss. We concluded that human activities greatly influenced soil erosion rates, being higher in the AG lands, followed by BF and F. Therefore, the current study could be very useful to policymakers and planners for proposing immediate conservation or restoration plans in a less studied area which has been shown to be vulnerable to soil erosion processes.

Keywords: soil management; land cover changes; Syria; soil erosion; hillslopes

---

1. Introduction

Soils are vital components of environmental systems and supply livelihoods, services and goods for humans and natural ecosystems [1,2]. Soils are formed by numerous factors such as parent material, topography, climate, water, organisms, and time; however, it is well-known that this process is slow and endangered by land degradation due to certain human activities [3–5]. Intensification of anthropogenic effects has become a key factor that causes negative structural shifts in the soil matrix and health; thus, there has been an acceleration of the erosional cycle from prehistoric times [6] to today [7]. Current soil erosion rates, caused by water or wind, are high and can be considered one of the most serious ecological threats to land sustainability globally [8], given that more than 75 billion tons per year of soil are lost due to soil erosion [9]. The problem associated with soil erosion by water is the result of spatial integration of physical and human factors, which vary significantly across scales (from pedon to watershed), and make any estimation difficult [10–12]. Soil erosion irremediably reduces the quality of the physico-chemical and biological properties, soil fertility and land productivity, which considerably affect cultivated areas [13,14]. Therefore, nature-based solutions to achieve land degradation neutrality can be a key to conserving ecosystem services [15,16]. However, for any ecological restoration, stakeholders and land managers must be fully motivated and convinced, and this is still a current challenge [17,18].

Soil erosion is progressively limiting the availability of resources, threatening biodiversity, and affecting food production, and is accelerated by specific drivers such as climate change, land use/land cover changes, overgrazing, inappropriate farming procedures, or armed conflicts [19–23]. Consequently, soil erosion is defined as a physical and anthropological challenge [24]. During the 1980s, statistics indicated that about two billion hectares of agricultural land had completely deteriorated since 1000 AD, and currently, the FAO estimates ~75 billion tons of agricultural soil loss, causing an annual cost of USD 400 billion [25]. Consequently, an increasing interest in soil stability and conservation is progressively evolving to deal with this worldwide environmental problem in the context of the landscape changes which have occurred in the current century [9,26,27].

Soil erosion is the outcome of the dynamic interaction between different ecosystem components, e.g., land use, inclination, rainfall intensity, and soil properties [28]. The mechanism of soil erosion by water includes splashing and detachment of soil particles due to the kinetic energy of raindrops, then the transportation of these particles by surface runoff [29]. However, due to its tremendous impact, recent research has been directed more towards erosion control techniques in many parts of the world, for example in Austria [30], Spain [31], China [32], Hungary [33], Germany [34], and France [35], among others.

The components of the Mediterranean environment are considered one of the most fragile around the world, especially as regards soils, which are exposed to severe degradation processes [36,37]. In several cases, rainfall and runoff have induced soil erosion, which is a well-known degradation challenge in terms of ecological mismanagement [8,38]. Rugged and dissected terrains, steep slopes, high rainfall intensities, shallow and skeletal soil thicknesses, receding and sparse vegetation, and chronic and severe drought stress in summer are among the most important physical factors which drive soil erosion [39,40]. Several authors have reported that the annual rates of soil erosion have reached dangerous levels, exceeding the allowable soil loss tolerance limits (2 to 12 Mg ha^{-1} year^{-1}) for agricultural and economic sustainability in the Mediterranean environment [41–45]; nevertheless, these numbers can vary depending on the main goal of the research and the specific area [46]. For example, Kouli et al. [47] determined that more than 1 Mg ha^{-1} year $^{-1}$ may be irreversible within a time dimension ranging from 50 to 100 years.

Syria is among the eastern Mediterranean countries which are seriously exposed to the problem of water-related soil erosion, especially in the coastal region of the country (CRoS). This area represents an appropriate terrestrial, structural, climatic, hydrological, and intense anthropological case of the acceleration of soil erosion by water [23,48]. In the CRoS, soil erosion by water is the first threat to agricultural activity, which is the pivot of economic life for 34.8% of the population [49]. Meanwhile, CRoS is considered the first agricultural stability zone in Syria, receiving more than 600 mm of rainfall and being used for rainfed agriculture, with a total agricultural land area of 2.7 million ha [50]. Accordingly, the issue of soil erosion in CRoS has been assessed by many local scholars at the administrative area or catchment area level, using different models such as the Revised Universal Soil Loss Equation (RUSLE) [51,52], the Water Erosion Prediction Project (WEPP) model [48,53], and the Coordination of Information on the Environment (CORINE) model [54]. On the other hand, a limited number of studies have dealt with soil erosion after wildfires. Al-Ali and Kheder [55] stressed the importance of monitoring soil erosion after wildfires, where the soil erosion from burnt forests reached 7.22 Mg ha^{-1} year^{-1}.

In the CRoS, as well as in the Mediterranean region in general, different anthropogenic activities (i.e., rapid changes in land use driven by intense population pressure or agricultural expansion) and climate change have rapidly exacerbated soil water erosion. However, information about soil erosion on the field-scale in the near-eastern Mediterranean remains limited compared to that in the western and northern Mediterranean. Some representative examples can be found in the territories of border countries, highlighting the importance of assessing land degradation processes from different points of view, e.g., [55–58]. Within this perspective, the main aim of this research was to bridge the gap in the common literature on soil water erosion in the coastal territories of Syria by measuring soil water erosion and runoff under three different land uses (agricultural land (AG), burnt forest (BF), forest/control plot (F)). Our hypotheses were the following: (i) agricultural areas are the main areas at risk of soil loss; (ii) burnt forests are endangered by the increased runoff and severe soil loss; (iii) the effect of inclination on erosion rates has a saturation curve, i.e., above a threshold inclination, the rate of erosion does not increase relevantly; and (iv) slopes and land cover have a significant interactive effect, thus, these two factors determine erosion hazard together.

2. Materials and Methods

2.1. Study Area

The study area is located in the western part of Syria (35°49′ to 36°31′ E; 34°49′ to 36°05′ N) within an area of 5274 km^2 (Figure 1). The elevation of the region ranges from 0 to 1700 m a.s.l. The Syrian coast area consists of three basic geomorphological units: the plain (0–100 m), the plateau (100–400) and the mountains (400–1700) [59]. The study area is characterized by narrow plains near the coast, followed by dissected mountains. The degree of inclination generally ranges from 0° to more than 60°. The coastal strip was affected by recent tectonics, which caused a fluctuation in the sea level from the Early Pleistocene to the recent "upper" Holocene. This is reflected in the diversity of rock formations such as sandstones, sands and conglomerates, which were laid down as sedimentary deposits with limestone and marls. Interestingly, these rocks were penetrated by basaltic rocks in the southern part of the coastline [60].

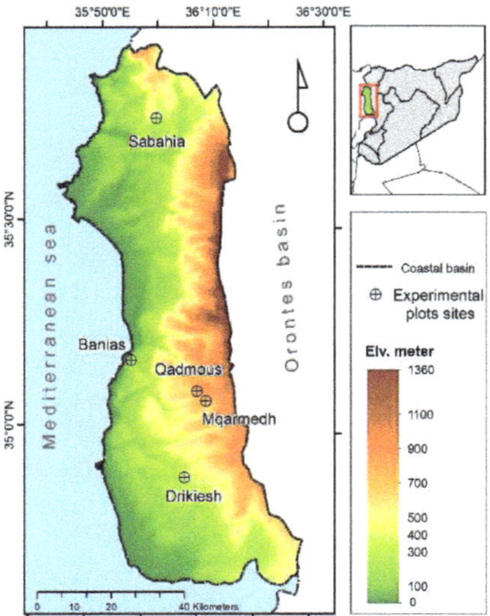

Figure 1. The coastal region of Syria and locations of the experimental plots.

According to the Köppen climate classification, the study area falls into two categories (Csa and Csb) with the main group being C, which follows the Mediterranean climate. The rainy winter season is mostly concentrated between November and March [61]. In general, the average rainfall ranges from 765 mm (near the coast) to 1250 mm (in the high mountains) [61]. The mean annual temperature in the plain areas is about 19.3 °C and in the mountains, it is about 14.8 °C [61]. The common soil orders are Inceptisols, Entisols, and Mollisols [62]. The study area includes the governorates of Tartous and Lattakia, with a population of approximately three million [61]. Syria is divided into five agricultural stability zones, according to distributed rainfall and the suitability for rainfed agriculture. The study area is located in the first agricultural stability zone, where rainfall exceeds 600 mm [63].

Traditional agriculture is the most essential economic axis for rural inhabitants, and most fields are cultivated with wheat, and olive and citrus orchards. Between 2010 and 2018, more than 800 wildfires were recorded in the coastal region of Syria. Wildfires usually occur between June and late August (summer season), and are typically induced by human activities. In this research, the experimental burnt sites were selected based on fire time and intensity.

2.2. Experimental Design

Based on a field survey conducted in the study area, five different locations (SY1, SY2, SY3, SY4, SY5) with different hillslope inclinations (38%, 45%, 15%, 29%, 10%) were chosen as representative sites for measuring soil erosion (Table 1). Three different land uses were selected at each location: (i) agricultural land (AG), where traditional cultivation, sowing, and harvesting occurs, with the absence of any mechanization; the common crops in AG lands are wheat (SY1, SY2), olives (SY3, SY4), and citrus orchards (SY5); (ii) burnt forest (BF), where soil cover varies from 30% to 55% with local natural vegetation; and (iii) forest land (F), which is characterized by mixed forest, and is used as a control plot without recently extensive human disturbance. The soil cover for all treatments was sampled without any disturbance.

Table 1. Experimental characteristics of the five locations studied.

Location	Code	X	Y	Slope (%)	Rainfall (mm)
Drikiesh	SY1	36°07′	34°53′	38%	965
Qadmous	SY2	36°09′	35°05′	45%	936
Banias	SY3	35°56′	35°10′	15%	914
Mqarmedh	SY4	36°10′	35°04′	29%	890
Sabahia	SY5	36°00′	35°45′	10%	765

Experimental plots of 2 × 1.6 m were installed with metal barriers of 0.5 m height (0.15 m into the soil) to collect runoff and soil loss. This method was previously adopted in Syria by [64], and applied by [61,65] and [61,66]. Nonetheless, the plots designed were similar to [61,67], but of a smaller size.

The amount of rainfall (mm) was measured on-site by placing a metal rainfall gauge at each location. Meanwhile, runoff (L/m^2) was recorded at each plot after each rainy event by recording the volume in each sediment collector. Soil loss (kg/m^2) was also determined by mixing the collected soil detachment and a representative sample of 5 L each. Finally, the samples were transported to the laboratory. In the laboratory, each sample was placed in a small container and dried in an oven (105 °C) for 24 h.

In addition, soil samples were collected at the beginning of the monitoring period from the topsoil (0–0.15 m) in each plot, and soil texture and soil erodibility factors were determined (Table 2). The design and performance of the chosen experimental plot with the sampling strategy were tested following [39] (Figure 2). Data were collected from October 2012 to December 2013 (i.e., the vegetation period). A total of 26 rainy storms were observed during the monitoring period.

Table 2. Soil texture and K value in the studied locations (SY1-SY5) for three land uses (agricultural land (AG), burnt forest (BF), and forest land (F)).

Code	Agricultural Land (AG)				Burnt Forest (BF)				Forest Land (F)			
	Sand	Silt	Clay	K	Sand	Silt	Clay	K	Sand	Silt	Clay	K
SY1	31.5	27.0	41.5	0.154	30.5	32.5	37.0	0.128	32.5	35.5	32.0	0.074
SY2	23.0	35.0	42.0	0.236	23.0	35.0	42.0	0.215	24.0	36.0	40.0	0.161
SY3	27.0	31.0	42.0	0.172	25.0	33.0	42.0	0.155	29.0	33.0	38.0	0.124
SY4	22.0	30.5	47.5	0.186	21.0	32.0	47.0	0.173	23.0	31.0	46.0	0.119
SY5	20.5	39.5	40.0	0.257	27.5	42.0	30.5	0.224	17.5	51.0	31.5	0.151

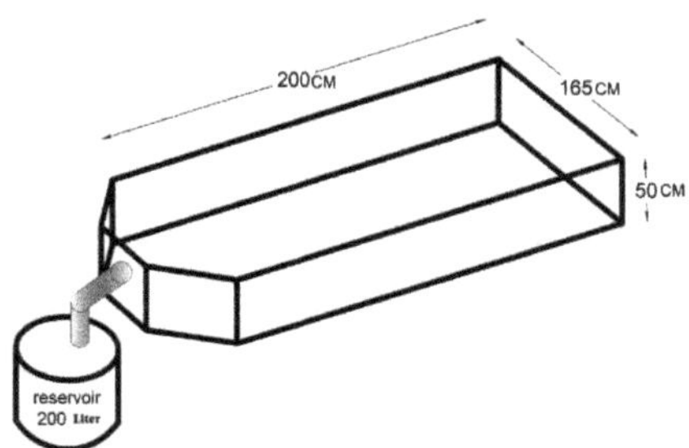

Figure 2. Sketch design for the experimental plot.

2.3. Data Analysis

Average, maximum, minimum, and median values were determined. Soil erosion and runoff data were depicted in boxplots, together with the linear regression among them in each land cover class. Normal distribution was checked by the Shapiro–Wilk test (S-W); as this failed, the non-parametric Kruskal–Wallis test (K-W) [68] was applied as an alternative to the one-way ANOVA. The K–W test aimed to detect the difference between the medians of the treatments with the following hypothesis:

H_0 was that the medians of the studied groups were from the same distribution, while H_1 represented the idea that the medians of the studied groups were different. As the K–W test did not show which plot is statistically different from any other, the pairwise comparison among slopes was performed with the Mann–Whitney test with Bonferroni correction. Pairwise analyses in the same hillslope but for different land uses (i.e., AG-F; AG-BF, BF-F) were neglected as we focused on the differences caused by inclination and did not analyze the obvious differences among land use types. Finally, to assess the relationships between the studied variables, a correspondence analysis was carried out. We applied a General Linear Model (GLM) to reveal the importance of rainfall, inclination and land use types. The inclination type was included as ordinal data and land use as a categorical dummy variable. We determined the model parameters, and the effect sizes expressed as partial η^2p, which expressed the contribution of each variable and the interaction of the factors as a standardized measure [69]. The effect can be very small ($\eta^2p < 0.01$), small ($0.01 > \eta^2p > 0.06$), medium ($0.06 > \eta^2p > 0.14$), and large ($\eta^2p > 0.14$). Differences among inclination degrees were analyzed with the t-test and ANOVA using the 1999 Monte-Carlo permutation. Statistical analyses were conducted with SPSS (v24; IBM, Chicago, IL, USA), the EViews software package (v10; [70] New York, NY, USA), and R 3.6.3 [71] with the gamlj package [72].

3. Results

3.1. Soil Water Erosion and Runoff

Observed runoff and soil erosion varied according to both inclination and land use, as can be observed in Appendix A (Figure A1). The total rainfall in the study area exceeded 750 mm, divided into 26 events. The average soil loss ranged between 0.74 ± 0.33 and 0.14 ± 0.07 kg/m^2, while runoff ranged between 42.14 ± 15.27 and 12.77 ± 5.84 L/m^2 in the AG (Table 3). Meanwhile, in the BF plots, mean soil loss ranged between 0.24 ± 0.10 and 0.03 ± 0.01 kg/m^2, and runoff from 22.95 ± 9.33 to 3.77 ± 1.62 L/m^2. The lowest amounts of soil loss and runoff were recorded in the F lands, where the ranges were between 0.07 ± 0.03 and 0.1 ± 0.001 kg/m^2, and 11.98 ± 4.73 and 1.78 ± 0.78 L/m^2, respectively.

Table 3. Univariate statistics of observed soil loss and runoff in the studied locations (SY1-SY5) under three land uses (AG: agricultural land, BF: burnt forest, F: forest).

System	Code	Soil Loss (kg/m^2)							Runoff (L/m^2)						
		Min.	Max.	Range	Median	Mean	SD	φ	Min.	Max.	Range	Median	Mean	SD	φ
AG	SY1	0.07	1.34	1.27	0.71	0.74	0.33	0.07	15.20	72.50	57.30	39.75	42.14	15.27	3.05
	SY2	0.23	1.17	0.94	0.66	0.69	0.29	0.06	13.00	72.50	59.50	41.75	41.42	16.88	3.38
	SY3	0.09	0.47	0.38	0.22	0.24	0.11	0.02	6.50	33.00	26.50	17.10	18.66	8.19	1.64
	SY4	0.01	0.94	0.93	0.45	0.50	0.25	0.05	5.20	50.00	44.80	25.25	28.07	12.61	2.52
	SY5	0.04	0.29	0.25	0.11	0.14	0.07	0.01	3.90	22.50	18.60	10.75	12.77	5.48	1.10
BF	SY1	0.08	0.43	0.35	0.20	0.22	0.09	0.02	7.50	37.00	29.50	18.75	20.35	7.80	1.56
	SY2	0.08	0.45	0.37	0.22	0.24	0.10	0.02	7.90	41.50	33.60	22.50	22.95	9.33	1.87
	SY3	0.04	0.22	0.18	0.10	0.11	0.05	0.01	1.80	12.00	10.20	5.10	6.06	2.92	0.58
	SY4	0.06	0.37	0.31	0.17	0.20	0.09	0.02	4.10	22.00	17.90	11.50	12.58	5.36	1.07
	SY5	0.01	0.05	0.04	0.02	0.03	0.01	0.00	1.10	7.20	6.10	3.25	3.72	1.62	0.32
F	SY1	0.02	0.10	0.08	0.05	0.05	0.02	0.001	4.10	21.50	17.40	11.20	11.98	4.73	0.95
	SY2	0.02	0.13	0.11	0.06	0.07	0.03	0.01	3.10	20.00	16.90	10.95	11.03	4.85	0.97
	SY3	0.01	0.05	0.04	0.02	0.02	0.01	0.001	1.10	6.90	5.80	3.40	3.79	1.79	0.36
	SY4	0.01	0.07	0.06	0.04	0.04	0.02	0.001	2.00	12.50	10.50	6.13	6.54	3.21	0.64
	SY5	0.00	0.02	0.02	0.01	0.01	0.00	0.001	0.50	3.50	3.00	1.50	1.78	0.78	0.16

Min: Minimum; Max: Maximum; SD (n): Standard deviation (n); φ: Standard error of the mean.

The highest soil loss was 1.34 ± 0.33 kg/m^2, registered in the AG lands with 38% inclination, and the lowest was in the gentlest slope (10%), reaching 0.29 ± 0.07 kg/m^2 (Figure 3a; Table 3). Soil loss was the highest in both BF and F with a hillslope inclination of 45%, reaching 0.45 ± 0.10 and 0.13 ± 0.03 kg/m^2, respectively (Figure 3b,c; Table 3).

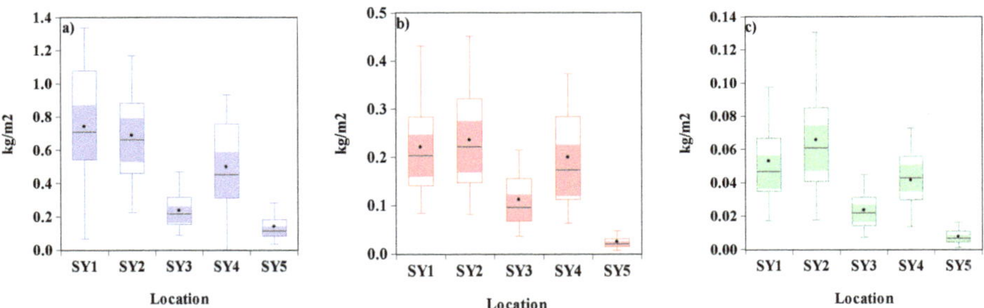

Figure 3. Box plots of soil erosion in each ecosystem (with respect to slope): (**a**) agricultural land; (**b**) !burnt forest, and (**c**) forest (median (——); mean (•); median 95% confidence (shaded).

Similarly, a maximum runoff was recorded in the AG lands with 72.5 L/m^2 in the steepest slopes (Figure 4a). In BF, the highest runoff was 41.51 L/m^2 with 45% inclination; meanwhile, the lowest was observed with the gentlest slopes, reaching 1.1 L/m^2 (Figure 4b). In F lands, the highest runoff was 21.50 L/m^2 (SY1) and the lowest was 3.50 L/m^2 (SY5) (Figure 4c).

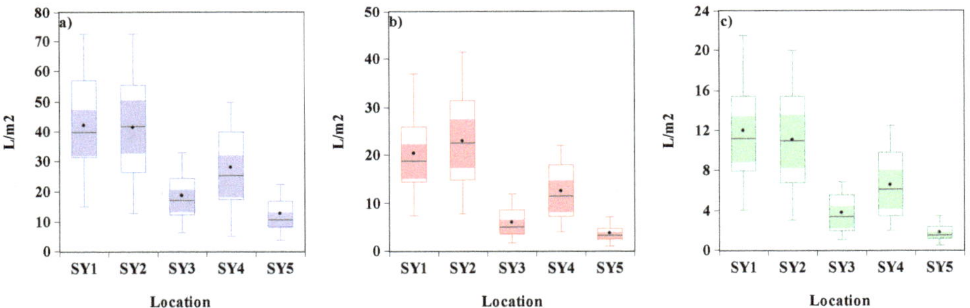

Figure 4. Box plots of runoff in each ecosystem (with respect to slope): (**a**) agricultural land; (**b**) burnt forest, and (**c**) forest (median (——); mean (•); median 95% confidence (shaded).

In each studied land-use type, regression analysis detected a high correlation between the generation of runoff and the activation of soil loss: R^2 values were 0.91, 0.87, and 0.89 ($p < 0.05$) in AG, BF, and F, respectively (Figure 5).

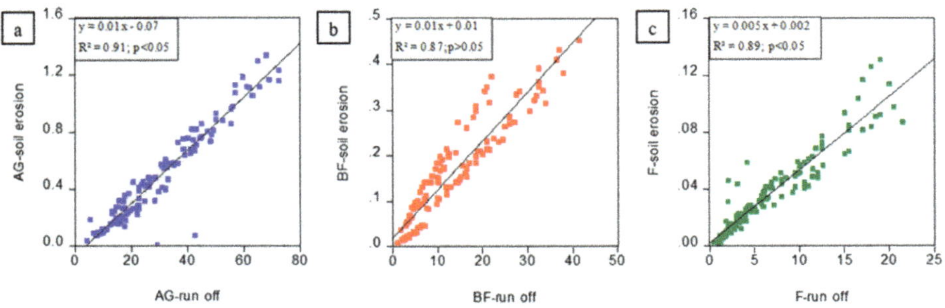

Figure 5. Correlation between soil loss (kg/m^2) and runoff (L/m^2) (regardless slope): (**a**) agricultural land, (**b**) burnt forest, and (**c**) forest.

3.2. *Impact of Inclination on Soil Water Erosion*

The Kruskal–Wallis test (K–W) showed that at least one of the studied plots was significantly ($p < 0.05$) different from other treatments in each slope inclination (SY1, SY2, SY3, SY4, SY5), and land use (AG, BF, F) (Table 4).

Table 4. Kruskal–Wallis analysis in each ecosystem for both soil water erosion and runoff ($p < 0.05$).

Kruskal–Wallis	Soil Loss		Runoff	
	H (chi^2)	p	H (chi^2)	p
SY1	63.41	0.00	48.99	0.00
SY2	62.47	0.00	45.69	0.00
SY3	59.44	0.00	43.81	0.00
SY4	55.52	0.00	40.89	0.00
SY5	65.83	0.00	59.14	0.00
Agricultural land	76.35	0.00	67.33	0.00
Burnt forest	78.57	0.00	92.08	0.00
Forest	83.09	0.00	86.17	0.00

The significance level is 0.05.

The pairwise comparison among the inclinations showed that there was a significant difference ($p < 0.05$) among them in the agricultural lands, both in the case of soil loss and runoff under different inclinations (Table 5). Differences were also significant ($p < 0.05$) between 15% (SY3) and 45% (SY2), and between 15% (SY3) and 38% (SY1). However, non-significant differences were noticed among the following plots: 10% vs 15%; 29% vs 45%; 29% vs 38%; and 45% vs 38% (Table 5). Just as with the AG, the F plots showed similar values with one exception: in the 29% vs the 38% plots, where the difference was significant regarding the runoff. In BF plots, significant differences were recorded in soil loss data among the following pairs: 10% vs 29%; 10% vs 38%; and 10% vs 45%; meanwhile, the pairwise analysis of runoff data showed identical results in the F plots.

Table 5. Pairwise comparisons between slopes for soil loss and runoff for three land uses (AG: agricultural land, BF: burnt forest, F: forest).

		Soil Loss					Runoff				
System	Inclination	Test Statistic	Std. Error	Std. Test Statistic	Sig.	Adj. Sig.[a]	Test Statistic	Std. Error	Std. Test Statistic	Sig.	Adj. Sig.[a]
AG	10–15%	18.64	10.45	1.78	0.07	0.75	18.96	10.45	1.82	0.07	0.70
	10–29%	52.89	10.45	5.06	0.00	**0.00**	42.44	10.45	4.06	0.00	**0.00**
	10–45%	70.35	10.45	6.73	0.00	**0.00**	66.96	10.45	6.41	0.00	**0.00**
	10–38%	72.65	10.45	6.95	0.00	**0.00**	70.00	10.45	6.70	0.00	**0.00**
	15–29%	−34.25	10.45	−3.28	0.00	**0.01**	−23.48	10.45	−2.25	0.03	0.25
	15–45%	51.71	10.45	4.95	0.00	**0.00**	48.00	10.45	4.60	0.00	**0.00**
	15–38%	54.02	10.45	5.17	0.00	**0.00**	51.04	10.45	4.89	0.00	**0.00**
	29–45%	17.46	10.45	1.67	0.10	0.95	24.52	10.45	2.35	0.02	0.19
	29–38%	19.77	10.45	1.89	0.06	0.59	27.56	10.45	2.64	0.01	0.08
	45–38%	2.31	10.45	0.22	0.83	1.00	3.04	10.45	0.29	0.77	1.00
BF	10–15%	38.98	10.45	3.73	0.00	**0.00**	16.96	10.45	1.62	0.10	1.00
	10–29%	67.44	10.45	6.46	0.00	**0.00**	50.56	10.45	4.84	0.00	**0.00**
	10–38%	74.71	10.45	7.15	0.00	**0.00**	75.73	10.45	7.25	0.00	**0.00**
	10–45%	77.71	10.45	7.44	0.00	**0.00**	80.21	10.45	7.68	0.00	**0.00**
	15–29%	−28.46	10.45	−2.72	0.01	0.06	−33.60	10.45	−3.22	0.00	**0.01**
	15–38%	35.73	10.45	3.42	0.00	**0.01**	58.77	10.45	5.63	0.00	**0.00**
	15–45%	38.73	10.45	3.71	0.00	**0.00**	63.25	10.45	6.05	0.00	**0.00**
	29–38%	7.27	10.45	0.70	0.49	1.00	25.17	10.45	2.41	0.02	0.16
	29–45%	10.27	10.45	0.98	0.33	1.00	29.65	10.45	2.84	0.01	0.05
	38–45%	−3.00	10.45	−0.29	0.77	1.00	−4.48	10.45	−0.43	0.67	1.00
F	10–15%	32.85	10.45	3.14	0.00	**0.02**	26.48	10.45	2.54	0.01	0.11
	10–29%	61.79	10.45	5.91	0.00	**0.00**	50.35	10.45	4.82	0.00	**0.00**
	10–38%	73.65	10.45	7.05	0.00	**0.00**	76.64	10.45	7.34	0.00	**0.00**
	10–45%	82.29	10.45	7.88	0.00	**0.00**	82.02	10.45	7.85	0.00	**0.00**
	15–29%	−28.94	10.45	−2.77	0.01	0.06	−23.87	10.45	−2.29	0.02	0.22
	15–38%	40.81	10.45	3.91	0.00	**0.00**	50.15	10.45	4.80	0.00	**0.00**
	15–45%	49.44	10.45	4.73	0.00	**0.00**	55.54	10.45	5.32	0.00	**0.00**
	29–38%	11.87	10.45	1.14	0.26	1.00	26.29	10.45	2.52	0.01	0.12
	29–45%	20.50	10.45	1.96	0.05	0.50	31.67	10.45	3.03	0.00	**0.02**
	38–45%	−8.64	10.45	−0.83	0.41	1.00	5.39	10.45	0.52	0.61	1.00

Each row tests the null hypothesis that Sample 1 and Sample 2 distributions are the same. Asymptotic significances (2-sided tests) are displayed. The significance level is 0.05. [a] Significance values have been adjusted by the Bonferroni correction for multiple tests. The bold numbers and bold color express the significance ($p < 0.05$).

The correspondence analysis revealed that erosion and runoff in the F plots can be discriminated and highly differentiated from both AG and BF, as it was located in a position further from the origin (x = 0, y = 0), whilst AG and BF were less distinct (Figure 6a). Similarly, erosion on 45% hillslope inclination, followed by 38%, was differentiated from other hillslope inclinations, while the 10%, followed by 45%, and 38% were differentiated from other slope inclinations in terms of runoff (Figure 6b).

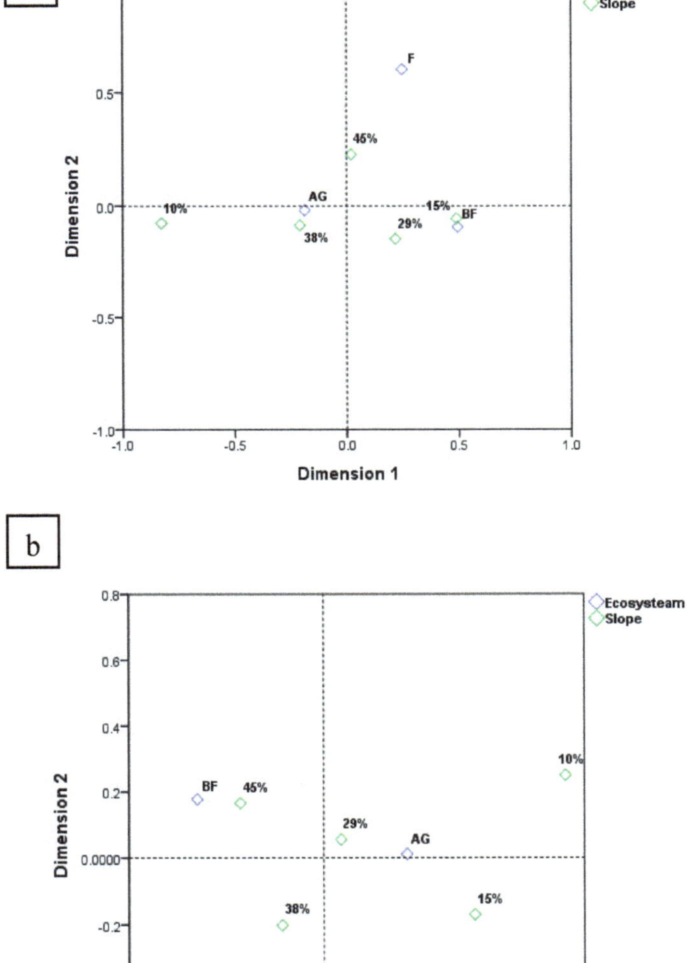

Figure 6. Correspondence analysis per plot: (**a**) soil loss and (**b**) runoff.

3.3. Multivariate Analysis of Factors and Covariates

The GLM revealed that all the factors involved (inclination and land use) and the covariate (rainfall) had a significant ($p < 0.001$) effect on the soil loss, and the explained variance was 85.1% (based on the adjusted $R^2 = 0.851$). Furthermore, the statistical interaction also obtained a significant ($p < 0.001$) effect (Table 6). Regarding the relevance of the predictors, land use registered the largest effect, while the effect of rainfall was 40% smaller, and the inclination effect was about half. The effect of the interaction of inclination and rainfall was similar to the rainfall effect.

Table 6. Summary of the General Linear Model (GLM) performed with soil erosion as the target variable (SS: sum of squares, df: the degree of freedom, F: F-statistic, p: significance, η^2p: effect size; $p < 0.001$ is highlighted in bold).

GLM	SS	df	F	p	η^2p
Model	24.33	15	142.8	<0.001	0.851
Inclination	2.30	4	50.7	<0.001	0.352
Land use	12.38	2	544.7	<0.001	0.744
Rainfall	3.69	1	324.5	<0.001	0.465
Inclination × Land use	3.41	8	37.5	<0.001	0.445
Residuals	4.25	374			
Total	28.58	389			

Generally, the soil loss rate of the AG lands was the greatest in all hillslopes, while the control areas (F) had the lowest rate. The erosion can be regarded as linear in these areas; locally estimated scatterplot smoothing (LOESS) curves were almost linear in all possible combinations (Figure 7). Visual evaluation of the data showed that inclination degrees can be divided into two different groups based on the soil loss: (i) inclination of 10 and 15%, and (ii) 29, 38 and 45%. In the case of group (i), the erosion rate was below 0.5 kg/m^2, although the difference between the AG lands was significant (mean difference: 0.096; $p_{M\text{-}C} < 0.0005$). Larger differences were caused by the heaviest rainfalls in the study area with 15% inclination. Erosion rates within group (ii) were similar regarding all the three land cover types, and the soil loss in the 45% inclination area was not significantly different ($p > 0.05$) from the 38% or 29% inclination degree areas according to the ANOVA test ($F = 2.059$, $df = 2$, $p = 0.129$).

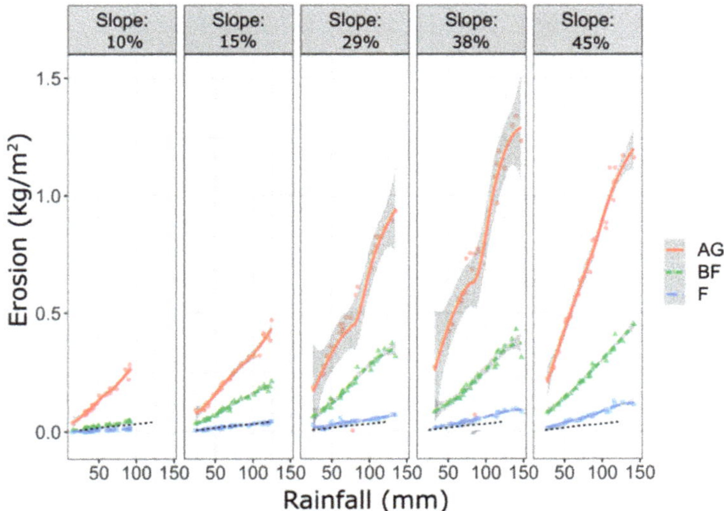

Figure 7. Relation between erosion rates and rainfall by slope and land cover type (agricultural land: AG; burnt forest: BF, and forest: F).

4. Discussion

Soil erosion by water is considered one of the most important agricultural sustainability challenges in the CRoS as a result of the following factors: heavy rainfall, severe inclinations, high erodibility, massive gushes of runoff, land-use changes, and non-sustainable agricultural practices [48]. Therefore, the assessment of water erosion derived from field analysis provides a detailed method of approaching the relationship between erosion, runoff and soil properties.

4.1. Criteria to Assess Current Erosion

4.1.1. The Role of Physical Features in Erosion

The climate of the study area is characterized by a high-intensity precipitation pattern with the intense kinetic energy of raindrops that hit the hillslopes with different land uses. Land use played an influential role in determining the quantities of eroded material and discharged runoff, which varied according to other physical features such as topography and soil properties [73]. Recently, forest lands in the CRoS were badly affected by severe wildfires, which increase the susceptibility to soil loss in the study area. In this regard, the importance of soil management was clear. In our research, soil loss and runoff were the highest in the AG and BF plots compared to the F plots. Within the study area, the cultivated land (AG plots) and burnt plots remained bare and exposed directly to raindrops, which could explain the high amount of soil erosion and runoff in comparison to the F plots, as other recent investigations in cultivated or abandoned fields have demonstrated [14,74], or in areas after recent wildfires [75].

4.1.2. The Role of Slope Steepness in Erosion

Of the five locations used for measuring soil loss and runoff, three of them were chosen with an inclination higher than 25%, i.e., SY1, SY2, SY4. Our statistical analysis revealed that from 29%, the critical limit was to be found above this value, similarly to a saturation curve, in that a greater slope gradient did not cause a relevant increase in the erosion rate (Figure 8). These results agree with other soil erosion and runoff reports presented by [76–79]. In the light of the high-intensity rainstorms in the study area, inclination was also a driving factor in the occurrence of high-velocity runoff events which enhance soil detachment. Additionally, this inclination could motivate both ponding depth and depressional storage [80–82]. Under the same land use, inclination degree could accelerate the erosion remarkably, as can be revealed from Table 3 and Figure 7. Land use had a relevant effect on the soil erosion rate, with the highest values observed in the AG plots, and the lowest ones in the F plots. In Syria, only a few studies have reported on soil erosion at the plot scale. Barneveld et al. [83] claimed that soil erosion in the NW part of Syria rarely exceeded 5 kg/m^2 in cultivated olive lands with average slopes of 25%. However, this difference in measuring soil erosion could be explained by the physiographic difference between each research location, especially the steepness of the slope, the form and development of terrain, precipitation intensity, soil characteristics, and agricultural practices. Notably, our results are higher than the erosion observed in Mediterranean mountains by [84] (147.3 g m^{-2}) and lower than results reported by [83] (5 kg/m^2).

4.1.3. Role of Human Activities in Erosion

If the physical factors are compared to human activities, the latter is the main driver of erosion through poor and unsustainable soil management and tampering with soil structure, and altering its physical, chemical and biological properties, especially its organic matter content [85]. However, these consequences should be considered as serious in fragile and vulnerable soils as in the Mediterranean environment. Intensive tillage and bare soils play a key role in accelerating soil erosion [86,87] by enhancing the separation of macro-aggregates, which negatively affects the soil aggregates' stability [88–92]. Soils in forest plots are protected by more vegetation and we hypothesize that soil aggregates are stronger and are not affected by the negative impacts of the kinetic energy of raindrops.

Some authors have observed that the collapse of soil aggregates can minimize soil porosity by blocking pores by fine particles (silt, clay) and can magnify soil sealing and crusting, and, subsequently, soil erosion can be enhanced [93]. As a consequence, some authors have even reported that the soil erodibility factor (K) is higher in AG plots for this reason, which indirectly indicates the susceptibility of AG plots to soil erosion [94]. In this regard, organic matter (OM) is expected to be higher in the F plots, which significantly enhances aggregate stability against rainy storms, while aggregates in AG and BF plots would be more vulnerable [95–97]. Our results are consistent with [98], who indicated that inappropriate agricultural practices in shallow topsoil can increase the susceptibility of runoff and erosion. This is extensive in various agricultural activities in the Mediterranean belt [99,100]. The relevance of OM content in mitigating erosion has been proved by several authors, i.e., soils with <2% OM are highly susceptible to erosion and runoff [101,102]. In addition, [36,103–105] highlighted the vital role of agricultural activities and the Mediterranean climate in accelerating soil erosion in semi-arid regions, while other studies stressed the importance of ground soil cover for preventing erosion and runoff [99,106–109]. As extensive fieldwork in the CRoS has revealed, in the AG plots there is an absence of most of the agricultural practices that conserve soil, especially crop rotation, maintaining tillage, contour and strip farming, grass water channels, and diversion structures.

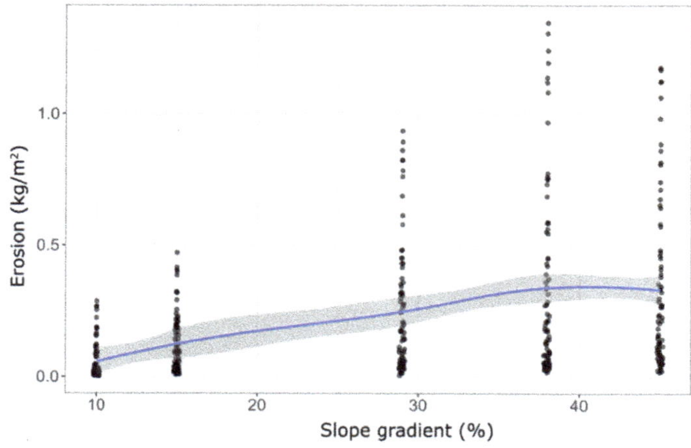

Figure 8. Slope gradient and erosion rate (——LOESS fit line with 95% confidence intervals).

4.2. *Dimensions of the Current Evaluation*

The CroS constitutes the first agricultural stability zone and the agricultural and economic backbone of the local population, and therefore the protection of its natural resources, especially soils from erosion, is a priority in the framework of agricultural sustainability. Thus, the implementation of some conservation practices (CP) or even the establishment of a national action plan for soil conservation to repair local ecosystems is a high priority. Some authors have recommended CP including soil mulching [110,111], tillage reduction [112,113], buffer strips and minimum cultivation [114] or a correct planification of soil terraces [115]. Nonetheless, field analysis of soil erosion is at the forefront of the measures that will develop strategies for preserving farmland, especially during the ongoing war that has negatively affected the agricultural and food system in the country. In the context of soil erosion, cultivated hillslopes in the CRoS are subject to intensive use pressure which includes poor maintenance technologies, and overuse of fertilizers. Consequently, soil aggregates are more dynamic once there are other agents of erosion, especially high-intensity raindrops. In this regard, the orographic precipitation model imposes high rainfall intensities, and consequently massive runoff which accelerates soil erosion. Unfortunately, in this research, the intensity and duration of rainfall could not be measured. However, further studies should address those elements instead of using the total rainfall amount per event. In addition, further research should be carried out to address appropriate measures for land conservation, especially with hillslopes of over 29% inclination.

5. Conclusions

In this research, soil loss and runoff were measured in five different locations (hillslopes) with three different land uses (AG, BF, F) in the coastal region of Syria. The main findings of this research are:

1. Observed soil loss and runoff were higher in the AG lands, followed by BF and F.
2. In the CRoS, land use has the greatest effect on soil erosion, followed by rainfall amount and hillslope inclination.
3. Concerning the inclination degree, SY1 (38%) and SY2 (45%) showed the greatest soil erosion and runoff amounts per event, followed by SY4 (29%), SY3 (15%), and SY5 (10%).
4. Regardless of the land use type, our results show an absence of statistical differences ($p < 0.05$) between 10 and 15% inclination, and between 38 and 45%.
5. Soil loss was 0.14 ± 0.07 kg/m^2 in the AG plots, while it did not exceed 0.1 ± 0.001 kg/m^2 in the F plots. Meanwhile, the highest runoff was recorded in the AG plots, which ranged between 3.77 ± 1.62 and 22.95 ± 9.33 L/m^2
6. In the CRoS, the pairwise comparison among the hillslopes revealed that 29% inclination can be the maximum tolerable threshold to apply urgent soil erosion control measures.

Few studies have dealt with soil erosion in Syria, and to our knowledge none of these have measured erosion per rainfall event at different hillslope positions comparing human disturbances under three different land uses. The outcome of this research could play an important role in setting up the first conservation plan in Syria. Moreover, the output of this research will contribute to bridging the gap in the common literature on soil water erosion in the near-eastern Mediterranean, and could be used for the improvement of erosion equations or soil protection policies not only in Syria, but all over the Mediterranean belt.

Author Contributions: Conceptualization, S.M., I.K. and J.I.; Data curation and collection, I.K. and J.I.; Resources, I.J.H.; Supervision, S.S. and J.R.-C.; Visualization, K.A., Q.B.P., N.T.T.L., D.T.A., A.M. and I.J.H.; Writing—original draft, S.M., and H.G.A.; Writing—review and editing, All authors. All authors have read and agreed to the published version of the manuscript.

Funding: This research received no external funding.

Acknowledgments: This paper is part of a research project of the first author (Safwan Mohammed) funded by the Tempus Public Foundation (Hungary) within the framework of the Stipendium Hungaricum Scholarship Programme. The research was supported by the Thematic Excellence Programme of the Ministry for Innovation and Technology in Hungary (ED_18-1-2019-0028) projects. Authors would like to thank Ministry of Local Administration and Environment (Syria), Tishreen University (Syria) and Debrecen University (Hungary) for their unlimited support.

Conflicts of Interest: The authors declare no conflict of interest.

Appendix A

AG (45%)
AG (38%)
AG (29%)
AG (15%)
AG (10%)
BF (45%)
BF (38%)
BF (29%)
kg/m2
Rainfal (mm)/Runoff (L/m2)
Soil erosion (kg/m2)
Rainfal event (mm)
Runoff (L/m2)

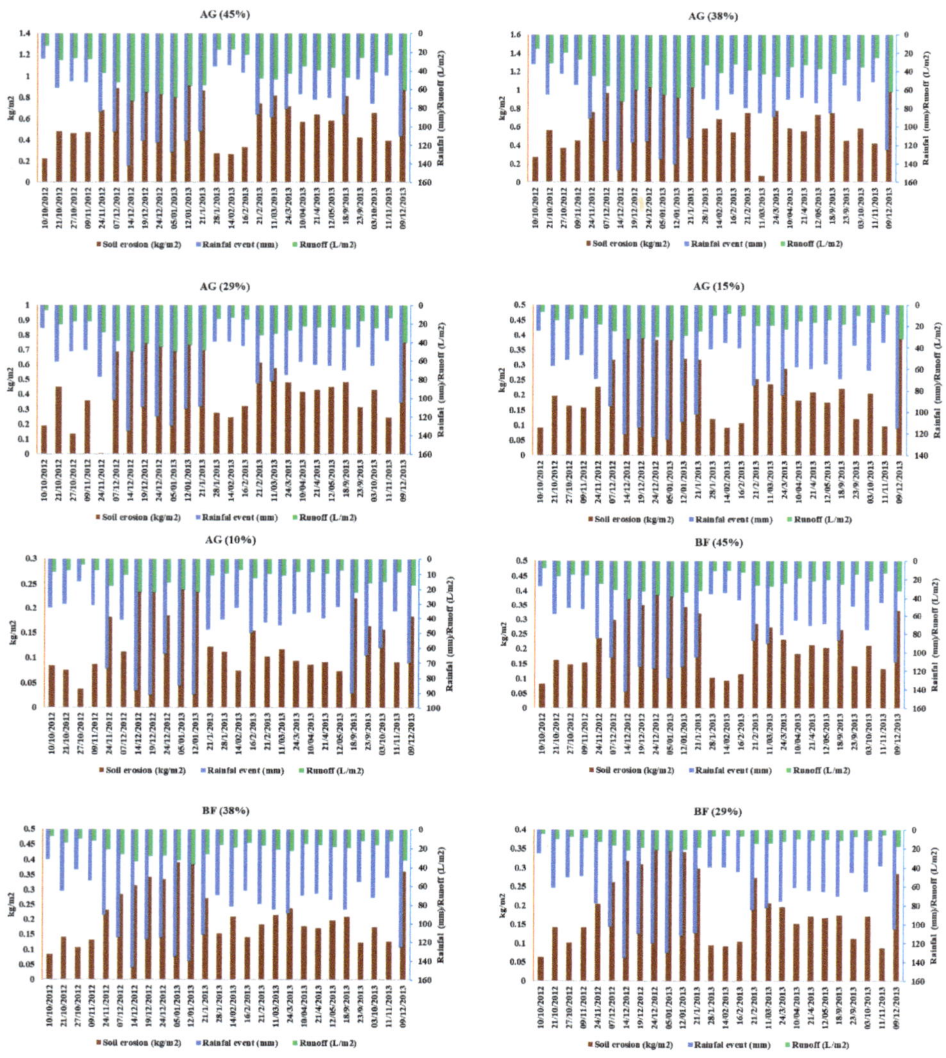

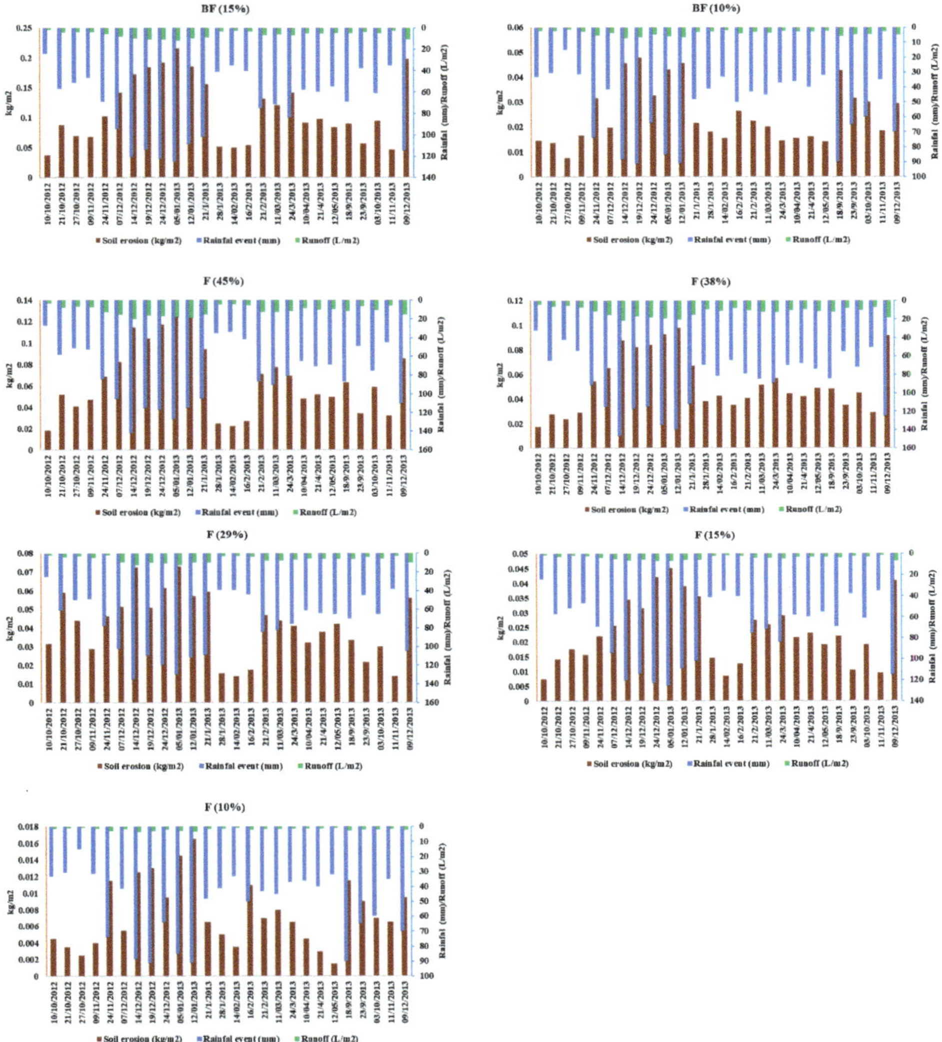

Figure A1. Dynamic interaction between erosion, runoff, and rainfall in each land use.

References

1. Baritz, R.; Wiese, L.; Verbeke, I.; Vargas, R. Voluntary guidelines for sustainable soil management: Global action for healthy soils. In *International Yearbook of Soil Law and Policy 2017*; Springer: Berlin, Germany, 2018; pp. 17–36.
2. Brevik, E.C.; Steffan, J.J.; Rodrigo-Comino, J.; Neubert, D.; Burgess, L.C.; Cerdà, A. Connecting the public with soil to improve human health. *Eur. J. Soil Sci.* **2019**, *70*, 898–910. [CrossRef]
3. Fang, H.; Sun, L.; Tang, Z. Effects of rainfall and slope on runoff, soil erosion and rill development: An experimental study using two loess soils. *Hydrolog. Process.* **2015**, *29*, 2649–2658. [CrossRef]
4. Alewell, C.; Egli, M.; Meusburger, K. An attempt to estimate tolerable soil erosion rates by matching soil formation with denudation in Alpine grasslands. *J. Soils Sedim.* **2015**, *15*, 1383–1399. [CrossRef]
5. Wang, L.; Li, X.A.; Li, L.C.; Hong, B.; Liu, J. Experimental study on the physical modeling of loess tunnel-erosion rate. *Bull. Eng. Geol. Environ.* **2019**, *78*, 5827–5840. [CrossRef]

6. Dotterweich, M.; Ivester, A.H.; Hanson, P.R.; Larsen, D.; Dye, D.H. Natural and human-induced prehistoric and historical soil erosion and landscape development in Southwestern Tennessee, USA. *Anthropocene* **2014**, *8*, 6–24. [CrossRef]
7. Romero-Díaz, A.; Ruiz-Sinoga, J.D.; Robledano-Aymerich, F.; Brevik, E.C.; Cerdà, A. Ecosystem responses to land abandonment in Western Mediterranean Mountains. *Catena* **2017**, *149*, 824–835. [CrossRef]
8. Panagos, P.; Borrelli, P.; Meusburger, K.; Yu, B.; Klik, A.; Lim, K.J.; Yang, J.E.; Ni, J.; Miao, C.; Chattopadhyay, N.; et al. Global rainfall erosivity assessment based on high-temporal resolution rainfall records. *Sci. Rep.* **2017**, *7*, 1–12. [CrossRef]
9. Borrelli, P.; Robinson, D.A.; Fleischer, L.R.; Lugato, E.; Ballabio, C.; Alewell, C.; Meusburger, K.; Modugno, S.; Schütt, B.; Ferro, V.; et al. An assessment of the global impact of 21st century land use change on soil erosion. *Nat. Commun.* **2017**, *8*, 1–13. [CrossRef]
10. Gholami, V.; Booij, M.; Tehrani, E.N.; Hadian, M. Spatial soil erosion estimation using an artificial neural network (ANN) and field plot data. *Catena* **2018**, *163*, 210–218. [CrossRef]
11. Prasannakumar, V.; Vijith, H.; Abinod, S.; Geetha, N. Estimation of soil erosion risk within a small mountainous sub-watershed in Kerala, India, using Revised Universal Soil Loss Equation (RUSLE) and geo-information technology. *Geosci. Front.* **2012**, *3*, 209–215. [CrossRef]
12. Efthimiou, N.; Lykoudi, E.; Karavitis, C. Comparative analysis of sediment yield estimations using different empirical soil erosion models. *Hydrol. Sci. J.* **2017**, *62*, 2674–2694. [CrossRef]
13. Ramos, M.; Martinez-Casasnovas, J. Soil moisture variability at different depths in land-levelled vineyards and its influence on crop productivity. *J. Hydrol.* **2006**, *321*, 131–146. [CrossRef]
14. Rodrigo-Comino, J. Five decades of soil erosion research in "terroir". The State-of-the-Art. *Earth Sci. Rev.* **2018**, *179*, 436–447. [CrossRef]
15. Sannigrahi, S.; Joshi, P.K.; Keesstra, S.; Paul, S.K.; Sen, S.; Roy, P.; Chakraborti, S.; Bhatt, S. Evaluating landscape capacity to provide spatially explicit valued ecosystem services for sustainable coastal resource management. *Ocean Coastal Manag.* **2019**, *182*, 104918. [CrossRef]
16. Sannigrahi, S.; Zhang, Q.; Pilla, F.; Joshi, P.K.; Basu, B.; Keesstra, S.; Roy, P.; Wang, Y.; Sutton, P.C.; Chakraborti, S. Responses of ecosystem services to natural and anthropogenic forcings: A spatial regression based assessment in the world's largest mangrove ecosystem. *Sci. Total Environ.* **2020**, *715*, 137004. [CrossRef]
17. Norman, L.M. Ecosystem services of riparian restoration: A review of rock detention structures in the madrean archipelago ecoregion. *Air Soil Water Res.* **2020**, *13*, 1178622120946337. [CrossRef]
18. Petrakis, R.E.; Norman, L.M.; Lysaght, O.; Sherrouse, B.C.; Semmens, D.; Bagstad, K.J.; Pritzlaff, R. Mapping perceived social values to support a respondent-defined restoration economy: Case Study in Southeastern Arizona, USA. *Air Soil Water Res.* **2020**, *13*, 1178622120913318. [CrossRef]
19. Cerdà, A.; Lavee, H.; Romero-Diaz, A.; Hooke, J.; Montanarella, L. Soil erosion and degradation in Mediterranean-type ecosystems. *Land Degrad. Dev.* **2010**, *21*, 71–74. [CrossRef]
20. Kaiser, J. *Wounding Earth's Fragile Skin*; American Association for the Advancement of Science: Washington, DC, USA, 2004.
21. Boardman, J.; Poesen, J. *Soil Erosion in Europe*; John Wiley & Sons: Hoboken, NJ, USA, 2007.
22. Pimentel, D.; Burgess, M. Soil erosion threatens food production. *Agriculture* **2013**, *3*, 443–463. [CrossRef]
23. Abdo, H.G. Impacts of war in Syria on vegetation dynamics and erosion risks in Safita area, Tartous, Syria. *Reg. Environ. Change* **2018**, *18*, 1707–1719. [CrossRef]
24. Falcão, C.J.L.M.; de Araújo Duarte, S.M.; da Silva Veloso, A. Estimating potential soil sheet Erosion in a Brazilian semiarid county using USLE, GIS, and remote sensing data. *Environ. Monit. Assess.* **2020**, *192*, 47. [CrossRef] [PubMed]
25. FAO. Status of the world's soil resources (SWSR)–main report. In *Food and Agriculture Organization of the United Nations and Intergovernmental Technical Panel on Soils*; FAO, I: Rome, Italy, 2015; Volume 650.
26. Hatna, E.; Bakker, M.M. Abandonment and expansion of arable land in Europe. *Ecosystems* **2011**, *14*, 720–731. [CrossRef]
27. Arnáez, J.; Lana-Renault, N.; Lasanta, T.; Ruiz-Flaño, P.; Castroviejo, J. Effects of farming terraces on hydrological and geomorphological processes. A review. *Catena* **2015**, *128*, 122–134. [CrossRef]
28. Dutta, S. Soil erosion, sediment yield and sedimentation of reservoir: A review. *Modeling Earth Syst. Environ.* **2016**, *2*, 123. [CrossRef]

29. Angulo-Martinez, M.; Beguería, S.; Navas, A.; Machin, J. Splash erosion under natural rainfall on three soil types in NE Spain. *Geomorphology* **2012**, *175*, 38–44. [CrossRef]
30. Klik, A.; Rosner, J. Long-term experience with conservation tillage practices in Austria: Impacts on soil erosion processes. *Soil Tillage Res.* **2020**, *203*, 104669. [CrossRef]
31. Martínez-Mena, M.; Carrillo-López, E.; Boix-Fayos, C.; Almagro, M.; Franco, N.G.; Díaz-Pereira, E.; Montoya, I.; de Vente, J. Long-term effectiveness of sustainable land management practices to control runoff, soil erosion, and nutrient loss and the role of rainfall intensity in Mediterranean rainfed agroecosystems. *Catena* **2020**, *187*, 104352. [CrossRef]
32. Chen, X.; Liang, Z.; Zhang, Z.; Zhang, L. Effects of soil and water conservation measures on runoff and sediment yield in red soil slope farmland under natural rainfall. *Sustainability* **2020**, *12*, 3417. [CrossRef]
33. Madarász, B.; Bádonyi, K.; Csepinszky, B.; Mika, J.; Kertész, Á. Conservation tillage for rational water management and soil conservation. *Hung. Geograph. Bull.* **2011**, *60*, 117–133.
34. Koch, H.J.; Stockfisch, N. Loss of soil organic matter upon ploughing under a loess soil after several years of conservation tillage. *Soil Tillage Res.* **2006**, *86*, 73–83. [CrossRef]
35. Armand, R.; Bockstaller, C.; Auzet, A.V.; van Dijk, P. Runoff generation related to intra-field soil surface characteristics variability: Application to conservation tillage context. *Soil Tillage Res.* **2009**, *102*, 27–37. [CrossRef]
36. García-Ruiz, J.M.; Nadal-Romero, E.; Lana-Renault, N.; Beguería, S. Erosion in Mediterranean landscapes: Changes and future challenges. *Geomorphology* **2013**, *198*, 20–36. [CrossRef]
37. Raclot, D.; le Bissonnais, Y.; Annabi, M.; Sabir, M. Challenges for mitigating Mediterranean soil erosion under global change. *Mediterr. Reg. Under Clim. Change* **2016**, 311.
38. Amate, J.I.; de Molina, M.G.; Vanwalleghem, T.; Fernández, D.S.; Gómez, J.A. Erosion in the Mediterranean: The case of olive groves in the south of Spain (1752–2000). *Environ. Hist.* **2013**, *18*, 360–382. [CrossRef]
39. Takken, I.; Govers, G.; Ciesiolka, C.; Silburn, D.; Loch, R. *Factors Influencing the Velocity-Discharge Relationship in Rills*; IAHS Publication: Oxfordshire, UK, 1998; pp. 63–70.
40. Bradford, J.; Foster, G. Interrill soil erosion and slope steepness factors. *Soil Sci. Soc. Am. J.* **1996**, *60*, 909–915. [CrossRef]
41. Nearing, M.; Deer-Ascough, L.; Laflen, J. Sensitivity analysis of the WEPP hillslope profile erosion model. *Trans. ASAE* **1990**, *33*, 839–0849. [CrossRef]
42. Rojo, L. *Plan nacional de restauración hidrológico-forestal y control de la erosión*; Memoria, Tomo I: Mapas Tomo II; ICONA: Madrid, Spain, 1990.
43. Irvem, A.; Topaloğlu, F.; Uygur, V. Estimating spatial distribution of soil loss over Seyhan River Basin in Turkey. *J. Hydrol.* **2007**, *336*, 30–37. [CrossRef]
44. Trabucchi, M.; Puente, C.; Comin, F.A.; Olague, G.; Smith, S.V. Mapping erosion risk at the basin scale in a Mediterranean environment with opencast coal mines to target restoration actions. *Reg. Environ. Change* **2012**, *12*, 675–687. [CrossRef]
45. Farhan, Y.; Zregat, D.; Farhan, I. Spatial estimation of soil erosion risk using RUSLE approach, RS, and GIS techniques: A case study of Kufranja watershed, Northern Jordan. *J. Water Res. Prot.* **2013**, *5*, 1247. [CrossRef]
46. Verheijen, F.G.; Jones, R.J.; Rickson, R.; Smith, C. Tolerable versus actual soil erosion rates in Europe. *Earth Sci. Rev.* **2009**, *94*, 23–38. [CrossRef]
47. Kouli, M.; Soupios, P.; Vallianatos, F. Soil erosion prediction using the revised universal soil loss equation (RUSLE) in a GIS framework, Chania, Northwestern Crete, Greece. *Environ. Geol.* **2009**, *57*, 483–497. [CrossRef]
48. Mohammed, S.; Khallouf, A.; Alshiehabi, O.; Pham, Q.B.; Linh, N.T.T.; Anh, D.T.; Harsányi, E. Predicting soil erosion hazard in Lattakia governorate (W Syria). *Int. J. Sediment Res.* **2020**. In press.
49. Al Bakeer, H. *Report About the Agricultural Situation in Syria*; Institute of Development Studies: Brighton, UK, 2018.
50. GCSAR: General Commission for Scientific Agricultural Research. *Natural Resources in the Costal Region of Syria*; Ministry of Agriculture: Damascus, Syria, 2013; p. 159. (In Arabic)
51. Abdo, H.; Salloum, J. Mapping the soil loss in Marqya basin: Syria using RUSLE model in GIS and RS techniques. *Environ. Earth Sci.* **2017**, *76*, 114. [CrossRef]

52. Hazem, G.A. Geo-modeling approach to predicting of erosion risks utilizing RS and GIS data: A case study of Al-Hussain Basin, Tartous, Syria. *J. Environ. Geol.* **2017**, *1*, 1–4.
53. Mohammed, S.; Kbibo, I.; Alshihabi, O.; Mahfoud, E. Studying rainfall changes and water erosion of soil by using the WEPP model in Lattakia, Syria. *J. Agric. Sci. Belgrad.* **2016**, *61*, 375–386. [CrossRef]
54. Barakat, M.; Mahfoud, I.; Kwyes, A. Study of soil erosion risk in the basin of Northern Al-Kabeer river at Lattakia-Syria using remote sensing andGIS techniques. *Mesop. J. Mar. Sci.* **2014**, *29*, 29–44.
55. Al-Ali, Y.A.Z.; Kheder, R. Studying the effect of forest fire on soil erosion and loss of some mineral elements in the forest of ein al-jaouz/tartous. *Biol. Sci. Ser.* **2014**, *36*, 277–290. (In Arabic)
56. Sarah, P. Soil organic matter and land degradation in semi-arid area, Israel. *Catena* **2006**, *67*, 50–55. [CrossRef]
57. Stavi, I.; Ragolsky, G.; Shem-Tov, R.; Shlomi, Y.; Ackermann, O.; Rueff, H.; Lekach, J. Ancient through mid-twentieth century runoff harvesting agriculture in the hyper-arid Arava Valley of Israel. *Catena* **2018**, *162*, 80–87. [CrossRef]
58. Lavee, H.; Poesen, J.; Yair, A. Evidence of high efficiency water-harvesting by ancient farmers in the Negev Desert, Israel. *J. Arid Environ.* **1997**, *35*, 341–348. [CrossRef]
59. Abdo, H.G. Evolving a total-evaluation map of flash flood hazard for hydro-prioritization based on geohydromorphometric parameters and GIS–RS manner in Al-Hussain river basin, Tartous, Syria. *Nat. Hazards* **2020**, *104*, 681–703. [CrossRef]
60. Ministry of Oil and Natural Resoures. Geology Map of Syria. 2009. Available online: https://geology-sy.org/ (accessed on 9 September 2020).
61. Directoriet of Meteorology. *Weather Data of Syria*; Syrian Ministry of Defense: Damascus, Syria, 2019.
62. Mohammed, S.; Khallouf, A.; Kiwan, S.; Alhenawi, S.; Ali, H.; Harsányi, E.; Kátai, J.; Habib, H. Characterization of Major Soil Orders in Syria. *Eurasian Soil Sci.* **2020**, *53*, 420–429. [CrossRef]
63. Ministry of Agriculture and Agrarian Reform (MoAAR). *The Agricultural Investment Map in the Syrian Arab Republic*; Ministry of Agriculture and Agrarian Reform (MoAAR): Damascus, Syria, 2020.
64. Kbibo, I.; Ibrahim, J.; Bou-Issa, A. Studying the effect of soil erosion for eight different systems with different slopes in the coastal area under forests, burned forest and planted soil system. *Tishreen Univ. J. Res. Sci. Stud. Biol. Sci. Ser* **2017**, *39*, 25–38.
65. Mohammed, S.; Al-Ebraheem, A.; Holb, I.J.; Alsafadi, K.; Dikkeh, M.; Pham, Q.B.; Linh, N.T.T.; Szabo, S. Soil management effects on soil water erosion and runoff in central Syria—A comparative evaluation of general linear model and random forest regression. *Water* **2020**, *12*, 2529. [CrossRef]
66. Safwan, M.; Alaa, K.; Omran, A.; Quoc, B.P.; Nguyen, T.T.L.; Van, N.T.; Duong, T.A.; Endre, H. Predicting soil erosion hazard in Lattakia Governorate (W Syria). *Int. J. Sediment Res.* **2020**. [CrossRef]
67. Wischmeier, W.H.; Smith, D.D. *Predicting Rainfall Erosion Losses: A Guide to Conservation Planning*; Department of Agriculture, Science and Education Administration: Washington, DC, USA, 1978.
68. McDonald, J.H. *Handbook of Biological Statistics*; Sparky House Publishing: Baltimore, MD, USA, 2009; Volume 2.
69. Field, A. *Discovering Statistics Using IBM SPSS Statistics*; Sage: Thousand Oacs, CA, USA, 2013.
70. McCullough, B.D. *Econometric Software Reliability: EViews, LIMDEP, SHAZAM and TSP*; JSTOR: New York, NY, USA, 1999.
71. Team, R.C. *R: A Language and Environment For Statistical Computing.(Version 3.6)*. 2019. Available online: http://www.r-project.org/index.html (accessed on 9 May 2020).
72. Gallucci, M. *R Package Version 2.0.5*. 2020. Available online: https://gamlj.github.io/ (accessed on 9 May 2020).
73. Feng, T.; Wei, W.; Chen, L.; Rodrigo-Comino, J.; Die, C.; Feng, X.; Ren, K.; Brevik, E.C.; Yu, Y. Assessment of the impact of different vegetation patterns on soil erosion processes on semiarid loess slopes. *Earth Surf. Process. Landf.* **2018**, *43*, 1860–1870. [CrossRef]
74. Rodrigo-Comino, J.; Brevik, E.C.; Cerdà, A. The age of vines as a controlling factor of soil erosion processes in Mediterranean vineyards. *Sci. Total Environ.* **2018**, *616*, 1163–1173. [CrossRef]
75. DeLong, S.B.; Youberg, A.M.; de Long, W.M.; Murphy, B.P. Post-wildfire landscape change and erosional processes from repeat terrestrial LIDAR in a steep headwater catchment, Chiricahua Mountains, Arizona, USA. *Geomorphology* **2018**, *300*, 13–30. [CrossRef]
76. Cerdà, A.; Morera, A.G.; Bodí, M.B. Soil and water losses from new citrus orchards growing on sloped soils in the western Mediterranean basin. *Earth Surface Process. Landf. J. Br. Geomorphol. Res. Group* **2009**, *34*, 1822–1830. [CrossRef]

77. Chaplot, V.; le Bissonnais, Y. Field measurements of interrill erosion under different slopes and plot sizes. *Earth Surface Process. Landf. J. Br. Geomorphol. Res. Group* **2000**, *25*, 145–153. [CrossRef]
78. Comino, J.R.; Sinoga, J.R.; González, J.S.; Guerra-Merchán, A.; Seeger, M.; Ries, J. High variability of soil erosion and hydrological processes in Mediterranean hillslope vineyards (Montes de Málaga, Spain). *Catena* **2016**, *145*, 274–284. [CrossRef]
79. Kairis, O.; Karavitis, C.; Kounalaki, A.; Salvati, L.; Kosmas, C. The effect of land management practices on soil erosion and land desertification in an olive grove. *Soil Use Manag.* **2013**, *29*, 597–606. [CrossRef]
80. Aksoy, H.; Kavvas, M.L. A review of hillslope and watershed scale erosion and sediment transport models. *Catena* **2005**, *64*, 247–271. [CrossRef]
81. Assouline, S.; Ben-Hur, M. Effects of rainfall intensity and slope gradient on the dynamics of interrill erosion during soil surface sealing. *Catena* **2006**, *66*, 211–220. [CrossRef]
82. Defersha, M.; Quraishi, S.; Melesse, A.M. The effect of slope steepness and antecedent moisture content on interrill erosion, runoff and sediment size distribution in the highlands of Ethiopia. *Hydrol. Earth Syst. Sci.* **2011**, *15*, 2367–2375. [CrossRef]
83. Barneveld, R.; Bruggeman, A.; Sterk, G.; Turkelboom, F. Comparison of two methods for quantification of tillage erosion rates in olive orchards of north-west Syria. *Soil Tillage Res.* **2009**, *103*, 105–112. [CrossRef]
84. Fonseca, F.; de Figueiredo, T.; Nogueira, C.; Queirós, A. Effect of prescribed fire on soil properties and soil erosion in a Mediterranean mountain area. *Geoderma* **2017**, *307*, 172–180. [CrossRef]
85. Jenny, J.P.; Koirala, S.; Gregory-Eaves, I.; Francus, P.; Niemann, C.; Ahrens, B.; Brovkin, V.; Baud, A.; Ojala, A.E.; Normandeau, A.; et al. Human and climate global-scale imprint on sediment transfer during the Holocene. *Proc. Nat. Acad. Sci. USA* **2019**, *116*, 22972–22976. [CrossRef]
86. Lieskovský, J.; Kenderessy, P. Modelling the effect of vegetation cover and different tillage practices on soil erosion in vineyards: A case study in Vráble (Slovakia) using WATEM/SEDEM. *Land Degrad. Dev.* **2014**, *25*, 288–296. [CrossRef]
87. Gao, Y.; Dang, X.; Yu, Y.; Li, Y.; Liu, Y.; Wang, J. Effects of tillage methods on soil carbon and wind erosion. *Land Degrad. Dev.* **2016**, *27*, 583–591. [CrossRef]
88. Paul, B.K.; Vanlauwe, B.; Ayuke, F.; Gassner, A.; Hoogmoed, M.; Hurisso, T.; Koala, S.; Lelei, D.; Ndabamenye, T.; Six, J. Medium-term impact of tillage and residue management on soil aggregate stability, soil carbon and crop productivity. *Agric. Ecosyst. Environ.* **2013**, *164*, 14–22. [CrossRef]
89. Kasper, M.; Buchan, G.; Mentler, A.; Blum, W. Influence of soil tillage systems on aggregate stability and the distribution of C and N in different aggregate fractions. *Soil Tillage Res.* **2009**, *105*, 192–199. [CrossRef]
90. Wang, Y.; Zhang, J.; Zhang, Z. Influences of intensive tillage on water-stable aggregate distribution on a steep hillslope. *Soil Tillage Res.* **2015**, *151*, 82–92. [CrossRef]
91. Bayat, F.; Monfared, A.B.; Jahansooz, M.R.; Esparza, E.T.; Keshavarzi, A.; Morera, A.G.; Fernández, M.P.; Cerdà, A. Analyzing long-term soil erosion in a ridge-shaped persimmon plantation in eastern Spain by means of ISUM measurements. *Catena* **2019**, *183*, 104176. [CrossRef]
92. Bogunovic, I.; Telak, L.J.; Pereira, P. Experimental comparison of runoff generation and initial soil erosion between vineyards and croplands of eastern Croatia: A case study. *Air Soil Water Res.* **2020**, *13*, 1178622120928323. [CrossRef]
93. Lin, Q.; Xu, Q.; Wu, F.; Li, T. Effects of wheat in regulating runoff and sediment on different slope gradients and under different rainfall intensities. *Catena* **2019**, *183*, 104196. [CrossRef]
94. Ayoubi, S.; Mokhtari, J.; Mosaddeghi, M.R.; Zeraatpisheh, M. Erodibility of calcareous soils as influenced by land use and intrinsic soil properties in a semiarid region of central Iran. *Environ. Monit. Assess.* **2018**, *190*, 192. [CrossRef]
95. Boix-Fayos, C.; Calvo-Cases, A.; Imeson, A.; Soriano-Soto, M. Influence of soil properties on the aggregation of some Mediterranean soils and the use of aggregate size and stability as land degradation indicators. *Catena* **2001**, *44*, 47–67. [CrossRef]
96. Kayet, N.; Pathak, K.; Chakrabarty, A.; Sahoo, S. Evaluation of soil loss estimation using the RUSLE model and SCS-CN method in hillslope mining areas. *Int. Soil Water Conserv. Res.* **2018**, *6*, 31–42. [CrossRef]
97. Novara, A.; Gristina, L.; Bodí, M.; Cerdà, A. The impact of fire on redistribution of soil organic matter on a Mediterranean hillslope under maquia vegetation type. *Land Degrad. Dev.* **2011**, *22*, 530–536. [CrossRef]

98. Toubal, A.K.; Achite, M.; Ouillon, S.; Dehni, A. Soil erodibility mapping using the RUSLE model to prioritize erosion control in the Wadi Sahouat basin, North-West of Algeria. *Environ. Monit. Assess.* **2018**, *190*, 210. [CrossRef] [PubMed]
99. Cerdà, A.; Jurgensen, M.; Bodi, M. Effects of ants on water and soil losses from organically-managed citrus orchards in eastern Spain. *Biologia* **2009**, *64*, 527–531. [CrossRef]
100. Laudicina, V.A.; Novara, A.; Barbera, V.; Egli, M.; Badalucco, L. Long-term tillage and cropping system effects on chemical and biochemical characteristics of soil organic matter in a Mediterranean semiarid environment. *Land Degrad. Dev.* **2015**, *26*, 45–53. [CrossRef]
101. Fullen, M.A.; Catt, J.A. *Soil Management: Problems and Solutions*; Routledge: London, UK, 2004; p. 269.
102. Conforti, M.; Buttafuoco, G.; Leone, A.P.; Aucelli, P.P.; Robustelli, G.; Scarciglia, F. Studying the relationship between water-induced soil erosion and soil organic matter using Vis–NIR spectroscopy and geomorphological analysis: A case study in southern Italy. *Catena* **2013**, *110*, 44–58. [CrossRef]
103. Giménez-Morera, A.; Sinoga, J.R.; Cerdà, A. The impact of cotton geotextiles on soil and water losses from Mediterranean rainfed agricultural land. *Land Degrad. Dev.* **2010**, *21*, 210–217. [CrossRef]
104. Cerda, A.; Rodrigo-Comino, J.; Novara, A.; Brevik, E.C.; Vaezi, A.R.; Pulido, M.; Gimenez-Morera, A.; Keesstra, S.D. Long-term impact of rainfed agricultural land abandonment on soil erosion in the Western Mediterranean basin. *Prog. Phys. Geogr. Earth Environ.* **2018**, *42*, 202–219. [CrossRef]
105. Schmid, T.; Rodríguez-Rastrero, M.; Escribano, P.; Palacios-Orueta, A.; Ben-Dor, E.; Plaza, A.; Milewski, R.; Huesca, M.; Bracken, A.; Cicuéndez, V.; et al. Characterization of soil erosion indicators using hyperspectral data from a Mediterranean rainfed cultivated region. *IEEE J. Sel. Topics Appl. Earth Obs. Remote Sens.* **2015**, *9*, 845–860. [CrossRef]
106. Zhao, G.; Mu, X.; Wen, Z.; Wang, F.; Gao, P. Soil erosion, conservation, and eco-environment changes in the Loess Plateau of China. *Land Degrad. Dev.* **2013**, *24*, 499–510. [CrossRef]
107. Ochoa, P.; Fries, A.; Mejia, D.; Burneo, J.; Ruíz-Sinoga, J.; Cerdà, A. Effects of climate, land cover and topography on soil erosion risk in a semiarid basin of the Andes. *Catena* **2016**, *140*, 31–42. [CrossRef]
108. Dunjó, G.; Pardini, G.; Gispert, M. The role of land use–land cover on runoff generation and sediment yield at a microplot scale, in a small Mediterranean catchment. *J. Arid Environ.* **2004**, *57*, 239–256. [CrossRef]
109. Bajocco, S.; de Angelis, A.; Perini, L.; Ferrara, A.; Salvati, L. The impact of land use/land cover changes on land degradation dynamics: A Mediterranean case study. *Environ. Manag.* **2012**, *49*, 980–989. [CrossRef] [PubMed]
110. Lucas-Borja, M.E.; Zema, D.A.; Carrà, B.G.; Cerdà, A.; Plaza-Alvarez, P.A.; Cózar, J.S.; Gonzalez-Romero, J.; Moya, D.; de las Heras, J. Short-term changes in infiltration between straw mulched and non-mulched soils after wildfire in Mediterranean forest ecosystems. *Ecol. Eng.* **2018**, *122*, 27–31. [CrossRef]
111. Rodrigo-Comino, J.; Giménez-Morera, A.; Panagos, P.; Pourghasemi, H.; Pulido, M.; Cerdà, A. The potential of straw mulch as a nature-based solution in olive groves treated with glyphosate. A biophysical and socio-economic assessment. *Land Degrad. Dev.* **2020**, *31*, 1877–1889. [CrossRef]
112. Alliaume, F.; Rossing, W.; Tittonell, P.; Jorge, G.; Dogliotti, S. Reduced tillage and cover crops improve water capture and reduce erosion of fine textured soils in raised bed tomato systems. *Agric. Ecosyst. Environ.* **2014**, *183*, 127–137. [CrossRef]
113. Seitz, S.; Goebes, P.; Puerta, V.L.; Pereira, E.I.P.; Wittwer, R.; Six, J.; van der Heijden, M.G.; Scholten, T. Conservation tillage and organic farming reduce soil erosion. *Agron. Sustain. Dev.* **2019**, *39*, 4. [CrossRef]
114. Wauters, E.; Bielders, C.; Poesen, J.; Govers, G.; Mathijs, E. Adoption of soil conservation practices in Belgium: An examination of the theory of planned behaviour in the agri-environmental domain. *Land Use Policy* **2010**, *27*, 86–94. [CrossRef]
115. Zuazo, V.D.; Ruiz, J.A.; Raya, A.M.; Tarifa, D.F. Impact of erosion in the taluses of subtropical orchard terraces. *Agric. Ecosyst. Environ.* **2005**, *107*, 199–210. [CrossRef]

Article

The Impact of Vegetation Successional Status on Slope Runoff Erosion in the Loess Plateau of China

Enhao Chang [1,2], Peng Li [1,2,*], Zhanbin Li [1,3], Yuanyi Su [1,2], Yi Zhang [1,2], Jianwen Zhang [1,2], Zhan Liu [1,2] and Zhineng Li [1,2]

1 State Key Laboratory of Eco-hydraulics in Northwest Arid Region, Xi'an University of Technology, Xi'an 710048, China; 123ceh@163.com (E.C.); zhanbinli@126.com (Z.L.); suyuanyi666@163.com (Y.S.); 18202915856@163.com (Y.Z.); zhangjianwen95@126.com (J.Z.); lipeng74@163.com (Z.L.); lizhineng01@126.com (Z.L.)

2 Key Laboratory National Forestry Administration on Ecological Hydrology and Disaster Prevention in Arid Regions, Xi'an 710048, China

3 State Key Laboratory of soil Erosion and Dry-land Farming on the Loess Plateau, Institute of Soil and Water Conservation, Chinese Academy of science and Ministry of Water Resources, Yangling 712100, China

* Correspondence: ttzlp@xaut.edu.cn; Tel./Fax: +86-29-8231-2651

Received: 4 November 2019; Accepted: 7 December 2019; Published: 11 December 2019

Abstract: Slope vegetation restoration is known to influence erosion in the Loess Plateau region in China. The ability of vegetation to mitigate soil erosion under extreme runoff, however, has not been studied in great detail in this region. Here, we examine five typical vegetation communities in the Loess Plateau region that originated from restoration efforts enacted at different times (1, 11, 15, 25, and 40 years). Water scouring experiments were carried out to monitor vegetation community succession and its effects on erosion. These results indicate that the sum of plant importance values increased from 260.72 to 283.06, species density increased from 2.5 to 4.5 per m^2, and the amount of litter and humus increased from 24.50 to 605.00 g/m^2 during the 1 to 40 years of vegetation community succession. Root biomass and root diameter reached a maximum of approximately 10.80 mg·cm^{-3} and 0.65 mm at 40 years of recovery. Slope runoff velocity decreased by 47.89% while runoff resistance increased by 35.30 times. The runoff power decreased by 19.75%, the total runoff volume decreased by 2.52 times, and the total sediment yield decreased by 11.60 times in the vegetation community. Slope runoff velocity and power had the largest correlation with aboveground vegetation (0.76, 0.74), total runoff had the largest correlation with underground roots (0.74), and runoff resistance was most strongly correlated with soil structure (0.71). Studies have shown that the succession of vegetation communities can enhance the aboveground ecological functions of plants, thereby significantly reducing the runoff velocity and power. The development of plant root system significantly reduces the runoff volume; the improved soil structure significantly increased the runoff resistance coefficient.

Keywords: vegetation community; vegetation importance value; root system; soil erosion; grey correlation analysis

1. Introduction

In recent years, there has been a series of studies conducted on soil and water conservation focusing on silt-dam gully engineering, terraced fields of slope engineering, and the Grain for Green Project on the Loess Plateau [1–3]. These efforts play an important role in the ecological restoration of Loess areas. The average annual sediment in the Yellow River has been reduced from an estimated 1530 million tons in the 1950s to 166 million tons in the 2010s [4]. About 40% to 50% of the reduction of the average annual sediment in the Yellow River comes from soil and water conservation measures in the Loess Plateau [5]. Since 1999, the Chinese government has carried out the Grain for Green Project, which has restored slope vegetation along the Loess Plateau. Observations from remote sensing show that the vegetation coverage of the Loess Plateau increased from 31.6% in 1999 to 59.6% in 2013 [6]. Restored forest and grassland areas accounted for about 56.7% of the total area of the Loess Plateau [7]. Some have suggested that vegetation restoration on the Loess Plateau has resulted in a 50% reduction in sedimentation along the Yellow River [8,9]. Ecological restoration of vegetation thus plays an important role in reducing slope soil erosion in the region [10]. Rainfall, however, does not automatically generate runoff. Erosion caused by a few short-duration heavy rainstorms can account for more than 60% to 90% of the total annual erosion [11]. Loess slopes can be damaged by slope runoff when extreme rainstorm events are frequent [12]. The slope is also the pioneer path of sediment production, which has a great contribution to the total amount of sediment observed in the outlet section of the watershed in the Loess hilly region [13]. As such, a better understanding of how vegetation can mitigate slope runoff and sediment under the erosion action of high-intensity slope runoff on the Loess Plateau is urgently required.

The Grain for Green Project facilitates a great change in landscape patterns in a certain sense, with the vegetation changing from annual crops to perennial native plants on the Loess Plateau [1]. The vegetation community, in the process of recovery without intervention, began to take place as species succession because of the extension of the years. Successional dynamics in the Loess Plateau typically follow a pattern of Artemisia plants, giving way to perennial rhizome grasses, and finally to perennial arbuscular herbs [14]. Gramineae, Legume, and Compositae occupy an important position in the natural succession of plant communities [15]. Previous studies have found that leguminous plants, such as *Lespedeza davurica*, can improve the soil organic matter, total nitrogen, total phosphorus, and available nitrogen content [16]. The Legume plants gradually became the dominant species, and soil organic carbon, in the 0–50 cm depth soil, gradually recovered after vegetation succession for about 20 years on Loess Plateau [17]. The Asteraceae and Gramineae plants are mainly represented by *Artemisia capillaris*, *Artemisia sacrorum*, and *Bothriochloa ischaemum*. They can grow naturally in the early stage of vegetation restoration, and their root systems have a strong ability to retain surface soil [18–20]. During succession, plant biomass, ground coverage, root structure, and function will change. The soil physical and chemical properties undergo predictable dynamics as well [21–23]. The number of species and individuals increases rapidly from 1 to 10 years in disturbed vegetation communities on the Loess Plateau [24]. The maximum root length density reached 31.04 mm/cm^3 in the 0–20 cm depth soil at 15 years after abandonment, and maximum root biomass density reached 3.35 mg/cm^3 after 21 years. Likewise, the water absorption capacity and the turnover frequency of root systems gradually increased in the process of vegetation restoration [17]. The slope is a basic unit of erosion, and erosion is primarily driven by the hydrodynamic index of the runoff and the erodibility of the topsoil [25]. The flow rate, flow velocity, and runoff depth are key factors directly affecting the slope separation process, and are also the basic parameters used for calculating other hydrodynamic indicators, such as runoff shear, runoff power, Reynolds number, and Froude number. These, in turn, are affected by factors, such as the underlying surface and vegetation [26,27]. Therefore, studying slope runoff hydrodynamics across vegetation communities in varying degrees of succession can help to reveal the role that the successional status has on mitigating slope erosion.

To study this, we studied slope runoff and erosion in five areas that had vegetation in varying degrees of succession (i.e., from 1 to 40 years after restoration) by the method of artificial simulation. Our goal was to analyze the effects of the vegetation successional status on slope soil erosion under water flushing conditions. We expect that results from this study will provide scientific guidance for future research on water erosion dynamic mechanisms and vegetation regulation principles in the Loess Plateau.

2. Materials and Methods

2.1. Study Area

The study area is located in the Xindan watershed on the Loess Plateau in China (E 110°15′–110°20′, N 37°27′–37°32′; 810–1120 m a.s.l.) (Figure 1). The soil type is to yellow loess soil, and the plough layer is thin (20–30 cm). Xindian watershed is a national soil erosion control area, which has been banned for about 60 years. The average annual temperature and precipitation in the study area is 9.7 °C and 486 mm, respectively. Precipitation does, however, vary widely across years and space. In recent years, there has been a notable decrease in precipitation days, an increase in the frequency of heavy rains, and an increase in both droughts and floods.

The successional sites included the *Artemisia capillaris* for 1 year since restoration, *A. sacrorum* for 11 years, *Bothriochloa ischaemum* for 15 years, *Lespedeza davurica* for 25 years, and *Ziziphus jujube* for 40 years. These species are dominant because of succession. It is also an inevitable sequence of vegetation succession in the Loess Plateau. This method of selection can be regarded as a method of a spatial sequence equivalent to a vegetation succession time series.

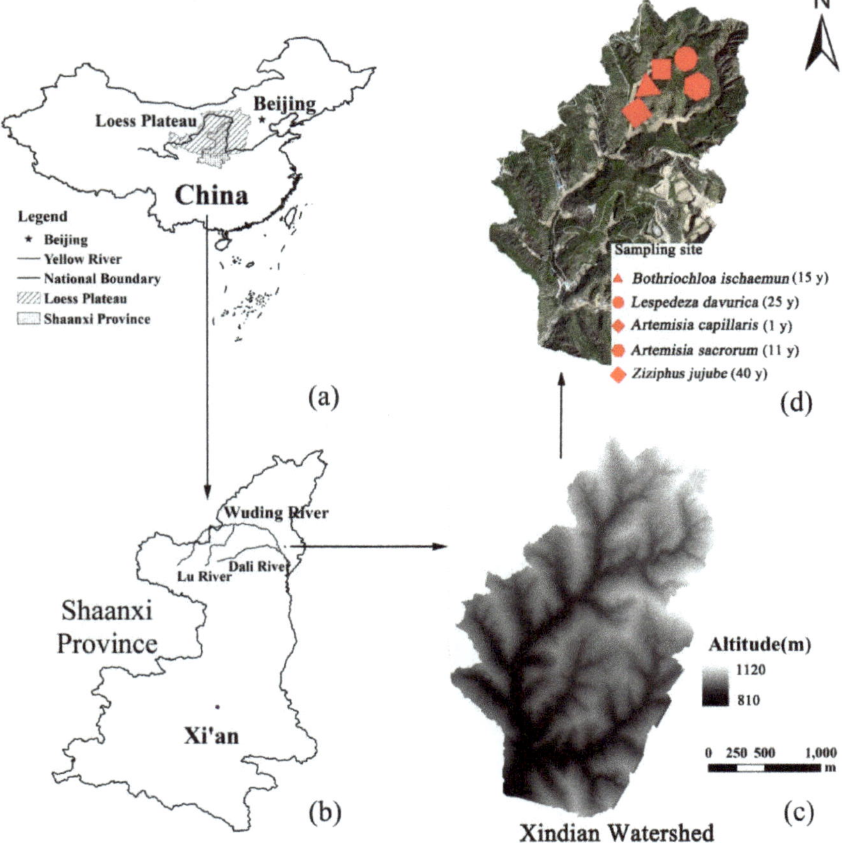

Figure 1. (**a**,**b**) Study area in Shaanxi Province, China; (**c**) Digital elevation model; (**d**) Digital image and sampling sites.

2.2. Experiments and Tests

The five experimental plot areas were 4 m × 0.5 m, and slope steepness ranged from 8 to 9°. In order to prevent lateral seepage of the slope flow during the test, plots were separated by a 2-mm thick steel plate. We installed flow-stabilizing devices and jet grooves at either ends of the plots, and dug a circular pit under the catchment groove where the sample collecting barrel was placed (Figure 2). Each scouring experiment lasted for 30 min, and the observation sections were set at 1, 2, 3, and 4 m of the plots to observe the runoff in sections. According to the series of precipitation data of the hydrological station in the past 30 years, the P–III frequency of rainstorms was calculated for each duration (10, 20, and 60 min; 3, 6, 12, and 24 h; and 3 d). In the torrential rains of different durations, the 20-year return period short-duration (60, 20, and 10 min) rainstorm intensities reached 0.9, 1.9, and 2.7 mm/min, respectively (Figure 3). As such, we used a flow rate of 4, 8, and 16 L/min for scouring, which is similar to a rain intensity of 2, 4, and 8 mm/min according to the catchment area of the plot.

Runoff and sediment samples were collected every minute, and runoff data was measured every 2 min. Sediment samples were left to settle first, then dried, and measured for sediment yield. We cut off the vegetation on the ground in each plot, retaining the stem of a certain height (5 cm) and the root system (Figure 2). We used the potassium permanganate stain tracing method to measure the runoff velocity on the slope. Runoff depth and width were measured using an artificial ruler. Runoff depth was used as a reference, and the hydrodynamic calculation uses the formula derivation value. Each experimental plot was scoured three times with different flow rates. A total of 45 experiments were conducted in this study.

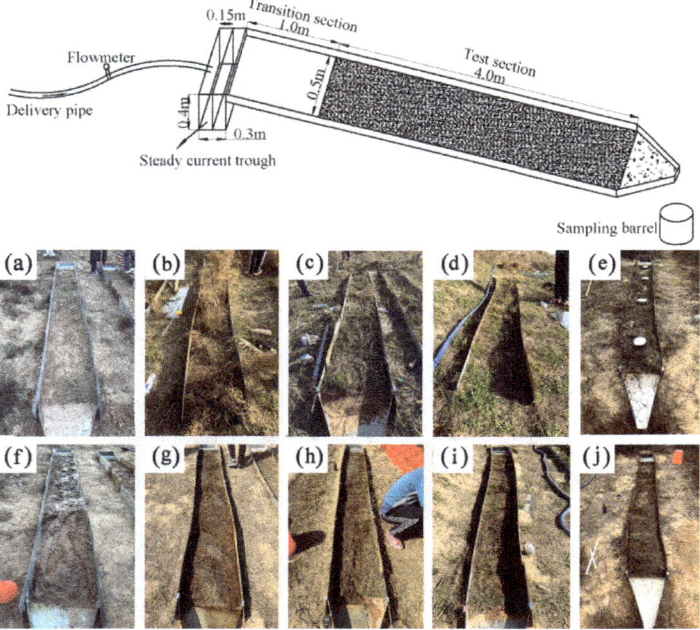

Figure 2. Device schematic diagram and field photos of the experimental process in the experimental area. (**a–e**) Before soil erosion; (**f–j**) After soil erosion; (**a**,**f**) *Artemisia capillaris* for 1 year since restoration; (**b**,**g**) *A. sacrorum* for 11 years; (**c**,**h**) *Bothriochloa ischaemum* for 15 years; (**d**,**i**) *Lespedeza davurica* for 25 years; (**e**,**j**) *Ziziphus jujube* for 40 years.

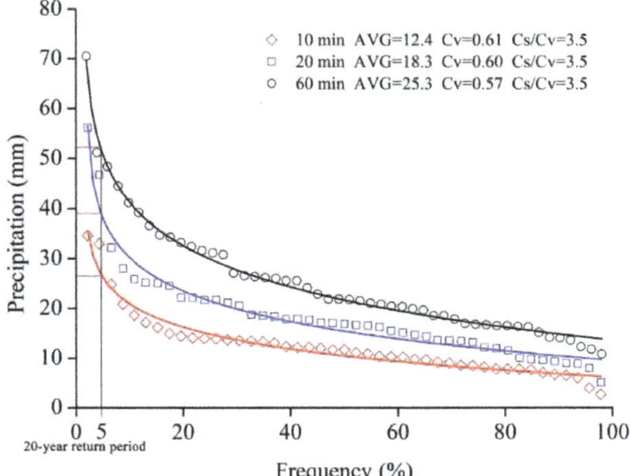

Figure 3. Short-term rainstorm frequency P–III curve in the study area.

2.3. Vegetation Community Survey and Sample Collection

In addition to experimental plots, three 2 × 2 m vegetation survey plots were established at each site. We sampled the vegetation by documenting the number of plants, estimating the percent ground cover, vegetation height, and humus layer thickness (Table 1). We harvested all aboveground plant tissue to measure aboveground vegetation biomass. We collected litter and humus layers for mass calculation. Root samples were collected using a root drill that was 9 cm in diameter with a barrel length of 10 cm from 0–20-, 20–40-, 40–60-, 60–80-, and 80–100-cm soil layers, based on previous reports of the root systems of vegetation on the Loess Plateau [1,28]. We used an iron box with an area of 0.2 × 0.15 m buckled into the soil to obtain undisturbed soil for measuring soil aggregates. A soil wreath knife with a volume of 100 cm^3 was used to obtain undisturbed soil for measuring the saturated hydraulic conductivity.

Table 1. Status of the experimental plots.

Years after Restoration/y	Dominant Species	Vegetation Type	Altitude/m	Slope Aspect	Slope/(°)	Vegetation Coverage/%	Humus Thickness/cm	Associated Species
1	*Artemisia capillaris*	Semi-shrub, herb	977	Shady slope	8	25	0.18	*Artemisia atrovirens, Artemisia scoparia*
11	*Artemisia sacrorum*	Semi-shrub, herb	965	Half-sunny slope	9	73	0.58	*Artemisia atrovirens, Artemisia scoparia, Tripolium vulgare, Lespedeza davurica, Ziziphus jujuba*
15	*Bothriochloa ischaemum*	Perennial herb	961	Half-sunny slope	8	86	0.73	*Artemisia sacrorum, Lespedeza davurica, Artemisia atrovirens, Taraxacum mongolicum*
25	*Lespedeza davurica*	Herbaceous subshrub	951	Half-shady slope	8	67	0.57	*Artemisia sacrorum, Lespedeza davurica, Bothriochloa ischaemum, Setaria viridis, Tripolium vulgare, Artemisia atrovirens, Artemisia scoparia, Taraxacum mongolicum*
40	*Ziziphus jujuba*	Deciduous arbors	963	Half-shady slope	9	76	0.69	*Artemisia sacrorum, Lespedeza davurica, Bothriochloa ischaemum, Setaria viridis, Tripolium vulgare, Clerodendrum mandarinorum, Asparagus cochinchinensis, Artemisia scoparia, Taraxacum mongolicum*

2.4. Sample Testing

All root samples were cleaned and scanned using a root scanner (EPSON, TWAIN PRO, Suwa City, Japan). We used the root-system analysis program WinRHIZO (QC., Quebec City, Canada) to analyze output from the root scanner. The program was used to estimate the root length, surface area, root tips, and diameter. Root biomass was measured by weighing the dried roots. Soil aggregate samples were air-dried naturally and then sifted into three grain classes to calculate the percentage, which were 0–0.25, 0.25–2, and >2 mm, respectively. Soils extracted from the drill core were used to measure the soil particle size using a laser particle size analyzer (Malvern, Mastersizer 2000, Birmingham, Britain). We only utilized soil median diameter d50 data in the grey correlation analysis. The saturated hydraulic conductivity of undisturbed soils collected with a wreath knife was measured by the constant head method.

2.5. Data Analysis

2.5.1. Vegetation Community Index

We measured the root length density (RLD), root weight density (RWD), and root tip density (RTD) for all samples according to Equations (1)–(3):

$$RLD = \frac{L}{V_s}, \tag{1}$$

$$RWD = \frac{M}{V_s}, \tag{2}$$

$$RTD = \frac{N}{V_s}, \tag{3}$$

where L is the sum of all root lengths per unit soil volume (mm); M is the dry weight of all roots per unit soil volume (mg); N is the sum of all the root tips per unit soil volume; and V_s is the volume per unit of soil.

Next, we measured the soil saturated hydraulic conductivity as an indicator of soil permeability. The higher the saturated hydraulic conductivity, the higher the soil permeability. Increases in soil permeability can increase the infiltration of runoff and play a better role in soil and water conservation. This was calculated according to Equation (4):

$$K = \frac{10QL}{A\Delta HT}, \tag{4}$$

where K is the soil saturated hydraulic conductivity (mm/min); Q is the outflow (mL) in time T; L is the linear distance (cm) of the water flow path; A is the cross-sectional area (cm^2) through which the water flows; ΔH is the total head difference (cm) of the start and end sections of the percolation path; and T is the outflow time (min). We only used the soil saturated hydraulic conductivity as one of the factors of the correlation analysis and did not analyze the results in this study.

We used the Shannon–Wiener index (H), Margalef index (R), and the vegetation importance value (Z) as three ecological indicators that can reflect the aboveground structure of vegetation communities. The Z reflects the ecological status of a certain plant in a vegetation community and can play a normalization role for the study of more complex vegetation communities. Equations (5)–(7) were as follows:

$$H = -\sum_{i=1}^{s} \frac{N_i}{N}\left(\ln \frac{N_i}{N}\right), \tag{5}$$

$$R = \frac{S-1}{\ln N}, \tag{6}$$

$$Z = RD + RF + RC, \tag{7}$$

where N is the sum of the number of plots; N_i is the number of plants of the i-th species; S is the total number of species per plot; and RD = (density of a species/total density of all species) × 100%; RF = (frequency of a species/total frequency of all species) × 100%; RC = (cover of a species/total coverage of all species) × 100%.

2.5.2. Calculation of Hydrodynamic Parameters

Hydrodynamic parameters included the Darcy–Weisbach (f), runoff shear (τ), runoff power (P), Reynolds number (Re), and Froude number (Fr) following Equations (8)–(15):

$$f = 8R{\cdot}J{\cdot}g/v^2, \tag{8}$$

$$J = [L{\cdot}\sin\theta - (v^2/2g)]/L, \tag{9}$$

$$h = \frac{Q}{dv}, \tag{10}$$

$$Re = \frac{vR}{\gamma}, \tag{11}$$

$$\gamma = \frac{0.01775}{1 + 0.0337t + 0.000221t^2}, \tag{12}$$

$$\mathrm{Fr} = v/\sqrt{gh}, \tag{13}$$

$$\tau = \rho RJ, \tag{14}$$

$$\omega = \tau v, \tag{15}$$

where R is the hydraulic radius in m; J is the hydraulic gradient in m/m; g is the gravitational acceleration constant of 9.8 m/s^2; v is the runoff velocity in m/s; θ is the slope in degrees; L is the slope length in m; h is the depth of runoff in m; Q is the flow rate in m^3/s; d is the runoff width in m; R is the hydraulic radius in m, which is approximately equal to the runoff depth, h; t is the water flow temperature in °C; and ρ is the water flow density in kg/m^3.

2.5.3. Grey Correlation Analysis

The gray correlation analysis formula is as follows, Equations (16) and (17):

$$\xi_{0i} = \frac{\min_i\min_k\left|x_0'(k) - x_i'(k)\right| + \rho\max_i\max_k\left|x_0'(k) - x_i'(k)\right|}{x_0'(k) - x_i'(k) + \rho\max_i\max_k\left|x_0'(k) - x_i'(k)\right|}, \tag{16}$$

$$\gamma(x_0, x_i) = \frac{1}{n}\sum_{i=1}^{n}\xi_{0i}, \tag{17}$$

where ξ_{0i} is the correlation coefficient, $\gamma(x_0, x_i)$ is the correlation degree, $x_0'(k) - x_i'(k)$ is the difference sequence, $\max_i\max_k\left|x_0'(k) - x_i'(k)\right|$ is the maximum difference, and $\min_i\min_k\left|x_0'(k) - x_i'(k)\right|$ is the minimum difference.

3. Results

3.1. Vegetation Succession Sequence and Structural Characteristics of Ground/Underground Parts

We showed the distribution of Z for each species (Figure 4). We also summed the Z of each species in the five experimental plots. The rankings of the plant species in terms of Z across the experimental plots was: *Artemisia sacrorum* (356.72) > *Artemisia capillaris* (214.36) > *Bothriochloa ischaemum* (189.31) > *Lespedeza davurica* (177.97) > *Artemisia atrovirens* (123.67) > *Ziziphus jujuba* (100.74) > *Artemisia scoparia* (81.39) > *Tripolium vulgare* (58.32) > *Setaria viridis* (36.34) > *Taraxacum mongolicum* (10.27) > *Clerodendrum mandarinorum* (9.66) > *Asparagus cochinchinensis* (5.36). The sum of Z of annual plants was 419.42, mainly *Artemisia capillaris*, *Artemisia scoparia*, and *Artemisia atrovirens*. Perennial herbaceous plants had a Z of 665.98, mainly *Bothriochloa ischaemum*, *Artemisia sacrorum*, and *Tripolium vulgare*. Small trees and semi-shrub plants had a Z of 278.71, which was primarily driven by *Lespedeza davurica* and *Ziziphus jujuba*. Therefore, it can be explained that perennial herbaceous plants are the main biological species for the natural restoration of vegetation in the area.

The Z across all species increased from 260.72 to 283.06 during the 40 years of succession that we analyzed. The Z of annual plants was 147.11, the perennial herbaceous plants was 93.12, and semi-shrubs and small trees was 20.49 in the 1-year vegetation community. Thus, annual plants were dominant in the 1-year vegetation community. Perennial herbaceous plants were dominant (153.23) in the 11- and 15-year vegetation communities by the same calculation method. Semi-shrubs and small trees were dominant (81.02) in the 25- and 40-year vegetation communities. It can be concluded that annual plants are the dominant species in the early stage of vegetation community succession, perennial grasses are the dominant species in the middle stage, and semi-shrubs and small trees are the dominant species in the later stage.

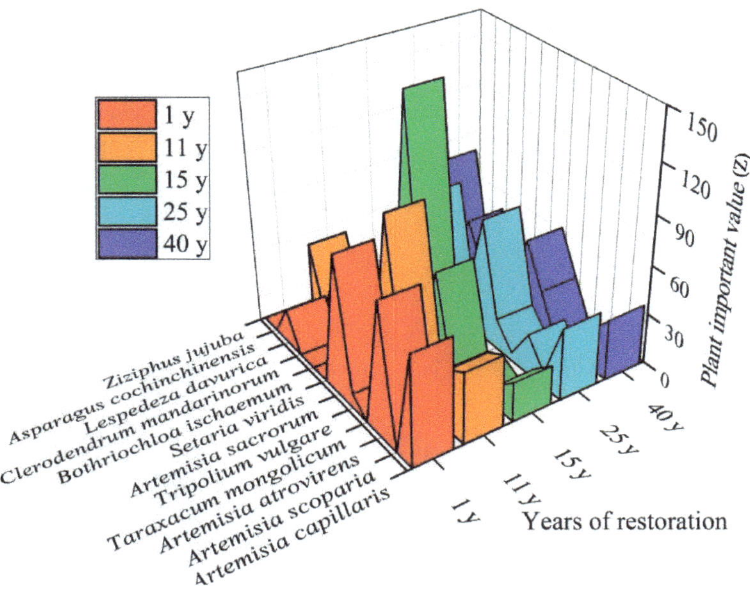

Figure 4. Distribution characteristics of the vegetation importance value.

The species density gradually increased from 2.5 to 4.5/m^2 from 1 to 40 years of succession (Table 2). Plant density increased first and then decreased and then stabilized. Plants density peaked at 135.5/m^2 at 15 years of succession, and then stabilized between 25 and 40 years (Table 2). Plant height increased from 25.25 to 159.32 cm/m^2 over the 40 years. Aboveground biomass increased from 28.83 to 753.33 g/m^2. The amount of litter and humus increased from 24.50 to 605.00 g/m^2.

Table 2. The ecological structure development of vegetation community succession from 1 to 40 years.

Years after Restoration/Year	Species Density (n/m^2)	Plants Density (n/m^2)	Plant Height (cm/m^2)	Aboveground Plant Biomass (g/m^2)	Litter and Humus (g/m^2)
1	2.5	14.33	25.25	28.83	24.50
11	3.5	44.5	58.33	195.83	115.33
15	3	135.5	54.17	165.55	151.33
25	4	34.67	54.67	195.33	220.5
40	4.5	34.35	159.32	753.33	605.00

Roots are highly sensitive to the soil environment and occupy an important position in the succession of vegetation communities. The more closely the root system is integrated with the soil, the more obvious its effect on the soil's physical and chemical properties, and the stronger the soil erosion resistance [29]. RWD and root diameter gradually increased from 1 to 40 years of succession (Figure 5). At 40 years, the maximum RWD and root diameter were 10.80 mg/cm^3 and 0.65 mm, respectively. RLD and RTD increased from 1 to 15 years, then decreased from 15 to 25 years before stabilizing. The average RLD and RTD reached a maximum of 7.72 mm/cm^3 and 2.80/cm^3 in the 15-year successional community. According to the results above, it can be explained that slender and thin are the main root morphology of the vegetation community in early succession. Perennial plants, however, increased in dominance with increasing successional age. Specifically, we found an increase in the dominance of semi-shrubs and small trees after 25 years. At this time, the RWD and root diameter was bigger. The RWD, RLD, root diameter, and RTD decreased with soil depth at each successional stage, and there were significant differences in the root index between some soil layers ($p < 0.05$) (Figure 5).

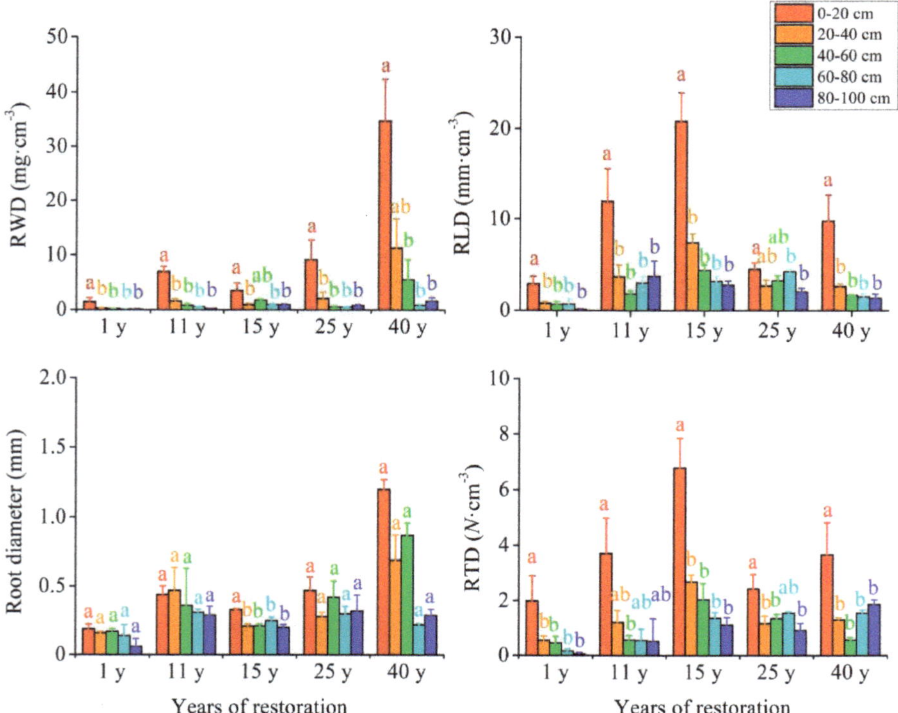

Figure 5. Distribution characteristics of root morphology in vegetation communities. Note: The same letter (a,b,c) indicates that there was no significant difference in the indexes of root system among groups, the significant level P = 0.05.

3.2. Dynamic Characteristics of Slope Runoff in Different Vegetation Communities

Soil erosion of vegetation communities on slopes is primarily affected by the hydrodynamic characteristics of runoff and by the composition of the underlying material. Runoff velocity ranged from 0.078 to 0.266 m/s across the whole experiment (Table 3). The runoff velocity range was 0.203 to 0.266 m/s under different flow conditions during the 1 to 15 years of succession. Runoff velocity ranged from 0.078 to 0.180 $m \cdot S^{-1}$ under different flow conditions during the 25 to 40 years of succession. In the later stage of succession, the runoff velocity was 52.37%, 38.77%, and 52.52% lower under the condition that the discharge rate was 4, 8, and 16 L/min, respectively. It can be concluded that vegetation community succession is a vegetation self-restoration process that can effectively reduce runoff velocity.

The Darcy–Weisbach (f) metric is used to indicate the resistance of the underpad to runoff. Generally, the larger the resistance coefficient, the more energy that is required for the water to overcome the resistance, and the smaller the sediment yield. The resistance coefficients ranged from 0.462 to 21.792 in the whole experiment (Table 3). The average resistance coefficient under different flow conditions was 0.827 in the 1 to 15 years of vegetation succession. The resistance coefficient was 11.223 during the 25 to 40 years of succession. The resistance coefficient of the slope runoff increased from 0.458 to 16.166 during the 25 to 40 years of vegetation succession.

Generally, the greater the flow shear stress, the greater the effective shear stress acting on the soil surface, and the greater the soil erosion intensity on the slope. The runoff shear stress increases with the increase of the discharge flow. When the scouring flow increased by 2.00 times, the shear stress increased by 1.63 times correspondingly in the whole experiment process. The runoff shear stress, however, also showed an increasing trend with the increase of the vegetation community succession years. The shear stress increased from 3.560 to 8.177 Pa during 1 to 40 years of vegetation community succession.

The runoff power can reflect the comprehensive influence of hydrodynamic characteristics on slope erosion. With the same trend of the shear stress, the runoff power of each vegetation community increased with the increase of the scouring flow (Table 3). This change in runoff power was due to the greater scouring force, faster runoff velocity, and greater shear stress, and the susceptibility to rill erosion. The maximum runoff power was 0.788 to 1.327 N/(m/s) in the 1-year vegetation community, and the minimum was 0.589 to 1.108 N/(m/s) in the 40-year vegetation community (Table 3).

Table 3. Characteristics of the dynamic parameters of slope runoff in different vegetation communities.

Years after Restoration/Year	Scouring Flow (L/min)	Water Temperature T (°C)	Runoff Depth h (m)	Runoff Width d (m)	Runoff Velocity v (m/s)	Reynolds Number Re	Froude Number Fr	Darcy–Weisbach f	Runoff Shear Stress τ (Pa)	Runoff Power P (N/(m/s))
1	4	8	0.002	0.124	0.238	409.455	1.567	0.462	3.318	0.788
	8	8	0.002	0.349	0.218	301.638	1.590	0.448	2.675	0.580
	16	8	0.001	0.282	0.211	689.693	1.581	0.463	4.686	1.327
11	4	22	0.001	0.241	0.211	290.719	1.838	1.007	2.099	0.442
	8	16	0.003	0.208	0.260	588.889	1.665	0.937	3.964	1.025
	16	14	0.003	0.358	0.266	676.453	1.553	0.909	4.742	1.256
15	4	18	0.002	0.159	0.256	432.667	1.950	0.770	2.861	0.725
	8	16	0.002	0.295	0.203	416.903	1.352	1.188	3.218	0.654
	16	16	0.004	0.338	0.238	751.111	1.278	1.338	4.954	1.178
25	4	22	0.003	0.156	0.146	464.339	0.847	4.432	4.336	0.628
	8	14	0.003	0.262	0.180	465.662	1.049	1.854	4.240	0.759
	16	16	0.007	0.371	0.110	670.753	0.430	12.554	9.457	1.038
40	4	17	0.005	0.189	0.078	346.970	0.362	21.792	7.611	0.589
	8	16	0.005	0.338	0.094	374.978	0.466	14.938	7.228	0.652
	16	16	0.006	0.412	0.115	636.482	0.467	11.767	9.693	1.108

3.3. Runoff and Sediment Yield under Different Vegetation Communities

The runoff volume and sediment yield on the slopes of the different vegetation communities were significantly different. Both the runoff volume and sediment yield decreased significantly with increasing successional age at a scouring flow of 4, 8, and 16 L/min (Table 4). Compared with the early successional community, the total runoff volume at 40 years decreased 3.52, 2.74, and 2.29 times, respectively, under the scouring flow of 4, 8, and 16 L·min^{-1}. The total sediment yield decreased 16.83, 9.31 and 11.65 times on average. It can be seen from the multiple of reducing runoff and sediment that the effect of vegetation restoration on reducing runoff and sediment decreases with the increase of the erosion discharge. In addition, the total runoff volume and sediment yield under different scouring flows were averaged and the following results were calculated: During 1 to 40 years of vegetation succession, the runoff volume decreased by an average of 2.52 times and the sediment yield decreased by an average of 11.60 times. Therefore, it can be concluded that the contribution of Loess slope vegetation succession to sediment reduction during water erosion is much greater than that for runoff reduction.

Table 4. Total runoff volume and sediment yield under different vegetation communities.

Scouring Flow/(L/min)	Years after Restoration/Years	Runoff Volume/L	Sediment Yield/kg
4	1	109.59	2.02
	11	41.93	0.36
	15	93.96	0.56
	25	73.39	0.09
	40	31.12	0.12
8	1	202.8	2.7
	11	147.29	0.58
	15	194.07	0.73
	25	171.72	0.15
	40	73.97	0.29
16	1	459.57	10.83
	11	363.73	1.38
	15	407.5	0.99
	25	399.53	0.31
	40	200.78	0.93

The soil erosion rate of the 1-year vegetation community reached a maximum value of 1.35 g/(m^2·s) when the scouring flow rate was 16 L/min. The minimum soil erosion rate was 0.01 g/(m^2·s) in the 25-year vegetation community under a scouring flow rate of 4 L/min, followed by the 40-year vegetation community at 0.02 g/(m^2·s) (Figure 6). The soil erosion rate decreased by 25.36 times from 1 to 40 years of succession under the same scouring flow rate. This indicates that the successional status of vegetation in the Loess Plateau has a significant effect on reducing the erosion rate of runoff.

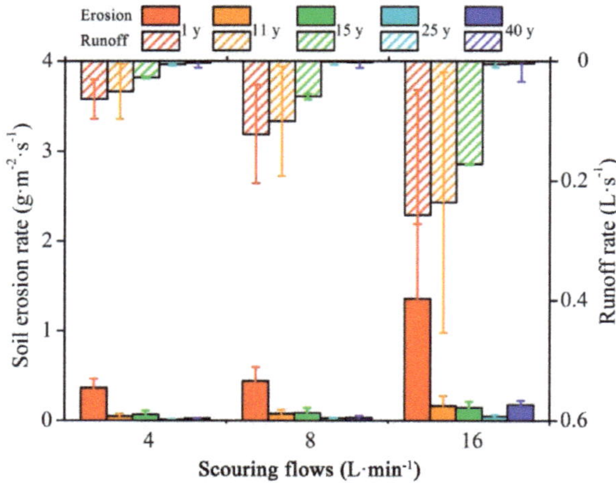

Figure 6. Average soil erosion and runoff rate under different vegetation communities.

3.4. Grey Correlation Analysis between Hydrodynamic Parameters and Ecological Factors of Vegetation Communities

The correlation between these hydrodynamic parameters and soil erosion rate can be illustrated (Table 5). The erosion rate was positively correlated with runoff velocity and power ($p < 0.01$). In contrast, the erosion rate was negatively correlated with resistance ($p < 0.01$), and was not correlated with shear force ($p > 0.05$).

The hydrodynamic parameters and the soil erosion rate showed a significant Pearson correlation, but the correlation coefficient does not indicate which hydrodynamic factors are most relevant to the erosion rate. Therefore, we identified the hydrodynamic factors most closely related to the erosion rate by the grey correlation analysis method: Runoff power (Table 6).

Table 5. Correlations between the soil erosion rate and hydrodynamic factors.

Scouring Flow/(L/min)	Sample Size	Runoff Velocity	Runoff Resistance	Runoff Power	Shear Stress
4	50	0.456 *	−0.319 *	0.412 **	−0.246
8	75	0.285 *	−0.213 *	0.235 *	−0.394 **
16	70	0.491 **	−0.367 **	0.376 **	−0.388 **
Total	195	0.414 **	−0.217 **	0.326 **	−0.112

* Indicates a significant correlation at $p < 0.05$; ** Indicates a significant correlation at $p < 0.01$.

Table 6. Correlation coefficient between the soil erosion rate and hydrodynamic factors and grey correlation degree.

Characteristic Indicators	Correlation Coefficient	1 Year	11 Years	15 Years	25 Years	40 Years	Correlation Degree γ (x_0, x_i)
Runoff velocity	ξ_1	0.37	0.65	0.67	0.72	1.00	0.68
Runoff resistance	ξ_2	0.38	0.99	1.00	0.70	0.43	0.70
Shear stress	ξ_3	0.39	1.00	0.99	0.67	0.62	0.73
Runoff power	ξ_4	0.48	0.96	1.00	0.86	0.99	0.86

In order to establish the relationship between the hydrodynamic parameters with the vegetation characteristics, we chose the hydrodynamic parameters as the characteristic indicators, and selected the vegetation community as the sequence index. The correlation degree of the total slope runoff volume and ecological factors of the vegetation community were ranked as follows: Underground part (0.74) > aboveground part (0.69) = soil structure (0.69) (Table 7). From the grey correlation analysis of the slope runoff power, it can be concluded that the aboveground part (0.74) > soil texture (0.73) > underground part (0.69) (Table 8). The grey correlation analysis of the slope runoff velocity shows that the aboveground part (0.76) > soil texture (0.70) > underground part (0.68) (Table 9). The grey correlation analysis of the runoff resistance shows that soil texture (0.71) > underground part (0.70) > aboveground part (0.65) (Table 10).

Table 7. Grey correlation between the total slope runoff volume and vegetation communities.

Sequence Index		Correlation Coefficient	1 Year	11 Years	15 Years	25 Years	40 Years	Correlation Degree	Mean
Underground part	RWD	ξ_5	0.66	0.99	0.81	1.00	0.42	0.78	0.74
	RLD	ξ_6	0.50	1.00	0.63	1.00	0.93	0.81	
	Root diameter	ξ_7	0.47	0.93	0.64	1.00	0.37	0.68	
	RTD	ξ_8	0.42	1.00	0.55	0.92	0.54	0.69	
Aboveground part	Vegetation important value	ξ_9	0.52	0.96	0.70	1.00	0.40	0.72	0.69
	Number of species	ξ_{10}	0.46	0.97	0.63	1.00	0.38	0.69	
	Number of plants	ξ_{11}	0.46	1.00	0.37	0.70	0.94	0.69	
	Shannon–Wiener index	ξ_{12}	0.57	0.77	0.42	1.00	0.35	0.62	
	Margalef index	ξ_{13}	0.41	1.00	0.38	0.70	0.76	0.65	
	Plant biomass	ξ_{14}	0.58	1.00	0.79	0.90	0.39	0.73	
	Litter and humus	ξ_{15}	0.52	0.75	1.00	0.80	0.67	0.75	
Soil structure	saturated hydraulic conductivity	ξ_{16}	0.52	0.79	1.00	0.56	0.33	0.64	0.69
	d_{50}	ξ_{17}	0.36	0.66	0.68	1.00	0.36	0.61	
	Macro-aggregate	ξ_{18}	0.48	1.00	0.65	0.97	0.36	0.69	
	Micro-aggregate	ξ_{19}	0.69	0.99	0.88	1.00	0.52	**0.82**	

Table 8. Grey correlation between the slope runoff power and vegetation communities.

Sequence Index		Correlation Coefficient	1 Year	11 Years	15 Years	25 Years	40 Years	Correlation Degree	Mean
Underground part	RWD	ξ_{20}	0.63	0.82	0.79	1.00	0.40	0.73	0.69
	RLD	ξ_{21}	0.43	0.94	0.40	0.94	1.00	0.74	
	Root diameter	ξ_{22}	0.43	1.00	0.60	0.91	0.34	0.66	
	RTD	ξ_{23}	0.39	0.64	0.36	1.00	0.70	0.62	
Aboveground part	Vegetation important value	ξ_{24}	0.52	0.67	1.00	0.61	0.48	0.66	0.74
	Number of species	ξ_{25}	0.48	1.00	0.76	0.67	0.46	0.67	
	Number of plants	ξ_{26}	0.64	1.00	0.43	0.94	0.96	0.79	
	Shannon–Wiener index	ξ_{27}	0.72	1.00	0.38	0.79	0.37	0.65	
	Margalef index	ξ_{28}	0.58	0.93	0.41	0.94	1.00	0.77	
	Plant biomass	ξ_{29}	0.62	0.91	0.88	1.00	0.41	0.76	
	Litter and humus	ξ_{30}	0.62	0.90	0.91	0.96	1.00	0.88	
Soil structure	saturated hydraulic conductivity	ξ_{31}	0.64	0.80	1.00	0.65	0.40	0.70	0.73
	d_{50}	ξ_{32}	0.37	0.90	1.00	0.61	0.58	0.69	
	Macro-aggregate	ξ_{33}	1.00	0.57	0.86	0.60	0.67	0.74	
	Micro-aggregate	ξ_{34}	0.67	0.99	0.86	1.00	0.50	0.80	

Table 9. Grey correlation between the slope runoff velocity and vegetation communities.

Sequence Index		Correlation Coefficient	1 Year	11 Years	15 Years	25 Years	40 Years	Correlation Degree	Mean
Underground part	RWD	ξ_{35}	0.56	0.65	0.64	1.00	0.34	0.64	0.68
	RLD	ξ_{36}	0.37	0.90	0.46	1.00	0.66	0.68	
	Root diameter	ξ_{37}	0.63	0.99	0.74	1.00	0.45	0.76	
	RTD	ξ_{38}	0.46	0.62	0.60	1.00	0.54	0.65	
Aboveground part	Vegetation important value	ξ_{39}	1.00	0.86	0.94	0.94	0.62	0.87	0.76
	Number of species	ξ_{40}	0.81	1.00	0.91	0.91	0.59	0.84	
	Number of plants	ξ_{41}	0.50	0.69	0.39	1.00	0.96	0.71	
	Shannon–Wiener index	ξ_{42}	0.91	1.00	0.55	0.89	0.48	0.76	
	Margalef index	ξ_{43}	0.54	0.69	0.68	1.00	0.35	0.65	
	Plant biomass	ξ_{44}	0.54	0.69	0.68	1.00	0.35	0.65	
	Litter and humus	ξ_{45}	0.50	0.99	1.00	1.00	0.60	0.82	
Soil structure	saturated hydraulic conductivity	ξ_{46}	0.58	0.65	1.00	0.69	0.35	0.66	0.70
	d_{50}	ξ_{47}	0.71	1.00	0.98	0.85	0.59	0.82	
	Macro-aggregate	ξ_{48}	1.00	0.48	0.74	0.55	0.52	0.66	
	Micro-aggregate	ξ_{49}	0.50	1.00	0.55	0.96	0.34	0.67	

Table 10. Grey correlation between the slope runoff resistance and vegetation communities.

Sequence Index		Correlation Coefficient	1 Year	11 Years	15 Years	25 Years	40 Years	Correlation Degree	Mean
Underground part	RWD	ξ_{50}	1.00	0.44	0.57	0.38	0.68	0.61	0.70
	RLD	ξ_{51}	1.00	0.62	0.47	0.84	0.38	0.66	
	Root diameter	ξ_{52}	0.93	0.60	0.82	1.00	0.45	0.76	
	RTD	ξ_{53}	1.00	0.81	0.53	0.98	0.43	0.75	
Aboveground part	Vegetation important value	ξ_{54}	0.68	0.70	0.72	1.00	0.40	0.70	0.65
	Number of species	ξ_{55}	0.68	0.61	0.68	1.00	0.37	0.67	
	Number of plants	ξ_{56}	1.00	0.76	0.40	0.78	0.38	0.66	
	Shannon–Wiener index	ξ_{57}	0.64	0.60	0.95	1.00	0.41	0.72	
	Margalef index	ξ_{58}	1.00	0.79	0.46	0.86	0.40	0.70	
	Plant biomass	ξ_{59}	1.00	0.35	0.43	0.36	0.41	0.51	
	Litter and humus	ξ_{60}	1.00	0.42	0.44	0.57	0.35	0.56	
Soil structure	saturated hydraulic conductivity	ξ_{61}	1.00	0.75	0.49	0.45	0.68	0.67	0.71
	d_{50}	ξ_{62}	0.72	0.64	0.68	1.00	0.39	0.69	
	Macro-aggregate	ξ_{63}	0.63	0.71	0.68	1.00	0.39	0.68	
	Micro-aggregate	ξ_{64}	1.00	0.62	0.97	0.91	0.52	0.80	

4. Discussion

4.1. Effects of Vegetation Community Restoration on Soil Structure and Erosion

Vegetation construction is an important measure for soil erosion control on the Loess Plateau. Specifically, it plays an important role in controlling soil erosion and reducing sediment along the Yellow River. Vegetation restoration can effectively improve soil properties, affecting the relationship between the plant–soil interface through various factors, such as stems, leaves, roots, and root exudates [30]. Improving soil properties takes time, and there are differences in the degree and efficiency of soil improvement between different succession development directions and different vegetation types [31]. Therefore, the growth and decline of dominant species is the basic unit of the succession and development of a vegetation community. The composition of the community and the spatial distribution of individuals constitute the structural characteristics of the vegetation community. Vegetation community structure includes diversity, species composition, community floristic composition, and community age structure [32]. The natural restoration process of abandoned farmland vegetation on the Loess Plateau can be roughly divided into the rapid recovery period, primary succession period, advanced succession period, and stable period [33].

Vegetation communities have the ability to control soil erosion, in part because they increase soil anti-erodibility [34]. The soil erodibility is affected by vegetation as they can increase soil organic matter and soil aggregate stability. In this study, the shear stress increased from 3.560 to 8.177 Pa from 1 to 40 years of vegetation succession (Table 3). At the same time, the erosion rate decreased from 1.35 to 0.02 g/(m^2·s) (Figure 5). This demonstrates that communities at later stages of succession have better anti-erosion properties than younger communities. The effects of vegetation on soil erosion reduction can be divided into three parts, namely, the aboveground interception of vegetation, the fixation of underground roots, and the resistance of the soil interface [35]. The vegetation types and cover on the Loess Plateau have changed dramatically over the past 30 years [36], and the regional vegetation ecosystem has been significantly improved. These improvements have effectively alleviated the serious effects of soil erosion in this area. Jiao et al. showed that the average soil erosion intensity in the early stages of vegetation succession was between 3087.6 and 4408.4 t/km^2/a, and the vegetation succession period was between 1245.2 and 1827.8 t/km^2/a [37]. This is consistent with the finding that vegetation restoration can effectively reduce the rate of soil erosion (Figure 5). However, with the occurrence of extreme rainstorm events or the increase of scouring intensity, places with better vegetation coverage are more prone to small gravity erosion, such as landslides and collapses. [38]. It can be assumed that when the scouring flow rate is great enough, the vegetation will gradually lose its effectiveness in reducing flow and reducing sediment, and can even induce extreme soil erosion events, such as landslides and collapses. Previous studies have shown that when grasslands on slopes scour under 5 L/min of discharge, the effect of sediment reduction on vegetated slopes is significantly greater than that on bare slopes while when the scouring flow increases to 8 L/min, the difference of sediment reduction between different grasslands is small, which indicates that the ability of vegetation to prevent and control runoff erosion on slopes is weakened with the increased discharge [39]. The total runoff volume decreased by an average of 3.52, 2.74, and 2.29 times at 4, 8, and 16 L/min of the scouring flow and the total sediment yield decreased by 16.83, 9.31, and 11.65 times on average in our data report (Figure 5). It was also just starting from the scouring flow of 8 L/min, and the difference in the control effect of vegetation communities on soil erosion became smaller. Therefore, the predecessors and our research can at least prove that when the scouring flow is large enough, the soil and water conservation of the vegetation gradually decreases. Whether it will aggravate slope erosion or induce gravity erosion remains to be further studied.

4.2. Effects of Vegetation Community Restoration on the Hydrodynamics of Slope Runoff

Vegetation can effectively reduce water erosion on slopes. and some of the more important reasons are the hydrodynamic parameters affecting the runoff, which will change the runoff velocity, flow regime, and erosion energy of the slope, thus affecting the soil erosion process [25,40]. Vegetation can increase the critical conditions of slope erosion by increasing runoff resistance, and reducing runoff velocity and power to improve the critical conditions of slope erosion [41]. Their research shows that among many hydrodynamic factors, the runoff power on the slope is the most closely related to the average sediment transport rate of runoff, and the runoff power is the factor that can best reflect the soil erosion rate. It is easier to analyze and simulate the soil erosion process by using runoff kinetic energy and power theory [42,43]. Previous studies have shown that vegetation type, vegetation coverage, vegetation litter, humus, and roots are important factors affecting soil erosion on slopes. Vegetation communities regulate the runoff of slopes through the interaction of aboveground and underground parts [44]. On the one hand, the ecological structure of aboveground plants and underground roots in vegetation communities can increase runoff resistance and reduce runoff power [45]. On the other hand, the correct succession of vegetation communities can improve the soil properties, greatly enhance the soil anti-erodibility, so that the formation of rills cannot be fully developed, and the runoff hydraulic energy slope always changes little [46,47]. In view of this problem, we continue to discuss the correlation degree of slope runoff volume, velocity, resistance coefficient, power, and other factors that affect the soil erosion rate. The average slope runoff volume decreased by 2.85 times with the succession and development of the vegetation community (1–40 years) (Table 4). This contribution came primarily from the root system of the vegetation community, and the average correlation degree was 0.74 (Table 7). The RLD was the most effective indicator of the runoff volume for roots. Runoff power was reduced by 19.75% from 1 to 40 years of succession (Table 3). The contribution came primarily from aboveground vegetation, with an average correlation degree of 0.74 (Table 8). The litter and humus quality were the most effective factors affecting the aboveground part. Vegetation community succession reduced the slope runoff velocity by 47.89%, and its deceleration effect came primarily from aboveground tissue, with the average correlation degree reaching 0.76 (Table 9). The vegetation importance value (0.87), species density (0.84), and Shannon–Wiener index (0.76) played key roles in the aboveground part. Therefore, the vegetation community complicates the composition of species through the development of succession, which is more effective in controlling soil erosion than the vegetation coverage of a single species [44]. The succession and development of the vegetation community also increased the content of soil aggregates, improved soil structure, and thus increased the resistance of surface soil to slope runoff. The correlation degree of soil structure, a series of the sequence index, to runoff resistance reached 0.71 (Table 10). Among them, the content of soil micro-aggregate was a key factor for the increased runoff resistance, and the correlation degree reached 0.8.

4.3. Implications for the Relationship between Vegetation Community Restoration and Slope Erosion

Around 50 years ago, the main contradiction in China's Loess Plateau was food production and ecological restoration. Strong soil erosion conditions have not allowed humans to cultivate on slopes [48,49]. Therefore, the Chinese government has adopted a series of eco-economic compensation measures in the hopes of resolving this important problem. However, after seeing significant increases in vegetation cover, the global climate changed, and the Loess Plateau still experiences strong soil erosion under extreme precipitation events [50]. Therefore, the new contradiction points to the benefit and mechanism of vegetation in controlling soil erosion. There were differences in the development direction of vegetation community succession, which leads to differences in the underlying surface. In the future, the difference analysis and quantitative description of slope erosion patterns should be examined more carefully. On the one hand, quantitative discussion of the effects of aboveground parts, underground parts, and soil structure of vegetation communities on slope erosion is required, and on the other hand, it is necessary to deeply analyze the relationship between the anti-erodibility of plant communities and the improved soil anti-erodibility, and to establish an evaluation model. This information will help us to understand the mechanism of a vegetation community regulating runoff and sediment on a slope more comprehensively. At the same time, the coupling and feedback between the slope hydrodynamic process and ecological vegetation processes can also be clarified from a scientific point of view.

5. Conclusions

We comprehensively analyzed the successional development of vegetation community restoration on the Loess Plateau in China. We found that vegetation communities in later stages of succession have the ability to control runoff erosion on slopes. In the early stages of vegetation community succession, communities were dominated by wormwood plants. Perennial grasses were dominant in the middle stages of succession, and semi-shrub and small trees became dominant in the later stages. The plant aboveground and underground parts, such as plant density, the number of species, root length density, and root biomass, increased gradually with the development of vegetation succession. After 40 years of natural succession of vegetation communities, the slope runoff velocity decreased by 47.89%, the runoff resistance coefficient increased by approximately 35.30 times, the runoff power decreased by approximately 19.75%, the total runoff volume decreased by approximately 2.52 times, and the total sediment yield was reduced by approximately 11.60 times. On the one hand, the role of vegetation in preventing and controlling slope water erosion indicated that the vegetation important value, number of species, vegetation diversity (Shannon–Wiener index), and litter humus layer of the aboveground part significantly reduced the runoff velocity and power. The total amount of runoff was significantly reduced by the development of vegetation roots. On the other hand, vegetation communities improved the soil structure, in which runoff resistance was significantly increased by the restoration of soil micro-aggregate. Our results are important for vegetation restoration in the Loess erosion slope. They provide a scientific basis for the study of the influence of the vegetation community on the resistance and control of soil erosion and can be used to benefit an evaluation of water and soil conservation in the "Grain for Green Project" on the Chinese Loess Plateau.

Author Contributions: Conceptualization, E.C. and P.L.; methodology, Z.L. (Zhanbin Li) and Y.S.; formal analysis, Y.Z. and J.Z.; software, Z.L. (Zhan Liu) and Z.L. (Zhineng Li); E.C. wrote the manuscript and all authors contributed to improving the paper.

Funding: This study was supported by the National Key Research and Development Program of China (2016YFC0402407), the National Basic Research Program of China (No. 2016YFC0402404), the Natural Science Foundations of China (No. 51779204, 41701603), Shaanxi Province Innovation Talent Promotion Plan Project Technology Innovation Team (2018TD-037) and Shaanxi Provincial Technology Innovation Guidance Project (2017CGZH-HJ-06).

Acknowledgments: We thank the reviewers for their useful comments and suggestions. We would like to thank Murphy Stephen at Yale University for his assistance with the English language and grammatical.

Conflicts of Interest: The authors declare no conflict of interest.

References

1. Chang, E.; Li, P.; Li, Z.; Xiao, L.; Zhao, B.; Su, Y.; Feng, Z. Using water isotopes to analyze water uptake during vegetation succession on abandoned cropland on the Loess Plateau, China. *CATENA* **2019**, *181*, 104095. [CrossRef]
2. Shi, P.; Zhang, Y.; Ren, Z.; Yu, Y.; Li, P.; Gong, J. Land-use changes and check dams reducing runoff andsediment yield on the Loess Plateau of China. *Sci. Total Environ.* **2019**, *664*, 984–994. [CrossRef] [PubMed]
3. Zhao, B.; Li, Z.; Peng, L.; Xu, G.; Gao, H.; Cheng, Y.; Chang, E.; Yuan, S.; Yi, Z.; Feng, Z. Spatial distribution of soil organic carbon and its influencing factors under the condition of ecological construction in a hilly-gully watershed of the Loess Plateau, China. *Geoderma* **2017**, *296*, 10–17. [CrossRef]
4. Ministry of Water Resources. *Chinese River Sediment Bulletin*; China Science Press: Beijing, China, 2018; pp. 3–4.
5. Yellow River Conservancy Commission of the Ministry of Water Resources. *Yellow River Basin Comprehensive Planning (2012–2030)*; Yellow River Water Conservancy Press: Zhengzhou, China, 2013; pp. 15–25.
6. Chen, Y.P.; Wang, K.B.; Lin, Y.S.; Shi, W.Y.; Song, Y.; He, X.H. Balancing green and grain trade. *Nat. Geosci.* **2015**, *8*, 739–741. [CrossRef]
7. Zhang, X.P.; Lin, P.F.; Chen, H.; Yan, R.; Zhang, J.J.; Yu, Y.P.; Liu, E.J.; Yang, Y.H.; Zhao, W.H.; Lv, D.; et al. Understanding land use and cover change impacts on run-off and sediment load at flood events on the Loess Plateau, China. *Hydrol. Process.* **2018**, *32*, 576–589. [CrossRef]
8. Gao, P.; Jiang, G.T.; Wei, Y.P.; Mu, X.M.; Wang, F.; Zhao, G.J.; Sun, W.Y. Streamflow regimes of the Yanhe River under climate and land use change, Loess Plateau, China. *Hydrol. Process.* **2015**, *29*, 2402–2413. [CrossRef]
9. Chen, H.; Cai, Q. Impact of hillslope vegetation restoration on gully erosion induced sediment yield. *Sci. China Ser. D* **2006**, *49*, 176–192. [CrossRef]
10. Zhou, P. A Study on Rainstorm Causing Soil Erosion in the Loess Plateau. *J. Soil Water Conserv.* **1992**, *6*, 1–5.
11. Schiavon, S.; Zecchin, R. Climate Change 2007: The Physical Science Basis. South African Geographical Journal Being A Record of the Proceedings of the South African. *Geogr. Soc.* **2007**, *92*, 86–87.
12. Gao, H.; Li, Z.; Li, P.; Ren, Z.; Yang, Y.; Wang, J. Paths and prevention of sediment during storm-runoff on the Loess Plateau: Based on the rainstorm of 2017-07-26 in Wuding River. *Sci. Soil Water Conserv.* **2018**, *16*, 66–72.
13. Zhang, J.T. Succession analysis of plant communities in abandoned croplands in the eastern Loess Plateau of China. *J. Arid Environ.* **2005**, *63*, 458–474. [CrossRef]
14. Zhou, J.; Fu, B.J.; Gao, G.Y.; Lu, Y.H.; Liu, Y.; Lu, N.; Wang, S. Effects of precipitation and restoration vegetation on soil erosion in a semi-arid environment in the Loess Plateau, China. *CATENA* **2016**, *137*, 1–11. [CrossRef]
15. Xiao, L.; Yao, K.; Li, P.; Liu, Y.; Chang, E.; Zhang, Y.; Zhu, T. Increased soil aggregate stability is strongly correlated with root and soil properties along a gradient of secondary succession on the Loess Plateau. *Ecol. Eng.* **2020**, *143*, 105671. [CrossRef]
16. Xiao, L.; Liu, G.; Li, P.; Xue, S. Direct and indirect effects of elevated CO_2 and nitrogen addition on soil microbial communities in the rhizosphere of Bothriochloa ischaemum. *J. Soils Sediments* **2019**, *19*, 3679–3687. [CrossRef]
17. Rutigliano, F.A.; D'Ascoli, R.; De Santo, A.V. Soil microbial metabolism and nutrient status in a Mediterranean area as affected by plant cover. *Soil Biol. Biochem.* **2004**, *36*, 1719–1729. [CrossRef]
18. Chang, E.; Li, P.; Xiao, L.; Xu, G.; Zhao, B.; Su, Y.; Feng, Z. The characteristics of root system behavior in vegetation succession in Loess Hilly and Gully Region. *Acta Ecol. Sin.* **2019**, *39*, 2090–2100.
19. Wang, T.; Li, P.; Liu, Y.; Hou, J.M.; Li, Z.B.; Ren, Z.P.; Cheng, S.D.; Zhao, J.H.; Reinhard, H. Experimental investigation of freeze-thaw meltwater compound erosion and runoff energy consumption on loessal slopes. *Catena* **2020**, *185*, 104310. [CrossRef]
20. Wang, B.; Zhang, G.H. Quantifying the Binding and Bonding Effects of Plant Roots on Soil Detachment by Overland Flow in 10 Typical Grasslands on the Loess Plateau. *Soil Sci. Soc. Am. J.* **2017**, *81*, 1567–1576. [CrossRef]
21. Shi, P.; Qin, Y.; Liu, Q.; Zhu, T.; Li, Z.; Li, P.; Ren, Z.; Liu, Y.; Wang, F. Soil respiration and response of carbonsource changes to vegetation restoration in the Loess Plateau, China. *Sci. Total Environ.* **2019**, in press. [CrossRef]

22. Jiao, F.; Wen, Z.-M.; An, S.-S. Changes in soil properties across a chronosequence of vegetation restoration on the Loess Plateau of China. *Catena* **2011**, *86*, 110–116. [CrossRef]
23. Wang, T.; Li, P.; Hou, J.M.; Li, Z.B.; Ren, Z.P.; Cheng, S.D.; Xu, G.C.; Su, Y.Y.; Wang, F.C. Response of the Meltwater Erosion to Runoff Energy Consumption on Loessal Slopes. *Water* **2018**, *10*, 1522. [CrossRef]
24. Martorell, C.; Almanza-Celis, C.A.I.; Pérez-García, E.A.; Sánchez-Ken, J.G. Co-existence in a species-rich grassland: Competition, facilitation and niche structure over a soil depth gradient. *J. Veg. Sci.* **2015**, *26*, 674–685. [CrossRef]
25. Meng, K.; Jiao, J.; Yin, Q.; Wang, N.; Wang, Z.; Li, Y.; Yu, W.; Wei, Y.; Yan, F.; Cao, B. Successional Trajectory Over 10 Years of Vegetation Restoration of Abandoned Slope Croplands in the Hill-Gully Region of the Loess Plateau. *Land Degrad. Dev.* **2016**, *27*, 919–932.
26. Nearing, M.A.; Simanton, J.R.; Norton, L.D.; Bulygin, S.J.; Stone, J. Soil erosion by surface water flow on a stony, semiarid hillslope. *Earth Surf. Process. Landf.* **2015**, *24*, 677–686. [CrossRef]
27. Liu, X.; Wang, F.; Yang, S.; Li, X.; Ma, H.; He, X. Sediment reduction effect of level terrace in the hilly-gully region in the Loess Plateau. *J. Hydraul. Eng.* **2014**, *45*, 793–800.
28. Zhang, G.H.; Luo, R.T.; Ying, C.; Shen, R.C.; Zhang, X.C. Correction factor to dye-measured flow velocity under varying water and sediment discharges. *J. Hydrol.* **2010**, *389*, 205–213. [CrossRef]
29. Zhang, B.Q.; He, C.S.; Burnham, M.; Zhang, L.H. Evaluating the coupling effects of climate aridity and vegetation restoration on soil erosion over the Loess Plateau in China. *Sci. Total Environ.* **2016**, *539*, 436–449. [CrossRef]
30. Peng, X.; Shi, D.; Dong, J.; Wang, S.; Li, Y. Runoff erosion process on different underlying surfaces from disturbed soils in the Three Gorges Reservoir Area, China. *CATENA* **2014**, *123*, 215–224. [CrossRef]
31. Zuazo, V.H.D.; Pleguezuelo, C.R.R. Soil-erosion and runoff prevention by plant covers. A review. *Agron. Sustain. Dev.* **2008**, *28*, 65–86. [CrossRef]
32. Li, P.; Zhao, Z.; Li, Z. Vertical root distribution characters of Robinia pseudoacacia on the Loess Plateau in China. *J. For. Res.* **2004**, *15*, 87–92.
33. Jordan, A.; Zavala, L.M.; Gil, J. Effects of mulching on soil physical properties and runoff under semi-arid conditions in southern Spain. *CATENA* **2010**, *81*, 77–85. [CrossRef]
34. Wang, C.; Ouyang, H.; Maclaren, V.; Yin, Y.; Shao, B.; Boland, A.; Tian, Y. Evaluation of the economic and environmental impact of converting cropland to forest: A case study in Dunhua county, China. *J. Environ. Manag.* **2007**, *85*, 746–756. [CrossRef] [PubMed]
35. Zhi, D.; Hongli, L.I.; Guoyong, R.E.N.; Gang, L.I.U.; Lin, M.A. Study on Soil Amelioration Effect of Planting Grasses in Wind—Sandy Land of Yellow River Floodplain. *Chin. J. Grassl.* **2008**, *30*, 84–87.
36. Bai, Y.F.; Xingguo, H.; Jianguo, W.; Zuozhong, C.; Linghao, L. Ecosystem stability and compensatory effects in the Inner Mongolia grassland. *Nature* **2004**, *431*, 181–184. [CrossRef]
37. Li, Y.Y.; Shao, M.A. Change of soil physical properties under long-term natural vegetation restoration in the Loess Plateau of China. *J. Arid Environ.* **2006**, *64*, 77–96. [CrossRef]
38. Chen, L.; Huang, Z.; Jie, G.; Fu, B.; Huang, Y. The effect of land cover/vegetation on soil water dynamic in the hilly area of the loess plateau, China. *CATENA* **2007**, *70*, 200–208. [CrossRef]
39. Lu, R.; Liu, Y.F.; Jia, C.; Huang, Z.; Liu, Y.; He, H.H.; Liu, B.R.; Wang, Z.J.; Zheng, J.Y.; Wu, G.L. Effects of mosaic-pattern shrub patches on runoff and sediment yield in a wind-water erosion crisscross region. *CATENA* **2019**, *174*, 199–205. [CrossRef]
40. Xin, Z.B.; Xu, J.X.; Zheng, W. Spatiotemporal variations of vegetation cover on the Chinese Loess Plateau (1981–2006): Impacts of climate changes and human activities. *Sci. China* **2008**, *51*, 67–78. [CrossRef]
41. Jiao, J.; Wang, Z.; Wei, Y.; Yuan, S.; Cao, B.; Li, Y. Characteristics of erosion sediment yield with extreme rainstorms in Yanhe Watershed based on field measurement. *Trans. Chin. Soc. Agric. Eng.* **2017**, *33*, 159–167.
42. Sasaki, Y.; Fujii, A.; Asai, K. Soil creep process and its role in debris slide generation—Field measurements on the north side of Tsukuba Mountain in Japan. *Eng. Geol.* **2000**, *56*, 163–183. [CrossRef]
43. Wang, L.; Yao, W.; Shen, Z.; Yang, C. Effects of grass coverage on shallow flow hydraulic parameters and sediment reduction. *Sci. Soil Water Conserv.* **2009**, *7*, 80–83.
44. Govers, G. Relationship between discharge, velocity and flow area for rills eroding loose, non-layered materials. *Earth Surf. Process. Landf.* **2010**, *17*, 515–528. [CrossRef]

45. Quinton, J.N.; Edwards, G.M.; Morgan, R.P.C. The influence of vegetation species and plant properties on runoff and soil erosion: Results from a rainfall simulation study in SE Spain. *Soil Use Manag.* **2010**, *13*, 143–148. [CrossRef]
46. Yan, Y.C.; Xin, X.P.; Xu, X.L.; Wang, X.; Yang, G.X.; Yan, R.R.; Chen, B.R. Quantitative effects of wind erosion on the soil structure and soil nutrients under different vegetation coverage in a semiarid steppe of northern China. *Plant Soil* **2013**, *369*, 585–598. [CrossRef]
47. Zhang, P.; Yao, W.Y.; Liu, G.B.; Xiao, P.Q. Experimental study on soil erosion prediction model of loess slope based on rill morphology. *CATENA* **2019**, *173*, 424–432. [CrossRef]
48. Pimentel, D. Soil Erosion: A Food and Environmental Threat. *Environ. Dev. Sustain.* **2006**, *8*, 119–137. [CrossRef]
49. Zhu, L.Q. Study on soil erosion and its effects on agriculture sustainable development in west Henan province loess hilly areas. *J. Food Agric. Environ.* **2013**, *11*, 906–908.
50. Zhao, B.H.; Li, Z.B.; Li, P.; Cheng, Y.T.; Gao, B. Effects of ecological construction on the transformation of different water types on Loess Plateau, China. *Ecol. Eng.* **2019**, in press. [CrossRef]

Article

Evaluation of Rainfall Erosivity Factor Estimation Using Machine and Deep Learning Models

Jimin Lee [1], Seoro Lee [1], Jiyeong Hong [1], Dongjun Lee [1], Joo Hyun Bae [2], Jae E. Yang [3], Jonggun Kim [1] and Kyoung Jae Lim [1,*]

1 Department of Regional Infrastructure Engineering, Kangwon National University, Chuncheon-si 24341, Korea; jiminlee217@gmail.com (J.L.); seorolee91@gmail.com (S.L.); jiyeong.hong.1@gmail.com (J.H.); dj90lee@gmail.com (D.L.); kimjg23@gmail.com (J.K.)
2 Korea Water Environment Research Institute, Chuncheon-si 24408, Korea; baegop@pusan.ac.kr
3 Department of Biological Environment, Kangwon National University, Chuncheon-si 24341, Korea; yangjay@kangwon.ac.kr
* Correspondence: kjlim@kangwon.ac.kr

Abstract: Rainfall erosivity factor (R-factor) is one of the Universal Soil Loss Equation (USLE) input parameters that account for impacts of rainfall intensity in estimating soil loss. Although many studies have calculated the R-factor using various empirical methods or the USLE method, these methods are time-consuming and require specialized knowledge for the user. The purpose of this study is to develop machine learning models to predict the R-factor faster and more accurately than the previous methods. For this, this study calculated R-factor using 1-min interval rainfall data for improved accuracy of the target value. First, the monthly R-factors were calculated using the USLE calculation method to identify the characteristics of monthly rainfall-runoff induced erosion. In turn, machine learning models were developed to predict the R-factor using the monthly R-factors calculated at 50 sites in Korea as target values. The machine learning algorithms used for this study were Decision Tree, K-Nearest Neighbors, Multilayer Perceptron, Random Forest, Gradient Boosting, eXtreme Gradient Boost, and Deep Neural Network. As a result of the validation with 20% randomly selected data, the Deep Neural Network (DNN), among seven models, showed the greatest prediction accuracy results. The DNN developed in this study was tested for six sites in Korea to demonstrate trained model performance with Nash–Sutcliffe Efficiency (*NSE*) and the coefficient of determination (R^2) of 0.87. This means that our findings show that DNN can be efficiently used to estimate monthly R-factor at the desired site with much less effort and time with total monthly precipitation, maximum daily precipitation, and maximum hourly precipitation data. It will be used not only to calculate soil erosion risk but also to establish soil conservation plans and identify areas at risk of soil disasters by calculating rainfall erosivity factors.

Keywords: rainfall erosivity factor; USLE R; machine learning; Deep Neural Network

Citation: Lee, J.; Lee, S.; Hong, J.; Lee, D.; Bae, J.H.; Yang, J.E.; Kim, J.; Lim, K.J. Evaluation of Rainfall Erosivity Factor Estimation Using Machine and Deep Learning Models. *Water* **2021**, *13*, 382. https://doi.org/10.3390/w13030382

Academic Editor: Su-Chin Chen
Received: 17 December 2020
Accepted: 28 January 2021
Published: 1 February 2021

1. Introduction

Climate change and global warming have been concerns for hydrologists and environmentalists [1–3]. Hydrologic change is expected to be more aggressive as a result of rising global temperature, that consequently results in a change in the current rainfall patterns [4]. Moreover, the Intergovernmental Panel on Climate Change (IPCC) [5] report showed that increasing rainfall events and rainfall intensity are expected to occur in the coming years [6]. Due to the frequent occurrence of greater intensity rainfall events, rainfall erosivity will increase, thus topsoil will become more vulnerable to soil erosion [7]. Soil erosion by extreme intensive rainfall is a significant issue from agricultural and environmental perspectives [8]. A decrease in soil fertility, the inflow of sediment into the river ecosystem, reduction of crop yields, etc., will occur due to soil erosion [9,10]. Therefore, effective best management practices should be implemented for better sustainable management of soil erosion. Furthermore, there is a need for a regional estimate of soil loss to proper decision-making related to appropriate control practice, since erosion occurs diversely over space and time [11].

During the last few decades, various empirical, physically based, and conceptual computer models [12] such as Soil and Water Assessment Tool (SWAT) [13], European Soil Erosion Model (EUROSEM) [14], Water Erosion Prediction Project (WEPP) [15], Sediment Assessment Tool for Effective Erosion Control (SATEEC) [16], Agricultural Non-Point Source Pollution Model (AGNPS) [17], Universal Soil Loss Equation (USLE) [18], Revised Universal Soil Loss Equation (RUSLE) [19] have been developed. Among the models, the USLE model is one of the most popular and widely used empirical erosion models to predict soil erosion because of its easy application and simple structures [20,21]. The USLE model [18] calculates the annual average amount of soil erosion by taking into account soil erosion factors, such as rainfall erosivity factors, soil erodibility factor, slope and length, crop and cover management factor, and conservation practice factor.

The Ministry of Environment of Korea has supported for use of USLE in planning and managing sustainable land management in Korea. To these ends, the USLE has been extensively used to predict soil erosion and evaluate various soil erosion best management practices (BMPs) in Korea. Various efforts have been made for the development of site-specific USLE parameters over the years [22]. Yu et al. [23] suggested monthly soil loss prediction at Daecheong Dam basin in order to improve the limitation of annual soil loss prediction. They found that over 50% of the annual soil loss occurs during July and August. The rainfall erosivity factor (R-factor) is one of the factors to be parameterized in the evaluation of soil loss in the USLE. The R-factor values are affected by the distribution of rainfall amount and its intensity over time and space.

Rainfall erosivity has been widely investigated due to its impact on soil erosion studies worldwide. Rainfall data at intervals of less than 30 min are required to calculate USLE rainfall erosivity factors. The empirical equations related to R-factor based on rainfall data, such as daily, monthly, or yearly, available in various spatial and temporal extents, have been developed using numerous data [24,25].

Sholagberu et al. [26] proposed a regression equation based on annual precipitation because it is difficult to collect sub-hourly rainfall data to calculate maximum 30-min rainfall intensity. Risal et al. [27] proposed a regression equation that can calculate monthly rainfall erosivity factors from 10-min interval rainfall data. In addition, the Web ERosivity Module (WERM), web-based software that can calculate rainfall erosivity factor, was developed and made available at http://www.envsys.co.kr/~werm. In the study by Risal et al. [27] on the R-factor calculation for South Korea, 10-min interval rainfall data, which cannot give the exact estimate of maximum 30-min rainfall intensity, was used. The Korea Meteorological Administration (KMA) provides 1-min rainfall data for over 50 weather stations in Korea. Estimation of R-factor values for South Korea using a recent rainfall dataset is needed for present and future uses because climate change causes changes in precipitation pattern and intensity to some degrees. However, the process of calculation of R-factor from rainfall data is time-consuming, although the Web ERosivity Module (WERM) software can calculate rainfall erosivity factor [27]. Furthermore, the radar rainfall dataset can be used to calculate spatial USLE R raster values using Web Erosivity Model-Spatial (WERM-S) [28]. These days, Machine Learning/Deep Learning (ML/DL) has been suggested as an alternative to predict and simulate natural phenomena [29]. Thus, ML/DL has been used for the prediction of flow, water quality, and ecosystem services [30–34]. These studies have implied that ML/DL is an efficient and effective way to calculate R factor values using recent rainfall time-series data provided by the KMA.

The objective of this study is to develop machine learning models to predict the monthly R-factor values, which are comparable with those calculated by the USLE method. For this aim, we calculated R-factor using 1-min interval rainfall data to estimate the maximum 30-min rainfall intensity of the target values, which is monthly R factor values at 50 stations in S. Korea. In the previous study by Risal et al. [27], the R-factor values for South Korea were calculated using 10-min interval rainfall data, which cannot give an exact estimate of maximum 30-min rainfall intensity. The procedure used in this study is shown in Figure 1.

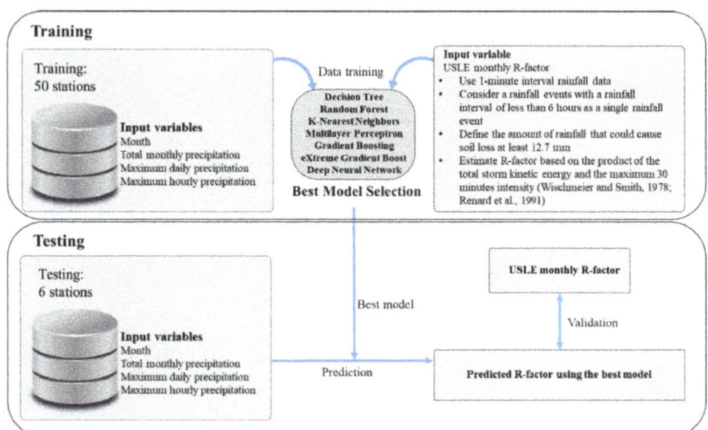

Figure 1. Study procedures.

2. Methods

2.1. Study Area

Figure 2 shows the location of weather stations where 1-min rainfall data have been observed over the years in South Korea. The fifty points marked in circles are observational stations that provide data used for training and validation to create machine learning models predicting rainfall erosivity factors, while six stations marked in green on the right map—Chuncheon, Gangneung, Suwon, Jeonju, Busan, and Namhae—represent the stations for the final evaluation of the results predicted by machine learning models selected through validation. Thiessen network presented using red lines of the map on the right shows a range of the weather environment around the weather stations.

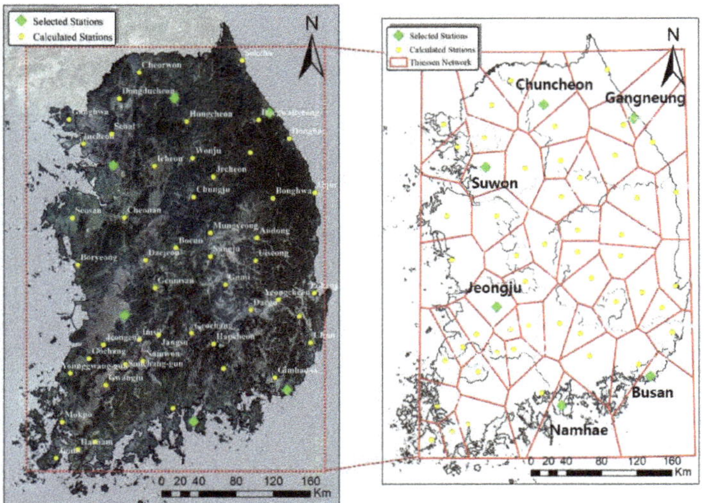

Figure 2. Weather stations in the study area.

2.2. Monthly Rainfall Erosivity Calculation

Monthly rainfall erosivity (R-factor) was calculated for each of the 50 weather stations in South Korea from 2013 to 2019. It was calculated based on the equation given in the USLE users' manual in order to calculate the R-factor value [18]. According to Wischmeier and Smith [18], a rainfall interval of fewer than six hours is considered a single rainfall event. In addition, the least amount of rainfall that could cause soil loss is at least 12.7 mm or more as specified in the USLE users' manual [35].

However, if the rainfall is 6.25 mm during 15 min, it is defined as a rainfall event that can cause soil loss. The calculations for each rainfall event are as follows.

$$\text{IF I} \leq 76\ \text{mm/hr} \rightarrow \text{e} = 0.119 + 0.0873\log_{10}\text{I} \quad (1)$$

$$\text{IF I} > 76\ \text{mm/hr} \rightarrow \text{e} = 0.283 \quad (2)$$

$$\text{E} = \Sigma\,(\text{e} \times \text{P}) \quad (3)$$

$$\text{R} = \text{E} \times \text{I}_{30max} \quad (4)$$

where I (mm h^{-1}) is the intensity of rainfall, e (MJ mm ha^{-1}) is unit rainfall energy, P (mm) is the rainfall volume during a given time period, E (MJ ha^{-1}) is the total storm kinetic energy, I_{30max} (mm h^{-1}) is the maximum 30-min intensity in the erosive event, and R (MJ mm ha^{-1} h^{-1}) is the rainfall erosivity factor. In this study, the monthly R-factor (MJ mm ha^{-1} h^{-1} $month^{-1}$) was estimated by calculating monthly E and multiplying it by I_{30max}. In addition, the monthly rainfall erosivity factor was calculated using Equations (1)–(4) [18] using the 1-min precipitation data provided on the Meteorological Data Open Portal site of the KMA (Korea Meteorological Administration).

2.3. Machine Learning Models

Machine learning can be largely divided into supervised learning, unsupervised learning, and reinforcement learning [36,37]. In this study, supervised learning algorithms were used. A total of seven methods (Table 1) were used to build models to estimate R-factor. Table 1 shows the information on machine learning models utilized in this study.

Table 1. Description of machine learning models.

Machine Learning Models	Module	Function	Notation
Decision Tree	Sklearn.tree	DecisionTreeRegressor	DT
Random Forest	Sklearn.ensemble	RandomForestRegressor	RF
K-Nearest Neighbors	Sklearn.neighbors	KNeighborsRegressor	KN
Gradient Boosting	Sklearn.ensemble	GradientBoostingRegressor	GB
eXtreme Gradient Boost	xgboost.xgb	XGBRegressor	XGB
Multilayer Perceptron	Sklearn, neural_network	MLPRegressor	MLP
Deep Neural Network	Keras.models.Sequential	Dense, Dropout	DNN

Decision Tree, Random Forest, K-Nearest Neighbors, Gradient Boosting, and Multilayer Perceptron imported and used the related functions from the Scikit-learn module (Version: 0.21.3), while eXtreme gradient boost was taken from the XGboost library (License: Apache-2.0) and used the regression functions. Deep Neural Network is trained by taking Dense and Dropout functions from "Keras.models.Sequential" module of TensorFlow (Version: 2.0.0) and Keras (Version: 2.3.1) framework. In this study, the standardization method was used during the pre-process for raw data. Moreover, the "StandardScaler" function, a preprocessing library of Scikit-learn, was used.

2.3.1. Decision Tree

The Decision Tree (DT) model uses hierarchical structures to find structural patterns in data for constructing decision-making rules to estimate both dependent and independent variables [38]. It first learns by continuing the yes/no question to reach a decision [39]. In this study, the DT model in the Scikit-learn supports only the pre-pruning. Entropy was based on classification and 2 for min_samples_split was given in Table 2.

Table 2. Critical hyperparameters in machine learning models.

Machine Learning Models	Hyperparameter
Decision Tree	criterion = "entropy", min_samples_split = 2
Random Forest	n_estimators = 52, min_samples_leaf = 1
K-Nearest Neighbors	n_neighbors = 3, weights = 'uniform', metric = 'minkowski'
Gradient Boosting	learning_rate = 0.01, min_samples_split = 4
eXtreme Gradient Boost	Booster = 'gbtree', max_depth = 10
Multilayer Perceptron	hidden_layer_sizes = (50,50,50), activation = "relu", solver = 'adam'
Deep Neural Network	kernel_initializer = 'normal', activation = "relu"

A model hyperparameter is a value that is set directly by the user when modeling. Table 2 shows the hyperparameter settings of the regressors used in this study.

2.3.2. Random Forest

Random Forest (RF) is a decision tree algorithm developed by Breiman [40] that applies the Bagging algorithm among the Classification and Registration Tree (CART) algorithm and the ensemble technique. RF creates multiple training data from a single dataset and performs multiple training. It generates several decision trees and improves predictability by integrating the decision trees [41]. Detailed tuning of the hyperparameter in RF is easier than an artificial neural network and support vector regression [42].

In this study, the hyperparameters in the RF are the following: 52 for n_estimators, and 1 for min_samples_leaf.

2.3.3. K-Nearest Neighbors

K-Nearest Neighbors (KNN) is a non-parametric method which can be used for regression and classification [43]. In this study, KNN was used for regression. KNN is an algorithm that finds the nearest "K" neighborhood from the new data in training data and uses the most frequent class of these neighbors as a predicted value [44]. In this study, the number of the nearest neighbors in KNN's hyperparameter was set as 3. The weights were calculated using a simple mean, and the distance was calculated by the Minkowski method [45].

2.3.4. Gradient Boosting and eXtreme Gradient Boost

Gradient Boosting (GB) is an ensemble algorithm belonging to the boosting family that can perform classification and regression analysis [46,47]. In GB, the gradient reveals the weaknesses of the model that have been learned so far, whereas other machine learning models (e.g., DT and RF) focus on it to boost performance [48]. In other words, the advantage of gradient boosting is that the other loss functions can be used as much as possible. Therefore, the parameters that minimize the loss function that quantifies errors in the predictive model can found for better R-factor prediction. In this study, the hyperparameters in the GB are the following: 0.01 for learning_rate, 4 for min_samples_split.

The eXtreme Gradient Boost (XGB) model is faster in training and classifying data than GB using parallel processing. It also has a regulatory function that prevents overfitting, which results in better predictive performance [49]. XGB is trained only by important features so that it calculates faster and performs better when compared to other algorithms [50,51]. The hyperparameters in the XGB are the following: gbtree for booster, and 10 for max_depth.

2.3.5. Multilayer Perceptron

Multilayer Perceptron (MLP) is a neural network that uses a back-propagation algorithm to learn weights [52]. MLP network consists of an input layer, a hidden layer, and an output layer (the R-factor). In this study, the hidden layer consisted of 50 nodes.

The hidden layers receive the signals from the nodes of the input layer and transform them into signals that are sent to all output nodes, transforming them into the last layer of outputs [53]. The output is used as input units in the subsequent layer. The connection between units in subsequent layers has a weight. MLP learns its weights by using the backpropagation algorithm [52].

2.3.6. Deep Neural Network

Deep Neural Network (DNN) is a predictive model that uses multiple layers of computational nodes for extracting features of existing data and depending on patterns learn to predict the outcome of some future input data [54]. The invention of the new optimizers enables us to train a large number of hyperparameters more quickly. In addition, the regularization and dropout allow us to avoid overfitting. The package used to build DNN in this study was TensorFlow developed by Google. In this study, the DNN model structure consisted of 7 dense layers and 1 dropout (Figure 3). Additional details about DNN can be found in Hinton et al. [55].

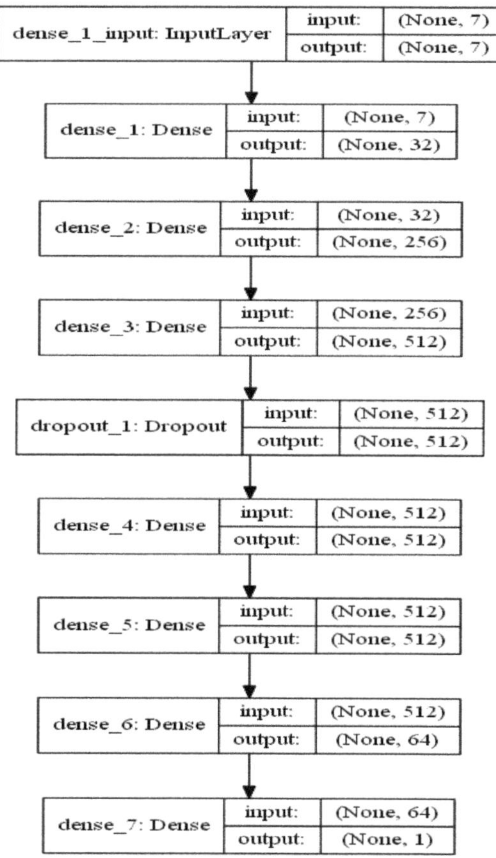

Figure 3. Illustration of the proposed Deep Neural Network (DNN) for rainfall erosivity (R-factor) prediction.

2.4. Input Data and Validation Method

Input data were compiled as shown in Table 3 to develop machine learning models to assess the R factor. The corresponding month from Jan. to Dec. was altered to numerical values, because rainfall patterns and their intensity may vary every month over space. Total monthly precipitation, maximum daily precipitation, and maximum hourly precipitation were calculated monthly and selected as the independent variables. The data can be easily downloaded in the form of monthly and hourly data among the Automated Synoptic Obstruction System (ASOS) data from the Korea Meteorological Administration (KMA)'s weather data opening portal site and organized as input data.

Table 3. The input data for machine learning models.

Description			Count	Mean	std	Min	25%	50%	75%	Max
Input variable	month	month (1~12)	4087	6.49	3.45	1	3	6	9	12
	m_sum_r	the total amount of monthly precipitation	4087	96.45	97.01	0	30.90	66.20	126.15	1009.20
	d_max_r	maximum daily precipitation	4087	39.39	38.10	0	14.50	27.10	51.35	384.30
	h_max_r	maximum hourly precipitation	4087	11.84	12.69	0	4.00	7.50	15.50	197.50
Output variable	R-factor	R-factor	4087	419.10	1216.79	0	15.99	77.84	326.24	43,586.61

The monthly R-factors data in the manner presented in the USLE for the 50 selected sites from 2013 to 2019 were designated as target values, and as the features are given in Table 4, month (1–12), total monthly precipitation, maximum daily precipitation, and maximum hourly precipitation were designated as the features. Among the data, 80% of randomly selected data were trained, the model was created, and then the remaining 20% of data were used for the validation of the trained model.

To assess the performance of each machine learning model, Nash–Sutcliffe efficiency (*NSE*), Root Mean Squared Errors (*RMSE*), the Mean Absolute Error (*MAE*), and coefficient of determination (R^2) was used. Numerous studies indicated the appropriateness of these measures to assess the accuracy of hydrological models [56–58]. *NSE*, *RMSE*, *MAE*, and R^2 for evaluation of the model accuracy can be calculated from Equations (5)–(8).

$$NSE = 1 - \frac{\sum (O_t - M_t)^2}{\sum (O_t - \overline{O_t})^2} \tag{5}$$

$$RMSE = \sqrt{\frac{\sum (O_t - M_t)^2}{n}} \tag{6}$$

$$MAE = \frac{1}{n}\sum \left| M_t - O_t \right| \tag{7}$$

$$R^2 = \frac{\left[\sum (O_t - \overline{O_t})(M_t - \overline{M_t})\right]^2}{\sum (O_t - \overline{O_t})^2 \sum (M_t - \overline{M_t})^2} \tag{8}$$

where O_t is the actual value of t, $\overline{O_t}$ is the mean of the actual value, M_t is the estimated value of t, $\overline{M_t}$ is the mean of the estimated value, and n is the total number of data.

Table 4. Monthly R-factor calculated by the Universal Soil Loss Equation (USLE).

Station Number	Station Name	R-Factor ($MJ\ mm\ ha^{-1}\ h^{-1}\ month^{-1}$)												R-Factor ($MJ\ mm\ ha^{-1}\ h^{-1}\ year^{-1}$)
		January	February	March	April	May	June	July	August	September	October	November	December	Annual
90	Sokcho	21	29	32	95	47	159	1039	2860	368	494	507	44	5694
95	Cheolwon	2	37	30	119	257	235	5867	2769	403	239	65	24	10,046
98	Dongducheon	2	69	31	91	287	455	3031	1364	424	228	47	24	6053
100	Daegwallyeong	4	19	20	79	577	194	1472	1669	453	237	40	8	4772
106	Donghae	17	29	34	207	27	157	592	1317	469	1461	128	8	4447
108	Seoul	0	31	37	95	284	266	2813	988	191	90	50	16	4861
112	Incheon	3	50	69	83	224	192	2193	897	406	480	88	27	4712
114	Wonju	2	22	31	83	284	699	2654	999	303	65	40	8	5189
127	Chungju	4	36	22	79	171	270	2075	1240	909	117	38	17	4978
129	Seosan	7	72	50	135	147	562	747	635	330	146	127	42	3000
130	Uljin	86	16	47	292	24	227	590	470	360	2816	143	30	5100
133	Daejeon	8	47	50	182	72	602	1658	1293	494	181	96	24	4707
136	Andong	1	32	53	101	48	325	1142	808	368	176	33	14	3100
137	Sangju	6	18	67	105	45	361	1098	1143	420	273	51	19	3605
138	Pohang	7	46	106	139	40	233	417	1051	910	1478	51	23	4500
143	Daegu	1	8	57	80	67	313	548	1322	238	340	26	12	3013
152	Ulsan	15	36	122	141	154	287	751	1499	727	1709	77	74	5591
156	Gwangju	8	59	92	156	74	927	1249	2458	703	361	99	49	6236
165	Mokpo	15	112	117	227	177	630	1023	944	2094	493	85	127	6044
172	Gochang	12	24	137	151	77	399	1768	2235	614	273	77	21	5788
175	Jindo	18	43	231	559	511	598	738	1323	799	636	113	36	5606
201	Ganghwa	1	26	35	60	193	59	1922	1255	648	654	48	20	4921
203	Icheon	2	34	103	83	211	207	2284	1068	450	162	45	24	4673
212	Hongcheon	1	11	23	81	461	162	2220	934	223	51	29	8	4204
217	Jeongseon	1	20	18	75	117	126	2165	654	355	101	36	16	3686
221	Jecheon	3	26	21	90	158	265	1616	1162	405	80	43	12	3881
226	Boeun	8	19	42	106	62	482	2016	1102	583	163	77	15	4675
232	Cheonan	2	17	21	86	106	248	3408	1002	249	110	68	15	5333
235	Boryeong	5	73	47	142	127	322	878	849	1014	184	149	29	3820
238	Guemsan	5	17	52	154	48	483	1126	1059	447	148	37	17	3591
244	Imsil	3	9	83	106	67	369	2329	1416	632	224	44	16	5297
245	Jeongeup	11	18	106	160	265	318	1679	1930	521	207	46	27	5287
247	Namwon	8	19	78	159	52	704	2988	2304	479	586	88	49	7512
248	Jangsu	5	34	85	151	80	246	1997	1812	715	308	53	30	5516
251	Gochanggoon	7	14	127	191	69	352	1448	2066	521	175	37	23	5029
252	Younggwang	7	15	130	178	114	292	994	2008	596	491	63	39	4928
253	Ginhae	7	43	220	200	339	385	1036	1600	1216	734	44	48	5872
254	Soonchang	4	14	93	194	89	456	1724	1304	629	363	58	16	4945
259	Gangjin	10	34	204	344	425	666	1156	9781	903	444	187	18	14,170
261	Haenam	10	15	223	206	177	595	965	1142	650	1250	83	91	5406
263	Uiryoong	4	24	121	184	156	334	629	1961	805	643	39	36	4936
266	Gwangyang	8	76	120	218	591	686	827	2488	2195	555	86	73	7924
271	Bonghwa	0	9	36	86	95	415	1154	706	327	98	28	13	2968
273	Mungyeong	7	18	54	139	102	331	1724	742	529	180	61	21	3908
278	Uiseong	1	6	69	101	74	220	632	647	326	106	31	7	2220
279	Gumi	3	12	68	103	95	380	1296	1442	554	364	32	14	4363
281	Yeongcheon	2	11	81	162	52	316	1259	1082	409	230	28	12	3643
283	Gyeongju	3	11	53	101	55	125	573	759	681	686	21	13	3081
284	Geochang	3	17	54	105	60	434	1184	1491	619	2246	33	55	6299
285	Hapcheon	2	22	57	173	105	700	1114	1646	810	676	47	24	5378

3. Results and Discussion

3.1. USLE R-Factor

For the selected 50 sites, monthly rainfall erosivity factors for each year from 2013 to 2019 were calculated, and the average monthly rainfall erosivity factors for seven years were obtained. Then, the seven-year average monthly rainfall erosivity, R-factor, was generated and shown in Table 1. Moreover, to give a comprehensive look at the degree of rainfall patterns by site, the average annual rainfall erosivity factor for each site is also presented in Table 4.

In this study, rainfall erosivity factor maps were generated to examine patterns of monthly R-factor calculated by USLE using rainfall data from 50 selected sites for evaluation. The R-factor distributions were mapped reflecting the geographical characteristics in South Korea (Figure 4). The high R-factor distribution in all regions during the summer months of July and August can be confirmed.

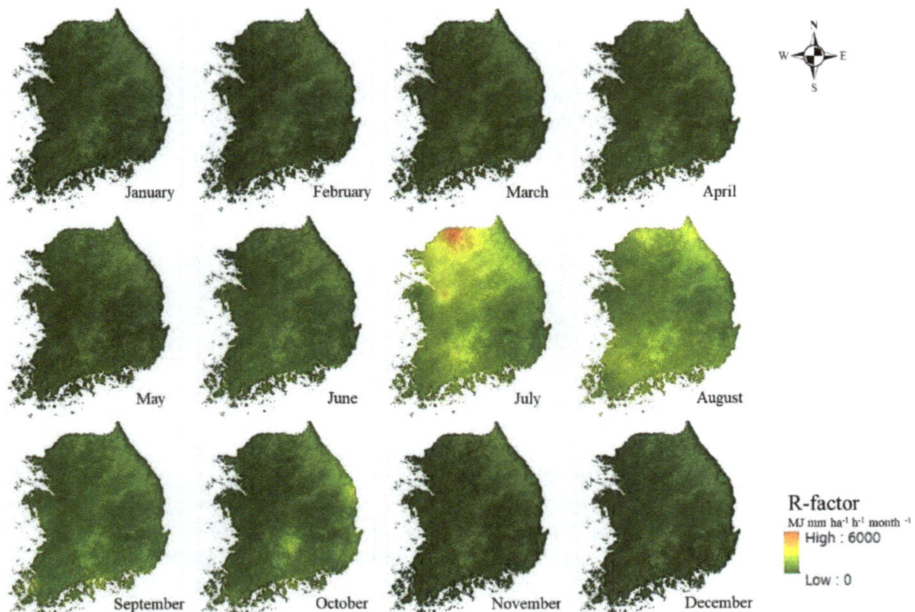

Figure 4. Spatial distribution of monthly R-factor calculated by USLE, using rainfall data from 50 weather stations for the period 2013–2019.

The monthly R-factors for two months from July to August contribute more than 50% of the total average annual R factor value of Korea. The rainfall occurs mainly in the wet season and the likelihood of erosion is very high compared to the dry season. In such a case, using the average annual R-factor value can give a misleading amount of soil erosion. For these reasons, the monthly R-factor would be helpful in analyzing the impact of the rainfall on soil erosion rather than the average annual R-factor.

3.2. Validation of Machine Learning Models

Table 5 shows the prediction accuracy results (*NSE*, *RMSE*, *MAE*, R^2) of seven machine learning models, by comparing the predicted R-factor. The results from the Deep Neural Network (DNN) showed the highest prediction accuracy with *NSE* 0.823, *RMSE* 398.623 MJ mm ha^{-1} h^{-1} month^{-1}, *MAE* 144.442 MJ mm ha^{-1} h^{-1} month^{-1}, and R^2 0.840.

Table 5. Prediction accuracy results of seven machine learning models.

Machine Learning Models	*NSE*	*RMSE* (MJ mm ha-$^{-1}$ h^{-1} month^{-1})	*MAE* (MJ mm ha-$^{-1}$ h^{-1} month^{-1})	R^2
Decision Tree	0.518	657.672	217.408	0.626
Multilayer Perceptron	0.732	490.055	158.847	0.783
K-Nearest Neighbors	0.817	405.327	149.923	0.794
Random Forest	0.800	423.345	148.147	0.799
Gradient Boosting	0.702	516.956	161.259	0.722
eXtreme Gradient Boost	0.791	433.230	159.275	0.788
Deep Neural Network	0.823	398.623	144.442	0.840

When comparing the results of DNN and the other machine learning models (Decision Tree, Random Forest, K-Nearest Neighbors, Multilayer Perceptron, Gradient Boosting, and eXtreme Gradient Boost), we can see that DNN provided more accurate prediction results over other machine learning algorithms. Moreover, the highest value of *NSE*, *RMSE*, *MAE*, and R^2 was found when the DNN was employed for the prediction R-factor values.

DNN had been proven for its good performance in a number of studies about the environment. In the study conducted by Liu et al. [59], the DNN showed better results, compared with results obtained by other machine learning algorithms, in predicting streamflow at Yangtze River. Nhu et al. [60] reported the DNN has the most impactful method in machine learning for the prediction of landslide susceptibility compared to other machine learning such as decision trees and logistic regression. In the study by Lee et al. [61], a DNN-based model showed good performance as a result of evaluating the heavy rain damage prediction compared to the recurrent neural network (RNN) in deep learning. Sit et al. [62] reported the DNN can be helpful in time-series forecasting for flood and support improving existing models. For these reasons, it has been shown that DNN performs better in various studies.

In this study, the second best-predicted model is the K-Nearest Neighbors (KNN). The result from the KNN model showed prediction accuracy with *NSE* 0.817, *RMSE* 405.327 MJ mm ha^{-1} h^{-1} month^{-1}, *MAE* 149.923 MJ mm ha^{-1} h^{-1} month^{-1}, and R^2 0.794 which indicates that the KNN is the most effective, aside from DNN, in predicting R-factor. According to Kim et al. [63], KNN has good performance results in predicting the influent flow rate and four water qualities like chemical oxygen demand (COD), suspended solids (SS), total nitrogen (TN), and total phosphorus (TP) at a wastewater treatment plant.

On the other hand, Decision Tree has prediction accuracy, with *NSE* 0.518, *RMSE* 657.672 MJ mm ha^{-1} h^{-1} month^{-1}, *MAE* 217.408, MJ mm ha^{-1} month-$^{-1}$, and R^2 0.626. This means that Decision Tree is less predictable than other machine learning models (Random Forest, K-Nearest Neighbors, Multilayer Perceptron, Gradient Boosting, eXtreme Gradient Boost, and Deep Neural Network). Hong et al. [37] also reported Decision Tree has less accuracy for the prediction of dam inflow compared to other machine learning models (Decision tree, Multilayer perceptron, Random forest, Gradient boosting, Convolutional neural network, and Recurrent neural network-long short-term memory).

Figure 5 shows the scattering graphs of the R-factors predicted by the seven machine learning models and calculated by the USLE method. All machine learning results represent a rather distracting correlation with less agreement. However, in Figure 5h, the Deep Neural Network algorithms predicted USLE R values calculated using the method suggested by USLE users' manual with higher accuracy, *NSE* value of 0.823.

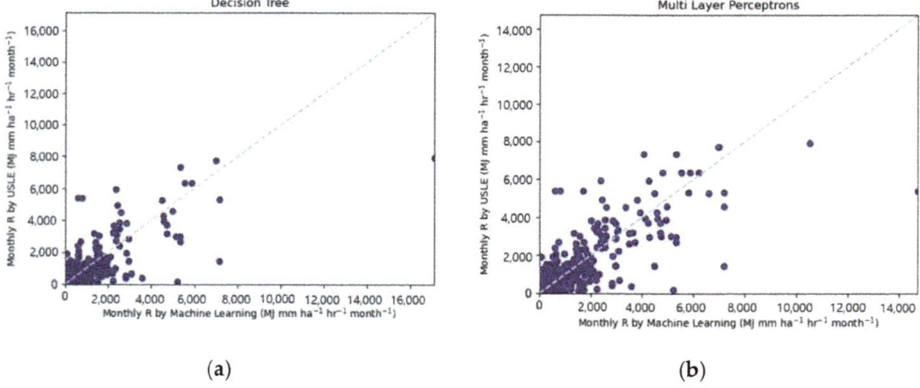

(a) (b)

Figure 5. *Cont.*

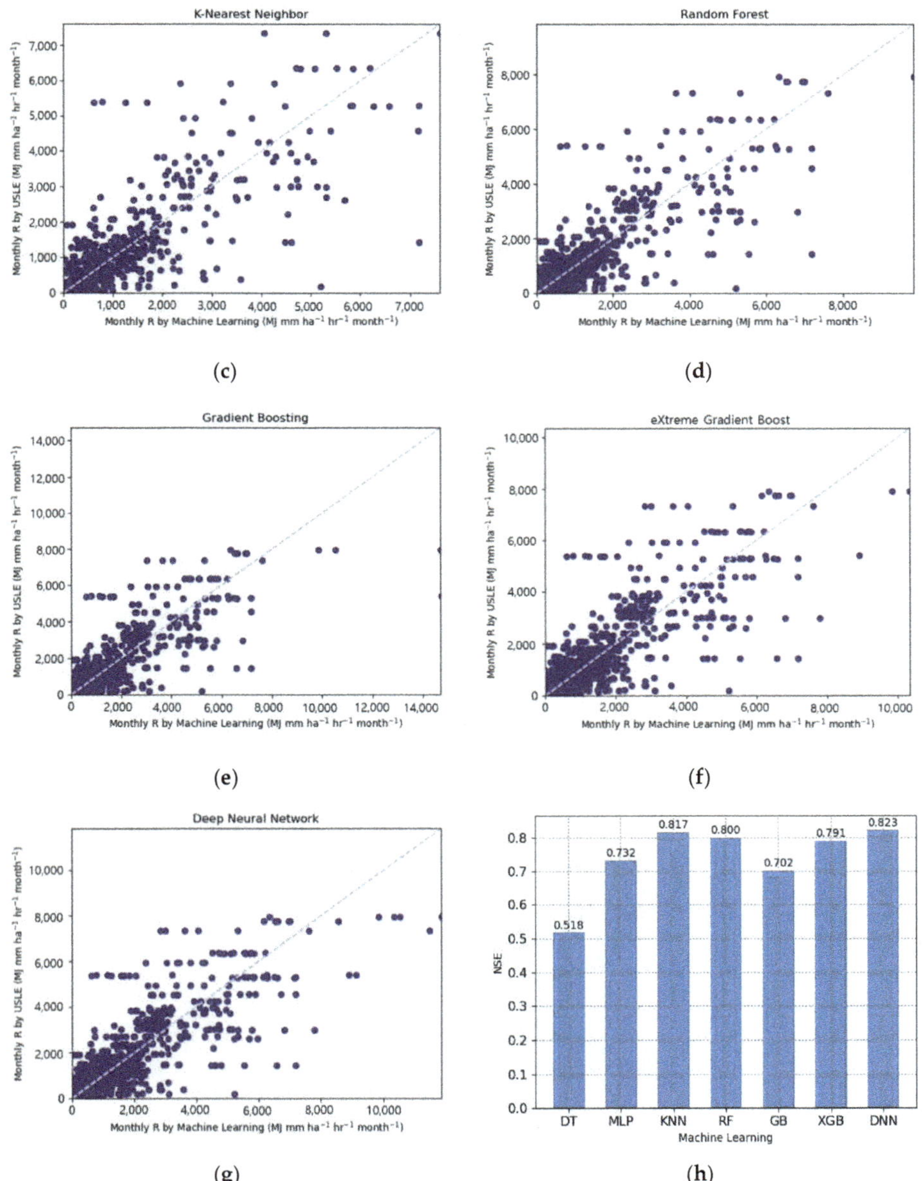

Figure 5. Comparison of (**a**) Decision Tree, (**b**) Multilayer Perceptron, (**c**) K-Nearest Neighbor, (**d**) Random Forest, (**e**) Gradient Boosting, (**f**) eXtreme Gradient Boost, and (**g**) Deep Neural Network calculated R-factor with validation data, and (**h**) comparison of machine learning accuracy.

Among the data, 80% of randomly selected data were trained, the model was created, and then the remaining 20% of data were used for the validation of the trained model. To prevent overfitting, the K-fold cross-validation was implemented for R^2 as shown in Table 6. As a result of the five attempts of K-fold cross-validation, the DNN showed the best results with an average R^2 of 0.783.

Table 6. K-fold cross validation results of seven machine learning models.

Fold	Coefficient of Determination (R^2)						
	Decision Tree	Multi-Layer Perceptron	K-Nearest Neighbors	Random Forest	Gradient Boosting	eXtreme Gradient Boost	Deep Neural Network
1	0.631	0.781	0.818	0.817	0.730	0.801	0.821
2	0.598	0.806	0.705	0.686	0.648	0.737	0.733
3	0.544	0.759	0.705	0.682	0.635	0.717	0.759
4	0.592	0.714	0.780	0.774	0.653	0.644	0.762
5	0.626	0.783	0.794	0.799	0.722	0.788	0.840
Average	0.598	0.769	0.760	0.752	0.678	0.737	0.783

Figure 6 shows the results of the prediction of the five machine learning models (i.e., Multilayer Perceptron, K-Nearest Neighbor, Random Forest, eXtreme Gradient Boost, and Deep Neural Network) at six sites for the testing of the selected models, as well as the time series comparison graph for 2013–2019 of the monthly R-factor values calculated on the USLE basis. At most sites, it showed that the time series trend fits well with a pattern similar to the USLE calculation value. In particular, looking at the distribution in Figure 6b Gangneung, the value of 9303 MJ mm ha^{-1} h^{-1} month^{-1} in October 2019, which represented the peak value of the rainfall erosivity factor, was generally well predicted by all machine learning models. Among the models, the result of the Random Forest model estimated a similar value with 8133 MJ mm ha^{-1} h^{-1} month^{-1}.

On the other hand, among six sites, the time series distribution values of the model prediction result in Busan showed a slightly different pattern from the USLE calculation R-factor. In particular, the result was overestimated as the values of 8241 MJ mm ha^{-1} h^{-1} month^{-1} in August 2014, and Multilayer Perceptron was almost twice overestimated at 16,725 MJ mm ha^{-1} h^{-1} month^{-1}.

However, the Random Forest (8188 MJ mm ha^{-1} h^{-1} month^{-1}) and eXtreme Gradient boost (8395 MJ mm ha^{-1} h^{-1} month^{-1}) algorithms were predicting very similar values. Therefore, the machine learning results could be seen as good at predicting the peak value.

A comparison of the machine learning model accuracies of *NSE* and R^2 of the test (validation) results at the six sites is shown in Figure 7. All five models had a coefficient of determination of 0.69 or higher, and the simulated values of the USLE method calculation and machine learning models showed high accuracy prediction. However, compared to Deep Neural Network, the *NSE* results of the four models (Multilayer Perceptron, K-Nearest Neighbor, Random Forest, eXtreme Gradient Boost) were less than 0.58, and the Deep Neural Network model showed 0.87 in both *NSE* and R^2. Therefore, the monthly average value of the R-factor, predicted by the DNN would be a good candidate algorithm for USLE R factor prediction (Table 5 and Figure 7).

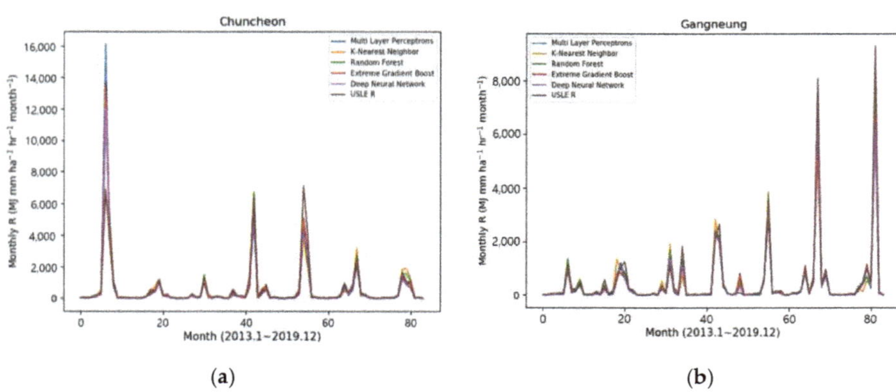

Figure 6. *Cont.*

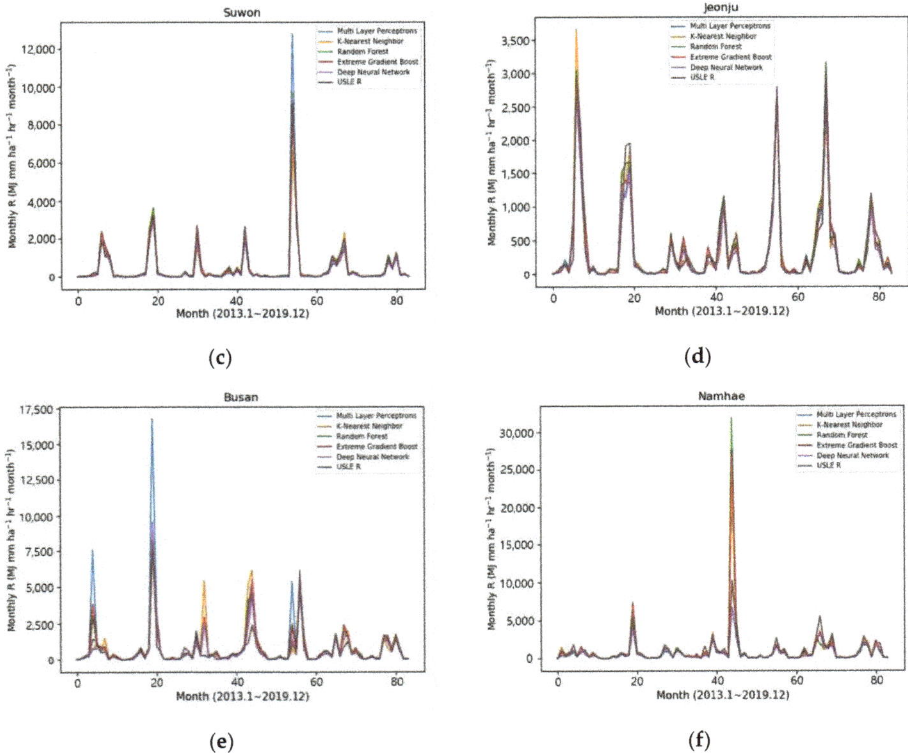

Figure 6. The comparisons of forecasting results of R-factor using machine learning in (**a**) Chuncheon, (**b**) Gangneung, (**c**) Suwon, (**d**) Jeonju, (**e**) Busan, and (**f**) Namhae.

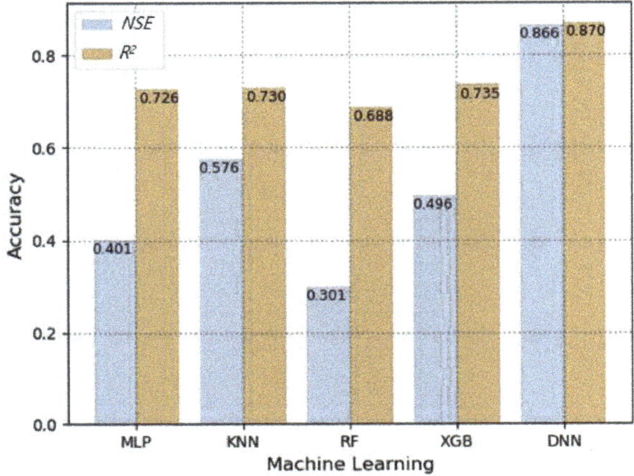

Figure 7. Comparison of prediction accuracy results by machine learning models in test sites.

Table 7 shows average monthly rainfall erosivity factor values at the six sites for testing, Chuncheon, Gangneung, Suwon, Jeonju, Busan, and Namhae, along with the USLE calculation and Deep Neural Network (DNN) prediction.

Table 7. Monthly R-factor calculated (C) by the previous method and predicted (M) by Deep Neural Network.

Station Number	Station Name	Method	R-Factor (MJ mm ha^{-1} h^{-1} month^{-1})												R-Factor (MJ mm ha^{-1} h^{-1} year^{-1})	*NSE*
			January	February	March	April	May	June	July	August	September	October	November	December	Annual	
101	Chuncheon	C	2	99	27	110	195	320	3466	1844	355	182	45	24	6670	0.814
		M	5	59	26	84	166	356	3543	1608	323	154	31	16	6372	
105	Gangneung	C	25	36	31	138	150	113	742	2476	390	1598	325	16	6040	0.874
		M	57	24	38	84	154	100	774	2189	287	1129	207	14	5055	
119	Suwon	C	3	29	80	115	278	165	2944	1463	388	107	84	27	5683	0.981
		M	5	36	56	88	210	156	2708	1352	312	85	52	19	5076	
146	Jeonju	C	7	17	104	105	108	526	1315	1511	311	240	67	21	4330	0.911
		M	7	16	80	103	95	581	1086	1261	310	153	50	19	3760	
159	Busan	C	32	79	199	408	472	781	1021	1729	1764	616	266	113	7479	0.883
		M	22	56	148	243	317	639	860	2205	2514	458	184	91	7736	
295	Namhae	C	19	407	366	946	892	1016	1608	1822	2169	1634	151	131	11,159	0.584
		M	14	161	198	748	607	867	1201	1164	1633	1010	85	110	7798	

Among average annual vales, the results for Busan showed a good performance with the Deep Neural Network (DNN) resulting in the average annual value of the rainfall erosivity factor of 257 MJ mm ha^{-1} h^{-1} year^{-1} difference over the USLE calculation result. In the case of Chuncheon, DNN also showed a good performance with an average annual rainfall erosivity factor difference of 298 MJ mm ha^{-1} h^{-1} year^{-1} difference over the USLE calculation result. On the other hand, the USLE calculation results for Namhae showed an average annual value of the rainfall erosivity factor difference of 3361 MJ mm ha^{-1} h^{-1} year^{-1} greater than the DNN result.

This is because, in the case of Namhae, the rainfall tendency lasted for a long period in the dry season from February to June compared to the other testing sites like Chuncheon, Gangneung, Suwon, Jeonju, and Busan. Moreover, the monthly R-factor calculation of Namhae in dry seasons was two to four times more than other testing sites. In particular, the monthly R-factor for February in Namhae figure being about five times higher than the monthly R-factor in Busan. This means that if the single set of learning data has a huge deviation or variation from other sets, it may result in the uncertainty of the entire result data. Therefore, the monthly R-factor of Namhae in the dry season from is containing uncertainty. In the future study, when predicting the R-factor of the Namhae, DNN model analysis will be implemented in consideration of rainfall trends by supplement the historical rainfall data.

R-factor can be calculated by machine learning algorithms with high accuracy and time benefit. The spatio-temporal calculation of the rainfall erosivity factor using machine learning techniques can be utilized for the estimation of the soil erosion due to rainfall at the target value. The DNN will be incorporated into the WERM website in the near future after further validation.

4. Conclusions

The main objective of this study is to develop machine learning models to predict monthly R-factor values which are comparable with those calculated by the USLE method. For this, we calculated R-factor using 1-min interval rainfall data for improved accuracy of the target value. The machine learning and deep learning models used in this study were Decision Tree, K-Nearest Neighbors, Multilayer Perceptron, Random forest, Gradient boosting, eXtreme Gradient boost, and Deep Neural Network. All of the models except Decision Tress showed *NSE* and R^2 values of 0.7 or more, which means that most of the machine learning models showed high accuracy for predicting the R-factor. Among these, the Deep Neural Network (DNN) showed the best performance. As a result of the validation with 20% randomly selected data, DNN, among the seven models, showed the greatest prediction accuracy results with *NSE* 0.823, *RMSE* 398.623 MJ mm ha^{-1} h^{-1} month^{-1}, *MAE* 144.442 MJ mm ha^{-1} h^{-1} month^{-1}, and R^2 0.840. Furthermore, the DNN developed in this study was tested for six sites (Chuncheon, Gangneung, Suwon, Jeonju, Busan, and Namhae) in S. Korea to demonstrate a trained model performance with *NSE* and R^2 of both 0.87. As a result of the comparative analysis of R-factor prediction through various models, the DNN was proven to be the best model for R-factor prediction in S. Korea with readily available rainfall data. The model accuracy and simplicity of machine learning and deep learning models insist that the models could replace traditional ways of calculating/estimating USLE R-factor values.

We found that the maximum 30 min intensity derived from 1-min interval rainfall data in this study is more accurate than that estimated from previous research. These methods can provide more accurate monthly, yearly, and event-based USLE R-factor for the entire period. Moreover, if the user has input data (month, the total amount of monthly precipitation, maximum daily precipitation, maximum hourly precipitation) as described in Table 3, the monthly R-factor can be easily calculated for the 50 specific stations in S. Korea by using the machine and deep learning models. Since the updated R-factor in this study reflected the recent rainfall data, which have high variability, it can improve the accuracy of the usage of the previous R-factor proposed by the Korean Ministry of

Environment [64] for future study. The results from this study can help the policymakers to update their guideline (Korean Ministry of Environment) [64] regarding the updated version of R-factors values for S. Korea.

It is expected that it will be used not only to calculate soil erosion risk but also to establish soil conservation plans and identify areas at risk of soil disasters by calculating rainfall erosivity factors at the desired temporal-spatial areas more easily and quickly.

However, this study evaluated the R-factor using machine learning models in S. Korean territory, under the monsoon region. Although deep learning models such as Deep Neural Network's applicability in S. Korea has been confirmed in this study, few studies have investigated and benchmarked the performances of a Deep Neural Network model-based USLE R-factor prediction trained. Therefore, future studies should be carried out for the diverse conditions of the other countries such as European countries, the United States, and African countries to broaden the applicability of machine learning technology in USLE R-factor (erosivity factor) analysis.

Author Contributions: Conceptualization and methodology: J.L., J.E.Y., J.K., and K.J.L.; formal analysis: J.H.B. and D.L.; data curation: J.L., J.H., and S.L.; writing—original draft preparation: J.L.; writing—review and editing: J.K.; supervision: K.J.L. All authors have read and agreed to the published version of the manuscript.

Funding: This research was funded by the Ministry of Environment of Korea as The SS (Surface Soil conservation and management) projects [2019002820003].

Institutional Review Board Statement: Not applicable.

Informed Consent Statement: Not applicable.

Data Availability Statement: Not applicable.

Conflicts of Interest: The authors declare no conflict of interest.

References

1. Diodato, N.; Bellocchi, G. Estimating monthly (R) USLE climate input in a Mediterranean region using limited data. *J. Hydrol.* **2007**, *345*, 224–236. [CrossRef]
2. Fu, B.J.; Zhao, W.W.; Chen, L.D.; Liu, Z.F.; Lü, Y.H. Eco-hydrological effects of landscape pattern change. *Landsc. Ecol. Eng.* **2005**, *1*, 25–32. [CrossRef]
3. Renschler, C.S.; Harbor, J. Soil erosion assessment tools from point to regional scales the role of geomorphologists in land management research and implementation. *Geomorphology* **2002**, *47*, 189–209. [CrossRef]
4. Christensen, O.; Yang, S.; Boberg, F.; Maule, C.F.; Thejll, P.; Olesen, M.; Drews, M.; Sorup, H.J.D.; Christensen, J. Scalability of regional climate change in Europe for high-end scenarios. *Clim. Res.* **2015**, *64*, 25. [CrossRef]
5. Stocker, T.; Gin, D.; Plattner, G.; Tignor, M.; Allen, S.; Boschung, J.; Nauels, A.; Xia, Y.; Bex, V.; Midgley, P.E. *Climate Change 2013: The Physical Science Basis: Contribution of Working Group I to the Fifth Assessment Report of the Intergovernmental Panel on Climate Change*; Cambridge University Press: Cambridge, UK; New York, NY, USA, 2013; p. 1535.
6. Achite, M.; Buttafuoco, G.; Toubal, K.A.; Lucà, F. Precipitation spatial variability and dry areas temporal stability for different elevation classes in the Macta basin (Algeria). *Environ. Earth Sci.* **2017**, *76*, 458. [CrossRef]
7. Shi, Z.; Yan, F.; Li, L.; Li, Z.; Cai, C. Interrill erosion from disturbed and undisturbed samples in relation to topsoil aggregate stability in red soils from subtropical China. *Catena* **2010**, *81*, 240–248. [CrossRef]
8. Kinnell, P.I.A.; Wang, J.; Zheng, F. Comparison of the abilities of WEPP and the USLE-M to predict event soil loss on steep loessal slopes in China. *Catena* **2018**, *171*, 99–106. [CrossRef]
9. Lee, J.; Park, Y.S.; Kum, D.; Jung, Y.; Kim, B.; Hwang, S.J.; Kim, H.B.; Kim, C.; Lim, K.J. Assessing the effect of watershed slopes on recharge/baseflow and soil erosion. *Paddy Water Environ.* **2014**, *12*, 169–183. [CrossRef]
10. Pandey, A.; Himanshu, S.; Mishra, S.K.; Singh, V. hysically based soil erosion and sediment yield models revisited. *Catena* **2016**, *147*, 595–620. [CrossRef]
11. Sigler, W.A.; Ewing, S.A.; Jones, C.A.; Payn, R.A.; Miller, P.; Maneta, M. Water and nitrate loss from dryland agricultural soils is controlled by management, soils, and weather. *Agric. Ecosyst. Environ.* **2020**, *304*, 107158. [CrossRef]
12. Lucà, F.; Buttafuoco, G.; Terranova, O. GIS and Soil. In *Comprehensive Geographic Information Systems*; Huang, B., Ed.; Elsevier: Oxford, UK, 2018; Volume 2, pp. 37–50.
13. Arnold, J.G.; Srinivasan, R.; Muttiah, R.S.; Williams, J.R. Large area hydrologic modeling and assessment part I: Model development1. *J. Am. Water Resour. Assoc.* **1998**, *34*, 73–89. [CrossRef]

14. Morgan, R.P.C.; Quinton, J.N.; Smith, R.E.; Govers, G.; Poesen, J.W.A.; Auerswald, K.; Chisci, G.; Torri, D.; Styczen, M.E. The European soil erosion model (EUROSEM): A dynamic approach for predicting sediment transport from fields and small catchments. *Earth Surf. Process. Landf.* **1998**, *23*, 527–544. [CrossRef]
15. Flanagan, D.; Nearing, M. USDA-water erosion prediction project: Hillslope profile and watershed model documentation. *NSERL Rep.* **1995**, *10*, 1–12.
16. Lim, K.J.; Sagong, M.; Engel, B.A.; Tang, Z.; Choi, J.; Kim, K.S. GIS-based sediment assessment tool. *Catena* **2005**, *64*, 61–80. [CrossRef]
17. Young, R.A.; Onstad, C.; Bosch, D.; Anderson, W. AGNPS: A nonpoint-source pollution model for evaluating agricultural watersheds. *J. Soil Water Conserv.* **1989**, *44*, 168–173.
18. Wischmeier, W.H.; Smith, D.D. *Predicting Rainfall Erosion Losses: A Guide to Conservation Planning*; Department of Agriculture, Science, and Education Administration: Wahsington, DC, USA, 1978; pp. 1–67.
19. Renard, K.G.; Foster, G.R.; Weesies, G.A.; McCool, D.K.; Yorder, D.C. Predicting Soil Erosion by Water: A Guide to Conservation Planning with the Revised Universal Soil Loss Equation (RUSLE). In *Agriculture Handbook*; U.S. Department of Agriculture: Washington, DC, USA, 1997; Volume 703.
20. Kinnell, P.I.A. Comparison between the USLE, the USLE-M and replicate plots to model rainfall erosion on bare fallow areas. *Catena* **2016**, *145*, 39–46. [CrossRef]
21. Bagarello, V.; Stefano, C.D.; Ferro, V.; Pampalone, V. Predicting maximum annual values of event soil loss by USLE-type models. *Catena* **2017**, *155*, 10–19. [CrossRef]
22. Park, Y.S.; Kim, J.; Kim, N.W.; Kim, S.J.; Jeong, J.; Engel, B.A.; Jang, W.; Lim, K.J. Development of new R C and SDR modules for the SATEEC GIS system. *Comput. Geosci.* **2010**, *36*, 726–734. [CrossRef]
23. Yu, N.Y.; Lee, D.J.; Han, J.H.; Lim, K.J.; Kim, J.; Kim, H.; Kim, S.; Kim, E.S.; Pakr, Y.S. Development of ArcGIS-based model to estimate monthly potential soil loss. *J. Korean Soc. Agric. Eng.* **2017**, *59*, 21–30.
24. Park, C.W.; Sonn, Y.K.; Hyun, B.K.; Song, K.C.; Chun, H.C.; Moon, Y.H.; Yun, S.G. The redetermination of USLE rainfall erosion factor for estimation of soil loss at Korea. *korean J. Soil Sci. Fert.* **2011**, *44*, 977–982. [CrossRef]
25. Panagos, P.; Ballabio, C.; Borrelli, P.; Meusburger, K. Spatio-temporal analysis of rainfall erosivity and erosivity density in Greece. *Catena* **2016**, *137*, 161–172. [CrossRef]
26. Sholagberu, A.T.; Mustafa, M.R.U.; Yusof, K.W.; Ahmad, M.H. Evaluation of rainfall-runoff erosivity factor for Cameron highlands, Pahang, Malaysia. *J. Ecol. Eng.* **2016**, *17*. [CrossRef]
27. Risal, A.; Bhattarai, R.; Kum, D.; Park, Y.S.; Yang, J.E.; Lim, K.J. Application of Web Erosivity Module (WERM) for estimation of annual and monthly R factor in Korea. *Catena* **2016**, *147*, 225–237. [CrossRef]
28. Risal, A.; Lim, K.J.; Bhattarai, R.; Yang, J.E.; Noh, H.; Pathak, R.; Kim, J. Development of web-based WERM-S module for estimating spatially distributed rainfall erosivity index (EI30) using RADAR rainfall data. *Catena* **2018**, *161*, 37–49. [CrossRef]
29. Lucà, F.; Robustelli, G. Comparison of logistic regression and neural network models in assessing geomorphic control on alluvial fan depositional processes (Calabria, southern Italy). *Environ. Earth Sci.* **2020**, *79*, 39. [CrossRef]
30. Noymanee, J.; Theeramunkong, T. Flood forecasting with machine learning technique on hydrological modeling. *Preced. Comput. Sci.* **2019**, *156*, 377–386. [CrossRef]
31. Vantas, K.; Sidiropoulos, E.; Evangelides, C. Rainfall erosivity and Its Estimation: Conventional and Machine Learning Methods. In *Soil Erosion—Rainfall Erosivity and Risk Assessment*; Intechopen: London, UK, 2019.
32. Zhao, G.; Pang, B.; Xu, Z.; Peng, D.; Liang, X. Assessment of urban flood susceptibility using semi supervised machine learning model. *Sci. Total Environ.* **2019**, *659*, 940–949. [CrossRef] [PubMed]
33. Vu, D.T.; Tran, X.L.; Cao, M.T.; Tran, T.C.; Hoang, N.D. Machine learning based soil erosion susceptibility prediction using social spider algorithm optimized multivariate adaptive regression spline. *Measurement* **2020**, *164*, 108066. [CrossRef]
34. Xiang, Z.; Demir, I. Distributed long-term hourly streamflow predictions using deep learning—A case study for state of Iowa. *Environ. Modeling Softw.* **2020**, *131*, 104761. [CrossRef]
35. Renard, K.G.; Foster, G.R.; Weesies, G.A.; Porter, J.P. RUSLE: Revised universal soil loss equation. *J. Soil Water Conserv.* **1991**, *46*, 30–33.
36. Alpaydin, E. *Introduction to Machine Learning*, 4th ed.; MIT Press: Cambridge, MA, USA, 2014; pp. 1–683.
37. Hong, J.; Lee, S.; Bae, J.H.; Lee, J.; Park, W.J.; Lee, D.; Kim, J.; Lim, K.J. Development and evaluation of the combined machine learning models for the prediction of dam inflow. *Water* **2020**, *12*, 2927. [CrossRef]
38. Wang, L.; Guo, M.; Sawada, K.; Lin, J.; Zhang, J.A. Comparative Study of Landslide Susceptibility Maps using Logistic Regression, Frequency Ratio, Decision Tree, Weights of Evidence and Artificial Neural Network. *Geosci. J.* **2016**, *20*, 117–136. [CrossRef]
39. Safavian, S.R.; Landgrebe, D. A survey of decision tree classifier methodology. *IEEE Trans. Syst. Man Cybern.* **1991**, *21*, 660–674. [CrossRef]
40. Breiman, L. Random forests. *Mach. Learn.* **2001**, *45*, 5–32. [CrossRef]
41. Moon, J.; Park, S.; Hwang, E. A multilayer perceptron-based electric load forecasting scheme via effective recovering missing data. *Kips Trans. Softw. Data Eng.* **2019**, *8*, 67–78.
42. Bae, J.H.; Han, J.; Lee, D.; Yang, J.E.; Kim, J.; Lim, K.J.; Neff, J.C.; Jang, W.S. Evaluation of Sediment Trapping Efficiency of Vegetative Filter Strips Using Machine Learning Models. *Sustainability* **2019**, *11*, 7212. [CrossRef]

43. Qu, W.; Li, J.; Yang, L.; Li, D.; Liu, S.; Zhao, Q.; Qi, Y. Short-term intersection Traffic flow forecasting. *Sustainability* **2020**, *12*, 8158. [CrossRef]
44. Yao, Z.; Ruzzo, W.L. A regression-based K nearest neighbor algorithm for gene function prediction from heterogeneous data. *BMC Bioinform.* **2006**, *7*, S11. [CrossRef]
45. Ooi, H.L.; Ng, S.C.; Lim, E. ANO detection with K-Nearest Neighbor using minkowski distance. *Int. J. Process. Syst.* **2013**, *2*, 208–211. [CrossRef]
46. Natekin, A.; Knoll, A. Gradient boosting machines, a tutorial. *Front. Neurorobot.* **2013**, *7*, 21. [CrossRef]
47. Friedman, J.H. Greedy function approximation: A gradient boosting machine. *Ann. Stat.* **2001**, *29*, 1189–1232. [CrossRef]
48. Ngarambe, J.; Irakoze, A.; Yun, G.Y.; Kim, G. Comparative Performance of Machine Learning Algorithms in the Prediction of Indoor Daylight Illuminances. *Sustainability* **2020**, *12*, 4471. [CrossRef]
49. Georganos, S.; Grippa, T.; Vanhuysse, S.; Lennert, M.; Shimoni, M. Very high resolution object-based land use–land cover urban classification using extreme gradient boosting. *IEEE Geosci. Remote Sens. Lett.* **2018**, *15*, 607–611. [CrossRef]
50. Babajide Mustapha, I.; Saeed, F. Bioactive molecule prediction using extreme gradient boosting. *Molecules* **2016**, *21*, 983. [CrossRef]
51. Lavecchia, A. Machine-learning approaches in drug discovery: Methods and applications. *Drug Discov. Today.* **2015**, *20*, 318–331. [CrossRef]
52. Rumelhart, D.E.; Hinton, G.E.; Williams, R.J. Learning representations by back-propagating errors. *Nature* **1986**, *323*, 533–536. [CrossRef]
53. Wang, Q.; Wang, S. Machine learning-based water level prediction in Lake Erie. *Water* **2020**, *12*, 2654. [CrossRef]
54. Siniscalchi, S.M.; Yu, D.; Deng, L.; Lee, C.H. Exploiting deep neural networks for detection-based speech recognition. *Neurocomputing* **2013**, *106*, 148–157. [CrossRef]
55. Hinton, G.E.; Osindero, S.; Teh, Y.W. A fast learning algorithm for deep belief nets. *Neural Comput.* **2006**, *18*, 1527–1554. [CrossRef]
56. Cancelliere, A.; Di Mauro, G.; Bonaccorso, B.; Rossi, G. Drought forecasting using the standardized precipitation index. *Water Resour. Manag.* **2007**, *21*, 801–819. [CrossRef]
57. Legates, D.R.; McCabe, G.J., Jr. Evaluating the use of "goodness-of-fit" measures in hydrologic and hydroclimatic model validation. *Water Resour. Res.* **1999**, *35*, 233–241. [CrossRef]
58. Moghimi, M.M.; Zarei, A.R. Evaluating the performance and applicability of several drought indices in arid regions. *Asia-Pac. J. Atmos. Sci.* **2019**, 1–17. [CrossRef]
59. Liu, D.; Jiang, W.; Wang, S. Streamflow prediction using deep learning neural network: A case study of Yangtze River. *Inst. Electr. Electron. Eng.* **2020**, *8*, 90069–90086. [CrossRef]
60. Nhu, V.; Hoang, N.; Nguyen, H.; Ngo, P.T.T.; Bui, T.T.; Hoa, P.V.; Samui, P.; Bui, D.T. Effectiveness assessment of Keras based deep learning with different robust optimization algorithms for shallow landslide susceptibility mapping at tropical area. *Catena* **2020**, *188*, 104458. [CrossRef]
61. Lee, K.; Choi, C.; Shin, D.H.; Kim, H.S. Prediction of heavy rain damage using deep learning. *Water* **2020**, *12*, 1942. [CrossRef]
62. Sit, M.A.; Demir, I. Decentralized flood forecasting using deep neural networks. *arXiv* **2019**, arXiv:1902.02308.
63. Kim, M.; Kim, Y.; Kim, H.; Piao, W. Evaluation of the k-nearest neighbor method for forecasting the influent characteristics of wastewater treatment plant. *Front. Environ. Sci. Eng.* **2015**, *10*, 299–310. [CrossRef]
64. Korea Ministry of Environment. *Notice Regarding Survey of Topsoil Erosion*; Ministry of Environment: Seoul, Korea, 2012; pp. 1–41.

Article

Impacts of the Degraded Alpine Swamp Meadow on Tensile Strength of Riverbank: A Case Study of the Upper Yellow River

Haili Zhu [1], Peng Gao [2,*], Zhiwei Li [3], Jiangtao Fu [4], Guorong Li [1], Yabin Liu [1], Xilai Li [5] and Xiasong Hu [1]

1 Department of Geological Engineering, Qinghai University, Xining 810016, China; qdzhuhaili@163.com (H.Z.); qdliguorong@163.com (G.L.); liuyabingqh@sina.com (Y.L.); huxiasong@sina.com (X.H.)
2 Department of Geography, Syracuse University, Syracuse, NY 13244, USA
3 State Key Laboratory of Water Resources and Hydropower Engineering Science, Wuhan University, Wuhan 430072, China; lizw2003@whu.edu.cn
4 School of Civil Engineering and Architecture, Shaanxi University of Technology, Hanzhong 723000, China; fujiangtao865@sina.com
5 College of Agriculture and Animal Husbandry, Qinghai University, Xining 810016, China; Xilai-li@163.com
* Correspondence: pegao@maxwell.syr.edu; Tel.: +1-315-436-5895

Received: 1 July 2020; Accepted: 18 August 2020; Published: 21 August 2020

Abstract: In the meandering riverbank of the Upper Yellow River (UYR), the native alpine swamp meadow (AS) has continuously degenerated into an alpine meadow (AM) due to climate change and intensified grazing. Its implication on river morphology is still not well known. This study examined this effect by in situ measurings of (1) physical properties of roots and their distribution in the soil-root mixture of the upper bank layer, and (2) the tensile strength in terms of excavating tests for triggering cantilever collapses of AS and AM riverbanks. The results showed that the root number in AS was significantly greater than that in AM, though the root distribution in both was similar. Also, the average tensile strength of individual roots in AS was 31,310 kPa, while that in AM was only 16,155 kPa. For the soil-root mixture, it decreased from 67.39 to 21.96 kPa. The weakened mechanical property was mainly ascribed to the lessened root number and the simpler root structure in the soil-root mixture of AM that reduces its ability to resist the external force. These findings confirmed that healthy AS can enhance bank stability and delay the development of tensile cracks in the riverbank of the meandering rivers in the UYR.

Keywords: alpine swamp meadow; alpine meadow; degradation of riparian vegetation; root distribution; tensile strength; tensile crack

---

1. Introduction

Banks of meandering rivers are often composed of silts and sand that have significantly higher compressive strength than tensile strength or cohesion [1–4]. There has been a consensus that riparian vegetation can reinforce bank strength [5–9] and stability through the interaction between soil and roots [10–16]. Mechanisms of riverbank failure are quite different from those found on hillslopes because steeper and shorter riverbanks tend to have a more variable profile and relatively small size of the failed block [17]. Accordingly, vegetation types and their root distributions throughout the bank profile play a critical role in resisting riverbank failure. In addition, the lower soil layer of the composite riverbank is subject to fluvial erosion, often resulting in a cantilever upper layer that includes the mixture of cohesive soil and vegetation. Once the gravity moment generated by the upper cantilever layer exceeds its tensile moment, vertical cracks are developed and continuously extend through the upper layer, causing cantilever bank failure [10,18]. The ability of the soil-root mixture to resist external forces is a key factor in evaluating the bank stability.

Many studies have analyzed the tensile strength of fiber-reinforced soils and explored its relationship with soil physical indices using laboratory experiments [19–23]. These studies found that the benefits of natural or synthetic fiber reinforcement include (1) improved ductility in tension compared with pure earth blocks and (2) inhibition of tensile crack propagation [21–24]. Since the root system is intertwined in the soils, artificial soil blocks remolded in these laboratory experiments cannot reflect the actual root branching structure and the complex interaction between roots and soils. It is thus imperative to perform in situ tensile tests using naturally rooted soils for revealing the tensile properties of riparian vegetation.

In recent decades, a series of studies have been carried out for examining the shear strength of rooted soils using either in situ or indoor shear-test experiments. These studies helped better understand the enhancement of the vertically extending root system through soils by revealing that coarse roots tend to stabilize the soil-root mixture and fine roots may enhance its mechanical strength [25–27]. Other studies have confirmed that the number of roots passing through the potential shear plane, root system distribution and its strength [27,28], the initial water content of the soils [12,29], and the soil-root friction [30], contribute to soil reinforcement. However, there is still a lack of studies for quantifying the tensile strength of the rooted soil and the reinforced traction in the soil-root mixture.

The Upper Yellow River (UYR) watershed is located in the hinterland of the Qinghai-Tibet Plateau and includes numerous rivers, lakes, and wetlands. It is an important source of fresh water but a fragile ecosystem in the western region of China. When the UYR flows from the source into the open terrain with low-lying hills and valleys on the eastern edge of the Qinghai-Tibet Plateau, it forms a unique curved shape in plane form, commonly called the first bend of the UYR [31]. Meandering rivers are widely developed in this area, accompanied by the vegetation cover of the alpine swamp meadow and its degraded type, the alpine meadow. Because of global climate change and anthropogenic disturbances, the alpine swamp meadow on the eastern Qinghai-Tibet Plateau has been undergoing severe degradation [32–35]. The degradation succession of the alpine swamp meadow caused changes in the composition of the plant community. Dicotyledonous plants with a straight root system replaced sedges and gramineous plants with a dense clump root system, leading to changes in the underground biomass of the plant community, reduction of the spatial distribution of the root system, and a significant decrease of root activity and bulk density [35,36]. The biological degradation affects not only the root distribution of riparian vegetation but also the tensile and shear strength of the vegetated riverbanks. This means that degraded riparian vegetation could decrease bank stability and affect lateral evolution trends of meandering rivers in the UYR. Therefore, quantifying the influence of alpine swamp meadow degradation on the strength of riverbank soils in the UYR is an important issue that needs to be addressed. Most previous studies have examined the shear strength of the vegetated soils [8–13,16,25,26,37]. Little has been done on quantifying the effect of degraded riparian vegetation on the tensile strength of the soil-vegetation mixture.

In this study, we addressed this issue by focusing on the riverbanks of a meandering reach that has initially formed a cantilever arm under fluvial erosion. The vegetation community is featured by a healthy alpine swamp meadow (AS) and moderately degraded alpine meadow (AM). We measured root numbers and their vertical distributions in the soil-root mixture within a depth range of 0.3 m. Furthermore, by excavating the sandy layer below the top soil-vegetation layer of the bank in situ for artificially initiating cantilever bank collapse, we measured the tensile strength of individual roots using in situ pullout tests and subsequently calculated that for the soil-root mixture of the collapsed bank blocks. By comparing the differences of these results between AS and AM mixtures, we revealed the effects of meadow degradation on bank tensile strength and the role of roots in slowing bank crack development.

2. Experiences at Field Scale and Procedure for Determining the Cantilanver Bank Collapse

2.1. Study Sites and Degradation of Alpine Swamp Meadow

The Upper Yellow River (UYR) watershed is within the Qinghai-Tibet Plateau located in western China (Figure 1a). In its downstream reach, the main channel is joined by a relatively small tributary, the Lanmucuo River (Figure 1b). Our study sites are in the upstream reach of this tributary (34°26′ N–35°02′ N; 101°29′ E–101°35′ E) (Figure 1c). This reach has elevations ranging between 3400 and 4200 m a.s.l. with a mean channel gradient of 0.19%. The UYR region is subject to the alpine monsoon climate that features a long, cold dry season from October to early May, and a short, warm wet season from middle May to September. The annual mean precipitation is 560.5 mm [38], most of which is concentrated in the period from June to September, accounting for more than 83% of the total. The annual mean evaporation and temperature are 1278 mm and −0.16 °C, respectively [39]. Under this climate, the ground consists of seasonally frozen soils.

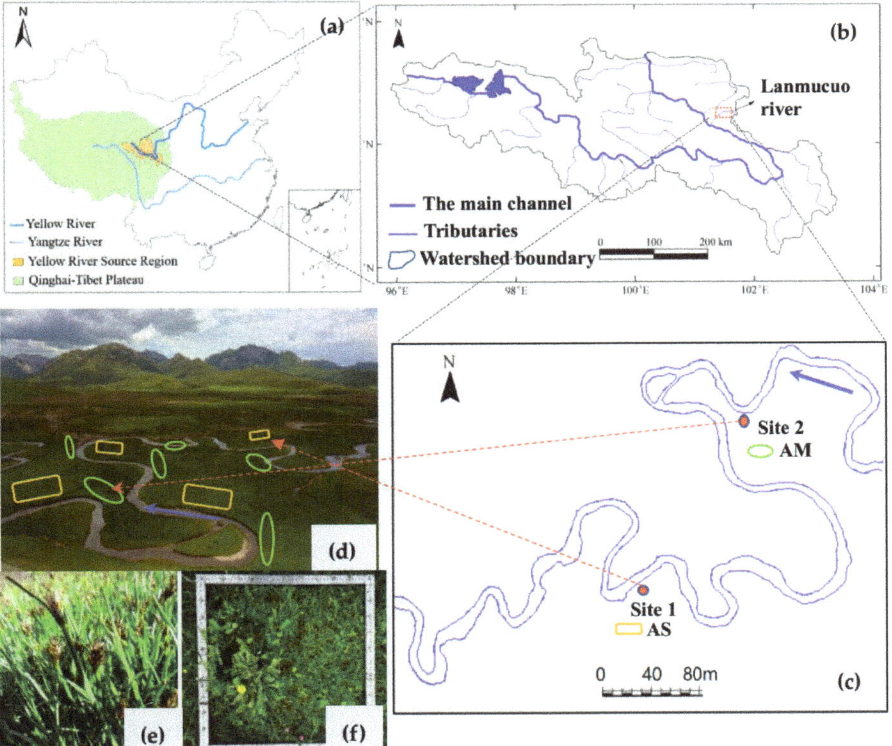

Figure 1. (**a**) The geography of the Qinghai-Tibet Plateau and Upper Yellow River; (**b**) the location of the study area in the Lanmucuo river; (**c**) the specific locations of the two selected sites; (**d**) the spatially distributed AS and AM along the riparian zone of the Lanmucuo River; (**e**) illustration of *Blysmus sinocompressus*, the dominant species of AS; (**f**) demonstration of the AM that is rich in species composition (some degraded species are discernable).

Grassland is the major type of land cover in the region. It is dominated by alpine swamp meadow (AS), which provides important ecosystem services to the regional environment. Affected by global warming and intensified grazing, AS in the region has gradually degraded to alpine meadows (AM) and alpine steppe meadows (ASM) [40]. Spatially, AS and AM are often seen along riverbanks, the Lanmucuo River, while ASM is mostly distributed on the upper sloping parts of the piedmonts (Figure 1d). Because the area covered by ASM is far away from river banks, it is not a concern in this study. Based on a set of qualitative and semi-qualitative indicators used for pasture degradation classification [33,35], a survey for riparian vegetation along a 22 km reach of the upper Lanmucuo River was conducted in the 2017–2019 period [16,41]. During the survey, the coverage and number of species, dominant species, and underground biomass of AS and AM were determined within each sampling plot with the size of 1 × 1 m. The underground biomass was represented by the root mass within the soil layer that is 0.3 m in depth. The results (Table 1) showed that the values of all measured metrics were different with statistical significance between AS and AM.

Table 1. Characteristics of the surveyed AS and AM communities in the study area.

Vegetation Type	Coverage (Mean ± SD *, %)	Number of Species (Mean ± SD)	Under-Ground Biomass (g m^{-2})	Dominant Species	Number of Surveyed Sites
AS	96.3 ± 2.7 ^{a}	3.5 ± 2.7 ^{b}	378 ± 74 ^{a}	*Blysmus sinocompressus*	40
AM	74.5 ± 11.6 ^{b}	14.8 ± 5.1 ^{a}	193 ± 56 ^{b}	*Kobresia pygmaea* *Elymus nutans* *Potentilla saundersiana*	44

* standard deviation; AS: alpine swamp meadow; AM: alpine meadow; Different superscripts of a and b denote significant differences ($p < 0.05$) between different vegetation types.

In general, distributions of surficial plant species, coverage, and rooting depth and lateral root spread in underground root systems are determined by the prevailing physical habitats [42]. In particular, AS tends to appear in areas with topographic lows and occupied by seasonally saturated water (i.e., in swales), whereas AM is typically developed around the apex of river bends (Figure 1d). Vegetation communities of AS are mainly composed of cold-tolerant hygrophytes and hydromesophytes, which have a simple community structure. *Blysmus sinocompressus* is the dominant and healthy specie (Figure 1e and Table 1) and its coverage reaches nearly 98%. The remaining 2% is contributed from other species, such as *Ranunculus nephelogenes* and *Pedicularis longiflora*. The dominant species of AM are *Kobresia pygmaea*, *Elymus nutans*, and *Potentilla saundersiana*, accounting for 30% of the coverage (Table 1). Other herbaceous species, such as *Poa annua*, *Nardostachys jatamansi*, *Saxifraga montana*, *Aconitum tanguticum*, and degraded species, such as *Leontopodium pusillum*, *Oxytropis ochrocephala*, take up to 30% of the coverage. This means that AM only takes 60% of the habitat (Figure 1f). Compared with AS, the number of species and the mesophytes in AM are significantly high (Table 1). These surficial ecological differences between AS and AM must affect their underground properties and the associated mechanical characteristics, which were investigated in this study.

2.2. Field Experiments and Measurements

Field experiments were conducted at the two selected sites that are about 200 m apart from each other in the study area (Figure 1c). These sites have natural vegetation communities and are not disturbed by livestock and human activities. Therefore, they are representative of the general vegetation distribution in the study area. Site 1 is covered by AS, while site 2 is topped by AM. The height of the riverbank to the water level of the channel flow at site 1 is about 0.2–0.4 m lower than that at site 2, suggesting that the groundwater level at site 1 is higher than that at site 2. This ·difference is the main factor that determines the spatial distribution of AS and AM in the study area. The riverbank at both sites had been undercut by river flows, forming a cantilever arm with a width of about 0.2 m and thickness of about 0.3 m at site 1, while 0.3 and 0.35 m, respectively at site 2 (Figure 2). Both cantilever arms were stable with no tensile cracks developed on the top. They were underlined by a layer of silt and sand with a thickness of approximately 0.3 to 0.6 m, whose bottoms were gradually intergraded into fine gravels deposited at the toes of both banks (Figure 2). Field experiments were conducted during an October storm event in 2018, such that the entire bank profiles were exposed, facilitating subsequent bank excavation and measurements.

Figure 2. Riverbanks and their vertical profiles at (**a**) site 1, which is covered by AS, and (**b**) site 2, which is covered by AM (not to scale).

At each site, experiments were performed at the face of a bank profile stretched longitudinally for 8 m (Figure 3). An area of 1 m long on each side (i.e., E1 and E2 in Figure 3) was selected for extending the existing cantilever arm by excavating the lower part of the bank profile. The two areas were selected apart from each other by 3 m to ensure that the excavation of the first one would not affect the stability of the second. During each experiment, excavation was executed gradually, such that the development of tensile cracks on top of the cantilever arm could be observed. Their emergence indicated the status of arm stability. Excavation ended when the cantilever arm reached the threshold that triggered the failure of the cantilever arm. Three types of measurements were subsequently performed after bank collapse, determining tensile strengths of individual roots, measuring root diameter, number, and distribution, and sampling soil for both in situ and laboratory analyses.

The tensile strength of a single root for a given plant species (T_r, MPa) may be determined by [2],

$$T_r = 4F/(\pi d^2) \tag{1}$$

where F is the maximum pullout force of measured individual roots (N) and d is the diameter of the corresponding single root (10^{-3} m). The value of F was measured using a HP-500 digital push-pull meter with a maximum load of 500 N and an indication error of ±0.5% (Leqing Aidebao Instruments Co. Ltd., Leqing, China). This measurement was taken at three plots within the experimental area; two (i.e., R1 and R3) were on the new face of the upper soil-root layer after cantilever failure and the other (i.e., R2) was at the face of the original upper bank (Figure 3). This design assured that the measured tensile strength accounted for spatial variability. At each plot, vegetation roots were separated from their surrounding soils by brushing soil particles away. Then, every single root was connected to the tension meter by a clamp. The value of F was recorded after applying a horizontal tensile force at a uniform speed until the root is broken or pulled out (Figure 4). The diameter of the same root was also measured using a Vernier caliper with an accuracy of 2×10^{-5} m. If the broken position of the root was close to the clamp, then the recorded F value was biased and not counted. There were more than 30 roots measured in each plot.

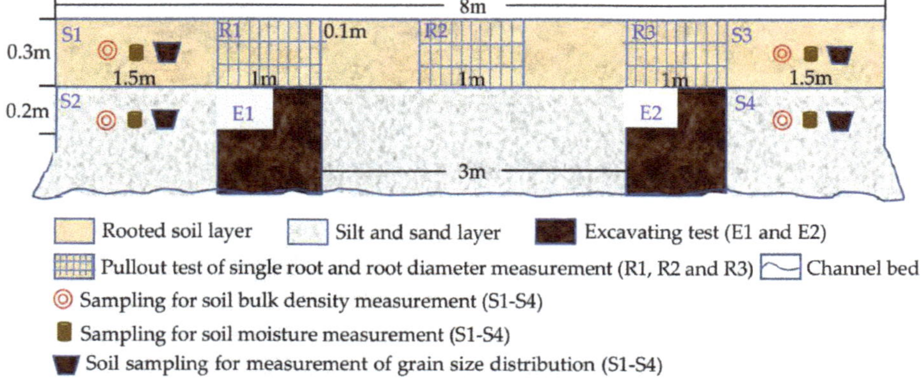

Figure 3. The design of bank excavation experiments and other measurements.

Figure 4. Measuring the maximum pullout force of individual roots using a tension meter.

Root diameters and numbers were measured within the front faces of the soil-root layer (0.3 m in thickness) in the collapsed bank block at R1 and R3, and the face of the same layer in the original bank at R2 (Figure 3). The surface of the face was washed by water to expose the root system and then divided into 30 small zones by laying a grid whose cells had the size of 0.10 × 0.10 m on the top. Within each of the 30 cells, root diameters were measured using the same Vernier caliper and the number of roots was recorded. All the roots that passed through the soil-root layer were measured. These values were classified into three groups based on the root diameter (i.e., <0.5, 0.5–1.0, and >1.0 × 10^{-3} m), and the depths with the soil-root layer (i.e., 0–0.10, 0.11–0.20, and 0.21–0.30 m). The root area ratio (RAR), defined as the ratio of the total area of all roots passing through the collapsed soil-root layer to the area of the plot (i.e., 1 × 0.3 m), was then calculated using the measured values. The measurements were repeated three times at each plot (i.e., R1, R2, and R3).

Soil samples were taken from two small plots in the soil-root layer around the depth of 0.10–0.20 m beneath the ground surface (i.e., S1 and S3 in Figure 3) and from two plots in the lower layer between 0.3 and 0.5 m from the ground surface (i.e., S2 and S4 in Figure 3) at both sites. These samples were carefully stored and transported back to the soil mechanics laboratory of Geological Engineering Department, Qinghai University for further analysis. The oven-drying method [43] was used to determine the soil moisture content, and the sieving method [43] was applied to determine the grain size distribution. Each measurement was repeated three times for accuracy. Additionally, a cutting ring with a capacity volume of 6.0 × 10^{-5} m^3 was used in situ to determine the soil bulk density at each plot (Figure 3).

2.3. Calculation

In the process of cantilever bank collapse, the weight of the cantilever body is originally balanced by the tensile strength from the soil-root mixture of the body. Extension of the cantilever arm (by artificial excavation in this case or by fluvial erosion under the natural condition) will increase its weight. Once the tensile strength is insufficient to balance the weight, tensile cracks develop from the top of the cantilever body. At this time, both tensile stress and compressive stress exhibit a triangle distribution, and the center axis of the failing cantilever block is located in the stress center of the cantilever body below the crack [44] (Figure 5). Further extension of the cantilever arm catalyzes the development of the tensile crack until the failure of the cantilever body occurs. Under the critical condition of the failure, the external moment of the cantilever body is balanced by the resistance moment of the soil-root mixture, which may lead to [44],

$$Wb_c/2 = \frac{l(d_1 - d_t)^2}{3(1+a)^2}\sigma_t + \frac{a^2 l(d_1 - d_t)^2}{3(1+a)^2}\sigma_c \tag{2}$$

where $W = \rho g b_c d_1 l$ (N) is the weight of the cantilever body; ρ is the bulk density of the soil-root mixture (kg m^{-3}); b_c is the critical width of the cantilever arm (m); d_1 is the thickness of the cantilever layer (m), d_t is the depth of the crack from the top of the bank (m); l is the unit length of the cantilever layer (m), that is 1 m; $a = \sigma_t / \sigma_c$, σ_t and σ_c are tensile and compressive stress of the soil-root mixture (N m^{-2}), following Ajaz's results [45], $a = 0.1$. Substituting $W = \rho g b_c d_1 l$ into Equation (2), the tensile strength σ_t of the soil-root mixture of the cantilever body may be expressed as

$$\sigma_t = 3(1+a)\rho g {b_c}^2 d_1 / 2(d_1 - d_t)^2 \tag{3}$$

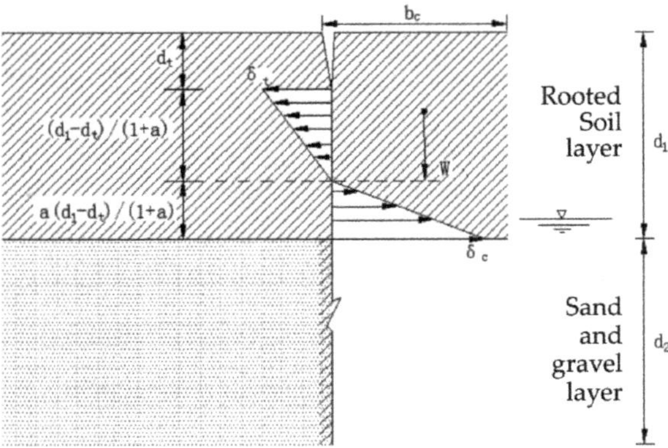

Figure 5. Stress analysis of the upper root-soil layer of a cantilever riverbank (modified from Figure 7 in Xia et al. [44], (not to scale)).

Using the measured d_1 and d_t in our tested AS and AM bank blocks, we calculated their σ_t values. The product of T_r and RAR also reflects the tensile strength of a soil-root mixture. We compared these two types of calculations for the tensile strength.

3. Results and Analysis

3.1. Characteristics of Soils and Root Distribution

Our results demonstrated that the composition and physical properties of the two layers in the vertical profile of the riverbank are significantly different between the two sites (Table 2). Generally, the upper soil-root layer was composed of silt, while the lower layer consisted of silty sand with some poorly graded fine gravel. In the upper layer, the average bulk density and moisture content at the two sites were 1260 kg·m^{-3} and 1560 kg·m^{-3}; 39.36% and 41.72%, respectively. They were about 180 kg·m^{-3} and 405 kg·m^{-3}; 17.12% and 31.42% higher than those in the lower layer. These two soil properties in the soil-root layer of AS were 21.96% and 3.57% higher than those in that of AM. For the lower layer, the average bulk density and moisture content at site 1 were greater than those at site 2 by 118.67% and 3.27%, respectively. It is obvious that the physical parameters (i.e., bulk density and moisture content) of the soil-root mixture of AS and AM are disparate. These differences could have different influences on the tensile strength of the root system in the soil [27], which lend the support for our analyses of tensile strength for both single roots and the soil-root mixture in this study.

Table 2. The physical properties of the two vertical layers at both sites.

Tested Site	Sampling Depth (m)	Soil Type	Bulk Density ρ (kg·m^{-3})	Moisture Content ω (%)	RAR (%)
1(AS)	0.10–0.20	Silt	1560	41.71	0.22
	0.30–0.40	Silty sand	1730	10.29	-
	0.10–0.20	Silt	1550	40.92	0.23
	0.30–0.40	Silty sand	1740	10.97	-
2(AM)	0.10–0.20	Silt	1260	40.42	0.12
	0.40–0.50	Silty sand	1710	24.25	-
	0.10–0.20	Silt	1290	39.36	0.11
	0.40–0.50	Silty sand	1650	22.24	-

Although the distribution of roots over different classes of root diameters is highly variable for different species [46], it may still be characterized by vertical patterns of root number and diameters along the depth of a riverbank. In this study, AS is mainly composed of *Blysmus sinocompressus*, belonging to the *Cyperaceous* family. This species has a typical dense and fibrous root system that may be up to 0.8 m long, and its rhizomes are typically about 0.25–0.60 m long. These roots mix with the surrounding soil, forming a soil-root layer that extends from the bank surface to the depth of 0.30 m. Our results showed that the total root number within the experimental block of this layer, which was 1.0 m long and 0.3 m deep (Figure 3), was 4345, with 2505, 1160, and 680 in the depth ranges of 0–0.10, 0.11–0.20, and 0.21–0.30 m, respectively. Among these roots, those with the root diameter <0.5, 0.5–1.0, and >1.0 × 10^{-3} m took about 66%, 25%, and 9%, respectively. AS is dominated by finer roots, which is evidenced by the fact that within each of the three depth ranges, they took 66%, 67%, and 76%, respectively (Figure 6a). Roots with medium diameters (0.5–1.0 × 10^{-3} m) in all three depth ranges took about 25%, while those with the diameters >1.0 × 10^{-3} m only existed in the depths of 0–0.10 and 0.11–0.20 m, taking merely 9% and 8%, respectively. More fine roots (<0.5 × 10^{-3} m) developed in the shallow depth range rather than in the deep depth range, supported by their distributions of 56%, 26%, and 18% in the depth ranges of 0–0.10, 0.11–0.20, and 0.21–0.30 m, respectively (Figure 6a). The roots with the mean diameter (0.5–1.0 × 10^{-3} m) followed a similar vertical distribution, featured by 58%, 27%, and 15% in the three depth ranges, respectively. Roots with the diameters >1.0 × 10^{-3} m only existed in the first two depth ranges with 71% and 29%, respectively.

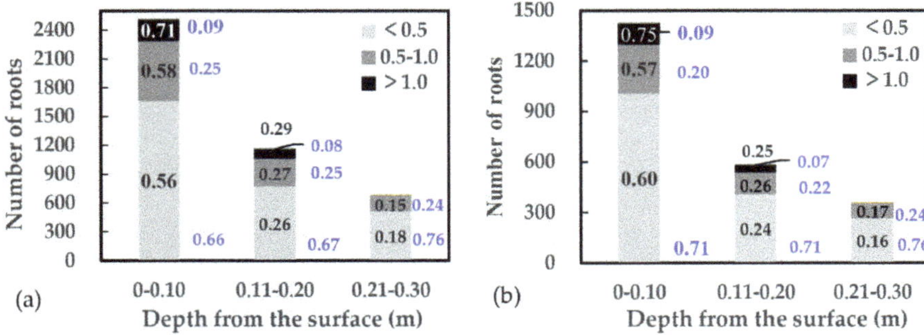

Figure 6. Root distributions in the three depth ranges of the experimental blocks for (**a**) AS and (**b**) AM. The number within each color of the three columns represents the percentage of roots in each sub-layer of the soil-root layer (i.e., the sum of the numbers in each color is 100%). The number outside of the columns represents the percentages of three root diameters in each sub-layer (i.e., the sum of these numbers along each column is 100%). The scaling of the y-axis is different between (**a**,**b**).

The AM is dominated by *Kobresia pygmaea*, *Elymus nutans*, and *Potentilla saundersiana*. Their root number took about 93% of the total in AM. The first two plants have dense and fibrous roots that are about 0.15–0.55 m long. Besides the three dominant plants, AM contains degraded species such as *Nardostachys chinensis*, *Oxytropis ochrocephala*, *Saxifraga montana*, and *Leontopodium pusillum*, whose roots are generally sparse and short. They have a typical tap root system with the length ranging between 0.04 and 0.13 m. The total number of roots in AS was only 2355 with 1420, 580, and 355 in the three depth ranges downward, respectively. Similarly to those in the AS, roots with the diameters $<0.5 \times 10^{-3}$ m dominated in each of the three depth ranges, taking 71%, 71%, and 76%, respectively (Figure 6b). The remaining roots at each depth were those with the mean diameters ($0.5–1.0 \times 10^{-3}$ m). They took roughly the same percentage of the total roots within each depth range, which was 20%, 22%, and 24%, respectively. Again, the coarse roots (i.e., diameter $>1.0 \times 10^{-3}$ m) only occupied the first two depth ranges, taking a small portion of the total roots in each (i.e., 9% and 7%, respectively). Along the vertical direction, roots with each diameter class demonstrated a similar distribution to their AS peers. For example, the fine roots (diameter $<0.5 \times 10^{-3}$ m) took 60%, 24%, and 16% from the top to the bottom depth ranges (Figure 6b). A similar vertical distribution appeared for the roots with the mean diameter (i.e., diameters were between 0.1 and 1.0×10^{-3} m) with 57%, 26%, and 17% in the three depth ranges, respectively. Also, the coarse roots only stayed in the first two depth ranges. Most of them (75%) occupy in the 0–0.10 m depth range (Figure 6b).

Although roots in AS and AM showed similar spatial and diameter distributions, which led to their similar average root diameters (i.e., $0.46 \pm 0.31 \times 10^{-3}$ m and $0.41 \pm 0.39 \times 10^{-3}$ m, respectively) in the entire mixed layer of 0.30 m, their root numbers were greatly different, which can be proved by a two-sample difference test ($p < 0.05$). The root number of AS was 46% higher than that of AM. Consequently, the RAR of the AS experimental block was 0.225% on average, which was about twice that of the AM block (Table 2). These results showed that the root number, which plays an important role in resisting tensile force imposed to the experimental block, is reduced greatly when AS is degraded to AM.

3.2. Tensile Trength of Individual Roots and Soil-Root Mixture

The effect of roots on soil strength of the riverbank does not only depend on the root number, but also the tensile strength of a single root (T_r) (i.e., Equation (1)). The average T_r of the dominant plant *Blysmus sinocompressus*, whose root diameters ranged between 0.20 and 1.76×10^{-3} m at site 1, was 31,310 kPa, and the maximum value can reach up to 128,000 kPa. The T_r for roots with the diameters ranging between 0.24 and 1.88×10^{-3} m at site 2 (i.e., AM) was 16,160 kPa. The average

T_r of the AS block was 48.4% higher than that of the AM block. There existed a strong relationship between the root diameter and T_r (Figure 7), which indicated that T_r decreases as the root diameter increases for both AS and AM. This finding is consistent with those reported in earlier studies [11,46–49]. This relationship may be described by a power function, whose exponent was different between the two sites (Figure 7). For the AS block, when the root diameter was less than 0.70×10^{-3} m, T_r decreased sharply as the root diameter increased, nearly following a linear relationship. Specifically, the value of T_r decreased drastically from 127,930 to 30,810 kPa, as the root diameter only increased from 0.20 to 0.70×10^{-3} m. As the root diameter continuously increased from 0.71 to 1.76×10^{-3} m, T_r only decreased from 26,480 to 7050 kPa. The differences of T_r were 97,120 kPa and 19,430 kPa before and after the threshold root diameter (i.e., 0.7×10^{-3} m). For the AM block, when the diameter was less than 0.4×10^{-3} m, T_r decreased greatly as the root diameter increased, and its value decreased from 73,380 to 23,590 kPa with the average of 39,050 kPa. However, the difference of T_r was only 25,860 kPa when the root diameter increased from 0.4 to 1.88×10^{-3} m. This change was much less than that for root diameters less than 0.4×10^{-3} m (i.e., 49,790 kPa). Therefore, the root diameter of 0.7 and 0.4×10^{-3} m can be taken as a threshold of the AS and AM blocks, respectively. Values of T_r were more sensitive to the changes of root diameters less than the threshold while remaining less variable for coarse roots whose diameters are greater than the threshold.

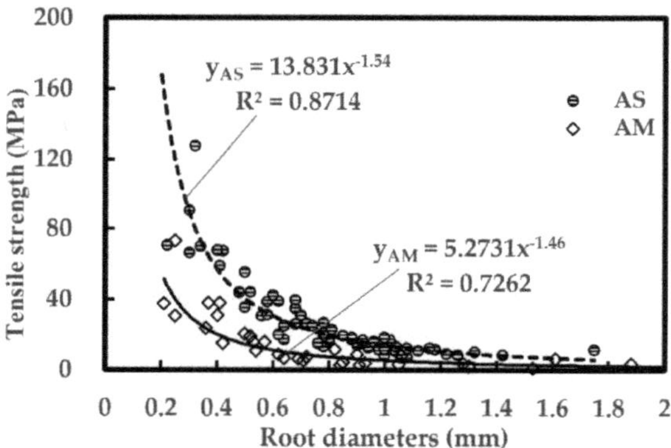

Figure 7. Root tensile strength vs. root diameters for the alpine swamp meadow (AS) and alpine meadow (AM).

Because there were about 93% and 90% fine roots with diameters less than 1×10^{-3} m in the mixed layer of AS and AM blocks, respectively, the difference in the root number should not be the main cause for the different T_r values between AS and AM blocks. Rather, their difference represented the true mechanical discrepancy between the two. In other words, the tensile strength of individual roots in healthy meadow plants (i.e., AS) is generally higher than that in the degraded meadow plants (i.e., AM).

The tensile strength of the soil-root mixture should be relevant to the physical properties of the mixture. At site 1, the moisture content and volume of the two tested bank blocks were different. They were 40.92% and 0.175 m^3 for block 1, and 41.71% and 0.187 m^3 fo block 2, respectively (Tables 2 and 3). The tensile strength of the soil-root mixture (σ_t) between the two tested sites was also slightly different. It was 66.86 kPa for Site 1 and 67.93 kPa for Site 2 (Table 3). At site 2, the moisture content and volume of the two tested experimental blocks were different by 1.06% and 0.004 m^3 (Tables 2 and 3), respectively. However, the variation of σ_t between the two was still minor, merely 1.34 kPa (Table 3). This shows that for either AS or AM, repetitive measurements of σ_t for different experimental blocks were consistent with each other, though the moisture content and volume of the experimental blocks might be slightly different.

Table 3. Physical and mechanical properties of the soil-root mixture at the two sites.

Tested Site	Tensile Strength of Root T_r (kPa)	Thickness of Slump Block d_1 (m)	Crack Depth d_t (m)	Width of Slump Block b_c (m)	Volume of Slump Block V (m^3)	Tensile Strength Based on Formula (3) σ_t (kPa)
1	31,670	0.25	0.035	0.70	0.175	66.86
	30,950	0.26	0.037	0.72	0.187	67.93
mean	31,310	0.255	0.036	0.71	0.181	67.395
2	15,420	0.35	0.048	0.52	0.182	21.29
	16,890	0.35	0.050	0.53	0.186	22.63
mean	16,155	0.35	0.049	0.525	0.184	21.96

These results suggest that variation of the moisture content and volume of the experimental blocks for the same type of plants has a negligible impact on σ_t. Nonetheless, between the AS and AM, the mean tensile strength of the rooted soil was significantly different, which was 67.39 and 21.96 kPa, respectively (Table 3). The value of σ_t for the AS at site 1 was about 3.07 times higher than that of the AM at site 2 (Table 3). The dramatic difference clearly reflected the differences in the mechanical characteristics of the root systems between the AS and AM blocks. It follows that the characteristics of root distribution and tensile strength of individual roots are the important factors that influence the tensile strength of the soil-root mixture.

The product of RAR and T_r is often used to evaluate the contribution of the root system to the soil tensile strength [9,27,49–54]. For a soil-vegetation mixture, the tensile strength is often viewed as that from the soil and root system, and the latter is much greater than the former [9,51]. In this study, the product of RAR and T_r for the AS and AM was 70.45 and 18.58 kPa, respectively. Their ratio was 3.79, which was greater than that of σ_t for the AS and AM blocks (i.e., 3.07) (Table 3). This indicates that conditions of the root system in a soil-root mixture are critical for determining the mechanical characteristic of the mixture.

In the two excavation tests for AS, the depths of the developed cracks in the collapsed block were 0.035 and 0.037 m, respectively, with the average of 0.036 m (Table 3). However, for AM, these depths were 0.048 and 0.050 m, respectively, giving rise to the average of 0.049 m (Table 3). The crack developed in the AM collapsed block was deeper by about 27% than that in the AS block.

4. Discussion

4.1. Effect of Degraded Riparian Vegetation on Tensile Strength of Individual Roots and the Soil-Root Mixture

Continuous vegetation degradation has forced alpine meadow, dominated by members of the Cyperaceae, to transform into bare land and subsequently Heitutan on the Qinghai-Tibet Plateau in western China. Areas of severely degraded alpine meadow on the Qinghai-Tibet Plateau are referred to as Heitutan and are characterized by increased proportions of bare land, reduced edible herbage, and commensurate increases in the dominance of less palatable species [55]. Specifically, when AS degenerates into AM, the distribution of dominant plants reduced significantly, from an original coverage of 98% to only about 30%. This change of the surface vegetation communities has further led to changes in the plant's underground biomass. Our results showed that the root number of AM block was 46% less than that of the AS block within the depth of 0–0.30 m (Figure 6). For plants of AS, many roots were much longer than this depth, some of which may have a length of up to 0.80 m (Figure 2a). For the plants of AM, however, the lengths of most of their roots are less than 0.30 m (Figure 2b). Thus, transforming from AS to AM means that deep-rooted plants with dense root systems are gradually replaced by plants with short and sparse root systems [35,55–58]. This change indicates that the shallow-rooted plants of AM are more vulnerable to high evaporation, livestock trampling, and human activities in the study area, resulting in further degradation [35].

In addition to their different densities and lengths, the two types of plants also have different root structures. The dominant plant of AS (i.e., *Blysmus sinocompressus*) has developed rhizomes. Roots derived from them often grow laterally (Figure 8a). Moreover, some roots have a wave shape (Figure 8a). In AM, however, most roots of the dominant plant (i.e., *Kobresia pygmaea*) grow vertically with no rhizome (Figure 8b). The other two dominant plants (i.e., *Elymus nutans* and *Potentilla saundersiana*) share a similar root structure. Compared with that of plant roots in AM, the structure of the dominant plant in AS effectively enhances the tensile strength of the soil-root mixture because (1) the laterally distributed roots may increase root density and (2) the wave-shaped roots have greater contact area with the surrounding soils and may resist higher external force by straightening its shape [59].

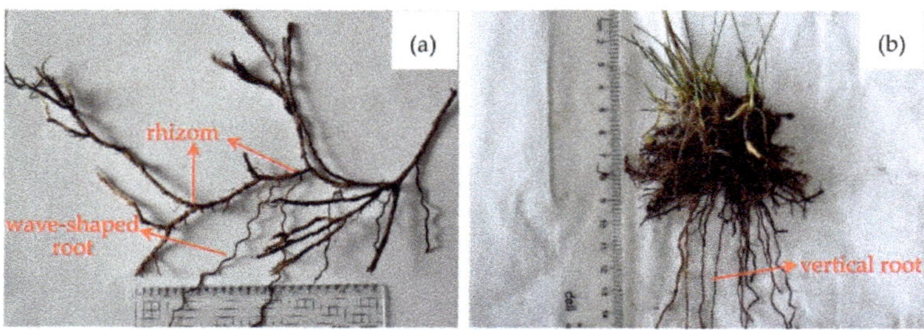

Figure 8. (**a**) The laterally distributed and wave-shaped root system of *Blysmus sinocompressus* in AS; (**b**) the distribution of straight roots in *Kobresia pygmaea* of AM.

The above-mentioned characteristics of root diameters and branching, and the tortuosity of plant roots, mainly affected the mechanical behavior of individual roots [28]. Our study showed that the influence of vegetation degradation is on not only the root number and structure but also on the mechanical characteristics of roots. In the study area, effective roots (i.e., roots with diameters less than 1×10^{-3} m) account for about 93% in the layer of 0–0.30 m (Figure 6) below the surface, indicating that most root systems mainly play the role of reinforcement [60]. When the cantilever arm of the meandering riverbank is formed, the soil and root system is subjected to the external load from the weight of the arm, giving rise to the deformation of the arm. The root system can convert part of the external load into the tensile stress and dissipate it to the surrounding soil through the soil-root interface. In this way, the root system and the surrounding soil particles can work together to balance the load and enhance the soil-root tensile resistance [61].

The tensile strength of the root system is a critical factor that directly reflects the effect of rooted soil consolidation. In our study, the average tensile strength of the root for the dominant plants in the AS reached 31,310 kPa, which was 48% higher than that of the AM (16,155 kPa). Mattia et al. [47] and Li et al. [62] measured the root tensile strength of Gramineae plants of *Lygeum spartum*, *Stipa purpurea* and sedge plant of *Kobresia pygmaea*. They found that their tensile strength ranged from 36,260 to 45,670 kPa, close to that of the sedge plant, *Blysmus sinocompressus* (i.e., 31,310 kPa) in our study. The tensile strength of the root for degraded plants of *Potentilla bifurca*, *Ajania tenuifolia* and *Saussurea salsa* decreases significantly, from 5110 to 25,610 kPa [62]. In our study, this value for the degraded meadow plants (i.e., the AM) was within this range. Evidently, the mechanical characteristics of roots changed because of degeneration. The root tensile strength of healthy meadow plants (i.e., the AS) is much higher than that of the degraded plants (i.e., the AM).

The contribution of the root system to the tensile strength of the rooted soil may be appropriately evaluated using the product of RAR and T_r [9,27,49,51,52]. To further analyze the reduction of the tensile strength in the soil-root mixture due to degradation, we compared our calculated values with those from earlier studies [12,47,62]. All of the products of RAR and root tensile strength of the four degraded herbaceous plants (i.e., 1, 2, 3, 5 in Figure 9) and the AM (i.e., 4) fell in the range of 2.04–38.4 kPa, which are lower than those of four healthy herbaceous plants (i.e., 7, 8, 9, 10 in Figure 9) and the AS (i.e., 6), which are within the range of 72.01–118.74 kPa. Clearly, there is a discrepancy of this product between healthy and degenerated plants, separated by the threshold of 50 kPa (Figure 9).

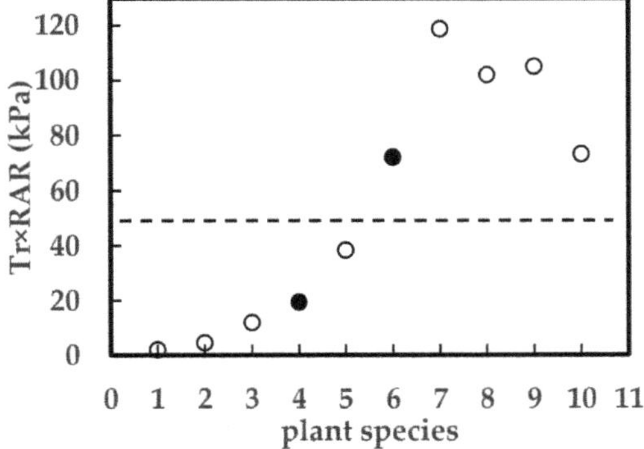

Figure 9. The product of T_r and RAR for healthy and degraded grassland plants. 1. *Saussurea salsa* [62]; 2. *Ajania tenuifolia* [62]; 3. *Potentilla bifurca* [62]; 4. *AM*; 5. *eontopodium nanum* [62]; 6. *AS*; 7. *Stipa purpurea* [62]; 8. *Lygeum spartum* [47]; 9. *Kobresia pygmaea* [62]; 10. *Helictotrichon filifolium* [46]. The hollow circles represent cited results, and the solid ones refer to the results in this study.

Because not all of the tensile strength of the roots is mobilized instantaneously at the moment of bank failure [9,50,51,63], the product of RAR and T_r overestimates the true tensile strength (Table 3). As more and more fiber materials have been used in engineering practices, their ability to improve the tensile strength of the composite has been tested for natural fibers [24,64–67] and synthetic fibers [18,24,58,68]. It is generally believed that fiber materials can reinforce tensile strength, and the degree of reinforcement varies for different fiber materials. Tang et al. [3] proposed that the tensile strength of soil-fiber composite ($\sigma_{composite}$) includes two parts, which are the tensile strength of natural soils (σ_{soil}) and the increase of the tensile strength due to the fiber ($\Delta\sigma_{fiber}$):

$$\sigma_{composite} = \sigma_{soil} + \Delta\sigma_{fiber} \tag{4}$$

According to Zhu et al. [69], the tensile strength of unsaturated clays is 0.7~0.8 times that of the shear strength. In the current study, the ratio of tensile strength to the shear strength of soil without root is taken as 0.75, and the shear strength is 9.29 kPa [70]. Using these values and Equation (4), the reinforced tensile strengths of plant roots in AS and AM may be calculated as 60.43 and 15.0 kPa, respectively. The reinforcement of the healthy plant root system of AS is about 4 times higher than that of the AM. Also, they are less than the products of RAR and T_r by 11.58 and 4.36 kPa, respectively.

The relationship between the value of $\Delta\sigma_{fiber}$ for short (0.03 m) and long (0.05 m) fibers and different fiber contents [24] (Figure 10) was established by setting the root content of AS and AM as 0.48% and 0.18%, respectively [70]. It demonstrated a positive correlation between $\Delta\sigma_{fiber}$ and the fiber content [1,20,21,67], and showed that the longer the fiber length, the greater the strengthening effect [3,24]. In Figure 10, the points representing AS and AM are in line with the established relationship for the fiber lengths of 0.05 and 0.03 m, respectively. This consistency indicates that both the root length and root content contribute to $\Delta\sigma_{fiber}$, and both variables are higher for plants of AS than those for plants of AM.

Furthermore, the tensile strength may also be affected by the initial water content and dry bulk density. In the fiber-reinforced soils, it is negatively related to water content and positively related to dry bulk density [3,24]. The average density of the natural soil-root mixture for AS is 18% higher than that of AM (Table 2). This may be attributed to the decreased root number and the dramatically increased pore volume of surface soil (i.e, soil in the 0–0.05 m layer) [58]. Higher bulk density should increase not only bonding forces between particles, which enhances the tensile strength, but also the interfacial contact area of the fiber-matrix structure, which improves the interfacial shear strength and associated friction [24]. In theory, the bonding forces between particles and friction of the soil-root mixture for AS is stronger than that of AM. Because the difference in the moisture content of the upper layer of soils between AS and AM is merely 1.43% (Table 2), its influence on the tensile strength of the soil-root mixture is relatively small. In this study, the influence of plant degradation on the soil tensile strength is mainly analyzed from a mechanical point of view, and the influences of soil properties (e.g., bulk density, moisture content, porosity, and particle size) on tensile strength and the impacts of degradation on hydraulic properties need to be further studied. It has been well known that the root system can change the hydraulic characteristics of soil [71–74], because it occupies the pore spaces of soil, thus reducing the porosity and increasing the water-holding capacity of soil [75]. In this study, the root number of AM decreased by about 46% compared with that of AS in the depth of 0–30 cm on the upper part of the riverbank, suggesting that the proportion of pores in the soil mass occupied by roots was reduced, which is evidenced by the fact that the density of soil mass was relatively lower than that of AS. Therefore, the water holding capacity of soil mass should be lower in AM than that in AS, implying that its suction of soil mass should also be lower [76].

According to Equation (4), the contributions of the root systems of AS and AM to the tensile strength of the soil-root mixture are 89.67% and 68.3%, respectively. This means that the influence of degradation on the tensile strength of the riverbank is mainly derived from the roots. In our study, when the AS degenerated to the AM, the rooted soil tensile strength decreased by 67.42% (Table 3). Li et al. [58] showed that the shear strength of the root-soil mixture decreases by 36.0% and 52.3% from severely degraded alpine meadows to moderately and slightly degraded alpine meadows, respectively. Thus, degradation of alpine swamp meadows obviously reduces the mechanical properties of the soils and weakens their ability to resist bank failure, wind and water erosion, and other external forces.

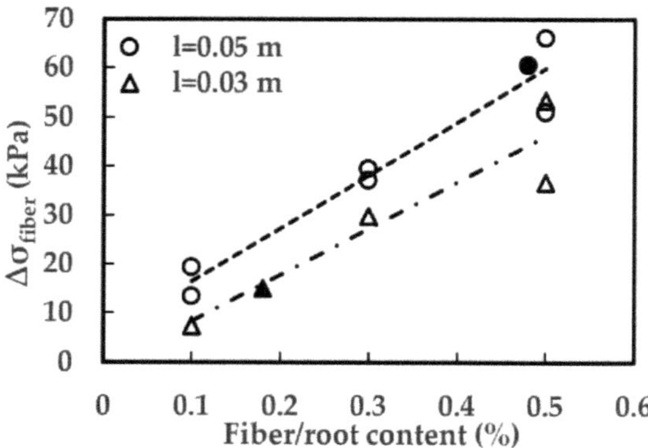

Figure 10. The relationship between fiber (root) content and reinforced tensile strength. The letter l in the diagram stands for the fiber length. The hollow circles and triangles are the data from Meriem et al. [24], while the solid circle and triangle refer to the reinforced tensile strength of AS and AM in this study, respectively.

4.2. The Role of Root System in Preventing Development of Riverbank Cracks

The development of cracks greatly destroys the integrity of the soil structure, weakens the mechanical properties of soils, reduces stability, increases permeability, intensifies evaporation, and increases soil erosion, resulting in a series of subsequent adverse effects on geotechnical engineering and the environment [77–80]. The tensile strength of the soil is an important mechanical parameter that controls the initiation and propagation of tensile cracks [3]. Therefore, enhancing the tensile strength of the soil and preventing the occurrence or slowing the expansion of cracks are critical for riverbank protection. Because plants in the AS have dense roots and strong single-root pullout resistance, they may enhance the tensile ductility of the rooted soils and inhibit the initial formation of tensile cracks. In our study, the expansion rate of tension cracks in cantilever arms due to weights of the AS blocks and the duration of block failure are generally slower than those for the AM blocks. This indicates that types of riparian vegetation, fiber root number within the soils, and the root tensile strength may considerably constrain the number of cracks and their propagation rates [81].

When a tested block of cantilever arms is about to fail, a penetrating crack forms from the surface of AS or AM by following the path of least resistance (Figure 11). For the tested block of AS, the crack had a zigzag shape, and its length was about 1.24 m, which is longer than the crack length of AM by 0.14 m. The crack propagated with a preference angle until it was interrupted by roots. This angle seemed greater in the block with a higher root content. This property is consistent with the result of Meriem et al. [24]. At the depth of 0–0.3 m, the root number of AS that passing through the collapsed profile was 1.84 times that of the root number of AM (Figure 6). As a result, a crack in AS propagated in a way that avoided most roots. Given that the length of the crack was longer than that of AM, the time needed to penetrate through the soil-root mixture is longer. Nonetheless, the crack pattern is smoother on the surface of the AM block because of its lower number of roots.

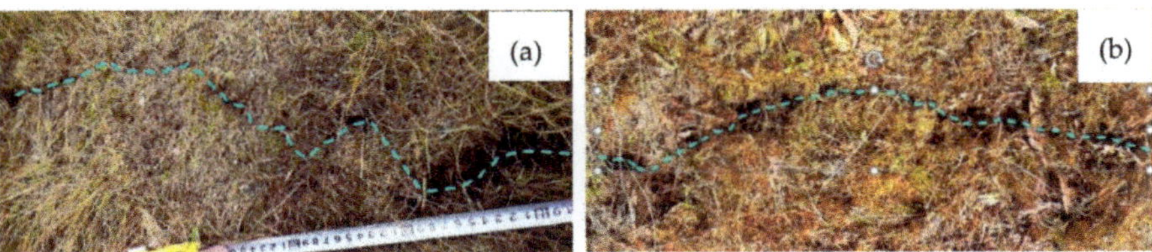

Figure 11. The planform crack pattern at the surface of the block in (**a**) AS and (**b**) AM.

Considering the impact of the degraded alpine swamp meadow on tensile strength of riverbank in the UYR region, the grassland management should practice regional rotation grazing, rational distribution of herds, and balanced use of the riparian alpine swamp meadow. In addition, anthropogenic disturbances, in particular engineering construction, should be reduced as much as possible.

5. Conclusions

Under the influence of climate change and increased anthropogenic disturbances, the alpine swamp meadow (AS), which is the main type of vegetation cover in the meandering riverbank of the Upper Yellow River (UYR), is subject to severe degradation and typically has transformed into the alpine meadow (AM). However, little is known about the tensile strength of the roots in soils and the influence of riparian vegetation degradation on the tensile strength of riverbanks. To reflect the actual interaction between soils and the roots with the natural root branching structure for AS and AM and its effect on bank strength, we measured properties of root vertical distribution and number and performed in situ root pullout and artificial excavation tests. Our results led to the following conclusions. First, though spatial and size distributions of roots in the soil-root mixture of AS and AM were similar, AS was characterized by a higher number of roots than AM. Second, the tensile strength of individual roots decreased with the diameter of roots in both AS and AM. Yet, the former always had higher tensile strength than the latter for any given root diameter. Similarly, the tensile strength of the soil-root mixture in AS was about three times higher than that in AM. This difference is mainly caused by the fact that the lateral extended and wave-shaped root structure in AS is more effective in enhancing the resistance of the soil-root mixture to the external force than the simple vertically distributed root structure in AM. Third, the tensile crack developed in the collapsed block for AM was deeper than that for AS, indicating the reduced resistance to the external force as AS gradually degraded to AM. The impact of the roots for the degraded vegetation (i.e., AM) was also reflected by the relatively smooth and shorter crack route on the surface of the collapsed block for AM. These findings call for better ecological management for preventing AS from being degraded to AM.

Although grassland degradation is a worldwide problem [82,83], how such degradation affects the mechanical properties of riparian riverbanks in the alpine environment has not been fully understood.

Our study provides firsthand evidence of weakened bank strength in the soil-root mixture due to degradation of vegetation and will serve as a benchmark for future investigation of riverbank strength, not only in the UYR, but also in other alpine regions in the world.

Author Contributions: Conceptualization, H.Z. and P.G.; Data curation, H.Z., G.L.; Formal analysis, H.Z., P.G.; Funding acquisition, H.Z., Z.L., G.L. and X.L.; Investigation, H.Z., Z.L., J.F., G.L., Y.L. and X.H.; Methodology, H.Z., P.G. and Z.L.; Project administration, H.Z.; Resources, X.L., G.L. and J.F.; Writing—original draft, H.Z.; Writing—review & editing, P.G. All authors have read and agreed to the published version of the manuscript.

Funding: This research was funded by the National Natural Science Foundation of China (Grant No. 41302258, 41762023, 41662023, and 51979012), the Project of Qinghai Science & Technology Department (Grant No. 2017-ZJ-776), the Natural Science Foundation Innovation Team Project of Qinghai Provincial Department of Science and Technology (Grant No. 2020-ZJ-904).

Acknowledgments: The authors thank undergraduate students B.L., B.X., R.Z. and Y.G. for data collection in the field. Thanks are extended to the anonymous reviewers and Editor for useful comments that improved the manuscript.

Conflicts of Interest: The authors declare no conflict of interest.

References

1. Ziegler, S.; Leshchinsky, D.; Ling, H.L.; Perry, E.B. Effect of short polymeric fibers on crack development in clays. *Soils Found.* **1998**, *38*, 247–253. [CrossRef]
2. Li, G.X.; Chen, L. Study of centrifugal model tests on texsol and cohesive soil slopes. *Chin. J. Geotech. Eng.* **1998**, *20*, 12–15. (In Chinese)
3. Tang, C.S.; Wang, D.Y.; Cui, Y.Y. Tensile Strength of Fiber reinforced soil. *Mater. Civ. Eng.* **2016**, *28*, 1–13. [CrossRef]
4. Divya, P.V.; Viswanadham, B.V.S.; Gourc, J.P. Evaluation of tensile strength-strain characteristics of fiber reinforced soil through laboratory tests. *J. Mater. Civ. Eng.* **2014**, *26*, 14–23. [CrossRef]
5. Simon, A.; Curini, A.; Darby, S.E.; Langendoen, E.J. Bank and near-bank processes in an incised channel. *Geomorphology* **2000**, *35*, 193–217. [CrossRef]
6. Schmidt, K.M.; Roering, J.J.; Stock, J.D.; Dietrich, W.E.; Montgomery, D.R.; Schaub, T. The variability of root cohesion as an influence on shallow landslide susceptibility in the Oregon Coast Range. *Can. Geotech. J.* **2001**, *38*, 995–1024. [CrossRef]
7. Roering, J.J.; Schmidt, K.M.; Stock, J.D.; Dietrich, W.E.; Montgomery, D.R. Shallow landsliding, root reinforcement, and the spatial distribution of trees in the Oregon Coast Range. *Can. Geotech. J.* **2003**, *40*, 237–253. [CrossRef]
8. Hubble, T.C.T.; Docker, B.B.; Rutherford, I.D. The role of riparian trees in maintaining riverbank stability: A review of Australian experience and practice. *Ecol. Eng.* **2010**, *36*, 292–304. [CrossRef]
9. Abernethy, B.; Rutherford, I.D. Where along a river's length will vegetation most effectivelystabilise stream banks? *Geomorphology* **1998**, *23*, 55–75. [CrossRef]
10. Thorne, C.R.; Tovey, N.K. Stability of composite river banks. *Earth Surf. Process. Landf.* **1981**, *6*, 469–484. [CrossRef]
11. Abernethy, B.; Rutherford, I.D. The distribution and strength of riparian tree roots in relation to riverbank reinforcement. *Hydrol. Process.* **2001**, *15*, 63–79. [CrossRef]
12. Simon, A.; Collison, A.J.C. Quantifying the mechanical and hydrologic effects of riparian vegetation on streambank stability. *Earth Surf. Process. Landf.* **2002**, *27*, 527–546. [CrossRef]
13. Yu, G.A.; Li, Z.; Yang, H.; Lu, J.; Huang, H.Q.; Yi, Y. Effects of riparian plant roots on the unconsolidated bank stability of meandering channels in the Tarim River, China. *Geomorphology* **2020**, *351*, 106958. [CrossRef]
14. Van De Wiel, M.J.; Darby, S.E. A new model to analyse the impact of woody riparian vegetation on the geotechnical stability of riverbanks. *Earth Surf. Proc. Landf.* **2007**, *32*, 2185–2198. [CrossRef]
15. Parker, G.; Shimizu, Y.; Wilkerson, G.V.; Eke, E.C.; Abad, J.D.; Lauer, J.W.; Paola, C.; Dietrich, W.E.; Voller, V.R. A new framework for modeling the migration of meandering rivers. *Earth Surf. Proc. Landf.* **2011**, *36*, 70–86. [CrossRef]
16. Zhu, H.L.; Hu, X.S.; Li, Z.W.; Song, L.; Li, K.; Li, X.L.; Li, G.R. The influences of riparian vegetation on bank failures of a small meadow-type meandering river. *Water* **2018**, *10*, 692. [CrossRef]

17. Abernethy, B.; Rutherford, I.D. The effect of riparian tree roots on the mass-stability of riverbanks. *Earth Surf. Process. Landf.* **2000**, *25*, 921–937. [CrossRef]
18. Samadi, A.; Amiri-Tokaldany, E.; Davoudi, M.H.; Darby, S.E. Experimental and numerical investigation of the stability of overhanging riverbanks. *Geomorphology* **2013**, *184*, 1–19. [CrossRef]
19. Lawton, E.C.; Khire, M.V.; Fox, N.S. Reinforcement of Soils by Multi-oriented Geo-synthetic Inclusions. *J. Geotech. Eng.* **1993**, *119*, 257–275. [CrossRef]
20. Plé, O.; Lê, T.N.H. Effect of polypropylene fiber reinforcement on the mechanical behavior of silty clay. *Geotext. Geomembr.* **2012**, *32*, 111–116. [CrossRef]
21. Mesbah, A.; Morel, J.C.; Walker, P.; Ghavami, K. Development of a direct tensile test for compacted earthblocks reinforced with natural fibers. *J. Mater. Civ. Eng.* **2004**, *16*, 95–98. [CrossRef]
22. Onur Akaya, A.; Özera, T.; Foxb, G.A.; Wilson, G.V. Application of fibrous streambank protection against groundwater seepage erosion. *J. Hydrol.* **2018**, *565*, 27–38. [CrossRef]
23. Das, N.; Singh, S.K. Geotechnical behaviour of lateritic soil reinforced with brown waste and synthetic fiber. *Int. J. Geotech. Eng.* **2019**, *13*, 287–297. [CrossRef]
24. Meriem, C.; Houda, G.; Mehrez, J. Tensile behaviour analysis of compacted clayey soil reinforced with natural and synthetic fibers: Effect of initial compaction conditions. *Eur. J. Environ. Civ. Eng.* **2020**, *24*, 354–380.
25. Wu, T.H.; McKinnell, W.P., III; Swanston, D.N. Strength of tree roots and landslides on Prince of Wales Island, Alaska. *Can. Geotech. J.* **1979**, *16*, 19–33. [CrossRef]
26. Waldron, L.J.; Dakessian, S. Soil reinforcement by roots: Calculation of increased soil shear resistance from root properties. *Soil Sci.* **1981**, *132*, 427–435. [CrossRef]
27. Pollen, N.; Simon, A. Estimating the mechanical effects of riparian vegetation on stream bank stability using a fiber bundle model. *Water Resour. Res.* **2005**, *41*, 1–11. [CrossRef]
28. Schwarz, M.; Lehmann, P.; Or, D. Quantifying lateral root reinforcement in steep slopes: From a bundle of roots to tree stands. *Earth Surf. Proc. Landf.* **2010**, *35*, 354–367. [CrossRef]
29. Li, Y.Z.; Fu, J.T.; Yu, D.M.; Hu, X.S.; Zhu, H.L.; Li, G.Y.; Hu, X.T. Mechanical effects of halophytes roots and optimal root content for slope protection in cold and arid environment. *Chin. J. Rock Mech. Eng.* **2015**, *34*, 1370–1383. (In Chinese)
30. Liu, Y.B.; Hu, X.S.; Yu, D.M.; Li, S.X.; Yang, Y.Q. Microstructural features and friction characteristics of the interface of shrub roots and soil in loess area of Xining Basin. *Chin. J. Rock Mech. Eng.* **2018**, *37*, 1270–1280. (In Chinese)
31. Li, Z.W.; Yu, G.A.; Xu, M.Z.; Hu, X.Y.; Yang, H.M.; Hu, S.X. Progress in studies on river morphodynamics in Qinghai-Tibet Plateau. *Adv. Water Sci.* **2016**, *27*, 617–628.
32. Li, J.; Zhang, F.W.; Lin, L.; Li, H.Q.; Du, Y.G.; Li, Y.K.; Cao, G.M. Response of the plant community and soil water status to alpine *Kobresia* meadow degradation gradients on the Qinghai–Tibetan Plateau, China. *Ecol. Res.* **2015**, *30*, 589–596. [CrossRef]
33. Wang, G.X.; Wang, Y.B.; Li, Y.S.; Cheng, H.Y. Influences of alpine ecosystem responses to climatic change on soil properties on the Qinghai-Tibet Plateau, China. *Catena* **2007**, *70*, 506–514. [CrossRef]
34. Wang, G.X.; Liu, G.S.; Li, C.J. Effects of changes in alpine grassland vegetation cover on hillslope hydrological processes in a permafrost watershed. *J. Hydrol.* **2012**, *444–445*, 22–33.
35. Zeng, C.; Zhang, F.; Wang, Q.J.; Chen, Y.Y.; Joswiak, D.R. Impact of alpine meadow degradation on soil hydraulic properties over the Qinghai-Tibetan Plateau. *J. Hydrol.* **2013**, *478*, 148–156. [CrossRef]
36. Pan, T.; Hou, S.; Wu, S.H.; Liu, Y.J.; Liu, Y.H.; Zou, X.T.; Herzberger, A.; Liu, J.G. Variation of soil hydraulic properties with alpine grassland degradation in the eastern Tibetan Plateau. *Hydrol. Earth Syst. Sci.* **2017**, *21*, 2249–2261. [CrossRef]
37. Krzeminska, D.; Kerkhof, T.; Skaalsveen, K.; Stolte, J. Effect of riparian vegetation on stream bank stability in small agricultural catchments. *Catena* **2019**, *172*, 87–96. [CrossRef]
38. Duan, S.Q. Runoff spatial difference of small-scale in Huangnan, Qinghai province and its cause. *Adv. Water Sci.* **2016**, *27*, 11–21.
39. Yu, Z.X. Investigation of variety resources of Oula sheep at Henan county in Qinghai province. *Anim. Husb. Feed Sci.* **2009**, *30*, 120–124. (In Chinese)
40. Ren, G.H.; Deng, B.; Hou, Y. Changes of community characteristics in the degradation process of the alpine swamp wetland in the Yellow River Source area. *Pratacult. Sci.* **2015**, *32*, 1222–1229. (In Chinese)

41. Xie, B.S.; Zhu, H.L.; Li, B.F.; Hu, X.S. Study on relationship between vegetation spatial distribution and soil properties in the meander riverside in source region of the Yellow River. *J. Sediment Res.* **2019**, *44*, 66–73. (In Chinese)
42. Schenk, H.J.; Jackson, R.B. Rooting depths, lateral root spreads and below-ground/above-ground allometries of plants in water-limited ecosystems. *J. Ecol.* **2002**, *90*, 480–494. [CrossRef]
43. Ministry of Construction. *GB/T 50123–1999. Standard for Soil Test Method*; Ministry of Construction: Beijing, China, 2010.
44. Xia, J.Q.; Zong, Q.L.; Xu, Q.X.; Deng, C.Y. Soil properties and erosion mechanisms of composite riverbanks in Lower Jingjiang Reach. *Adv. Water Sci.* **2013**, *24*, 810–820. (In Chinese)
45. Ajaz, A. Stress-Strain Behaviour of Compacted Clays in Tension and Compression. Ph.D. Thesis, Cambridge University, Cambridge, MA, USA, 1973.
46. Baets, S.D.; Poesen, J.; Reubens, B.; Wemans, K.; Baerdemaeker, J.D.; Muys, B. Root tensile strength and root distribution of typical Mediterranean plant species and their contribution to soil shear strength. *Plant Soil* **2008**, *305*, 207–226. [CrossRef]
47. Mattia, C.; Bischetti, G.B.; Gentile, F. Biotechnical characteristics of root systems of typical Mediterranean species. *Plant Soil* **2005**, *278*, 23–32. [CrossRef]
48. Bischetti, G.B.; Chiaradia, E.A.; Simonato, T.; Speziali, B.; Vitali, B.; Vullo, P.; Zocco, A. Root strength and root area ratio of forest species in Lombardy (Northern Italy). *Plant Soil* **2005**, *278*, 11–22. [CrossRef]
49. Docker, B.B.; Hubble, T.C.T. Quantifying root-reinforcement of river bank soils by four Australian tree species. *Geomorphology* **2008**, *100*, 401–418. [CrossRef]
50. Lawrence, C.J.; Rickson, R.J.; Clark, J.E. The effect of grass roots on the shear strength of colluvial soils in Nepal. In *Advances in Hillslope Processes*; Anderson, M.G., Brooks, S.M., Eds.; John Wiley and Sons: Chichester, UK, 1996; Volume 2, pp. 857–868.
51. Reubens, B.; Poesen, J.; Danjon, F.; Geudens, G.; Muys, B. The role of fine and coarse roots in shallow slope stability and soil erosion control with a focus on root system architecture: A review. *Trees* **2007**, *21*, 385–402. [CrossRef]
52. Schwarz, M.; Cohen, D.; Or, D. Soil-root mechanical interactions during pullout and failure of root bundles. *Geophys. Res.* **2010**, *115*, 1–19. [CrossRef]
53. Schwarz, M.; Preti, F.; Giadrossich, F.; Lehmann, P.; Or, D. Quantifying the role of vegetation in slope stability: A case study in Tuscany (Italy). *Ecol. Eng.* **2010**, *36*, 285–291. [CrossRef]
54. Zhu, J.Q.; Wang, Y.Q.; Wang, Y.J.; Zhang, H.L.; Li, Y.P.; Li, Y. An analysis on soil physical enhancement effects of root system of *PinusTabulaeformis* and *Acer Truncatum* based on two models. *Bull. Soil Water Conserv.* **2015**, *35*, 277–282. (In Chinese)
55. Li, X.L.; Perry, G.L.W.; Brierley, G.; Sun, H.Q.; Li, C.H.; Lu, G.X. Quantitative assessment of degradation classifications for degraded alpine meadows (heitutan), Sanjiangyuan, Western China. *Land Degrad. Dev.* **2014**, *25*, 417–427. [CrossRef]
56. Cheng, H.Y.; Wang, G.X.; Hu, H.C.; Wang, Y.B. The variation of soil temperature and water content of seasonal frozen soil with different vegetation coverage in the headwater region of the Yellow River, China. *Environ. Geol.* **2008**, *54*, 1755–1762. [CrossRef]
57. Niu, Y.J.; Zhou, J.W.; Yang, S.W.; Chu, B.; Zhu, H.M.; Zhang, B.; Fang, Q.G.; Tang, Z.S.; Hua, L.M. Plant diversity is closely related to the density of zokor mounds in three alpine rangelands on the Tibetan Plateau. *PeerJ* **2019**, *7*, e6921. [CrossRef]
58. Li, G.R.; Li, X.L.; Chen, W.T.; Li, J.F.; Zhu, H.L.; Hu, X.S.; Zhou, H.K.; Sun, H.Q. Effects of degradation severity on the physical, chemical and mechanical properties of topsoil in alpine meadow on the Qinghai-Tibet Plateau, west China. *Catena* **2020**, *187*, 104370. [CrossRef]
59. Tang, C.S.; Li, J.; Wang, D.Y.; Shi, B. Investigation on the interfacial mechanical behavior of wave-shaped fiber reinforced soil by pullout test. *Geotext. Geomembr.* **2016**, *44*, 872–883. [CrossRef]
60. Li, Y.; Zhu, X.M.; Tian, J.Y. The effectiveness of plant roots in improving soil anti-scourability on the loess plateau. *Sci. Bull.* **1991**, *12*, 935–938. (In Chinese)
61. Zhou, Y.; Watts, D.; Cheng, X.P.; Li, Y.H.; Luo, H.S.; Xiu, Q. The traction effect of lateral roots of Pinus yunnanensis on soil reinforcement: A direct in situ test. *Plant Soil* **1997**, *190*, 77–86. [CrossRef]

62. Li, G.Y.; Hu, X.T.; Li, X.L.; Yu, D.M.; Fu, J.T.; Zhu, H.L.; Hu, X.S. Mechanical Effects of Alpine Grassland Plants in Slope Protection in Maqin County of the Source Area of the Yellow River. *Mt. Res. Dev.* **2014**, *32*, 550–560. (In Chinese)
63. Comino, E.; Druetta, A. The effect of Poaceae roots on the shear strength of soils in the Italian alpine environment. *Soil Tillage Res.* **2010**, *106*, 194–201. [CrossRef]
64. Bessadok, A.; Roudesli, S.; Marais, S.; Follain, N.; Lebrun, L. Alfa fibers for unsaturated polyester composites reinforcement: Effects of chemical treatments on mechanical and permeation properties. *Compos. Part A* **2009**, *40*, 184–195. [CrossRef]
65. Tran, H.N.; Shinji, O.; Nguyen, H.T.; Satoshi, K. Effect of alkali treatment on interfacial and mechanical properties of coir fiber reinforced poly(butylene succinate) biodegradable composites. *Compos. Part B* **2011**, *42*, 1648–1656.
66. Parisi, F.; Asprone, D.; Fenu, L.; Prota, A. Experimental characterization of Italian composite adobe bricks reinforced with straw fibers. *Compos. Struct.* **2015**, *122*, 300–307. [CrossRef]
67. Khiem, Q.T.; Tomoaki, S.; Hiroshi, T. Improvement of mechanical behavior of cemented soil reinforced with waste cornsilk fibers. *Constr. Build. Mater.* **2018**, *178*, 204–210.
68. Tang, C.S.; Shi, B.; Cai, Y.; Gao, W.; Chen, F.J. Experimental study on polypropylene fiber improving soft soils. *Rock Soil Mech.* **2007**, *28*, 1796–1800. (In Chinese)
69. Zhu, C.H.; Liu, J.M.; Yan, B.W.; Ju, J.L. Experimental study on relationship between tensile and shear strength of unsaturation clay earth material. *Chin. J. Rock Mech. Eng.* **2008**, *27* (Suppl. S2), 3453–3458. (In Chinese)
70. Li, B.F.; Zhu, H.L.; Xie, B.S.; Luo, L.Y.; Li, G.R.; Hu, X.S. Study on tensile properties of root-soil composite of alpine meadow plants in the riparian zone of the Yellow River source region. *Chin. J. Rock Mech. Eng.* **2020**, *39*, 424–432. (In Chinese)
71. Leung, A.K.; Garg, A.; Ng, C.W.W. Effects of plant roots on soil–water retention and induced suction in vegetated soil. *Eng. Geol.* **2015**, *193*, 183–197. [CrossRef]
72. Ng, C.W.W.; Leung, A.K.; Woon, K.X. Effects of soil density on grass-induced suction distributions in compacted soil subjected to rainfall. *Can. Geotech. J.* **2014**, *51*, 311–321. [CrossRef]
73. Gabrm, A.; Akran, M.; Taylor, H.M. Effect of simulated roots on the permeability of silty soil. *Geotech. Test. J.* **1995**, *18*, 112–115.
74. Huat, B.B.K.; Alif, H.J.; Low, T.H. Water infiltration characteristics of unsaturated soil slope and its effect on suction and stability. *Geotech. Geol. Eng.* **2006**, *24*, 1293–1306. [CrossRef]
75. Ng, C.W.W. Atmosphere-plant-soil interactions: Theories and mechanisms. *Chin. J. Geotech. Eng.* **2017**, *39*, 1–47. (In Chinese)
76. Pollen, N.; Simon, A. Hydrologic and hydraulic effects of riparian root networks on streambank stability: Is mechanical root-reinforcement the whole story? *Geomorphology* **2010**, *116*, 353–362. [CrossRef]
77. Morris, P.H.; Graham, J.; Wiliams, D.J. Cracking in drying soils. *Can. Geotech. J.* **1992**, *29*, 263–277. [CrossRef]
78. Albrecht, B.A.; Benson, C.H. Effect of desiccation on compacted natural clays. *J. Geotech. Geoenviron.* **2001**, *127*, 67–75. [CrossRef]
79. Peron, H.; Hueckel, T.; Laloui, L.; Hu, L.B. Fundamentals of desiccation cracking of fine-grained soils: Experimental characterization and mechanisms identification. *Can. Geotech. J.* **2009**, *46*, 1177–1201. [CrossRef]
80. Tang, C.S.; Shi, B.; Liu, C.; Zhao, L.Z.; Wan, B.J. Influencing factors of geometrical structure of surface shrinkage cracks in clayey soils. *Eng. Geol.* **2008**, *101*, 204–217. [CrossRef]
81. Miller, C.J.; Rifai, S. Fiber reinforcement for waste containment soil liners. *J. Environ. Eng.* **2004**, *130*, 891–895. [CrossRef]
82. Muller, S.; Dutoit, T.; Alard, D.; Grévilliot, F. Restoration and rehabilitation of species-rich grassland ecosystems in France: A review. *Restor. Ecol.* **1998**, *6*, 94–101. [CrossRef]
83. Carrick, P.J.; Krüger, R. Restoring degraded landscapes in lowland Namaqualand: Lessons from the mining experience and from regional ecological dynamics. *J. Arid Environ.* **2007**, *70*, 767–781. [CrossRef]

Article

Static Liquefaction Capacity of Saturated Undisturbed Loess in South Jingyang Platform

Rui-Xin Yan [1,2], Jian-Bing Peng [2,*], Jin-Yuan Zhang [3,*] and Shao-kai Wang [2]

1. Department of Architecture and Civil Engineering, Xi'an Technological University, Xi'an 710021, China; yanruixin1985@163.com
2. Key Laboratory of Western China's Mineral Resource and Geological Engineering, Ministry of Education, Department of Geological Engineering and Surveying of Chang'an University, Xi'an 710054, China; 2015026020@chd.edu.cn
3. Department of Architecture and Civil Engineering, Xi'an University of Science and Technology, Xi'an 710054, China

* Correspondence: dicexy_1@chd.edu.cn (J.-B.P.); 18204058032@stu.xust.edu.cn (J.-Y.Z.); Tel.: +86-139-9288-1198 (J.-B.P.); +86-186-5318-3466 (J.-Y.Z.)

Received: 15 June 2020; Accepted: 11 August 2020; Published: 16 August 2020

Abstract: According to a previous geological investigation, high-speed and long-distance loess landslides in the South Jingyang platform in Shaanxi Province are closely related to the static liquefaction of loess. Considering the typical loess landslides in this area, isotropic consolidated undrained (ICU) triaxial tests and scanning electron microscopy analyses were conducted in this study. The main conclusions are as follows: (1) The stress-strain curves indicate strong strain softening under different confining pressures. The pore water pressure increases significantly and then remains at a high level; (2) The liquefaction potential index (LPI) shows an increasing trend followed by stabilization; the larger the LPI is, the smaller the state parameter (ψ) is. The steady-state points of the loess are in the instability region; however, the steady-state strength is not zero; (3) Based on the ICU test results, the average pore diameter decreases; the shape ratio remains essentially unchanged; and the fractal dimension and roundness show different trends. The proportions of the macropore and mesopore decrease; that of the small pore increases slightly; and that of the micropore increases significantly; (4) The compression deformation of the highly spaced pores causes rapid strain hardening. A rapid strain softening results from the pore throat blockage at the beginning of particle rearrangement and reorganization. A stable strain softening is related to the agglomeration blocking of the reconstructed pore throat in the gradually stable stage of particle rearrangement and reorganization.

Keywords: loess; ICU; static liquefaction; mechanical behavior; pore structure

---

1. Introduction

The phenomenon of static liquefaction was first discovered by Terzaghi and Peck [1] when they performed an experiment on saturated silty fine sand. The phenomenon was described as follows: saturated silty fine sand with uniform viscous liquid properties appears under a very small disturbance action; this was termed "spontaneous liquefaction". Subsequently, many scholars [2–4] have reported this phenomenon. They proposed that this phenomenon should be distinguished from the liquefaction of saturated sand under a dynamic load, and they studied it separately. Therefore, a more accurate conceptual description of "static liquefaction" was gradually formed. In other words, the stress-strain curve of the material element or the sample shows an obvious strain softening characteristic in the static loading process. This indicates that the deviator stress can only maintain a very low shear strength after reaching the peak value. When the external load continues to act, the soil shows strong instability. With the further cognition of the phenomenon of static liquefaction, many scholars [5–7] began to understand the influence of static liquefaction on the loss of the soil resistance, and performed detailed analysis of engineering disasters resulting from static liquefaction. Several laboratory tests were also conducted, and their results have provided a clear understanding of the mechanism of static liquefaction.

Due to the characteristics of loess, such as being porous, having weak cementation, and water sensitivity, it is prone to cause slope instability due to static liquefaction under the saturated condition. According to analyses of the characteristics of the loess landslides in the South Jingyang platform, Shaanxi Province, the loess landslides are closely related to static liquefaction of the saturated loess. Relevant scholars have carried out systematic research on the static liquefaction typical characteristics of loess landslides in this region. Leng et al. [8] speculated that the movement mechanism of the landslide was related to the liquefaction of loess, considering the water accumulation at the toe of the slope and the unique properties of the slip soil after the occurrence of the landslide in this area. In the process of shearing, the loess liquefaction led to a sharp decrease in the shear strength of the displaced landslide materials. Through a field investigation of the Western Miaodian and Zhaitou landslides and using a digital elevation model, Peng et al. [9] found that the static liquefaction of the sliding surface was caused by the loss of shear resistance and bond strength. Xu [10] conducted scores of field investigations and measurements of the loess landslides in the South Jingyang platform. It was determined that the shear opening of the flow landslides was low, and most of the landslides were in the loess saturated zone on the edge of the plateau and were caused by a rise in the groundwater level. The mechanism for these kinds of loess landslides is related to the static liquefaction of the saturated loess at the bottom. In addition, the slope of the flow slip (about 12°) is much lower than the internal friction angle of the saturated loess, which also preliminarily confirms loess liquefaction in the process of flowing. According to Li et al. [11], the pore water pressure produced by the undrained shear at the bottom of the sliding body in the loess tableland area is much higher than the pore water pressure inside the sliding body. Therefore, shearing occurred at the bottom. The high pore water pressure liquefied the bottom material, which resulted in a large difference between the driving force and resistance.

The phenomenon and mechanism of loess static liquefaction depends on the systematic experimental analysis. Therefore, relevant scholars have carried out laboratory experimental research and analyzed the mechanism of loess static liquefaction. For example, Ma et al. [12] determined that surface water infiltration led to water accumulation at the bottom of the loess layer through isotropic consolidated undrained (ICU) and ring shear tests. The loess in this layer was highly liquefied: a high pore water pressure was generated rapidly, and a deviatoric stress reached the maximum value under low deformation. Zhuang et al. [13] conducted ring shear tests on saturated loess in this region and determined that the pore water pressure and shear resistance immediately rose to the maximum value after shearing. Thereafter, the pore water pressure was maintained near the maximum value; however, half of the shear resistance was lost after a small shear displacement. The effective stress path showed that the saturated loess was prone to liquefaction under the undrained shear action. This indicates that the loess landslide had liquefied during the sliding process. Yan et al. [14] conducted ICU tests on the loess in this area and determined that it was easy to form a saturated softening zone on the top of the paleosol layer (relative aquiclude) at the toe of the slope. The relative sliding between the loess particles at this location can increase the pore water pressure sharply and reduce the effective stress. After reaching the steady state, the saturated loess was in a low-confining-pressure unstable state and was prone to plastic flow. Li and Jin [15] used the global digital systems (GDS) triaxial apparatus to conduct isotropic/anisotropic consolidated undrained (ICU/ACU) compression tests and constant-shear-drained (CQD) triaxial tests of the undisturbed loess in the Dongfeng landslide for the South Jingyang platform. The results showed that the stress-strain mode of the soil displayed a strong strain softening type, and the starting friction angle of the soil failure was far less than the steady-state friction angle. When the stress path passed through the instability area, incomplete drainage shear shrinkage failure occurred, which is a typical static liquefaction phenomenon. Liu et al. [16] carried out undrained monotonic triaxial shear tests on saturated undisturbed loess samples taken from Heping Town, Yuzhong County, Lanzhou City. They found that loess samples with confining pressure of 100 kPa and 150 kPa exhibited complete static liquefaction characterized by that the pore water pressure reached the level of the initial confining pressure and the effective confining pressure decreased to zero. Wang et al. [17] also determined that the final pore pressures in the undrained triaxial compression tests were as great as 65% of the initial confining pressures. Besides, through effective stress path passing through flow liquefaction line (FLL), it can be concluded that the loess is susceptible to flow liquefaction failure. Pei et al. [18] conducted a series of consolidated undrained triaxial compression tests on undisturbed loess obtained from the source area of Shibeiyuan landslide. The tests revealed that although the confining pressures were different (≤200 kPa), the effective stress decreased with the accumulation of excess pore pressure and the steady state strength decreased continuously to zero. Xu et al. [19,20] carried out ICU tests on undisturbed loess retrieved from the backwall of a loess landslide in Heifangtai plateau, Lanzhou City, Gansu Province. The loess reached peak shear strength with an axial strain of <2%, accompanied by a sharp increase in pore pressure and followed by monotonic strain softening. The results showed that the saturated loess had strong liquefaction potential. Based on the study of the mechanical behavior of the static liquefaction of the saturated loess, the mechanism of the static liquefaction of loess has been clearly summarized as follows: ① the explosion of the pore water pressure; ② the effective stress drops sharply, even to zero; ③ and the occurrence of flow plastic deformation. However, the internal mechanism is still unclear, such as: ① an unclear understanding of the mechanism of the explosion of the pore water pressure in the process of static liquefaction; ② incomplete cognition of the relationship between the occurrence of static liquefaction and the change in the internal structure of the saturated loess; and ③ the essence of the sudden drop in the strength of the saturated loess due to static liquefaction, which can result in a sharp drop in the effective stress. However, the mechanism of the sudden drop in strength that is caused by the loess structure is not clear. Therefore, analyzing the internal mechanism of loess liquefaction and scientifically understanding the three typical characteristics of static liquefaction from multiple perspectives are the key work in the next stage.

In view of the above shortcomings, we carried out the ICU tests under different confining pressures for the saturated undisturbed loess. In addition, scanning electron microscopy was performed before and after the ICU tests. The stress-strain characteristics of the saturated undisturbed loess were studied. The variation trend of the pore water pressure and of the corresponding mechanical mechanism was also analyzed. Finally, the evolution process and internal mechanism of static liquefaction were determined.

2. Development Characteristics of the Landslide and the Correlation Analysis with Static Liquefaction

The South Jingyang platform is located in the South Bank of the Jinghe River, Jingyang County, Shaanxi Province. The platform is 30–90 m higher than the Jinghe River, and the altitude is 450–500 m. It is almost parallel to the flow direction of the Jinghe River and is distributed in a belt. The South Jingyang platform presents the east–west extension trend as a whole; the total extension is 28 km; and it mainly spans the Taiping, Jiangliu, and Gaozhuang administrative townships [21]. Under the influence of the buried fault of the Jinghe River, the quaternary loess was uplifted and was deposited on the south of the Jinghe River, and the loess platform was formed [22]. It has been affected by large-scale agricultural irrigation since 1980. As of April 2018, there have been 92 loess landslides on the South Jingyang platform. Among them, sliding had occurred multiple times for 17 landslides, which have caused serious casualties and property losses [23]. Based on the triggering factors and landslide movement characteristics, loess landslides in the study area can be classified into four types [22]: irrigation flow landslides, irrigation slide landslides, erosion slide landslides, and engineering-induced landslides (Figure 1). In previous studies, 92 landslides have been investigated, which include 35 irrigation flow landslides, 37 irrigation slide landslides, 15 erosion slide landslides, and 5 engineering induced landslides (artificial loading, cutting slope toe, engineering disturbance, etc.). The statistical results show that the number of irrigation flow landslides and the irrigation slide landslides in the study area is equivalent, which accounts for 38.04% and 40.22% respectively, which is far more than erosion slide landslides and engineering induced landslides (only 21.73% in total). The irrigation flow landslides largely slide out from the Middle Pleistocene (Q2) loess layer at the slope toe. Since the water content of the slope toe and the silt (or sand gravel) layer at the terrace is large, the soil near the sliding surface can produce high pore water pressures during movement. This makes the soil in the sliding zone liquefy. Even if the terrace slide bed is gentle, it still presents significant high-speed and long-distance characteristics. The plane shape of the sliding mass is primarily semicircular or circular, and the sliding distance is generally 200–300 m. The volume is mostly between hundreds of thousands of square meters to one million square meters, which indicates that these landslides are a medium-sized landslide. The shear opening of irrigation slide landslides is relatively high and is predominantly in the unsaturated loess layer. The irrigation slide landslides are associated with the reduction of suction (shear strength) of the loess matrix caused by the infiltration of surface water such as farmland irrigation. From the perspective of landslide evolution, sliding occurs mainly in the second and third stages of the landslide, while some of the sliding, which results from the deep buried groundwater level, occurs in the first stage.

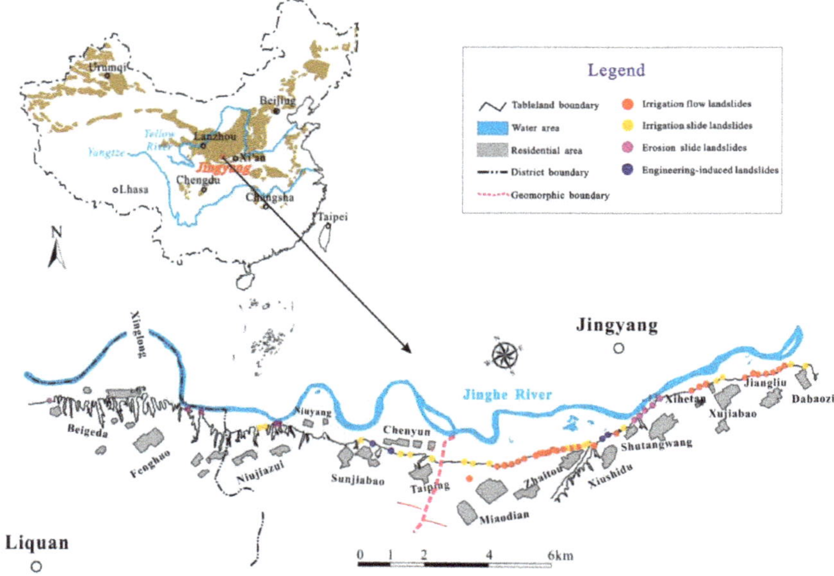

Figure 1. The loess landslides distribution in the South Jingyang platform.

The geological structure of the loess slope exhibits different types of structural planes, including joint fissures and weak planes. In the South Jingyang platform, the joint fissures are mainly the fissures on the edge of the platform, and the weak planes are mainly the loess-paleosol interface. According to a field geological structure survey, fissures within 7 m from the edge of the platform are intensively developed; most of these are unloading fissures. The fissures appear along the top of the platform or are near the steep slope, and they continue to extend and penetrate under the unloading effect. However, due to the cutting action of the steep cliff, the extended length of these fissures is often small. Fissures more than 15 m away from the edge of the platform are primarily collapsible fractures and are predominantly at the top of the platform. Under the action of uneven collapsibility, they extend deeply [12,14]. Although the number of collapsible fissures is less than unloading fissures, the plane extension length and joint opening are larger. Collapsible fissures are the dominant channel for the surface water to infiltrate the loess slope [12,24,25], which can cause the surface water to quickly invade the inside of the landslide body and to accelerate the rise of the groundwater level [21]. The South Jingyang platform presents typical interbedded sequence characteristics of the loess-paleosol [14], and the loess-paleosol contact interface is significant (Figure 2). However, because it is influenced by the tension of the fissures on the edge of the platform, the S1–S5 paleosol layer often breaks, which can easily cause water infiltration. Paleosol has a lower infiltration rate and water-holding capacity than loess does [26] and can be regarded as a natural aquiclude in comparison with loess [14]. The accumulation of surface water near the interface of a certain loess-paleosol is enhanced by the infiltration-promoting effect of the fissures on the edge of the platform. This causes the bottom of the Lishi loess to be saturated, which produces the saturated softening layer. Therefore, liquefaction may occur at the loess-paleosol interface, after which the saturated softening layer becomes the bottom sliding surface of the landslide [27].

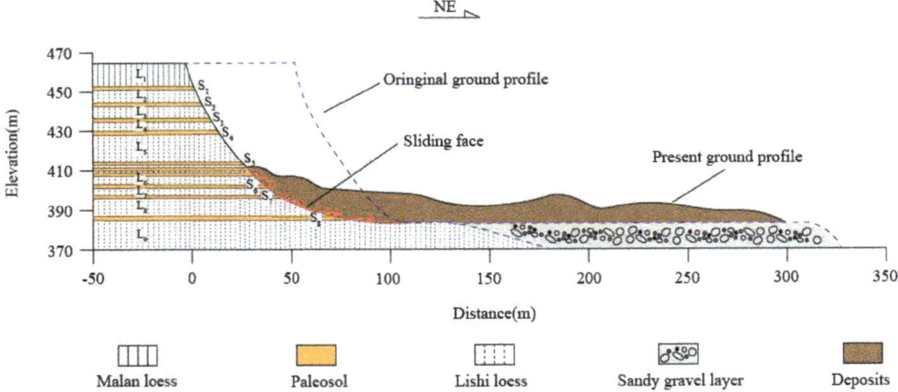

Figure 2. The profile of loess-paleosol contact interface in the South Jingyang platform.

At about 10:00 AM on 26 July 2014, a large-scale loess landslide occurred in Hetan village, Western Miaodian. This can be regarded as the first large-scale landslide because of the long-time interval from the Western Miaodian landslide in 1987. In fact, four landslides occurred on 26–28 May 2015 and 7–8 August 2015. However, with the increase in the sliding sequence, the volume in the landslide decreased gradually [12]. This indicates that the landslide that occurred on July 26, 2014 had a triggering effect. The main sliding direction of the landslide was NE61°, which is slightly different from that of the 1987 landslide (NE48°). A giant sliding body with a sliding distance of 278 m and a total volume of about 50 × 104 m^3 was formed (Figure 3). The posterior wall of the landslide was about 40 m high and 222 m wide, and it was characterized by a typical concave ring chair. The top 10–15 m was vertical and was controlled by the Late Pleistocene (Q3) and Q2 loess vertical joints. There were five paleosol layers that were exposed, and the landslide started from the vicinity of the fifth paleosol layer; moreover, the water retention effect of the paleosol layer was significant. In addition, the sliding surface of the posterior wall was smooth, and the water content was high.

Figure 3. Development characteristics of the "7.26" high-speed and long-distance landslide in Hetan village, Western Miaodian.

3. Experimental Samples and Process

Based on the above investigation and analysis, the occurrence of landslides in this region is closely related to static liquefaction. In order to further analyze the evolution process and the typical characteristics of the static liquefaction of loess in this region, a static liquefaction test was conducted. In addition, because the thickness of the Holocene (Q4) loess (generally ≤ 5 m), Late Pleistocene (Q3) Malan loess (generally < 10 m), and other new loess were not large, these were mainly located at the top of the loess slope. Their failure mode was mainly a tensile fracture, and shear liquefaction was unlikely to occur. Although the Middle Pleistocene (Q2) Lishi loess (10–70 m) was the main component of the loess slope in the study area, shear failure and sliding were the main failure modes in this layer. Therefore, the Q2 loess was selected for the ICU tests.

3.1. Experimental Samples

The undisturbed loess samples were acquired from the Zhaitou (ZT) and Shutangwang (STW) landslide. Among them, the ZT loess samples were obtained from the adit of the posterior wall of the landslide. The sampling location is about 21 m away from the top of the platform, and the corresponding stratum number is L2. The ZT loess samples were yellow-brown, slightly wet, and have slightly dense uniform soil. The samples were mainly composed of silt particles. In addition, needle pores, a small amount of snail shell fragments, and concretions were found in the samples. The STW loess samples were obtained from the surface of the posterior wall of the landslide. The sampling location is about 17 m away from the top of the platform, and the corresponding stratum number is L2. The STW loess samples were light yellow-brown and dry, and its other characteristics were like those of the ZT loess samples. The ZT and STW loess consisted of Q2 Lishi loess.

The physical properties of the samples were determined by referring to the relevant provisions of the standard for the soil test method (GB/T 50123-1999) [28]. The water content was measured using the drying method; the density was measured using the round knife method; and the dry density value was obtained via conversion. The specific gravity was measured using the pycnometer method, and the liquid limit and plastic limit water content were determined using the combined liquid plastic limit method. The basic physical properties of the two loesses are shown in Table 1.

Table 1. Basic physical properties of the Zhaitou (ZT) and Shutangwang (STW) loess.

Samples	Water Content (%)	Dry Density (g/cm^3)	Void Ratio	Specific Gravity (g·cm^{-3})	Liquid Limit (%)	Plastic Limit (%)	Liquid Index	Plastic Index
ZT	17	1.44	0.88	2.71	25.3	19.61	−0.46	5.73
STW	9.8	1.43	0.89	2.72	24.4	20.35	−2.59	4.08

It can be observed from Table 1 that the physical properties of the ZT and STW Q2 loess were close except for the water content. ZT loess samples were procured from the adit, while the STW loess samples were procured from the surface of the slope. Due to the water evaporation at the surface loess, the measured water content of the STW loess samples was relatively low. The liquid index of the two loess was less than zero and the plasticity index was low. This indicates that their plasticity in the natural state was low.

In addition, a laser particle size analyzer was used to analyze the grading of two Q2 loess samples, and a semi quantitative analysis of their mineral compositions was performed using an X-ray diffractometer. The selected diffraction angle was 0.5–30°. The specific results are presented in Table 2.

Table 2. The particle and mineral composition of the ZT and STW loess.

Samples	Particle Size(%)			Mineral Composition(%)					
	Clay (≤5 μm)	Silt (5–75 μm)	Fine Sand (75–250 μm)	Quartz	Carbonate	Illite	Kaolinite	Chlorite	Others
ZT	22.6	74.1	3.3	29	23	20	5	14	9
STW	10.6	82.2	7.2	31	20	19	3	17	10

As shown in Table 2, the particle composition of the ZT and STW Q2 loess were quite similar, with the proportion of the silt being the most, clay being the second, and fine sand being the least. Meanwhile, the corresponding mineral composition and content were highly similar. The highest content was quartz and carbonate, which constituted the main components of the cement and the skeleton of the loess. Clay minerals mainly consisted of illite, chlorite, and a relatively small amount of kaolinite. The hydrophilic characteristics of illite and chlorite were obvious, which makes it easy for the clay minerals to agglomerate and to form tuberculosis under water saturation conditions. The above mineral composition objectively revealed the typical structural characteristics of the loess in the area.

3.2. Experimental Process

In view of the typical characteristics of the static liquefaction phenomenon of the South Jingyang platform landslide: the pore water of the saturated loess caused by the groundwater immersion cannot be discharged in time during the shear process, which caused the pore water pressure to continue rising and to maintain a high level. Therefore, to accurately reflect the stress state of the saturated loess during the sliding of the loess slope, isotropic consolidated undrained (ICU) triaxial tests were conducted on ZT and STW loess. In addition, to further explore the internal mechanism of the static liquefaction of loess in this area and finding the internal relationship between the static liquefaction and the microstructure characteristics directly, it is necessary to observe and analyze the microstructure of the ZT and STW loess before and after the ICU tests. The specific experimental process is as follows:

(1) The undisturbed loess samples were molded into Φ50 mm × 100 mm cylinders according to the requirements of the standard for the soil test method (GB/T 50123-1999) [28]; and these cylinders were placed in the load cell of the Wykeham Farrance (WF, Wykeham Farrance, Milan, Italy) stress path triaxial apparatus (Figure 4). After connecting the pressure sensor, pore water pressure sensor, cell pressure controller, and the back pressure controller with the corresponding port of the load cell, the WF stress path triaxial apparatus was used to carry out the back pressure saturation treatment on the undisturbed loess. ① A 50 kPa cell pressure was applied to the

samples, and the back pressure was set at 10 kPa lower than the cell pressure to prevent structural damage of the samples caused by the pore water pressure, which is higher than the confining pressure. After the back pressure inflow was stable, the back pressure was kept constant for 6–8 h to make the water gradually infiltrate the samples. ② Under the premise of ensuring that the cell pressure was 10 kPa higher than the back pressure, the cell pressure and back pressure were applied to the samples step by step. Each stage was increased by 50 kPa and it remained for 6–8 h before conducting the next stage. ③ When the B value was higher than 0.95, the samples were considered to be saturated. Multiple tests showed that the undisturbed loess samples generally required an applied back pressure that ranged from 190 to 240 kPa, and in 1–2 days to ensure that the soil sample reached the saturation state.

(2) To reflect the discrepancy in the static liquefaction characteristics of the undisturbed loess under the different confining pressures (σ_c'), four confining pressures (150 kPa, 250 kPa, 350 kPa, and 450 kPa, corresponding to the numbers 1–4, respectively, for the ZT and STW loess) were selected. After the saturation of the loess (B ≥ 0.95), the cell pressure and back pressure (the cell pressures were 150 kPa, 250 kPa, 350 kPa, and 450 kPa higher than the back pressure) were set appropriately to ensure that the sample would be isotropically consolidated under four confining pressures. When the confining pressure attained the set value and remained stable, the drainage valve was opened to start the consolidation. In the consolidation process, the pore water pressure and volumetric strain changed rapidly at first and then it tended to be gradually stable. When the pore water pressure dissipated to more than 95% and the volumetric strain remained stable, the consolidation was completed.

(3) After the back pressure saturation and isotropic consolidation, undrained triaxial tests were conducted with a 0.1 mm/min shear rate, according to the relevant provisions of the standard for the soil test methods (GB/T 50123-1999) [28]. To observe the strength change characteristics after the peak value more clearly, a 25% axial strain was taken as the termination condition of these tests.

(4) For preparing the cylindrical samples for the ICU tests, cube samples with a side length of 1–2 cm had to be fabricated. Freeze-drying and section gold spraying treatments were performed successively on the cube samples. Then, the samples were placed under a field emission scanning electron microscope (JSM-7610F) and observed under different magnifications (500× and 2000×). The 500× magnification was used for observing the overall pore structure of the section, and 2000× magnification was used for observing of the occurrence and contact arrangement of the different particles. Five areas were randomly scanned at each magnification.

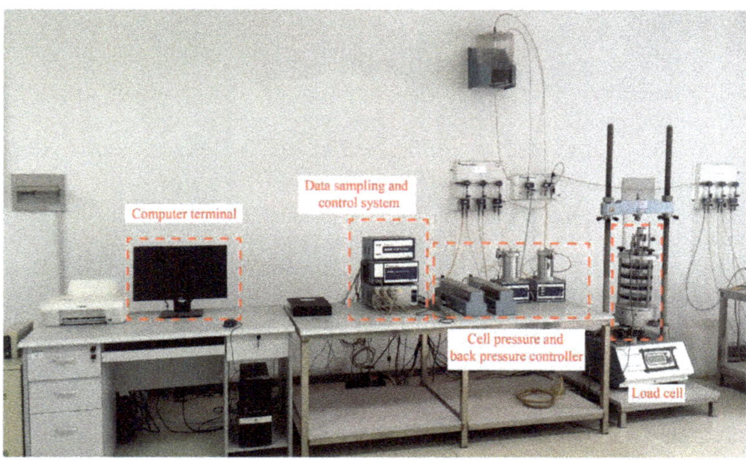

Figure 4. Wykeham Farrance (WF) stress path triaxial apparatus.

4. Results

4.1. Macroscopic Characteristics Analysis of the Shear Failure and Static Liquefaction

At first, the stress-strain characteristics under the different confining pressures were studied to obtain the liquefaction potential index (LPI) evolution process of the undisturbed loess. Moreover, the variation law of the pore water pressure and the influence of the confining pressures on the pore water pressure were analyzed. Finally, the evolution process of the static liquefaction was determined.

4.1.1. Stress-Strain Characteristics Analysis

Based on the experimental data from the ICU tests, the stress-strain curves of the ZT and STW undisturbed loess under the confining pressures of 150 kPa, 250 kPa, 350 kPa, and 450 kPa were plotted (Figure 5).

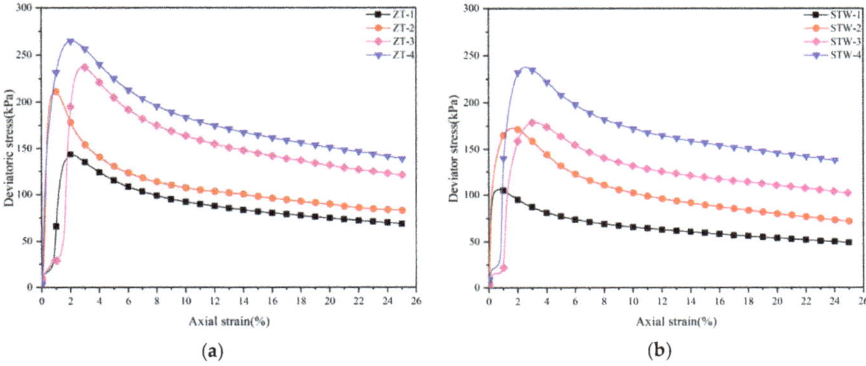

Figure 5. Stress-strain curves of the ZT and STW loess. (**a**) ZT. (**b**) STW.

It can be observed from Figure 5 that the stress-strain curves of the ZT and STW loess under the different confining pressures had a similar change in the trends, and all of them present obvious strain softening characteristics [12]. The stress-strain curves can be divided into three stages.

(1) Rapid strain hardening stage: At the initial stage of loading, when the axial strain is very small ($\varepsilon_a < 3\%$), the deviator stress rises rapidly and reaches the peak value. The stress-strain curve of this stage is close to the straight line. The larger the confining pressure is, the greater the deviator stress peak value is. The stress-strain curves in this stage indicate that the internal spaced pore structure is compressed under the action of the external load, and that the contact between the particles is closer [29].

(2) Rapid strain softening stage: When the peak strength is reached, the deviator stress decreases sharply, and the corresponding axial strain is about 1–3%. This indicates that the spaced pore structure may be damaged after compression with the increase in strain, and the ability to resist deformation is evidently weakened. There is a strain threshold (ε_{ap}) between the compression and the failure of the pore structure, which reflects that the structural loess requires a certain compression deformation to undergo structural failure. The stress-strain curves of the ZT and STW loess show that ε_{ap} is always in a certain range (generally between 1 and 3%). The stress-strain curves of the undisturbed loess in the South Jingyang platform provide similar results [9,12,15,30].

(3) Stable strain softening stage: When the axial strain reaches a certain value (ε_a = 10–15%), the stress tends to be stable with an increase in the strain. This shows that the strength of the soil structure softens after it has been destroyed. The clay aggregates, which are inlaid with the large framework of the silt particles, may gradually fall into the pore throat channel; this results in the reorganization of the loess structure [16]. With the continuous increase in the strain, the reorganization process

of the loess structure tends to be stable, and its initial structure is completely destroyed under the action of the shear stress. In addition, the new orientation arrangement has statistically reached a stable state [31]. Therefore, in the ICU tests of the saturated undisturbed loess, when the axial strain reaches 10–15%, the loess is considered to undergo steady-state deformation. When the loess reaches a steady state, its shear resistance is known as the steady-state strength [32]. Compared with its peak strength, the steady-state strength is about 0.4–0.6. Therefore, there is the possibility of static liquefaction. The above structural changes can also find the corresponding intuitive evidence in the later microstructure morphology analysis.

4.1.2. Variation Analysis of the Pore Water Pressure

Based on the ICU tests data, the pore water pressure-strain curves and effective confining pressure-strain curves of the ZT and STW loess were plotted, as shown in Figures 6 and 7.

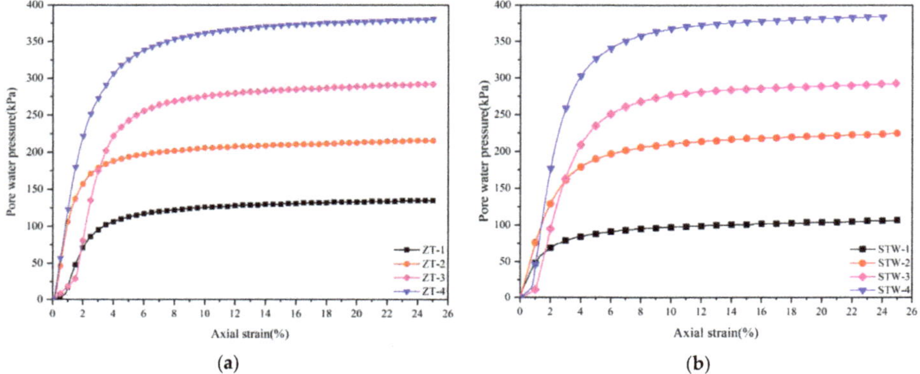

Figure 6. Pore water pressure-strain curves of the ZT and STW loess. (**a**) ZT. (**b**) STW.

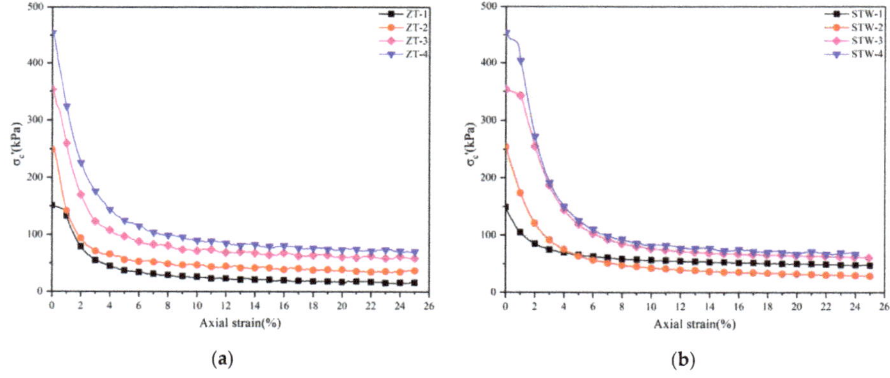

Figure 7. Effective confining pressure-strain curves of the ZT and STW loess. (**a**) ZT. (**b**) STW.

According to Figure 6, the pore water pressure curves of the ZT and STW loess exhibited almost the same change trend and were mainly divided into two stages: (1) Sharp rise stage: at the beginning of loading (the axial strain ε_a was less than 3%), the pore water pressure rose rapidly, close to the confining pressure level. (2) Stable stage: after reaching the curve inflection point, the increment of the pore pressure was very small with the increase in the strain, and it tended to be gradually stable. The inflection point of the curve can be considered as the critical point of the pore water pressure change and is closely associated with the structural yield stress being reached [33]. Under the undrained condition, the pore water pressure rose rapidly. Accompanied by the particle rearrangement, the spaced pore structure collapses, structural failure of the loess structure begins, and the effective stress gradually transfers to the pore water pressure [16]. When the collapse of the structure tended to end, the clay aggregates slipped and squeezed into the pore throat, which seriously hindered the discharge of the pore water. As a result, the pore water pressure remains at a relatively high level. The variation trend is accurately captured through the microstructure morphology analysis.

It can be observed from Figure 7 that under the different confining pressures, the effective confining pressure of the ZT and STW loess decreased sharply when the axial strain was less than 3%, and it tended to be gradually stable with the increase in the axial strain. The effective confining pressure of the ZT loess in the steady state increased with the increase in the initial confining pressure. Under the confining pressure of 250 kPa, the pore pressure of the STW loess increased more; hence, the effective confining pressure was the lowest in the steady state. Under the other confining pressure conditions, the effective confining pressure of the STW loess in the steady state also increased with the increase in the initial confining pressure. Combined with Figures 6 and 7, it is not difficult to determine that when the axial strain was less than 3%, the effective confining pressure decreased with the increase in the pore water pressure, and the pore water pressure and the effective confining pressure tend to be stable with the increase in the axial strain. Although the effective confining pressure at the end of the shear was a lower value (20–100 kPa), it did not reach zero. This means that the loess had not reached complete static liquefaction; however, it had decreased significantly in comparison to the initial confining pressure. At this point, the loess was in a very unstable state and had a strong flow plastic characteristic [14].

4.2. Analysis of the Microstructure Characteristics before and after the ICU Tests

Based on the aforementioned analysis of the ICU test results for the loess under the different confining pressures, the ZT and STW loess samples show prominent strain softening characteristics. After reaching the peak value, the partial stress decreased significantly; the pore water pressure increased sharply and remained high; and the effective confining pressure decreased rapidly. As a result, the ZT and STW loess have the possibility of static liquefaction. In order to find the relationship between the static liquefaction and the microstructure characteristics and to understand the internal mechanism of the static liquefaction in this area more clearly, the scanning electron microscope (SEM) images captured before and after the ICU tests were analyzed in this section.

In Figures 8 and 9(a1,a2) are SEM images of loess after the back pressure saturation treatment on the triaxial apparatus. Therefore, according to the experimental process, the SEM images of (a1) and (a2) reflect the micromorphology of loess before consolidated undrained triaxial tests. Moreover, (b)–(e) in Figures 8 and 9 are the SEM images of loess after consolidated undrained triaxial tests under different confining pressures. Through a comparative analysis of the microstructure changes before and after the ICU tests under the different confining pressures (Figures 8 and 9), the following results were determined. (1) Before the ICU tests, there were a large number of scattered spaced pores between the loess particles, which was confirmed to the microstructure characteristics of the loess [34]. After the ICU tests, the original pore throat channel of the spaced pores between the loess coarse particles was filled by fine particles, which can be also observed in the previous study [9]. Therefore, the pore throat channel was blocked, the boundary shape of the pore throat channel tended to be more complicated, and the intergranular pores were gradually dominated by mosaic pores. (2) Before the ICU tests, the contact between the coarse particles was mainly point-point and point-face. After the ICU tests, the appearance of fine particles, such as clay aggregates, in the contact position of the particles destroyed the contact state between the coarse particles of the loess. This resulted in the contact state between the coarse particles gradually transforming to face-face. The face-face contact relation mentioned in here includes direct face-face contact and indirect face-face contact [35,36]. (3) Under the confining pressures of 350 kPa and 450 kPa, the edges of the coarse particles peeled off from the fine particles. The same phenomenon was found in another study [37]. This indicates that the spaced framework formed by overlapping of the coarse particles becomes fully contacted under the compression deformation; however, the framework structure cannot bear a greater load. At the contact point of the coarse particles, the cracks would appear due to the stress concentration and they develop towards the inside of the particles, which gradually produces fragmentation.

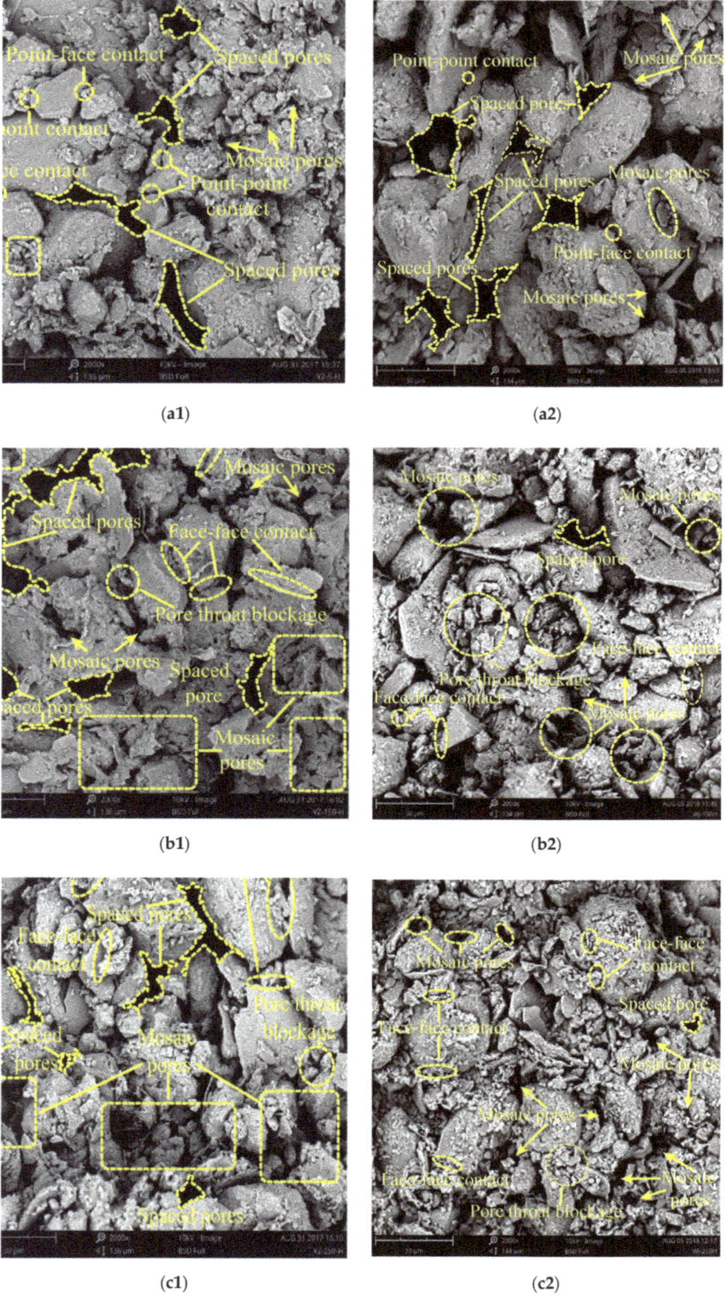

(**a1**) (**a2**)

(**b1**) (**b2**)

(**c1**) (**c2**)

Figure 8. *Cont.*

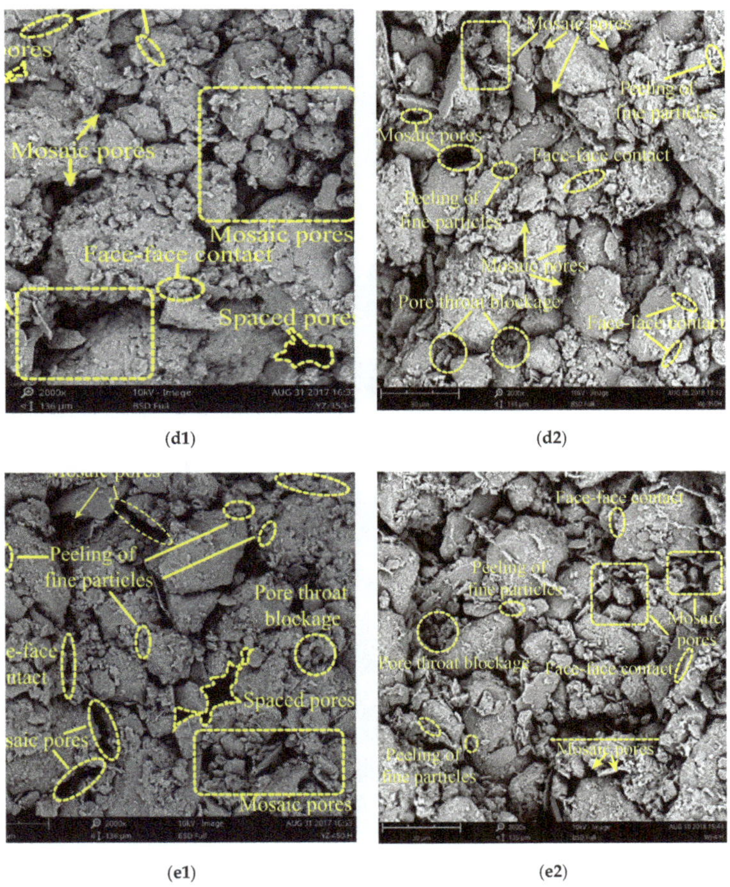

Figure 8. SEM images (2000×) of the ZT and STW loess before and after the isotropic consolidated undrained (ICU) tests. (**a1**) Saturated (ZT). (**a2**) Saturated (STW). (**b1**) 150 kPa (ZT). (**b2**) 150 kPa (STW). (**c1**) 250 kPa (ZT). (**c2**) 250 kPa (STW). (**d1**) 350 kPa (ZT). (**d2**) 350 kPa (STW). (**e1**) 450 kPa (ZT). (**e2**) 450 kPa (STW).

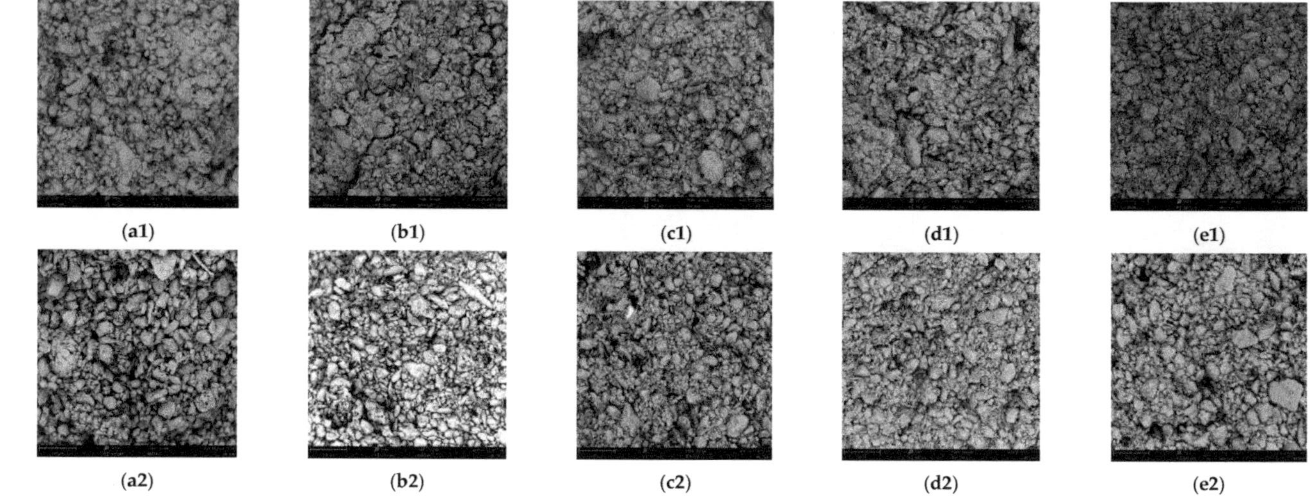

Figure 9. SEM images (500×) of the ZT and STW loess before and after the ICU tests. (**a1**) Saturated (ZT). (**a2**) Saturated (STW). (**b1**) 150 kPa (ZT). (**b2**) 150 kPa (STW). (**c1**) 250 kPa (ZT). (**c2**) 250 kPa (STW). (**d1**) 350 kPa (ZT). (**d2**) 350 kPa (STW). (**e1**) 450 kPa (ZT). (**e2**) 450 kPa (STW).

5. Discussion

5.1. Analysis of Capacity of Static Liquefaction

The stress-strain curves, pore water pressure-strain curves, and the effective confining pressure-strain curves of the ZT and STW loess under the different confining pressures were analyzed. The loess under the different confining pressures shows prominent strain softening characteristics, but the deviator stress at the end of the shear was not zero. In addition, the pore water pressure rose rapidly under a small deformation ($\varepsilon_a < 3\%$), and the effective confining pressure decreased with the increase in the pore water pressure and it reached a lower value (20–100 kPa) at the end of the shear. Therefore, the saturated undisturbed loess may undergo static liquefaction under the undrained shear action; however, the possibility of static liquefaction still needs further discussion.

5.1.1. Analysis Change Characteristics of the LPI

To further understand the effect of the confining pressure on the liquefaction resistance of the saturated undisturbed loess, the previous concept of the liquefaction potential index (LPI) was used for the analysis. Ng. C [38] defined the LPI as follows:

$$LPI = \frac{q_{\max} - q_{\min}}{q_{\max}} \quad (1)$$

where $q_{\max}$ is the peak shear stress, and $q_{\min}$ is the quasi-steady shear stress. This parameter is similar to the brittleness index (I_B) proposed by Bishop (1967) [39] for the strain softening materials. It can be used to characterize the degree of reduction in the deviatoric stress of the saturated undisturbed loess under the different confining pressures; thus, it reflects more accurately the ability of the loess to resist the static liquefaction. When the $LPI = 1$, this indicates that there is complete static liquefaction of the soil. When the $LPI = 0$, this indicates that there is no static liquefaction of the soil. When $0 < LPI < 1$, this indicates that the soil has a certain static liquefaction potential and the ability to resist static liquefaction weakens with the increase in the LPI. According to Formula (1), the LPI of the ZT and STW loess under the confining pressures of 150 kPa, 250 kPa, 350 kPa, and 450 kPa is calculated, and the results are shown in Figure 10a.

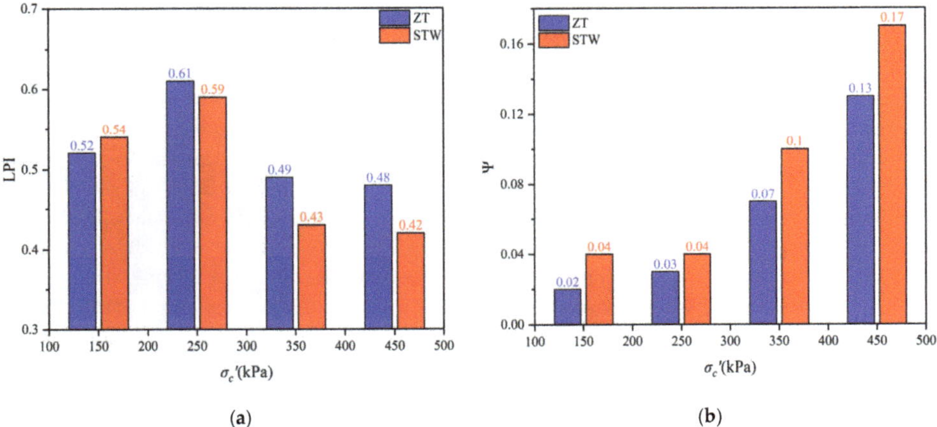

Figure 10. Measurement index of the liquefaction capacity. (**a**) liquefaction potential index (LPI). (**b**) ψ.

Been [40] defined the state parameter (ψ) as follows:

$$\psi = e_0 - e_{ss} \quad (2)$$

where e_0 and e_{ss} are the void ratios in the initial and steady states, respectively, for the same effective mean normal stress. When $\psi > 0$, the soil is in a "loose" state, and the undrained shear can cause shear shrinkage. The larger the state parameter ψ is, the more obvious the shear shrinkage behavior of soil is and the stronger the liquefaction capacity is. According to Formula (2), the ψ of the ZT and STW loess for the confining pressures of 150 kPa, 250 kPa, 350 kPa, and 450 kPa were calculated, and the results are shown in Figure 10b.

It can be observed from Figure 10a that the *LPI*s of the ZT and STW loess were almost equal under the different confining pressures and range between 0.4 and 0.6. In addition, with the increase in the confining pressure, no clear monotonic change rule; this is similar to the results of previous ICU tests on saturated undisturbed loess in the South Jingyang platform [9,12,30]. At first, with the increase in the confining pressure from 150 to 250 kPa, the *LPI* shows an upward trend. At this point, the *LPI*s of the ZT and STW loess were 0.61 and 0.59, respectively. Then, with the increase in the confining pressure from 250 to 350 kPa, the *LPI* decreased to some extent. Finally, with the increase in the confining pressure from 350 to 450 kPa, the *LPI* remained unchanged. It can be determined from Figure 10b that ψ of the ZT and STW loess increased with the increase in the confining pressure from 250 to 450 kPa; however, it remained unchanged with the increase in the confining pressure from 150 to 250 kPa. Moreover, for a confining pressure of 150 kPa and 250 kPa, the LPI was larger; yet ψ was smaller. When the confining pressure was 350 kPa and 450 kPa, the LPI was smaller, but ψ was larger.

Furthermore, the possibility of static liquefaction existed in different layers and the buried depth of the undisturbed loess. In addition, there was no obvious relationship between static liquefaction and confining pressure. In fact, according to the research on the buried depth of the sliding surface of the loess landslides in the South Jingyang platform, it is determined that there is no clear loess stratigraphic unit for the sliding of the high-speed and long-distance loess landslide; however, there is a prominent sliding face for the conventional sand landslides [41,42]. Although no prominent sliding face was observed in the loess landslides of the South Jingyang platform, the loess-paleosol of the South Jingyang platform presents typical interbedded sequence characteristics. Therefore, an evident water retention effect was easily produced at the paleosol interface. This is because the paleosol layer was equivalent to a relatively dense aquiclude. Some landslides in this area do have the possibility of sliding along the paleosol layer [12,24], but the specific formation was not regular.

5.1.2. Analysis of the Stress Path Evolution

The stress-strain curves and the pore pressure-strain curves of the ZT and STW loess indicate that the loess exhibited strong strain softening, and the pore water pressure remained relatively high. For the post peak state description and the steady-state evaluation of this kind of strain softening soil, the potential liquefaction state can be well recognized. Based on previous studies [43–45], the stability of the ZT and STW loess was evaluated by establishing a steady state line and an instability line. The stress paths of the ZT and STW loess were plotted using the experimental data from the ICU tests; the abscissa is the effective average stress $p' = 1/3(\sigma_1' + 2\sigma_3')$ (MPa), and the ordinate is the deviator stress $q' = \sigma_1' - \sigma_3'$ (MPa), as shown in Figure 11.

Figure 11 shows that the ZT and STW loess under the different confining pressures all have shear shrinkage phenomenon. With the increase in the confining pressure, the peak point of the effective stress path gradually increased. In addition, after the effective stress path of the ZT and STW loess reached the peak point, they all reached the steady state with an increase in the strain. The stress path was divided into the stable region and instability regions by the instability and steady-state lines. If the soil stress path entered the instability region between the instability line and the steady-state line, the soil was very likely to undergo static liquefaction. If the soil stress path is in the stable region below the instability line, static liquefaction of the soil is unlikely to occur [43,44,46]. Accordingly, it can be concluded that static liquefaction of the ZT and STW loess is more likely to occur under different confining pressures.

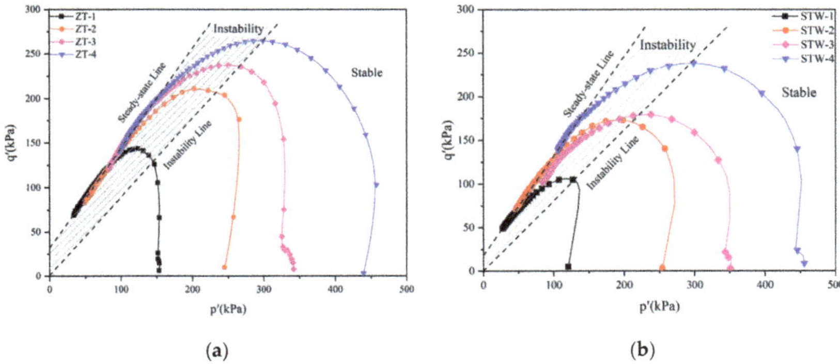

(**a**) (**b**)

Figure 11. The stress paths of the ZT and STW loess. (**a**) ZT. (**b**) STW.

For the purpose of making the steady-state line pass through the steady-state points of the loess as much as possible, after trying to draw the steady-state lines, it was determined that this effect can be achieved when the steady-state line is a straight line without passing through the origin. Based on the stress path that from the ICU tests of the loess, some scholars [17,47] observed that the steady-state line is a straight line that does not pass through the origin. Although the uniqueness of the steady-state line of sand is controversial [48,49], most of it is a straight line that passes through the origin [43–45]. Loess still has cohesion in the steady-state stage [50], while the cohesion of sand is usually negligible. In addition, the macroscopic mechanical behavior of loess is the result of the coupling effect of the particle cementation strength and the friction strength. After overcoming the cementation strength between the loess particles, the greater the friction strength between the particles, the weaker the strain softening phenomenon [50]. The friction strength between the particles depends on the shape, size, and the relative position of the particles after structural reorganization [51]. The difference in the friction strength between the particles results in a different static liquefaction phenomenon after the peak value for the loess and sand; however, both display shear shrinkage characteristics.

5.2. Correlation Analysis of the Macro and Micro Characteristics of Static Liquefaction

Through the analysis of the mechanical behavior of ICU tests and the structural evolution, the following conclusions can be obtained. (1) The most obvious feature of the static liquefaction is that the pore water pressure rises sharply and it remains high. (2) The stable-state points fall into the instability region; however, the steady-state strength is not zero, which reflects that there is still a certain residual strength, and it should belong to incomplete static liquefaction. Furthermore, the typical characteristics of static liquefaction are all derived from the change in the loess internal structure [52,53]. As demonstrated from the comparison of the SEM images of the loess before and after the ICU tests (Figure 8), significant changes had taken place in the internal morphology; however, a quantitative analysis of the changes in the pore structure is still needed to accurately understand the relationship between the changes in the macroscopic mechanical behavior and the changes in the microstructure.

To further analyze the changes in the pore structure of the loess before and after the ICU tests, image analysis software (Image-pro Plus (IPP, Media Cybernetics Inc. Rockville, MD, USA)) was used to process the SEM images at 500 times magnification (Figure 9). This software can effectively distinguish the image from the background by threshold segmentation of the original SEM image, and then it obtains the determined value by performing a statistical analysis of the corresponding quantitative parameters. The specific operation flow is as follows. (1) Each of the original SEM images is uniformly corrected (including improving the image brightness and contrast, using the median filter to denoise the image to accurately define the pore boundary, using image binary processing technology to accurately identify the pore structure, etc.). Among them, the median filter is a noise reduction method that the middle value of the gray of the neighbor pixels replaces the gray value of the central pixel, and is mainly used to eliminate mutational noise points and improve image quality. The above uniform corrections of SEM images are shown in Figure 12. (2) The quantitative parameters were selected for the statistical analysis, among which, the quantitative parameters involved in this study include the average pore diameter, shape ratio, roundness, and the pore contour fractal dimension. The concepts related to the quantitative parameters are described as follows, and the conceptual graph (Figure 12e) is shown for understanding adequately. (3) The quantitative characterization of the pore structure is achieved through the above parameters, and the detailed results are illustrated in Figure 13.

(1) Average pore diameter (μm): it is the average length of diameters (brown lines in Figure 12) measured at 2degree intervals and passing through the centroid of pore.
(2) Shape ratio: it is the ratio of the long axis to the short axis of ellipse equivalent to the pore, and the long axis and short axis are showed in Figure 12. The larger the shape ratio is, the more the pore shape resembles a long strip; the smaller the shape ratio, the more equiaxed the pore is. When the shape ratio is 1, the pore is square-shaped or circular [54].
(3) Roundness: $R = 4\pi A/P^2$, where P is the perimeter of the pore and A is the pore area. The perimeter is the length of the red contour, and the pore area is the white area inside the red contour (Figure 12). This index can better reflect the plane shape characteristics of the pore; the larger the roundness value is, the closer the pore is to the circle [55].
(4) Fractal dimension: it is used to describe the fractal characteristics of the pores, and it reflects the irregularity of the contact boundary between the pores and the solid particles in loess [56]. The calculation of the pore contour (blue contour in Figure 12) fractal dimension is based on the "silt-island" method, proposed by Mandelbrot [57]: $\lg P = D/2 \times \lg A + C$ where P is the perimeter of the pore; A is the pore area; and C is a constant. The pore contour fractal dimension can be determined by plotting the scatter map in the $\lg P - \lg A$ double logarithmic coordinate and performing a linear regression analysis. The larger the pore contour fractal dimension, the more complex the pore contour.

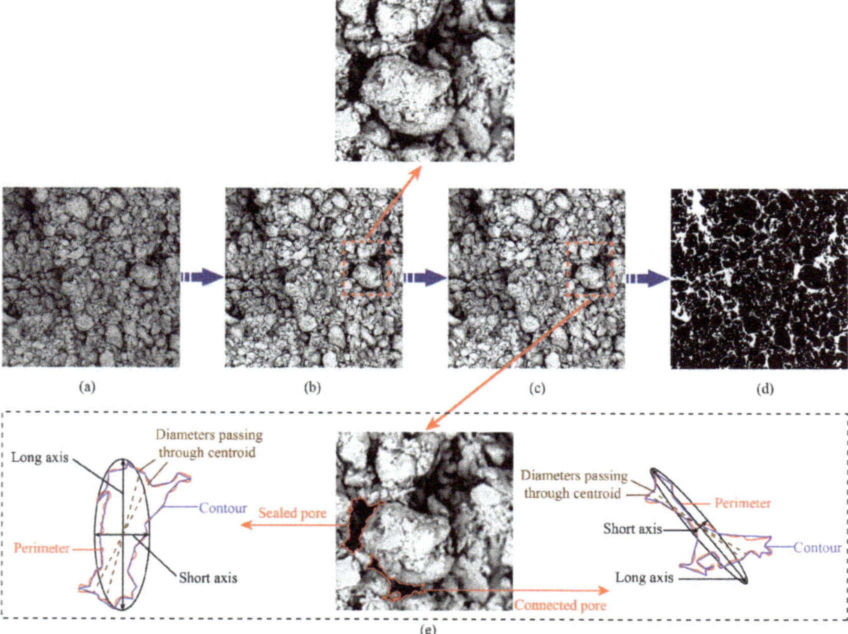

Figure 12. SEM images correction process. (**a**) Original SEM images. (**b**) Improving brightness and contrast. (**c**) Median filter. (**d**) Binary processing. (**e**) Quantitative parameters of different types pore structure.

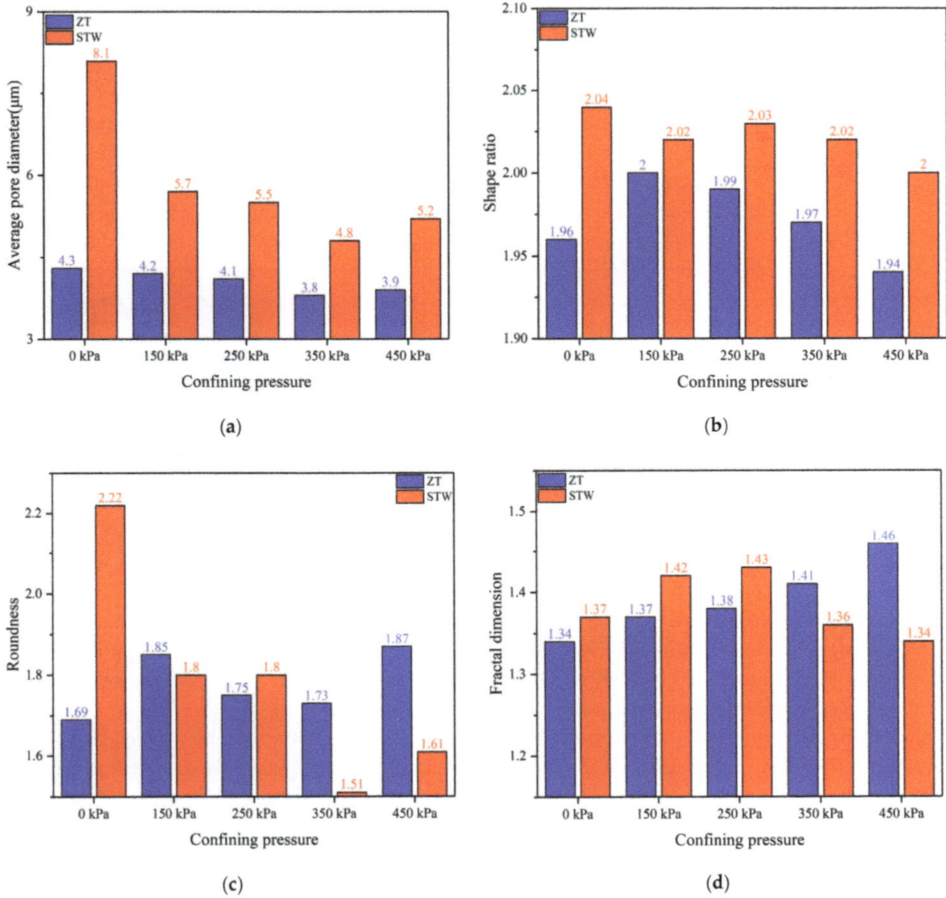

Figure 13. The quantitative parameters of the ZT and STW loess before and after the ICU tests. (**a**) Average pore diameter. (**b**) Shape ratio. (**c**) Roundness. (**d**) Pore contour fractal dimension.

It can be observed from Figure 13 that after the ICU tests, the average pore diameter decreased and the shape ratio was basically unchanged; however, the roundness and pore contour fractal dimension shows two different trends: the roundness and the pore contour fractal dimension of the ZT loess increased after the ICU tests. After the ICU tests, the roundness of the STW loess decreased, and the pore contour fractal dimension increased or decreased. It can be inferred that a large deformation (e.g., crushing, rotating, sliding, etc.) occurred in the aggregate spaced pore structure formed by overlapping of the coarse particles inside the loess; this resulted in the sudden and interlocking initial structural damage. The deformation of the pore structure essentially involves particle arrangement and reorganization [46,58]. In addition, the differences in the shape ratio, roundness, and pore contour fractal dimension before and after the ICU tests reflect the uncertainty rearrangement and reorganization of the particles. However, the difference in the average pore diameter before and after the ICU tests shows that the fine particle aggregates can enter the original pore throat in the process of rearrangement and reorganization of the particles. This can be observed intuitively in Figure 8b–e and is also verified by other studies [9,17,57]. At the beginning of the particles rearrangement and reorganization, the water transport channel formed by spaced pores is blocked to a certain extent. Hence, the macroscopic mechanical behavior is characterized by a rapid decrease in the deviatoric stress after the peak stress and a slow increase in the pore water pressure (rapid strain softening stage). Subsequently, the process of particle rearrangement and reorganization tends to be stable, and the migration and agglomeration of fine particles under the action of pore water pressure continue [26]. The synergistic effect of particle rearrangement and reorganization and the pore water pressure causes the agglomeration blocking of the reconstructed pore throat. Therefore, the macroscopic mechanical behavior shows that the deviatoric stress gradually tended to shift to the steady-state stage, and the pore water pressure was maintained at a high level (stable strain softening stage). In addition, the macroscopic mechanical behavior of the loess before reaching the peak strength displayed a rapid increase in the deviatoric stress and the pore water pressure (rapid strain hardening stage). It is reasonable to speculate that the widespread spaced pore structure in loess often requires a certain compression deformation to ensure full contact and the formation of a force chain [59]. This deformation process can also compress the pore water transport channel.

Based on the different pore diameter ranges of the loess (macropore > 32 μm, mesopore 8–32 μm, small pore 2–8 μm, and micropore < 2 μm) proposed by Lei [60], and the IPP was used to perform a statistical analysis on the different pore proportion, the detailed results are presented in Figure 14. As illustrated in Figure 14, after the ICU tests, the different pore proportion change characteristics of the ZT and STW loess were as follows: the macropore and mesopore decreased, the small pore increased slightly, and the micropore increased significantly. The micropores are mainly cement pores; the small pores are mainly mosaic pores and a small amount of cement pores; the mesopores are mainly spaced pores and some mosaic pores; and the macropores mainly includes root holes, wormholes, and fractures [60]. This shows that during the particles reorganization, the intergranular pores are gradually filled by fine particles such as clay aggregates; thus, resulting in an increase in the proportion of the micropore and small pore, which is also illustrated by the decrease in the average pore diameter and Figure 8b–e after the ICU tests. After the ICU tests, most of the above pore structure quantitative parameters (e.g., average pore diameter, shape ratio, roundness, pore contour fractal dimension, and the proportion of the different pores) did not show an obvious monotonic change law with the increase in the confining pressure. The static liquefaction characteristics of the loess come from the change in the internal structure of the loess, which again reflects that there was no clear relationship between the confining pressure and the static liquefaction of the loess in the South Jingyang platform.

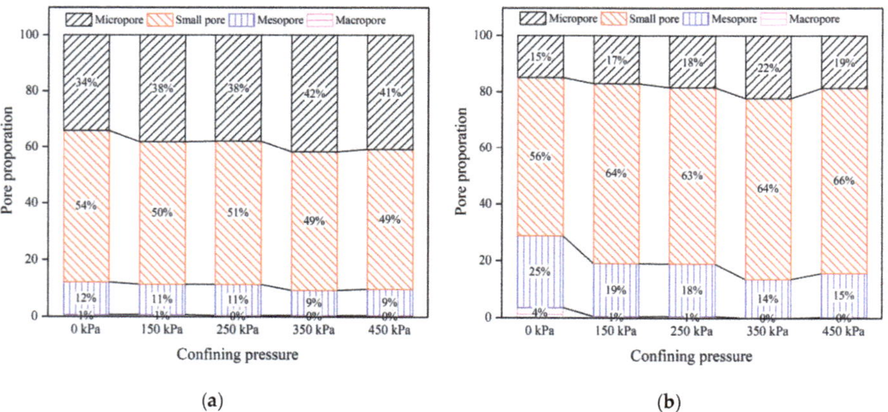

Figure 14. Different pore proportion of the ZT and STW loess before and after the ICU tests. (**a**) ZT. (**b**) STW.

The characteristics of the two grain phases (the gain composition was mainly silt and clay) of the ZT and STW loess shown in Table 2 and the pore proportions shown in Figure 14 indicate that the spaced structure formed by the accumulation and overlapping of silt constituted the entire structural framework of the loess. As the particle size of clay was sufficiently small compared with that of silt, the van der Waals force and repulsive force between the clay counteracted the filling effect due to gravity in the relatively loose silt skeleton structure. Therefore, the clay forms relatively closed agglomerates through agglomeration [46]; however, the high content of clay rendered the clay aggregates inlaid or resulted in the filling of the pores or contact areas in the spaced structure (Figures 8 and 9). Therefore, the structure of the loess, which is composed of silt and is filled with clay aggregates, was similar to that of sand, which is composed of coarse grains and is filled with fine grains such as silt [44]. According to a hypothetical model of the interaction between fine grains and coarse grains proposed by Lade and Yamamuro [44], when the fine grains occupy the pores between the coarse grains, they only increase the material density and have negligible influence on the behavior of the soil. When the fine grains occupy the location near the contact of the coarse particles, under the isotropic shear action, the fine grains tend to slide into the pores between the coarse grains. This promotes the rearrangement of the coarse particle structure, it increases the volume shrinkage of the soil, and it produces a greater liquefaction potential under the undrained condition. Therefore, the structure characteristics of the loess provides better structural conditions for the occurrence of static liquefaction phenomenon, which makes it easy to have a structural reorganization under the action of an external load. Clay aggregates would enter the spaced structure that is composed of overlapping silt and hinders the dissipation of pore water [9,17,57].

6. Conclusions

According to the stress-strain curves, pore water pressure-strain curves, and the regions of the stress path, the static liquefaction of the loess in the South Jingyang platform is incomplete. Through a comparative analysis of the microstructure characteristics before and after the ICU tests, the evolution process and internal mechanism of the static liquefaction were revealed. The main conclusions are as follows:

(1) The stress-strain curves of the ZT and STW loess show the characteristics of the strong strain softening under the different confining pressures, and there were stages of rapid strain hardening, rapid strain softening, and stable strain softening. The pore water pressure under the different confining pressures increased rapidly to the effective confining pressure level after a small axial strain ($\varepsilon_a < 3\%$). By having a continuous shear, it was always near this level; hence, its change also presents phased characteristics: a sharp rise stage and a stable stage. The effective confining pressure decreased with an increase in the pore water pressure and then it remained stable. Afterwards, it reached a lower value (20–100 kPa) at the end of the shear.

(2) The *LPI* of the ZT and STW loess shows a trend of first increasing and then stabilizing with the increase in the confining pressure. The larger LPI was, the smaller ψ was, which reflected that the loess in different layers and buried depths had the ability of static liquefaction. The steady-state points of the saturated loess were the instability region. However, the steady-state strength was not zero, which indicates that there was a certain residual strength and it belonged to the incomplete static liquefaction.

(3) After the ICU tests, the average pore diameter of the ZT and STW loess decreased, and the shape ratio essentially remained unchanged. The pore contour fractal dimension and the roundness show two different trends. The change in the proportion of the different pores is as follows: the macropore and mesopore decreased, the small pore increased slightly, and the micropore increased significantly. However, most of the above pore structure quantitative parameters did not show a monotonous change rule with the increase in the confining pressures.

(4) When $\varepsilon_a < 3\%$, the compression of the spaced pore structure in the loess resulted in full contact and the formation of a force chain, and it compressed the pore water transport channel. The macroscopic mechanical behavior shows rapid strain hardening. When $\varepsilon_a = 1$–3%, the deformation of the spaced pore structure occurred significantly, that is, the process of particle arrangement and recombination. At the beginning of this process, fine aggregates could enter the original pore throat channel, which caused the water transport channel formed by the spaced

pore to be blocked to a certain extent. The macroscopic mechanical behavior shows rapid strain softening. When ε_a = 10–15%, the gradually stable rearrangement and reorganization of the particles and the migration and agglomeration of finer grains under the effect of the pore water pressure caused agglomeration blocking of the reconstructed pore throat. The macroscopic mechanical behavior indicates stable strain softening.

Author Contributions: Writing—Original Draft Preparation, R.-X.Y.; Funding Acquisition, J.-B.P.; Writing—Review and Editing, R.-X.Y. and J.-Y.Z.; Data Curation, S.-k.W. All authors have read and agreed to the published version of the manuscript.

Funding: This work was supported by the Major Program of National Natural Science Foundation of China (Grant No. 41790441), the National Natural Science Foundation of Shaanxi Province, China (Grant No. 2018JQ5124), and the Foundation of Key Laboratory of Western Mineral Resources and Geological Engineering of Ministry of Education, Chang' an University (Grant No. 300102268503).

Acknowledgments: We thank the entire team for their efforts to improve the quality of the article. At the same time, we would like to thank editor for his timely handling of the manuscripts.

Conflicts of Interest: The authors declare no conflict of interest.

References

1. Terzaghi, K.; Peck, R.B. *Soil Mechanics in Engineering Practice*, 2nd ed.; John Wiley & Sons: Hoboken, NJ, USA, 1967; Chapter 2, p. 108, ISBN 0471852732.
2. Castro, G. *Liquefaction of Sands*; Harvard Soil Mechanics Series No. 81; Harvard University: Cambridge, MA, USA, 1969.
3. Kramer Steven, L.; Seed, H.B. Initiation of soil liquefaction under static loading conditions. *J. Geotech. Eng.* **1988**, *114*, 412–430. [CrossRef]
4. Mróz, Z.; Boukpeti, N.; Drescher, A. Constitutive model for static liquefaction. *Int. J. Geomech.* **2003**, *3*, 133–144. [CrossRef]
5. Konrad, J.M.; Watts, B.D. Undrained shear strength for liquefaction flow failure analysis. *Can. Geotech. J.* **1995**, *32*, 783–794. [CrossRef]
6. McKenna, G.; Luternauer, J.; Kostaschuk, R. Large-scale mass-wasting events on the Fraser River delta front near Sand Heads, British Columbia. *Can. Geotech. J.* **1992**, *29*, 151–156. [CrossRef]
7. Olson Scott, M.; Stark Timothy, D.; Walton William, H.; Castro, G. 1907 static liquefaction flow failure of the North Dike of Wachusett Dam. *J. Geotech. Geoenviron. Eng.* **2000**, *126*, 1184–1193. [CrossRef]
8. Leng, Y.; Peng, J.; Wang, Q.; Meng, Z.; Huang, W. A fluidized landslide occurred in the Loess Plateau: A study on loess landslide in South Jingyang tableland. *Eng. Geol.* **2018**, *236*, 129–136. [CrossRef]
9. Peng, J.; Zhuang, J.; Wang, G.; Dai, F.; Zhang, F.; Huang, W.; Xu, Q. Liquefaction of loess landslides as a consequence of irrigation. *Q. J. Eng. Geol. Hydrogeol.* **2018**, *51*, 330–337. [CrossRef]
10. Xu, L.; Dai, F.-C.; Min, H.; Kwong, A.K.L. Loess landslide types and topographic features at South Jingyang Plateau, China. *Diqiu Kexue-Zhongguo Dizhi Daxue Xuebao/Earth Sci.-J. China Univ. Geosci.* **2010**, *35*, 155–160.
11. Li, T.; Wang, P.; Xi, Y. *The Mechanisms for Initiation and Motion of Chinese Loess Landslides*; Springer: Berlin/Heidelberg, Germany, 2013; pp. 105–122.
12. Ma, P.; Peng, J.; Wang, Q.; Zhuang, J.; Zhang, F. The mechanisms of a loess landslide triggered by diversion-based irrigation: A case study of the South Jingyang Platform, China. *Bull. Eng. Geol. Environ.* **2019**, *78*, 4945–4963. [CrossRef]
13. Zhuang, J.; Peng, J.; Wang, G.; Javed, I.; Wang, Y.; Li, W. Distribution and characteristics of landslide in Loess Plateau: A case study in Shaanxi province. *Eng. Geol.* **2018**, *236*, 89–96. [CrossRef]
14. Yan, R.-X.; Peng, J.-B.; Huang, Q.-B.; Chen, L.-J.; Kang, C.-Y.; Shen, Y.-J. Triggering Influence of Seasonal Agricultural Irrigation on Shallow Loess Landslides on the South Jingyang Plateau, China. *Water* **2019**, *11*, 1474. [CrossRef]
15. Li, H.J.; Jin, Y.L. Initiation analysis of an irrigation-induced loess landslide. *Appl. Mech. Mater.* **2012**, *170–173*, 574–580. [CrossRef]
16. Liu, W.; Chen, W.; Wang, Q.; Wang, J.; Lin, G. Effect of pre-dynamic loading on static liquefaction of undisturbed loess. *Soil Dyn. Earthq. Eng.* **2020**, *130*, 105915. [CrossRef]

17. Wang, G.; Zhang, D.; Furuya, G.; Yang, J. Pore-pressure generation and fluidization in a loess landslide triggered by the 1920 Haiyuan earthquake, China: A case study. *Eng. Geol.* **2014**, *174*, 36–45. [CrossRef]
18. Pei, X.; Zhang, X.; Guo, B.; Wang, G.; Zhang, F. Experimental case study of seismically induced loess liquefaction and landslide. *Eng. Geol.* **2017**, *223*, 23–30. [CrossRef]
19. Xu, L.; Dai, F.C.; Tham, L.G.; Tu, X.B.; Min, H.; Zhou, Y.F.; Wu, C.X.; Xu, K. Field testing of irrigation effects on the stability of a cliff edge in loess, North-West China. *Eng. Geol.* **2011**, *120*, 10–17. [CrossRef]
20. Xu, L.; Dai, F.C.; Gong, Q.M.; Tham, L.G.; Min, H. Irrigation-induced loess flow failure in Heifangtai platform, North-West China. *Environ. Earth Sci.* **2012**, *66*, 1707–1713. [CrossRef]
21. Peng, J.; Qiao, J.; Leng, Y.; Wang, F.; Xue, S. Distribution and mechanism of the ground fissures in Wei River Basin, the origin of the Silk Road. *Environ. Earth Sci.* **2016**, *75*, 718. [CrossRef]
22. Peng, J.; Wang, G.; Wang, Q.; Zhang, F. Shear wave velocity imaging of landslide debris deposited on an erodible bed and possible movement mechanism for a loess landslide in Jingyang, Xi'an, China. *Landslides* **2017**, *14*, 1503–1512. [CrossRef]
23. Peng, J.; Ma, P.; Wang, Q.; Zhu, X.; Zhang, F.; Tong, X.; Huang, W. Interaction between landsliding materials and the underlying erodible bed in a loess flowslide. *Eng. Geol.* **2018**, *234*, 38–49. [CrossRef]
24. Zhuang, J.-Q.; Peng, J.-B. A coupled slope cutting—A prolonged rainfall-induced loess landslide: A 17 october 2011 case study. *Bull. Eng. Geol. Environ.* **2014**, *73*, 997–1011. [CrossRef]
25. Tu, X.B.; Kwong, A.K.L.; Dai, F.C.; Tham, L.G.; Min, H. Field monitoring of rainfall infiltration in a loess slope and analysis of failure mechanism of rainfall-induced landslides. *Eng. Geol.* **2009**, *105*, 134–150. [CrossRef]
26. Derbyshire, E.; Dijkstra, T.A.; Smalley, I.J.; Li, Y. Failure mechanisms in loess and the effects of moisture content changes on remoulded strength. *Quat. Int.* **1994**, *24*, 5–15. [CrossRef]
27. Xu, L.; Dai, F.C.; Tu, X.B.; Javed, I.; Woodard, M.J.; Jin, Y.L.; Tham, L.G. Occurrence of landsliding on slopes where flowsliding had previously occurred: An investigation in a loess platform, North-West China. *Catena* **2013**, *104*, 195–209. [CrossRef]
28. GB/T50123-1999. *Standard for Soil Test Method*; China Planning Press: Beijing, China, 1999.
29. Wang, H.N.; Ni, W.K.; Liu, G.Y. Ct images analysis of damage process in loess under triaxial conditions. *Adv. Mater. Res.* **2011**, *243–249*, 3175–3181. [CrossRef]
30. Jiang, M.; Hu, H.; Peng, J.; Leroueil, S. Experimental study of two saturated natural soils and their saturated remoulded soils under three consolidated undrained stress paths. *Front. Archit. Civ. Eng. China* **2011**, *5*, 225–238. [CrossRef]
31. Mandelbrot, B.B.; Passoja, D.E.; Paullay, A.J. Fractal character of fracture surfaces of metals. *Nature* **1984**, *308*, 721–722. [CrossRef]
32. Poulos, S.J. Steady state of deformation. *J. Geotech. Eng. Div.* **1981**, *107*, 553–562.
33. Wang, Y.; Xie, W.; Gao, G. Effect of different moisture content and triaxial test methods on shear strength characteristics of loess. *E3S Web Conf.* **2019**, *92*, 07007. [CrossRef]
34. GAO, G.-R. Classification of microstructures of loess in China and their collapsibility. *Sci. Sin.* **1981**, *24*, 962.
35. Li, P.; Vanapalli, S.; Li, T. Review of collapse triggering mechanism of collapsible soils due to wetting. *J. Rock Mech. Geotech. Eng.* **2016**, *8*, 256–274. [CrossRef]
36. Fang, X.; Shen, C.; Li, C.; Wang, L.; Chen, Z. Quantitative analysis of microstructure characteristics of Pucheng loess in Shaanxi Province. *Yanshilixue Yu Gongcheng Xuebao Chin. J. Rock Mech. Eng.* **2013**, *32*, 1917–1925.
37. Wang, F.W.; Sassa, K.; Wang, G. Mechanism of a long-runout landslide triggered by the August 1998 heavy rainfall in Fukushima Prefecture, Japan. *Eng. Geol.* **2002**, *63*, 169–185. [CrossRef]
38. Ng, C.W.W.; Fung, W.; Cheuk, C.Y.; Zhang, L. Influence of stress ratio and stress path on behavior of loose decomposed granite. *J. Geotech. Geoenviron. Eng.* **2004**, *130*. [CrossRef]
39. Bishop, A.W. Progressive failure with special reference to the mechanism causing it. In *Proceedings of the Geotechnical Conference*; Norwegian Geotechnical Institute: Oslo, Norway, 1968; pp. 142–150.
40. Been, K.; Jefferies, M.G. A state parameter for sands. *Géotechnique* **1985**, *35*, 99–112. [CrossRef]
41. Sladen, J.; D'Hollander, R.; Krahn, J.; Mitchell, D. Back analysis of the nerlerk berm liquefaction slides. *Can. Geotech. J.* **1985**, *22*, 579–588. [CrossRef]
42. Been, K.; Conlin, B.; Crooks, J.; Fitzpatrick, S.; Jefferies, M.; Rogers, B.; Shinde, S. Back analysis of the Nerlerk berm liquefaction slides: Discussion. *Can. Geotech. J.* **2011**, *24*, 170–179. [CrossRef]
43. Vaid, Y.P.; Chung, E.K.F.; Kuerbis, R.H. Stress path and steady state. *Can. Geotech. J.* **1990**, *27*, 1–7. [CrossRef]

44. Lade, P.; Yamamuro, J. Effects of nonplastic fines on static liquefaction of sands. *Can. Geotech. J.* **1997**, *34*, 918–928. [CrossRef]
45. Sladen, J.; D'Hollander, R.; Krahn, J. The liquefaction of sands, a collapse surface approach. *Can. Geotech. J.* **1985**, *22*, 564–578. [CrossRef]
46. Lade, P.V.; Yamamuro, J.A. Evaluation of static liquefaction potential of silty sand slopes. *Can. Geotech. J.* **2011**, *48*, 247–264. [CrossRef]
47. Zhou, Y.-X.; Zhang, D.-X.; Luo, C.-Y.; Chen, J. Experimental research on steady strength of saturated loess. *Yantu Lixue/Rock Soil Mech.* **2010**, *31*, 1486–1490.
48. Ishihara, K. Liquefaction and flow failure during earthquakes. *Géotechnique* **1993**, *43*, 351–451. [CrossRef]
49. Konrad, J.M. Undrained response of loosely compacted sands during monotonic and cyclic compression tests. *Géotechnique* **1993**, *43*, 69–89. [CrossRef]
50. Tian, K.-L.; Wang, P.; Zhang, H.-L. Discussion on stress-strain relation of intact loess considering soil structure. *Yantu Lixue Rock Soil Mech.* **2013**, *34*, 1893–1898.
51. Shen, Y.; Yang, H.; Xi, J.; Yang, Y.; Wang, Y.; Wei, X. A novel shearing fracture morphology method to assess the influence of freeze–thaw actions on concrete–granite interface. *Cold Reg. Sci. Technol.* **2020**, *169*, 102900. [CrossRef]
52. Hu, R.L.; Yeung, M.R.; Lee, C.F.; Wang, S.J. Mechanical behavior and microstructural variation of loess under dynamic compaction. *Eng. Geol.* **2001**, *59*, 203–217. [CrossRef]
53. Wang, Q.; Wang, L.M.; Wang, J.; Ma, H.P.; Zhong, X.M.; Wang, N.; Wang, J. Laboratory study on the liquefaction properties of the saturation structural loess. *Appl. Mech. Mater.* **2012**, *170–173*, 1339–1343. [CrossRef]
54. Xie, X.; Qi, S.; Zhao, F.; Wang, D. Creep behavior and the microstructural evolution of loess-like soil from Xi'an area, China. *Eng. Geol.* **2018**, *236*, 43–59. [CrossRef]
55. Li, X.A.; Hong, B.; Wang, L.; Li, L.; Sun, J. Microanisotropy and preferred orientation of grains and aggregates (POGA) of the Malan loess in Yan'an, China: A profile study. *Bull. Eng. Geol. Environ.* **2020**, *79*, 1893–1907. [CrossRef]
56. Zhao, S.; Zhao, Y.; Wu, J. Quantitative analysis of soil pores under natural vegetation successions on the Loess Plateau. *Sci. China Earth Sci.* **2010**, *53*, 617–625. [CrossRef]
57. Zhang, X.-C.; Huang, R.-Q.; Xu, M.; Pei, X.-J.; Han, X.-S.; Song, L.-J.; Zhang, F.-Y. Loess liquefaction characteristics and its influential factors of Shibeiyuan landslide. *Yantu Lixue Rock Soil Mech.* **2014**, *35*, 801–810.
58. Li, P.; Xie, W.; Pak, R.Y.S.; Vanapalli, S.K. Microstructural evolution of loess soils from the Loess Plateau of China. *Catena* **2019**, *173*, 276–288. [CrossRef]
59. Liu, Z.; Liu, F.; Ma, F.; Wang, M.; Bai, X.; Zheng, Y.; Yin, H.; Zhang, G. Collapsibility, composition, and microstructure of loess in China. *Can. Geotech. J.* **2016**, *53*, 673–686. [CrossRef]
60. Lei, X.-Y. Pore type of loess in China and collapsibility. *Sci. China Ser. B* **1987**, *17*, 1309–1316.

Article

Characteristics of a Debris Flow Disaster and Its Mitigation Countermeasures in Zechawa Gully, Jiuzhaigou Valley, China

Xing-Long Gong [1,2,3,†], Kun-Ting Chen [1,†], Xiao-Qing Chen [1,2,3,*], Yong You [1], Jian-Gang Chen [1,2], Wan-Yu Zhao [1] and Jie Lang [4]

1 Key Laboratory of Mountain Hazards and Earth Surface Processes, Institute of Mountain Hazards and Environment, Chinese Academy of Sciences, Chengdu 610041, China; gongxinglong@imde.ac.cn (X.-L.G.); kuntingchen@imde.ac.cn (K.-T.C.); yyong@imde.ac.cn (Y.Y.); chenjg@imde.ac.cn (J.-G.C.); wyzhao@imde.ac.cn (W.-Y.Z.)
2 CAS Center for Excellence in Tibetan Plateau Earth Sciences, Beijing 100101, China
3 University of Chinese Academy of Sciences, Beijing 100049, China
4 Jiuzhai Valley National Park Administration Bureau, Sichuan 623402, China; jielang157698955@163.com
* Correspondence: xqchen@imde.ac.cn; Tel.: +86-028-6671-3416
† These authors contributed equally to this work.

Received: 5 March 2020; Accepted: 26 April 2020; Published: 28 April 2020

Abstract: On 8 August 2017, an Ms 7.0 earthquake struck Jiuzhaigou Valley, triggering abundant landslides and providing a huge source of material for potential debris flows. After the earthquake debris flows were triggered by heavy rainfall, causing traffic disruption and serious property losses. This study aims to describe the debris flow events in Zechawa Gully, calculate the peak discharges of the debris flows, characterize the debris flow disasters, propose mitigation countermeasures to control these disasters and analyse the effectiveness of countermeasures that were implemented in May 2019. The results showed the following: (1) The frequency of the debris flows in Zechawa Gully with small- and medium-scale will increase due to the influence of the Ms 7.0 Jiuzhaigou earthquake. (2) An accurate debris flow peak discharge can be obtained by comparing the calculated results of four different methods. (3) The failure of a check dam in the channel had an amplification effect on the peak discharge, resulting in a destructive debris flow event on 4 August 2016. Due to the disaster risk posed by dam failure, both blocking and deposit stopping measures should be adopted for debris flow mitigation. (4) Optimized engineering countermeasures with blocking and deposit stopping measures were proposed and implemented in May 2019 based on the debris flow disaster characteristics of Zechawa Gully, and the reconstructed engineering projects were effective in controlling a post-earthquake debris flow disaster on 21 June 2019.

Keywords: debris flow; Zechawa Gully; mitigation countermeasures; Jiuzhaigou Valley

1. Introduction

A debris flow—a very to extremely rapid surging flow of saturated debris in a steep channel—is a widespread hazardous phenomenon in mountainous areas [1–3]. Because of their characteristics of high flow velocities, high impact forces and long run-out distances, debris flows pose a great threat to the safety of people, can cause catastrophic damage to infrastructure elements (such as roads and houses), and can even block rivers, leading to fatalities and property damage downstream [4–10]. In recent years, post-earthquake debris flow hazards have been widely investigated due to their long activity duration, high occurrence frequency and catastrophic damage [11–14]. Numerous studies have focused on rainfall thresholds and sediment supply to characterize the occurrence of post-earthquake debris flows. In the areas affected by the 1999 Chi-Chi earthquake and the 2008 Wenchuan earthquake, the thresholds for rainfall triggering post-earthquake debris flows were analysed, and it was recognized that the rainfall threshold in periods shortly after the earthquakes was markedly lower than that before the earthquake and gradually recovered over time [14–20]. In fact, a devastating earthquake generates a large sediment supply in the form of co-seismic collapses and landslides and changes the grain size of the material and the watershed permeability characteristics, thereby indirectly reducing the debris flow-triggering rainfall thresholds [18,21]. Because earthquakes tend to produce abundant loose material, if sufficient rainfall occurs soon after an earthquake, a catastrophic debris flow can be triggered. For example, influenced by the Wenchuan earthquake on 12 May 2008, a catastrophic debris flow event was triggered on 14 August 2010 in Hongchun Gully, claiming the lives of 32 people [8]. Similarly, five debris flow events were triggered in Wenjia Gully in the three rainy seasons after the Wenchuan earthquake, including a giant debris flow event on 13 August 2010 [9,13].

As an effective way to mitigate debris flow hazards, engineering countermeasures have attracted widespread attention [22–33], and the mitigation of debris flows is usually carried out by stabilizing, blocking, drainage and deposit stopping measures [11,23]. Check dams, which act to stabilize the bed, consolidate hillslopes, decrease the slope, and retain and control the transport of sediment, are commonly used engineering structures for controlling debris flows and can generally be divided into solid-body dams and open dams [25,28,29]. Because solid-body dams are associated with many drawbacks, such as the erosion of the dam foundation and changes in the hillslope-to-channel connectivity [26,27], open dams are more efficient at controlling debris flows [28,29]. After the Wenchuan earthquake, to protect people's lives and property and ensure smooth traffic, a large number of debris flow engineering structures, especially check dams, were built. However, due to the insufficient realization on the characteristics and formation mechanisms of post-earthquake debris flows, many newly-built engineering structures have failed to mitigate debris flows and have instead caused catastrophic damage. For example, due to the failure of check dams in Sanyanyu Valley on 8 August 2010, more than 200 buildings were damaged, and approximately 1700 people died [34]. Similarly, during the "8.13" Wenjiagou debris flow event, engineering structures failed, causing seven deaths and the burial of more than 497 houses [9,35]. Therefore, further research should be carried out to propose appropriate mitigation countermeasures for post-earthquake debris flows.

Recently, an Ms 7.0 earthquake struck Jiuzhaigou Valley on 8 August 2017, triggering abundant landslides and providing a vast source of material for debris flows. Due to the influence of heavy rainfall, post-earthquake debris flows were triggered in Jiuzhaigou Valley and heavily damaged infrastructure elements, such as pedestrian walkways and scenic roads, causing traffic disruption and serious property losses [36–38]. It is necessary to evaluate the characteristics of post-earthquake debris flows in Jiuzhaigou Valley, and to propose appropriate mitigation countermeasures to avoid catastrophic events, but only a few studies related to post-earthquake debris flow mitigation in this area have been published to date. In this paper, Zechawa Gully is taken as a case study to characterize a debris flow disaster and then discuss mitigation countermeasures. To improve the accuracy of parameter calculation, four different methods were used to calculate the debris flow peak discharge and quantify the debris flow magnitude. According to the survey and analysis, the destructive debris flow event in 2016 was caused by a dam breach. After the Ms 7.0 Jiuzhaigou earthquake on 8 August 2017, abundant loose solid material was available for debris flow activity, and at least one post-earthquake debris flow occurred in September 2017. The risk of dam breaches led to the implementation of engineering countermeasures with blocking and deposit stopping measures. Such works were finished on May 2019. On 21 June 2019, a post-earthquake debris flow was triggered by heavy rainfall, and the engineering countermeasures played a useful role in controlling the debris flow disaster even though the debris flow magnitude was greater than the design standard of the reconstruction engineering projects.

2. Background

2.1. Formation Conditions of the Zechawa Gully Debris Flow

Zechawa Gully, with gully mouth coordinates of 103°55′22.8″ E, 33°08′34.8″ N, is located in Jiuzhaigou Valley, Sichuan Province, China, and lies approximately 13.9 km from a scenic entrance (Figure 1a,b). The outlet of the Zechawa Gully debris flow coincides with the location of the only scenic road from Nuorilang Waterfall to Long Lake (Figure 1c). The study area is the transition zone from the Qinghai-Tibet Plateau to the Sichuan Basin and belongs to the peripheral mountainous area of the Sichuan Basin. The watershed covers an area of 1.96 km^2 and features five tributaries; the main channel is 2.57 km long and has a 61.1% longitudinal slope. The elevation difference of Zechawa Gully is approximately 1601 m, with a maximum elevation of 4040 m in the southwest of the watershed and a minimum elevation of 2439 m at the gully mouth near the scenic road. The topography of Zechawa Gully is steep, with 86.9% of the total area of the watershed having a slope exceeding 25°. The flow path of debris flow along Zechawa Gully can be divided into a formation zone, transport zone and deposition zone (Table 1). The formation zone is located in the upper reaches of Zechawa Gully (elevation above 3620 m), with an area of 0.26 km^2 and a channel length of 470 m. The transport zone is situated in the middle reaches, with the elevations ranging from 3620 m to 2600 m. The area of the transport zone is approximately 1.47 km^2, and the channel length is approximately 1530 m. The deposition zone, with an area of 0.23 km^2 and a channel length of approximately 570 m, is located in the area below an elevation of 2600 m.

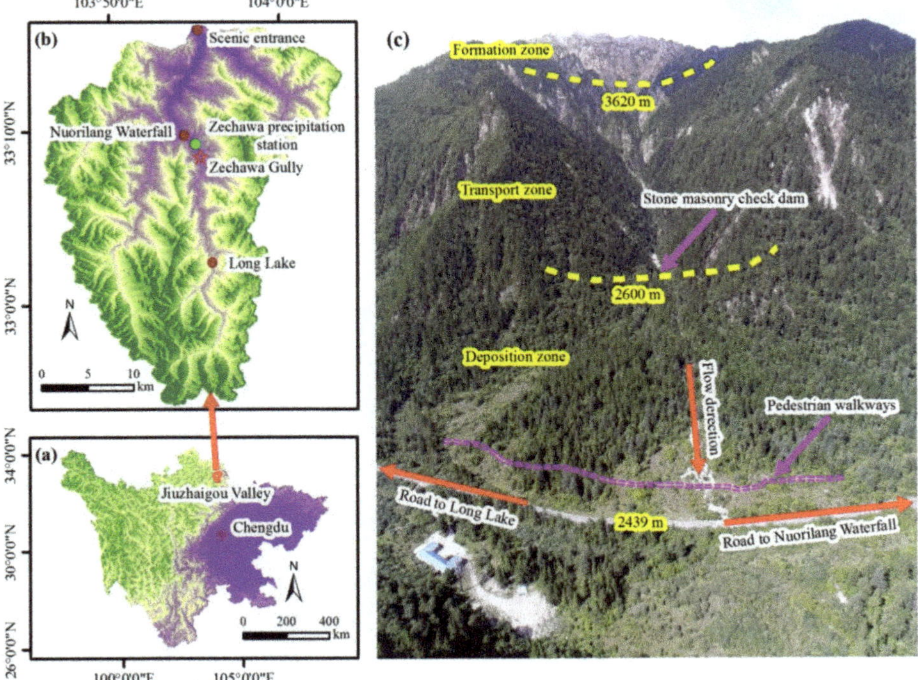

Figure 1. Location of Zechawa Gully and its full view. (**a**) Location of Jiuzhaigou Valley in Sichuan Province; (**b**) Location of Zechawa Gully in Jiuzhaigou Valley; (**c**) The full view of Zechawa Gully. The flow direction of the debris flow is perpendicular to the pedestrian walkways and the scenic road (from Nuorilang Waterfall to Long Lake).

Table 1. Zone division of Zechawa Gully.

Zone Division	Formation Zone	Transport Zone	Deposition Zone
Elevation (m)	4040–3620	3620–2600	2600–2439
Average gully gradient (‰)	708	415	244
Gully length (m)	470	1530	570
Gully characteristics	Steep slope (>50°), bare bedrock with severe frost weathering, low vegetation coverage and abundant collapsed regions	Steep slopes, a large number of landslides and high abundance of debris flow sediments on the gully bed	Gentle topography with no collapses or landslides

Compared with the characteristics of the formation zone and transport zone, the topography of the deposition zone is gentle, with no collapses and landslides, and debris flow material tends to be deposited in this area, forming a large debris flow fan. Zechawa Gully is generally a "v"-shaped channel with the characteristics of a narrow gully bed, steep lateral slopes and a high longitudinal slope, providing favourable topographic conditions for the formation of debris flows.

The study area is located in the Songpan-Ganzi Block, and the outcropping strata are mainly Quaternary and Mesozoic (Figure 2a). The lithology consists mainly of limestone and slate with a small amount of sandstone, which were intensely deformed by folding and thrusting during the Late Triassic and Early Jurassic [39,40]. In addition, since the Quaternary, the geological tectonic movement in this area has been intense due to the influence of the Tazang fault (the eastern part of the East Kunlun Fault Zone), Minjiang fault and Huya fault [41–45] (Figure 2b). Historically, seismicity has occurred on the Minjiang fault and Huya fault, including the 1960 Zhangla Ms 6.7 earthquake, the 1973 Huanglong Ms 6.5 earthquake, and the 1976 Songpan-Pingwu earthquake swarm (Ms = 7.2, 6.7, and 7.2). A recent earthquake was the Jiuzhaigou 7.0 earthquake, which occurred on 8 August 2017 on the north-western extension of the Huya fault; the rupture was dominated by left-lateral strike-slip motion [41,46–48]. On the whole, seismicity is frequent in the study area due to the geological conditions of the region, resulting in the fracture of the rock mass in the study area and triggering abundant collapses and landslides, which provide a rich source of loose material for incorporation into debris flows.

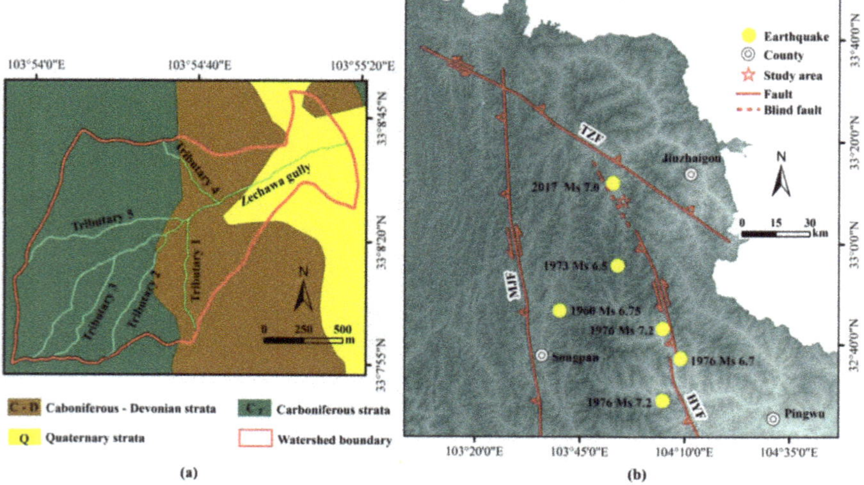

Figure 2. Study area maps. (**a**) Geologic map of the study area; (**b**) Topographic map of the Tazang fault (TZF), the Minjiang fault (MJF), the Huya fault (HYF) and the blind extension of the HYF (modified from Zhao et al. [41]).

The study area features a plateau cold temperate-subarctic monsoon climate. Due to the blocking effect of the Longmen Mountains to the southeast of the study area, most of the warm and humid air currents from the Pacific Ocean stay to the east of the Longmen Mountains. Therefore, the rainfall in Jiuzhaigou Valley west of the Longmen Mountains is relatively low, and the annual average precipitation is only 761.8 mm. The impact of cold air and high-pressure cold air currents from Mongolia in the winter is greatly weakened by the blocking of the Qinling Mountains to the north of the study area, causing this region to exhibit a mild climate, moderate precipitation and an annual average temperature of 7.3 °C [49]. There are more than 150 rainfall days annually in the study area, and the rainfall is concentrated mainly in May to September in the form of rainstorms. According to the rainfall data from the Jiuzhaigou Administration Bureau, the maximum rainfall over 24 h in Jiuzhaigou Valley is greater than 50 mm, and the precipitation increases with increasing elevation. The lowest average annual precipitation, at 696.6 mm, is found at the outlet of Jiuzhaigou Valley at an elevation of 1996 m. The highest annual average precipitation, at 957.5 mm, is found at Long Lake at an elevation of 3100 m. The snowpack period is from October to April, and the largest recorded snowpack depth exceeded 150 mm. The rainfall conditions of the study area are characterized by concentrated heavy rainfall, which is favourable for the formation of debris flows.

2.2. Description of the Debris Flow Events in Zechawa Gully

Due to the steep topography, adequate supply of loose material and intense precipitation in the study area, debris flows are active in Zechawa Gully. The earliest recorded debris flow event occurred in August 2006 and buried pedestrian walkways. In July 2008, another debris flow occurred again and blocked the scenic road. To prevent debris flows from causing further damage to the downstream pedestrian walkways and the only scenic road and to ensure the safety of residents and tourists in scenic areas, engineering countermeasures were taken in 2009. These countermeasures were designed to resist a debris flow with a 20-year return period. One stone masonry check dam 34.7 m long and 8 m high was constructed at the end of the transport zone of Zechawa Gully in 2009 (Figure 1c), and one auxiliary dam was constructed close to the stone masonry check dam. The stone masonry check dam was designed to be able to trap a volume of 2.24×10^4 m^3 of debris flow material [50].

On 4 August 2016, another destructive debris flow was triggered in Zechawa Gully. The rainfall data from the Zechawa precipitation station (103°55′04.8″ E, 33°09′18.0″ N, Figure 1b) showed that the preceeding rainfall that accumulated from 26 July 2016 to 3 August 2016 was only 8.8 mm, and the intraday rainfall was 6.7 mm on 4 August 2016. During this debris flow event, large amounts of sediment were trapped in front of the stone masonry check dam, resulting in a deposited thickness of 7 m and width of 30 m, and the length of the debris flow deposit behind the check dam was 44 m according to field measurements (Figure 3a). As sediments deposited, a breach formed in the check dam. Ultimately, the average width of the breach was 20.5 m, and the residual height of the check dam was 6 m (Figure 3b). The large kinetic energy of strong flow waves formed by the breach of check dam caused a high erosion of the downstream gully bed. During the movement of the debris flow material, the trees on both sides of the channel were impacted, leaving noticeable mud marks (Figure 3c). According to the field investigation, the total volume of the debris flow material transported downstream the failed check dam was approximately 1.39×10^4 m^3. Some of the material was deposited on the debris flow fan with a deposit area of 0.77×10^4 m^2, a thickness of 0.8–1.5 m and a volume of 0.89×10^4 m^3. Additional material with a volume of 0.5×10^4 m^3 was transported to the scenic road. During this debris flow event, the pedestrian walkways were buried again, and the only scenic road from Nuorilang Waterfall to Long Lake was blocked, causing traffic disruption and serious property loss [37,51].

Figure 3. Images of Zechawa Gully debris flow in different periods: **(a)**–**(c)** 6 August 2016, **(d)**–**(f)** 16 August 2017, **(g)**–**(i)** 23 October 2017; **(j)** large boulder transported by the debris flow that occurred in September 2017.

On 8 August 2017, the Ms 7.0 Jiuzhaigou earthquake struck the study area, and abundant landslides were triggered (Figure 3d), providing a vast source of material for debris flows. However, this earthquake had little influence on the breach shape of the check dam (Figure 3e) or the downstream topography of the check dam (Figure 3f). Subsequently, heavy rainfall occurred in the study area in September 2017. The rainfall data from Zechawa precipitation station showed that the total rainfall in September 2017 was 243.2 mm, accounting for approximately 32% of the total annual rainfall (Figure 4). Affected by the heavy rainfall in September 2017, a debris flow occurred, and the topography changed significantly. At the upstream check dam, the erosion caused by the debris flow was intense. An erosional trench approximately 1.0 m in depth was formed upstream of the dam (Figure 3g), and the breach in the dam was detectably deepened due to the erosion induced by the debris flow (Figure 3h). Due to the very high transport capacity of the debris flow, a large boulder with a long-axis length of 1.3 m, an intermediate-axis length of 1.1 m and a short-axis length of 0.7 m was transported to a point 20 m downstream of the check dam, and this boulder was composed of masonry (Figure 3j). Downstream of the check dam, the debris flow material was deposited in the channel. Additionally, trees on both sides of the channel were broken due to the very large destructive power of the debris flow, and new mud marks were left on the trees (Figure 3i). Fortunately, pedestrian walkways and scenic roads were not destroyed again. To reduce the disaster risk of the post-earthquake debris flow in Zechawa Gully, one concrete check dam, one concrete auxiliary dam and one concrete retaining wall were constructed in May 2019.

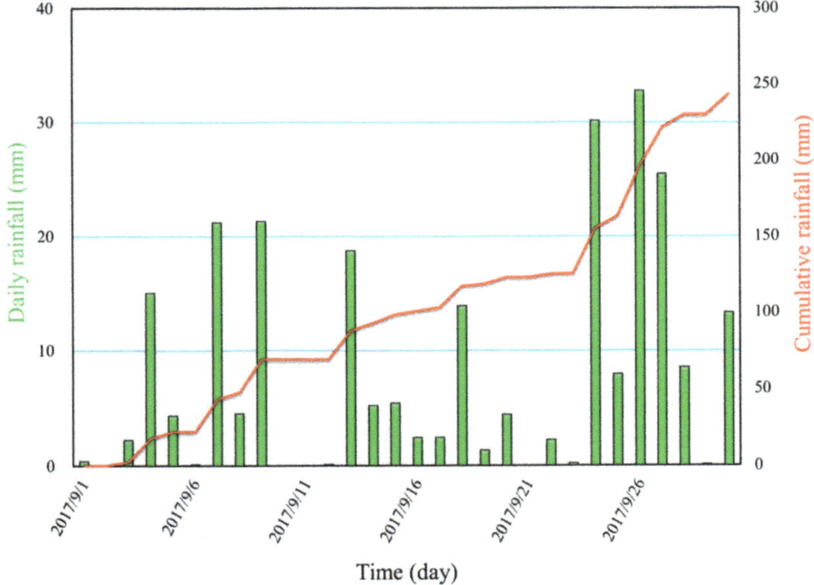

Figure 4. Rainfall distribution in September 2017 recorded by the Zechawa precipitation station.

A rainfall event started at 20:00 on 20 June 2019 and ended at approximately 08:00 on 21 June 2019 in Jiuzhaigou Valley. According to reports from patrol personnel, a post-earthquake debris flow was triggered by this storm at approximately 03:00 on 21 June 2019, and the rainfall data from the Zechawa precipitation station showed that the accumulated rainfall from 21:00 on 20 June 2019 to 02:00 on 21 June 2019 was 18.1 mm. According to the field investigation, the total volume of debris flow material was approximately 2.3×10^4 m^3. The debris flow material volume trapped by the concrete check dam was approximately 0.48×10^4 m^3 (Figure 5). Some of the other debris flow material was trapped behind the retaining wall with a deposit area of 0.3×10^4 m^2, a maximum deposit thickness of

4 m at the middle of the retaining wall and a deposit volume of 0.66×10^4 m^3 (Figure 6). The middle of the retaining wall was partially damaged, resulting in a breach with a width of 8.5 m, due to the high impact force of the debris flow. This breach allowed a portion of the debris flow material with a volume of 1.16×10^4 m^3 to be transported to the debris flow fan and scenic road (Figure 7). The material volume deposited on the fan was approximately 0.93×10^4 m^3 with a deposit area of 0.62×10^4 m^2 and an average deposit thickness of 1.5 m. The volume of the material blocking the scenic road was approximately 0.23×10^4 m^3, with a deposit length of 180 m and an average deposit thickness of 1.8 m.

Figure 5. Overview of the reconstructed check dams in Zechawa Gully (taken on 25 June 2019).

Figure 6. Overview of the reconstructed retaining wall in Zechawa Gully (taken on 23 June 2019).

Figure 7. The debris flow that occurred on 21 June 2019 buried pedestrian walkways and blocked the scenic road (taken on 22 June 2019).

3. Calculation of the Debris Flow Peak Discharge

In the mountainous areas of China, due to the lack of observation data, the rain-flood method and cross-section survey method have been widely used to calculate the debris flow peak discharge [52]. Under the assumption that the occurrence frequencies of rainstorms, floods and debris flows are the same, the rain-flood method is widely employed to calculate the debris flow peak discharge under different occurrence frequencies [53,54]. The cross-section survey method calculates the peak discharge of a debris flow that has occurred based on the mud mark and cross-sectional morphology of the channel [7,55].

For the debris flow event that occurred on 4 August 2016, two obvious typical cross-sections downstream of the stone masonry check dam are available for the calculation of the debris flow discharge through the cross-section survey method. Moreover, the pedestrian walkways were buried, and the scenic roads were blocked, and the stone masonry check dam in the channel was broken during this debris flow event. According to previous research, the amplification effect caused by dam breakage can contribute to debris flow damage in downstream towns [9,56]. Therefore, to characterize the relationship between dam failure and the occurrence of the debris flow on 4 August 2016, the dam-breaking peak discharges were estimated through the dam-breaking calculation method.

During the debris flow event that occurred in September 2017, the cross-section survey method was unavailable due to the lack of an available cross-section. A coarse boulder with dimensions of 1.3 m, 1.1 m and 0.7 m was transported 20 m downstream of the check dam by the debris flow in September 2017. According to previous studies, the largest transported particle reflects the maximum kinetic energy of flooding in mountain streams, and the maximum particle size parameters are widely used to reconstruct the velocity, depth and peak discharge of floods [57]. Thus, in this study, based on the assumption that the rainstorm, flood and debris flow frequencies were the same, the maximum particle size parameters were used to calculate the flood peak discharge, and the peak discharge of the debris flow in September 2017 was then estimated by using the methodology proposed by Lanzoni [58] according to the calculated flood peak discharge.

3.1. Rain-Flood Method

The debris flow peak discharges under different occurrence frequencies are computed by Ref. [54]:

$$Q_{df} = D_{df}\left(1 + \psi_{df}\right)Q_f \quad (1)$$

$$\psi_{df} = (\gamma_{df} - \gamma_w)/(\gamma_s - \gamma_{df}) \quad (2)$$

where D_{df} is the blockage coefficient, whose value varies with the degree of blockage, namely, very serious blockage (D_{df} = 3.0–2.6), serious blockage (D_{df} = 2.5–2.0), normal blockage (D_{df} = 1.9–1.5) and minor blockage (D_{df} = 1.4–1.1); ψ_{df} is the amplification coefficient of the debris flow peak discharge; γ_{df} is the density of the debris flow (t/m^3); γ_w is the density of water (t/m^3), usually taken as 1.00 t/m^3; γ_s is the density of the solid material (t/m^3), usually taken as 2.65 t/m^3; and Q_f is the flood peak discharge under different return periods (m^3/s), which is calculated by:

$$Q_f = 0.278\varphi\frac{S}{t^n}F \quad (3)$$

where φ is the runoff coefficient of the flood peak, which is related to the convergence of runoff; S is the rainfall intensity (mm); t is the runoff confluence time of the rainstorm (h); n is the attenuation index of the rainstorm; and F is the watershed area (m^2). Here, φ, S, t and n are calculated by the following empirical equations:

$$\varphi = 1 - 1.1\frac{\eta}{S}{t_0}^n \quad (4)$$

$$S = H_1K_1 \quad (5)$$

$$t = t_0\varphi^{-\frac{1}{4-n}} \quad (6)$$

$$n = 1 + 1.285(lg\frac{H_1K_1}{H_6K_6}) \quad (7)$$

where H_1 and H_6 are the 1-hour average rainfall and 6-hour average rainfall, respectively (mm), which are obtained from "The Rainstorm and Flood Calculation Manual of Medium and Small Basins in Sichuan Province" (published in 2010, with rainfall data from 1978 to 2004); K_1 and K_6 are the modulus coefficients corresponding to H_1 and H_6 under different return periods, respectively, which can be obtained from a Pearson type III distribution table; η is the runoff yield parameter, which reflects the average infiltration intensity (mm/h); t_0 is the runoff confluence time of the rainstorm when φ equals 1, which can be calculated by:

$$\eta = 3.6K_PF^{-0.19} \quad (8)$$

$$t_0 = [\frac{0.383}{mS^{1/4}/\theta}]^{\frac{4}{4-n}} \quad (9)$$

where K_p is the modulus coefficient when the variation coefficient is equal to 0.23, which is obtained from the Pearson type III distribution table; m is the runoff confluence parameter; and θ is the watershed characteristic parameter, which is obtained from:

$$m = 0.221\theta^{0.204} \quad (10)$$

$$\theta = \frac{L}{J^{1/3}F^{1/4}} \quad (11)$$

where L is the main channel length and J is the longitudinal slope of the channel.

3.2. Cross-Section Survey Method

Because natural channels have irregular channel bottoms, information on the channel roughness is not easy to obtain and measure. Therefore, an empirical formulation (Manning formula) was developed for turbulent flows in rough channels. It can be applied to calculate the discharge for fully rough turbulent flows and water flows. Although it is an empirical relationship, it has been found to be reasonably reliable [59,60]. Thus, the Manning formula was employed to obtain debris flow peak discharge when computing by the cross-section survey method. Based on the mud marks and cross-section morphology of the channel, the debris flow peak discharge Q_{df} (m^3/s) can be obtained by Ref. [54]:

$$Q_{df} = A_{df}V_{df} \tag{12}$$

where A_{df} is the area of the cross-section (m^2), and V_{df} is the average velocity of the debris flow (m/s), which can be calculated by:

$$V_{df} = \frac{1}{n_{df}}{R_{df}}^{2/3}{I_{df}}^{1/2} \tag{13}$$

where n_{df} is the roughness coefficient of the debris flow gully, R_{df} is the hydraulic radius of the debris flow (m), and I_{df} is the longitudinal slope gradient of the channel bed (m/m).

3.3. Dam-Breaking Calculation Method

Considering the scarcity of observational data in this study, three commonly used semi-empirical methods are employed to obtain the dam-breaking peak discharge during the debris flow event on 4 August 2016. The semi-empirical method of the Ministry of Water Resources of the People's Republic of China (MWR) [61] estimates the debris flow peak discharge Q_{df} through:

$$Q_{df} = \frac{8}{27}\sqrt{g}[B_0h_0/B_m]^{0.28}B_m(h_0 - h_d)^{1.22} \tag{14}$$

$$Q_{df} = \frac{8}{27}\sqrt{g}(\frac{B_0}{B_m})^{0.4}(\frac{h_0 + 10h_d}{h_0})^{0.3}B_m(h_0 - h_d)^{1.5} \tag{15}$$

where g is acceleration due to gravity (9.8 m^2/s); B_0 is the debris flow width before breakage (m); h_0 is the debris flow depth before breakage (m); B_m is the breach width (m), and h_d is the residual height of the dam.

The semi-empirical method of Dai and Wang [62] calculates the debris flow peak discharge Q_{df} by:

$$Q_{df} = 0.27\sqrt{g}(L_b/B_0)^{1/10}(B_0/B_m)^{1/3}B_m(h_0 - \kappa h_d)^{3/2} \tag{16}$$

where L_b is the deposit length of the debris flow material behind the check dam (m); κ is the influence factor that accounts for residual height, which is obtained by:

$$\kappa = \begin{cases} 1.4(B_mh_d/B_0h_0)^{1/3}, B_mh_d/B_0h_0 < 0.3 \\ 0.92, B_mh_d/B_0h_0 > 0.3 \end{cases} \tag{17}$$

3.4. Maximum Boulder Size Method

Based on the particle size parameters of the maximum-sized boulder, the debris flow peak discharge can be obtained through Ref. [58]:

$$Q_{df} = \frac{1}{1-C}Q_f \tag{18}$$

$$C = \frac{\rho_f \tan\beta}{(\rho_s - \rho_f)(tan\phi_{df} - \tan\beta)} \tag{19}$$

where C is the transported sediment concentration; ρ_f is the fluid density (kg/m^3); ρ_s is the sediment density; β is the bed slope angle (degrees), and the value of β is usually between 15° to 25° when using

Equation (19) [63]; φ_{df} is the quasi-static friction angle (degrees); and Q_f is the flood peak discharge (m^3/s), which was estimated by the methods of Schoklitsch, Helley, Williams and Clarke.

3.4.1. Method of Schoklitsch

This method estimates the flood peak discharge Q_f (m^3/s) by computing the unit width flux by Ref. [64,65]:

$$q_f = \frac{0.0194 d_I}{(\tan\beta)^{4/3}} \tag{20}$$

$$Q_f = q_f * B_f \tag{21}$$

where q_f is the unit width flux; d_I is the diameter of the boulder intermediate axis (m), and B_f is the channel width (m).

3.4.2. Method of Helley

This method computes the "bed velocity" for incipient motion (overturning) by equating the turning moments for fluid, drag, and lift with the resisting moment of the submerged particle weight. The critical velocity V_f (bed velocity) can be calculated by Ref. [66]:

$$V_f = 3.276\left[\frac{(\rho_b/1000 - 1)d_L(ds + d_I)^2 MR_L}{(C'_D d_S d_L MR_D + 0.178 d_I d_L MR_L)}\right]^{0.5} \tag{22}$$

$$MR_L = d_I cos\alpha/4 + \sqrt{\frac{3}{16} d_S{}^2} \sin\alpha \tag{23}$$

$$MR_D = 0.1 d_S \cos\alpha + \sqrt{\frac{3}{16} S_2^d} \cos\alpha - d_I \sin\alpha/4 \tag{24}$$

where ρ_b is the maximum boulder density (kg/m^3); d_L is the diameter of the boulder long axis (m); d_S is the diameter of the boulder short axis (m); C'_D is the drag coefficient; MR_D and MR_L are the drag turning arm and lift turning arm, respectively; and α is the original imbrication angle of the deposited boulder. During the calculation process, Equation (22) uses English units of feet, and the units of critical velocity calculated by Equation (22) need to be converted into metres per second.

The critical velocity V_f calculated by Equation (22) needs to be converted to the average velocity V_{avg} [57]:

$$V_{\text{avg}} = 1.2 V_f \tag{25}$$

The flood peak discharge Q_f can then be calculated as the product of the average velocity, mean depth and channel width by:

$$Q_f = V_{avg} h_f B_f \tag{26}$$

where h_f is the mean flood depth (m). Given that the channel width was much larger than the mean depth of flooding, the hydraulic radius obtained by the Manning formula can estimate the average depth; thus, h_f was obtained by the Manning formula:

$$h_f = \left(\frac{V_{avg} n_f}{\sqrt{\tan\beta}}\right)^{1.5} \tag{27}$$

where n_f is the roughness coefficient of a mountain stream.

3.4.3. Method of Williams

This approach calculates either the bed shear stress or the stream power needed to entrain the boulder. First, the intermediate axis diameter of the largest boulder d_I is obtained through field investigation, and then the empirical relationship between the unit stream power w, bed shear stress τ, average velocity V_{avg} and d_I is established by Ref. [67]:

$$w = 0.079{d_I}^{1.3} \quad (28)$$

$$\tau = 0.17d_I \quad (29)$$

$$V_{avg} = 0.065{d_I}^{0.5} \quad (30)$$

V_{avg}, h_f and Q_f based on the shear stress can be determined by Equations (30)–(32), respectively:

$$h_f = \frac{\tau}{\rho_b g \tan\beta} \quad (31)$$

$$Q_f = \frac{w * B_f}{\rho_b g \tan\beta} \quad (32)$$

V_{avg} and h_f based on the stream power can be obtained by:

$$V_{avg} = \frac{Q_f \rho_b g \tan\beta}{B_f \tau} \quad (33)$$

$$h_f = \frac{w}{\rho_b g \tan\beta * 0.065\sqrt{d_I}} \quad (34)$$

The value of Q_f in Equation (33) is obtained by Equation (32); then, Q_f based on the stream power can be obtained by inserting the calculated values of V_{avg} and h_f from Equations (33) and (34), respectively, into Equation (26).

3.4.4. Method of Clarke

This method assumes that the critical force (i.e., the minimum force needed to move the boulder) is equal to the resisting force and that the critical force is equal to the sum of the lift force and drag force. The critical velocity V_f (bed velocity) required to carry the maximum-sized boulder is solved by the following formula [68]:

$$V_f = \left\{2[(F_D/C_D)/\rho_f]/A_B\right\}^{0.5} \quad (35)$$

where C_D is the lift coefficient of the boulder, which is dependent on the shape of the largest boulder, with C_D = 1.18 for a cubic boulder and 0.20 for a spherical boulder; A_B is the cross-sectional area of the largest boulder; and F_D is the drag force, which is obtained by:

$$F_D = C_D F_C/(C_L + C_D) \quad (36)$$

where C_L is the lift drag coefficient, which is dependent on the shape of the largest boulder, with C_L = 0.178 for a cubic boulder and 0.20 for a spherical boulder; and F_C is the critical force, which is calculated by:

$$F_C = F_R \quad (37)$$

$$F_R = M_B[(\rho_b - \rho_f)/\rho_b]g(\mu cos\beta - sin\beta) \quad (38)$$

where μ is the shape coefficient, which is dependent on the shape of the largest boulder, with μ = 0.675 for a cubic boulder and 0.225 for a spherical boulder; and M_B is the boulder mass (kg). M_B can be obtained for a cubic boulder and a spherical boulder by Equations (39) and (40), respectively:

$$M_B = \rho_b D^3 \quad (39)$$

$$M_B = \rho_b[(\pi/6)D^3] \quad (40)$$

where D is the nominal diameter of the boulder (m), which is solved by:

$$D = (d_L d_I d_S)^{0.33} \quad (41)$$

The flood peak discharge Q_f can be obtained by inserting the calculated value of V_f into Equations (25)–(27).

4. Results

4.1. The Calculated Debris Flow Peak Discharge in 2016

With the data collected during the field investigation, the peak discharge of the debris flow that occurred on 4 August 2016 was estimated by the cross-section survey method and dam-breaking calculation method. Table 2 shows the calculation results for the debris flow peak discharge. The permissible debris flow peak discharges at the two typical mud mark cross-sections estimated by the cross-section survey method were 33.29 m^3/s and 36.69 m^3/s. The values of A_{df}, R_{df} and I_{df} were obtained through field investigation. The roughness coefficient of the debris flow gully (n_{df}) is related to the properties of the debris flow fluid and channel characteristics, and the value in this case is 0.1 according to a field survey [54].

Table 2. Calculation results of the debris flow peak discharge by using the cross-section survey method and dam-breaking calculation method.

Methods		Parameters					
		A_{df} (m^2)	R_{df} (m)	I_{df}	V_{df} (m/s)	n_{df}	Q_{df} (m^3/s)
cross-section survey method		6.45	0.75	0.391	5.16	0.1	33.29
		9.58	0.85	0.182	3.83	0.1	36.69
		B_O (m)	h_O (m)	B_m (m)	h_d (m)	L_b (m)	Q_{df} (m^3/s)
dam-breaking calculation method	Equation (14)	30.0	7.0	20.5	6.0	/	36.5
	Equation (15)	30.0	7.0	20.5	6.0	/	43.6
	Equation (16)	30.0	7.0	20.5	6.0	44.0	36.8

According to the calculation results in Table 2, the permissible maximum debris flow peak discharges resulting from the breach in the check dam varied from 36.5 m^3/s to 43.6 m^3/s. The calculation result by Equation (14) was the lowest (36.5 m^3/s), and the calculation result by Equation (15) was the highest (43.6 m^3/s). The values of B_O, h_O, B_m, h_d, and L_b were obtained by field investigation. Since the data inputs used in Equations (14)–(16) were the same, the differences among the results arose from the different combinations of data used for a given technique. The calculated values are reasonable and are similar to the debris flow peak discharge estimated by the cross-section survey method.

4.2. The Calculated Debris Flow Peak Discharge in 2017

With data collected during the field investigation, the peak discharge of the debris flow that occurred in September 2017 was calculated by the maximum boulder size method. Table 3 shows the calculation results. The calculated values of Q_f vary from 0.58 m^3/s to 6.05 m^3/s, and the calculated values of Q_{df} range from 1.76 m^3/s and 18.33 m^3/s. The minimum permissible debris flow peak discharge of 1.76 m^3/s is estimated through the method of Schoklitsch, and the maximum discharge of 18.33 m^3/s is estimated through the method of Helley. ρ_f is usually taken as 1150 kg/m^3 considering the turbidity of the flood waters [68]. ρ_s is usually taken as 2650 kg/m^3. Owing to the absence of information, a value of 36.5° was given for φ_{df} based on previous studies [58]. The values of d_L, d_I, d_S, ρ_b, B_f, β, and α were obtained through field investigation. The transported sediment concentration (C) is 0.67 by inserting the values of ρ_f, ρ_s, β and φ_{df} into Equation (19). The roughness coefficient of a mountain stream (n_f) is related to the channel characteristics, and a value of 0.05 was used here according to a field survey [69].

Table 3. Summary of the calculation results based on the maximum boulder size methods.

Basic parameters	d_L (m)	1.3	B_f (m)	6.5	ρ_s (kg/m^3)	2650	
	d_I (m)	1.1	β (degrees)	19	φ_{df} (degrees)	36.5	
	d_S (m)	0.7	α (degrees)	6	C	0.67	
	ρ_b (kg/m^3)	2250	n_f	0.05			
			Parameters				
Method			**V_{avg}(m/s)**	**h_f(m)**	**Q_f(m^3/s)**	**Q_{df}(m^3/s)**	
Schoklitsch [64]			/	/	0.58	1.76	
Helley [66]			4.26	0.22	6.05	18.33	
Williams [67]	Shear stress		2.16	0.03	0.61	1.85	
	Stream power		3.80	0.04	1.07	3.24	
Clarke [68]			2.49	0.10	1.59	4.82	

4.3. The Calculated Debris Flow Peak Discharge under Different Occurrence Frequencies

According to the magnitude of the debris flow, hazard degree and importance of the protection object, mitigation countermeasures in Zechawa Gully were required to resist a debris flow with a return period of 20–50 years [70]. Thus, the debris flow peak discharges under 10-, 20- and 50-year return periods were computed, and the calculated results of related parameters are listed in Table 4. The possible debris flow peak discharges under 10-year, 20-year and 50-year return periods are 22.27 m^3/s, 32.73 m^3/s and 48.27 m^3/s respectively. In the calculation sections, the values of F, L and J are different, resulting in different debris flow peak discharges estimated by the rain-flood method.

Table 4. Calculation results of the debris flow peak discharge by using the rain-flood method.

Calculation Content	Parameters	Unit	Return Periods		
			10-Year	**20-Year**	**50-Year**
The flood peak discharge	θ	//	2.14	2.14	2.14
	m	/	0.26	0.26	0.26
	H_1	mm	15	15	15
	H_6	mm	25	25	25
	K_1	/	1.72	2.10	2.58
	K_6	/	1.66	1.99	2.42
	K_P	/	1.31	1.42	1.56
	S	mm	25.8	31.5	38.7
	n	/	0.73	0.74	0.8
	η	mm/h	4.26	4.62	5.07
	t_0	h	1.52	1.43	1.34
	φ	/	0.75	0.79	0.82
	t	h	1.66	1.54	1.43
	Q_f	m^3/s	6.37	8.58	11.55
The debris flow peak discharge	γ_{df}	t/m^3	1.8	1.85	1.9
	D_{df}	/	1.8	1.85	1.9
	Q_{df}	m^3/s	22.27	32.73	48.27
	W_{df}	m^3	0.88×10^4	1.30×10^4	1.91×10^4

To better compare with the debris flow peak discharges calculated by the cross-section survey method, dam-breaking calculation method and maximum boulder size method, the calculation section located at the check dam site was selected to compute the debris flow peak discharges through the rain-flood method. The values of F, L and J were obtained from a topographic map with a scale of 1:5000. According to the results of the querying specification table and spot investigation, the average density of the debris flow was 1.8 t/m^3. Under given conditions, the debris flow density is positively related to the debris flow peak discharge [54,71], thus the densities of the debris flows γ_{df} under the three return periods (10-year, 20-year and 50-year) were 1.8 t/m^3, 1.85 t/m^3 and 1.9 t/m^3, respectively.

According to the site investigation, the blockage degree of the channel was normal, and the values of D_{df} were considered to be 1.8–1.9.

5. Discussion

5.1. The Applicability and Limitations of the Calculated Debris Flow Peak Discharge

The debris flow peak discharge is an important parameter for debris flow disaster prevention and risk assessment. As debris flows occur in remote mountain areas, it is difficult to measure the peak discharge and other parameters of debris flow under the conditions of severe weather and traffic delays. At present, the debris flow peak discharge is usually calculated by the rain-flood method and cross-section survey method based on certain assumptions, resulting in calculation results with low credibility. In this study, under certain assumptions, the peak discharge of debris flow was estimated by the rain-flood method, the cross-section survey method, the dam-breaking calculation method and the maximum boulder size method, and comparative analysis of the calculation results was conducted to obtain an accurate peak discharge. The limitations of the calculation results are explained as follows:

(1) Due to the complexity of debris flows and the measurement limitation, the values of relevant parameters are usually obtained by field surveys and querying the specifications. In this study, the roughness coefficient of the debris flow gully (n_{df}), the roughness coefficient of a mountain stream (n_f), the density of debris flow (γ_{df}) and the blockage coefficient (D_{df}) were obtained through field investigations and querying specifications.

(2) Considering the complexity of the debris flow and the operability of the calculation method, it is necessary to make certain assumptions and simplifications to obtain the peak discharge of the debris flow in the calculation process. The rain-flood method assumes that the occurrence frequencies of rainstorms, floods and debris flows are the same and that the calculated flood peak discharge is completely converted into the peak discharge of the debris flow [54]. Under such assumptions, important parameters such as debris flow peak discharge and total volume of debris flow material under different occurrence frequencies can be obtained, which provide important references for the design of engineering countermeasures. In addition, the breach in the check dam was idealized as a trapezoidal shape, and the average width of the breach was taken as the calculated value of B_m in the dam-breaking calculation.

(3) Four methods were used to estimate the peak discharge of the debris flow based on the maximum particle size parameters (Table 3), and the related issues in the calculation are as follows: Both Clarke and Helley solved for the critical velocity required to move the largest boulder, obtained the flow depth through the Manning formula, and finally calculated the peak discharge. Differences in the critical velocity result in differences in the flow depth and peak discharge. The method of Clarke idealizes the largest boulder as either cubic or spherical for the shape-dependent parameters, and the calculated velocities are averaged to provide the critical velocity. By setting the critical force $F_C = 0$, the downward gravitational component is balanced by the gravity-induced friction, and the extreme use condition of this method can be obtained. The limit bed slope angle (β) is equal to 34.1° for a cubic boulder and 12.7° for a spherical boulder when using the Clarke method; therefore, a spherical boulder is easier to move than the cubic boulder under the same conditions. According to the field investigation, β is equal to 19°, which exceeds the limit bed slope angle for a spherical boulder. Therefore, the selected boulder in this study was considered a cubic boulder, resulting in a calculated critical velocity that is higher than the actual value. Compared with the method of Clarke, the method of Helley neglects the bed slope, ignoring the downstream gravitational component. Generally, the bed slope of a stream is small; even for a stream with a channel longitudinal slope of 10%, the downstream gravitational component is negligibly small compared to the fluid drag and lift, so this component can be ignored [57]. However, the bed slope is 19° in this study, and neglecting the gravitational component results in a calculated critical velocity that is much higher than the actual value, ultimately resulting in

a higher calculated peak discharge. The methods of Schoklitsch and Williams estimate the peak discharge by establishing an empirical correlation based on boulder size parameters without considering the influence of the boulder shape on the calculation results. In addition, the values of w, τ and V_{avg} in the method of Williams represent the lowest values, and the actual values are higher than the calculated value.

(4) In summary, certain assumptions and simplifications were made in the calculation process, causing the peak discharge of the debris flow calculated by a single method to exhibit low accuracy. Thus, multiple methods should be used to comprehensively obtain the peak discharge, further quantifying the scale of debris flow disasters. It is worth noting that the method for calculating the debris flow peak discharge proposed in this study is mainly based on the specifications in China, especially the selection of some parameters. When calculating the debris flow peak discharge in other countries, local specifications should be considered.

5.2. The Scales of the Debris Flow Disasters in 2016 and 2017

To identify the disaster characteristics and the occurrences of debris flow events, the peak discharges of the debris flows occurring on 4 August 2016 and in September 2017 were estimated based on field investigations, and the calculation results were compared with the debris flow peak discharges under different occurrence frequencies to quantify the scale of the debris flow disasters. The related explanations are as follows:

(1) The debris flow peak flow obtained by the cross-section survey method and dam-breaking calculation method are essentially the same and are generally equivalent to the peak discharge of the debris flow with a 20-year return period (Tables 2 and 4). In addition, the total volume of the debris flow material W_{df} is estimated by Ref. [54]:

$$W_{df} = 0.264 Q_{df} T_{df} \tag{42}$$

where T_{df} is the duration time of the debris flow (s), and its value is approximately 1500 s based on the reports of patrol personnel. The value of Q_{df} is the average calculation result through the cross-section survey method and dam-breaking calculation method, and its value is 37.38 m^3/s. The total volume of debris flow material from Equation (42) is 1.48×10^4 m^3, which is consistent with the value of 1.39×10^4 m^3 based on the field investigation. Thus, it is reasonable that the scale of the debris flow on 4 August 2016 is equivalent to that of a debris flow with a 20-year return period. Moreover, based on the study above, the debris flow peak discharges calculated by Equations (14)–(16) were similar to the values obtained by the cross-section survey method. Thus, we conclude that the debris flow peak discharge on 4 August 2016 was amplified by the failure of the check dam, causing widespread damage, and this aspect also explains why the magnitude of the debris flow on 4 August 2016 was large even though the accumulated rainfall and rainfall intensity were extremely low. Similarly, check dam failures have led to catastrophic disasters in other regions, such as the "8.13" Wenjiagou debris flow event [72] and the "8.8" Zhouqu debris flow event [73,74].

(2) Based on the above analysis, the flood peak discharge estimated by the method of Helley is the largest, and is equivalent to that of a debris flow with a 10-year return period. Both of the peak discharges calculated by the methods of Clarke and Helley are larger than the actual value, while the value calculated by the method of Williams is smaller than the actual value. In addition, compared with the extensive destruction of the 2016 debris flow event with a 20-year return period, the destruction of the 2017 debris flow event was smaller, according to the field investigation. Therefore, it is reasonable that the magnitude of the debris flow in September 2017 was less than that of a debris flow with a 10-year return period.

(3) In the remote mountain areas of China, rainfall data are difficult to obtain, and the rainfall throughout a whole catchment usually cannot be recorded by precipitation stations due to the

influence of terrain, resulting in inconsistencies between the triggering rainfall and the scale of debris flow disasters. Thus, the relationships between the occurrence of debris flow disasters and the triggering rainfall are not researched in this paper.

5.3. Mitigation Countermeasures in Zechawa Gully

More than 23×10^4 m^3 of loose solid material was generated by the Ms 7.0 Jiuzhaigou earthquake and remains available as material for debris flows in Zechawa Gully in the near future [37,75]. Therefore, appropriate engineering countermeasures must be taken in a timely manner to mitigate post-earthquake debris flow disasters. According to the field investigation and calculation results above, the stone masonry check dam built in 2009 were broken, and the failure of the check dam amplified the debris flow peak discharge, resulting in a very large amount of damage during the debris flow event on 4 August 2016. Thus, the potential failure of a check dam should be fully taken into account during engineering design processes, and an integrated strategy including blocking measures and deposit stopping measures should be adopted for debris flow mitigation. On the one hand, the construction of deposit stopping structures (e.g., retaining walls) can increase the retention capacity of engineering structures; on the other hand, the debris flow material can be trapped by the deposit stopping structures even if the blocking structures (e.g., check dams) in the channel are damaged, thereby reducing the disaster risk downstream.

The engineering countermeasure taken in 2009 were designed to resist a debris flow with a 20-year return period but were damaged during the debris flow event in 2016. Considering the high-frequency and large-scale characteristics of post-earthquake debris flows, engineering countermeasures were designed to resist a debris flow with a 50-year return period after the Ms 7.0 Jiuzhaigou earthquake based on the scale, damage degree and threatened objects threatened by the subsequent debris flows. The total volume of debris flow material with a 50-year return period can be obtained by inserting the calculated value of Q_{df} into Equation (42), and the resulting value is 1.91×10^4 m^3 (Table 4). Thus, the designed engineering structures are required to trap at least 1.91×10^4 m^3 of debris flow material. In addition, the control principles of prevention projects should not only control the debris flow itself but also operate in harmony with the landscape and reduce the harm to landscape resources, as required in Jiuzhaigou Valley [76]. Under the guidance of these principles, in conjunction with the specific characteristics of the Zechawa debris flows, a concrete check dam and a concrete auxiliary dam were constructed in the channel, and a concrete retaining wall was constructed on the debris flow fan. The concrete check dam, 42.6 m long and 6 m high, was built close to but downstream of the broken stone masonry check dam in order to reduce the peak discharge, stabilize the gully bed, minimize scouring along the bottom and sides of the gully, and stabilize the debris flow material trapped behind the broken check dam. The downstream concrete auxiliary dam, 38.1 m long and 3 m high, was constructed close to the concrete check dam to protect the latter's foundation (Figure 5). Moreover, the reconstructed check dams were located somewhat upstream in the gully and were satisfactorily concealed. The retaining wall with a total length of 95.6 m was built 93 m away from the scenic road and is out of sight of tourists, and it can trap a volume of 2.27×10^4 m^3 of debris flow materials (Figure 6). In May 2019, new control works (the reconstructed check dam and the retaining wall) were finished.

5.4. Effectiveness of Mitigation Countermeasures and Evaluation of Debris Flow Impact Force

On 21 June 2019, one post-earthquake debris flow was triggered by heavy rainfall, and a volume of 2.3×10^4 m^3 of debris flow material was transported; this value was greater than the calculated total volume of debris flow material with a 50-year return period in Table 4. A volume of 0.48×10^4 m^3 of debris flow sediment was trapped by the concrete check dam (Figure 5), which contributed to stabilizing the gully bed and preventing entrainment of additional material. Moreover, a volume of approximately 0.66×10^4 m^3 debris flow sediment was trapped by the retaining wall (Figure 6), and a portion of material with a volume of 1.16×10^4 m^3 emerged from the breach in the middle of the retaining wall and was transported downstream. During the debris flow event on 21 June 2019, the prevention projects played a satisfactory role in controlling the debris flow disaster even though the flow magnitude exceeded the design standard.

In addition, studying the damage mechanism of mitigation structures is significant for effective debris flow mitigation. According to previous studies, the huge impact force of a debris flow can contribute significantly to the destruction of mitigation structures [34,77], and numerous impact models have been established [77–80]. Through comprehensive analysis of the existing debris flow impact models, a modified hydro-static model with a good prediction capability was proposed by Vagnon [77]. Therefore, the impact force of debris flow on the retaining wall was evaluated to study the damage mechanism by Ref. [77]:

$$P_{peak} = 2.07{F_r}^{1.64}\gamma_{df}gh_{df} \tag{43}$$

$$F_r = V_{df}/\sqrt{gh_{df}} \tag{44}$$

where P_{peak} is the peak impact pressure (kN/m^2); F_r is the Froude number; and h_{df} is the mean debris flow depth (m). Considering the large scale of the debris flow disaster on 21 June 2019, γ_{df} is taken as 1.9 t/m^3 according to Table 4. Based on field investigation, the average velocity of the debris flow (V_{df}) near the retaining wall was calculated through Equation (13), and related parameters are shown in Table 5.

Based on the related report, the designed resistance of the retaining wall is 51.34 KN/m^2 [75], which is far below the calculated value of the peak impact pressure (80.39 kN/m^2) in Table 5. The debris flow impact force was greater than the resistance of the retaining wall, causing partial failure of the retaining wall on 21 June 2019. Thus, the resistance of the retaining wall should be increased during the design processes. In general, considerable attention should be given to the post-earthquake debris flow disaster in Zechawa Gully in the future, and it is necessary to repair the broken retaining wall with a greater design resistance and remove the debris flow material deposited behind the retaining wall to prepare for the next post-earthquake debris flow in the near future.

Table 5. Calculation results of the debris flow impact force on the retaining wall on 21 June 2019.

γ_{df} (t/m^3)	h_f (m)	R_{df} (m)	I_{df}	n_{df}	F_r	P_{peak} (kN/m^2)
1.9	1.55	1.11	0.19	0.1	1.20	80.39

6. Conclusions

This study is intended to describe the debris flow events in Zechawa Gully, characterize the debris flow disaster, propose appropriate mitigation countermeasures and analyse the effectiveness of mitigation countermeasures that were already implemented in May 2019. Field investigations were conducted in a timely manner to determine the debris flow peak discharge, and the disaster characteristics and occurrence of debris flows in 2016 were analysed. The following conclusions can be drawn:

(1) In this study, the debris flow peak discharge was calculated using the rain-flood method, cross-section survey method, dam-breaking calculation method and maximum boulder size method. Based on our research, compared with previous results based on a single method, an accurate debris flow peak discharge can be obtained by comparing the results of each calculation method with each other, which increases the parameter accuracy for debris flow disaster prevention and risk assessment.

(2) According to the classification criterion of the debris flow scale, the debris flows in Zechawa Gully can be classified as small-scale events (with a total volume of debris flow material less than 1.0×10^4 m^3) and medium-scale events (with a total volume of debris flow material between 1.0×10^4 m^3 and 10×10^4 m^3) [81]. The scale of the debris flow event on 4 August 2016 was

equivalent to that of a debris flow with a 20-year return period. After the Ms 7.0 Jiuzhaigou earthquake, at least one debris flow with a scale less than that of a debris flow with a 10-year return period was triggered in September 2017, and a destructive debris flow with a scale greater than that of a debris flow with a 50-year return period was triggered in June 2019.

(3) The debris flow peak discharge on 4 August 2016 was amplified by the failure of the stone masonry check dam, causing widespread damage. Due to the disaster risk caused by dam breach incidents, an integrated strategy including blocking measures and deposit stopping measures should be adopted for debris flow mitigation.

(4) Based on the debris flow hazard characteristics of Zechawa Gully, optimized engineering countermeasures (including blocking measures and deposit stopping measures) with a design standard of a 50-year return period were proposed. Combined with the debris flow control principles for national parks, one satisfactorily concealed concrete check dam and one retaining wall out of view of tourists were constructed in Zechawa Gully in May 2019.

(5) On 21 June 2019, a post-earthquake debris flow was triggered by heavy rainfall, and the engineering countermeasure, including blocking and deposit stopping measures, were effective in mitigating the debris flow disaster even though the debris flow magnitude was greater than the design standard of the reconstructed engineering projects. More attention should be paid to the post-earthquake debris flow disaster in Zechawa Gully, and it is necessary to repair the broken retaining wall with greater design resistance and to remove the debris flow material deposited behind the retaining wall in a timely manner to prepare for upcoming post-earthquake debris flows in the near future.

Notation

A_B	Cross-sectional area of the largest boulder
A_{df}	Area of the cross-section
B_f	Channel width
B_m	Breach width
B_0	Debris flow width before breakage
C	Transported sediment concentration
C_D	Lift coefficient of the boulder, which is dependent on the shape of largest boulder
C'_D	Drag coefficient
C_L	Lift drag coefficient, which is dependent on the shape of the largest boulder
D	Nominal diameter of the boulder
D_{df}	Blockage coefficient
d_I	Diameter of the boulder intermediate axis
d_L	Diameter of the boulder large axis
d_S	Diameter of the boulder short axis
F	Watershed area
F_C	Critical force
F_D	Drag force
F_r	Froude number
g	Acceleration due to gravity
H_1	1-hour average rainfall
H_6	6-hour average rainfall
h_d	Residual height of check dam
h_{df}	Mean debris flow depth
h_f	Mean flood depth
h_0	Debris flow depth before breakage
I_{df}	Longitudinal slope gradient of the channel bed
J	Longitudinal slope of the channel

K_1	Modulus coefficients corresponding to H_1 under different return periods
K_6	Modulus coefficients corresponding to H_6 under different return periods.
K_p	Modulus coefficient when the variation coefficient is equal to 0.23
L	Main channel length
L_b	Deposit length of the debris flow material behind the check dam
M_B	Boulder mass
MR_D	Drag turning arm
MR_L	Lift turning arm
m	Runoff confluence parameter
n	Attenuation index of the rainstorm
n_{df}	Roughness coefficient of the debris flow gully
n_f	Roughness coefficient of a mountain stream
P_{peak}	Peak impact pressure
Q_{df}	Debris flow peak discharge
Q_f	Flood peak discharge
q_f	Unit width flux
R_{df}	Hydraulic radius of the debris flow
S	Rainfall intensity
T_{df}	Duration time of the debris flow
t	Runoff confluence time of the rainstorm
t_0	Runoff confluence time of the rainstorm when ϕ equals 1.
V_{avg}	Average velocity
V_{df}	Average velocity of the debris flow
V_f	Critical velocity (bed velocity)
W_{df}	Total volume of the debris flow material
w	Unit stream power
α	Original imbrication angle of the deposited boulder
β	Bed slope angle
γ_{df}	Density of the debris flow
γ_s	Density of the solid material
γ_w	Density of the water
θ	Watershed characteristic parameter
μ	Shape coefficient, which is dependent on the shape of the largest boulder
ρ_b	Maximum boulder density
ρ_f	Fluid density
ρ_s	Sediment density
τ	Bed shear stress
η	Runoff yield parameter, which reflects the average infiltration intensity
ϕ	Runoff coefficient of the flood peak, which is related to the convergence of runoff
φ_{df}	Quasi-static friction angle
ψ_{df}	Amplification coefficient of the debris flow peak discharge
κ	Influence factor that accounts for residual height

Author Contributions: X.-L.G. and K.-T.C. contributed to the conceptualization, methodology, analysis and manuscript writing of the study. X.-Q.C. proposed the main structure of this study and approved the final version. Y.Y. and J.-G.C. helped perform the analysis with constructive discussions. W.-Y.Z. and J.L. provided resources and participated in the field investigations. All authors have read and agreed to the published version of the manuscript.

Funding: This research was supported by the "8.8" Jiuzhaigou earthquake stricken area ecological disaster prevention and control of key scientific and technological support project of Land and Resources Department of Sichuan Province (Research on Prevention and Control Technology of Ecological Debris Flow Disasters, Grant No. KJ-2018-24), the National Natural Science Foundation of China (Grant No. 51709259), the Foundation for Young Scientists of the Institute of Mountain Hazards and Environment, CAS (Grant No. SDS-QN-1912), CAS "Light of West China" Program, the Youth Innovation Promotion Association CAS (2017426) and the Open Foundation of Key Laboratory of Mountain Hazards and Earth Surface Processes, CAS.

Conflicts of Interest: The authors declare no conflict of interest.

References

1. Huggett, R. *Fundamentals of Geomorphology*; Routledge: London, UK, 2007. [CrossRef]
2. Jakob, M.; Hungr, O. *Debris-Flow Hazards and Related Phenomena*; Springer: Berlin, Germany, 2005.
3. Hungr, O.; Leroueil, S.; Picarelli, L. The Varnes classification of landslide types, an update. *Landslides* **2014**, *11*, 167–194. [CrossRef]
4. Chen, K.-T.; Chen, X.-Q.; Hu, G.-S.; Kuo, Y.-S.; Huang, Y.-R.; Shieh, C.-L. Dimensionless assessment method of landslide dam formation caused by tributary debris flow events. *Geofluids* **2019**. [CrossRef]
5. Chen, K.-T.; Chen, X.-Q.; Niu, Z.-P.; Guo, X.-J. Early identification of river blocking induced by tributary debris flow based on dimensionless volume index. *Landslides* **2019**. [CrossRef]
6. Chen, K.-T.; Lin, C.-H.; Chen, X.-Q.; Hu, G.-S.; Guo, X.-J.; Shieh, C.-L. An assessment method for debris flow dam formation in Taiwan. *Earth Sci. Res. J.* **2018**, *22*, 37–43. [CrossRef]
7. Ni, H.; Zheng, W.; Song, Z.; Xu, W. Catastrophic debris flows triggered by a 4 July 2013 rainfall in Shimian, SW China: Formation mechanism, disaster characteristics and the lessons learned. *Landslides* **2014**, *11*, 909–921. [CrossRef]
8. Tang, C.; Zhu, J.; Ding, J.; Cui, X.F.; Chen, L.; Zhang, J.S. Catastrophic debris flows triggered by a 14 August 2010 rainfall at the epicenter of the Wenchuan earthquake. *Landslides* **2011**, *8*, 485–497. [CrossRef]
9. Yu, B.; Ma, Y.; Wu, Y. Case study of a giant debris flow in the Wenjia Gully, Sichuan Province, China. *Nat. Hazards* **2012**, *65*, 835–849. [CrossRef]
10. Chen, M.-L.; Liu, X.-N.; Wang, X.-K.; Zhao, T.; Zhou, J.-W. Contribution of Excessive Supply of Solid Material to a Runoff-Generated Debris Flow during Its Routing Along a Gully and Its Impact on the Downstream Village with Blockage Effects. *Water* **2019**, *11*, 169. [CrossRef]
11. Chen, X.; Cui, P.; You, Y.; Chen, J.; Li, D. Engineering measures for debris flow hazard mitigation in the Wenchuan earthquake area. *Eng. Geol.* **2015**, *194*, 73–85. [CrossRef]
12. Ni, H.; Tang, C.; Zheng, W.; Xu, R.; Tian, K.; Xu, W. An overview of formation mechanism and disaster characteristics of post-seismic debris flows triggered by subsequent rainstorms in Wenchuan earthquake extremely stricken areas. *Acta Geol. Sin.* **2014**, *88*, 1310–1328. [CrossRef]
13. Ni, H.Y.; Zheng, W.M.; Tie, Y.B.; Su, P.C.; Tang, Y.Q.; Xu, R.G.; Wang, D.W.; Chen, X.Y. Formation and characteristics of post-earthquake debris flow: A case study from Wenjia gully in Mianzhu, Sichuan, SW China. *Nat. Hazards* **2011**, *61*, 317–335. [CrossRef]
14. Yu, B.; Wu, Y.; Chu, S. Preliminary study of the effect of earthquakes on the rainfall threshold of debris flows. *Eng. Geol.* **2014**, *182*, 130–135. [CrossRef]
15. Guo, X.; Cui, P.; Li, Y.; Zhang, J.; Ma, L.; Mahoney, W.B. Spatial features of debris flows and their rainfall thresholds in the Wenchuan earthquake-affected area. *Landslides* **2016**, *13*, 1215–1229. [CrossRef]
16. Guo, X.; Cui, P.; Marchi, L.; Ge, Y. Characteristics of rainfall responsible for debris flows in Wenchuan earthquake area. *Environ. Earth Sci.* **2017**, *76*, 596. [CrossRef]
17. Li, T.-T.; Huang, R.-Q.; Pei, X.-J. Variability in rainfall threshold for debris flow after Wenchuan earthquake in Gaochuan River watershed, Southwest China. *Nat. Hazards* **2016**, *82*, 1967–1980. [CrossRef]
18. Ma, C.; Wang, Y.; Hu, K.; Du, C.; Yang, W. Rainfall intensity-duration threshold and erosion competence of debris flows in four areas affected by the 2008 Wenchuan earthquake. *Geomorphology* **2017**, *282*, 85–95. [CrossRef]
19. Shieh, C.L.; Chen, Y.S.; Tsai, Y.J.; Wu, J.H. Variability in rainfall threshold for debris flow after the Chi-Chi earthquake in central Taiwan, China. *Int. J. Sediment Res.* **2009**, *24*, 177–188. [CrossRef]
20. Zhou, W.; Tang, C. Rainfall thresholds for debris flow initiation in the Wenchuan earthquake-stricken area, Southwestern China. *Landslides* **2014**, *11*, 877–887. [CrossRef]
21. Tang, C.; van Asch, T.W.J.; Chang, M.; Chen, G.Q.; Zhao, X.H.; Huang, X.C. Catastrophic debris flows on 13 August 2010 in the Qingping area, Southwestern China: The combined effects of a strong earthquake and subsequent rainstorms. *Geomorphology* **2012**, *139–140*, 559–576. [CrossRef]
22. Zeng, Q.L.; Yue, Z.Q.; Yang, Z.F.; Zhang, X.J. A case study of long-term field performance of check-dams in mitigation of soil erosion in Jiangjia stream, China. *Environ. Geol.* **2008**, *58*, 897–911. [CrossRef]
23. Peng, C.; Yongming, L. Debris-Flow Treatment: The Integration of Botanical and Geotechnical Methods. *J. Resour. Ecol.* **2013**, *4*, 097–104. [CrossRef]

24. DeWolfe, V.G.; Santi, P.M.; Ey, J.; Gartner, J.E. Effective mitigation of debris flows at Lemon Dam, La Plata County, Colorado. *Geomorphology* **2008**, *96*, 366–377. [CrossRef]
25. Piton, G.; Carladous, S.; Recking, A.; Tacnet, J.M.; Liébault, F.; Kuss, D.; Quefféléan, Y.; Marco, O. Why do we build check dams in Alpine streams? An historical perspective from the French experience. *Earth Surf. Process. Landf.* **2016**, *42*, 91–108. [CrossRef]
26. Cucchiaro, S.; Cavalli, M.; Vericat, D.; Crema, S.; Llena, M.; Beinat, A.; Marchi, L.; Cazorzi, F. Geomorphic effectiveness of check dams in a debris-flow catchment using multi-temporal topographic surveys. *CATENA* **2019**, *174*, 73–83. [CrossRef]
27. Cucchiaro, S.; Cazorzi, F.; Marchi, L.; Crema, S.; Beinat, A.; Cavalli, M. Multi-temporal analysis of the role of check dams in a debris-flow channel: Linking structural and functional connectivity. *Geomorphology* **2019**. [CrossRef]
28. Piton, G.; Recking, A. Design of Sediment Traps with Open Check Dams. I: Hydraulic and Deposition Processes. *J. Hydraul. Eng.* **2016**, *142*. [CrossRef]
29. Bernard, M.; Boreggio, M.; Degetto, M.; Gregoretti, C. Model-based approach for design and performance evaluation of works controlling stony debris flow with an application to a case study at Rovina di Cancia (Venetian Dolomites, Northeast Italy). *Sci. Total Environ.* **2019**. [CrossRef]
30. Albaba, A.; Lambert, S.; Kneib, F.; Chareyre, B.; Nicot, F. DEM Modeling of a Flexible Barrier Impacted by a Dry Granular Flow. *Rock Mech. Rock Eng.* **2017**, *50*, 3029–3048. [CrossRef]
31. Leonardi, A.; Wittel, F.K.; Mendoza, M.; Vetter, R.; Herrmann, H.J. Particle-Fluid-Structure Interaction for Debris Flow Impact on Flexible Barriers. *Comput.-Aided Civ. Infrastruct. Eng.* **2015**, *31*, 323–333. [CrossRef]
32. Zhu, Z.-H.; Yin, J.-H.; Qin, J.-Q.; Tan, D.-Y. A new discrete element model for simulating a flexible ring net barrier under rockfall impact comparing with large-scale physical model test data. *Comput. Geotech.* **2019**, *116*, 103208. [CrossRef]
33. Albaba, A.; Lambert, S.; Nicot, F.; Chareyre, B. Relation between microstructure and loading applied by a granular flow to a rigid wall using DEM modeling. *Granul. Matter* **2015**, *17*, 603–616. [CrossRef]
34. Wang, G.L. Lessons learned from protective measures associated with the 2010 Zhouqu debris flow disaster in China. *Nat. Hazards* **2013**, *69*, 1835–1847. [CrossRef]
35. Liu, F.Z.; Xu, Q.; Dong, X.J.; Yu, B.; Frost, J.D.; Li, H.J. Design and performance of a novel multi-function debris flow mitigation system in Wenjia Gully, Sichuan. *Landslides* **2017**, *14*, 2089–2104. [CrossRef]
36. Chen, X.-Q.; Chen, J.-G.; Cui, P.; You, Y.; Hu, K.-H.; Yang, Z.-J.; Zhang, W.-F.; Li, X.-P.; Wu, Y. Assessment of prospective hazards resulting from the 2017 earthquake at the world heritage site Jiuzhaigou Valley, Sichuan, China. *J. Mt. Sci.* **2018**, *15*, 779–792. [CrossRef]
37. Hu, X.; Hu, K.; Tang, J.; You, Y.; Wu, C. Assessment of debris-flow potential dangers in the Jiuzhaigou Valley following the August 8, 2017, Jiuzhaigou earthquake, Western China. *Eng. Geol.* **2019**, *256*, 57–66. [CrossRef]
38. Zhao, B.; Wang, Y.S.; Luo, Y.H.; Li, J.; Zhang, X.; Shen, T. Landslides and dam damage resulting from the Jiuzhaigou earthquake (8 August 2017), Sichuan, China. *R. Soc. Open Sci.* **2018**, *5*, 171418. [CrossRef]
39. Chen, S.F.; Wilson, C.; Deng, Q.D.; Zhao, X.L.; Zhi, L.L. Active faulting and block movement associated with large earthquakes in the Min Shan and Longmen Mountains, Northeastern Tibetan Plateau. *J. Geophys. Res. Solid Earth* **1994**, *99*, 24025–24038. [CrossRef]
40. Yin, A.; Harrison, T.M. Geologic evolution of the Himalayan-Tibetan orogen. *Ann. Rev. Earth Planet. Sci.* **2000**, *28*, 211–280. [CrossRef]
41. Zhao, D.; Qu, C.; Shan, X.; Gong, W.; Zhang, Y.; Zhang, G. InSAR and GPS derived coseismic deformation and fault model of the 2017 Ms7.0 Jiuzhaigou earthquake in the Northeast Bayanhar block. *Tectonophysics* **2018**, *726*, 86–99. [CrossRef]
42. Liu, Z.; Tian, X.; Gao, R.; Wang, G.; Wu, Z.; Zhou, B.; Tan, P.; Nie, S.; Yu, G.; Zhu, G.; et al. New images of the crustal structure beneath eastern Tibet from a high-density seismic array. *Earth Planet. Sci. Lett.* **2017**, *480*, 33–41. [CrossRef]
43. Ren, J.; Xu, X.; Yeats, R.S.; Zhang, S. Millennial slip rates of the Tazang fault, the Eastern termination of Kunlun fault: Implications for strain partitioning in eastern Tibet. *Tectonophysics* **2013**, *608*, 1180–1200. [CrossRef]
44. Zhao, X.L.; Deng, Q.D.; Chen, S.F. Tectonic geomorphology of the Minshan uplift in Western Sichuan, Southwestern China. *Seismol. Geol.* **1994**, *16*, 429–439. (In Chinese)

45. Zhou, R.; Li, Y.; Densmore, A.L.; Ellis, M.A.; He, Y.; Wang, F.; Li, X. Active tectonics of the eastern margin of the Tibet Plateau. *J. Mineral. Petrol.* **2006**, *26*, 40–51. (In Chinese) [CrossRef]
46. Fan, X.; Scaringi, G.; Xu, Q.; Zhan, W.; Dai, L.; Li, Y.; Pei, X.; Yang, Q.; Huang, R. Coseismic landslides triggered by the 8th August 2017 M s 7.0 Jiuzhaigou earthquake (Sichuan, China): Factors controlling their spatial distribution and implications for the seismogenic blind fault identification. *Landslides* **2018**, *15*, 967–983. [CrossRef]
47. Nie, Z.; Wang, D.-J.; Jia, Z.; Yu, P.; Li, L. Fault model of the 2017 Jiuzhaigou Mw 6.5 earthquake estimated from coseismic deformation observed using global positioning system and interferometric synthetic aperture radar data. *Earth Planets Space* **2018**, *70*, 55. [CrossRef]
48. Xie, Z.; Zheng, Y.; Yao, H.; Fang, L.; Zhang, Y.; Liu, C.; Wang, M.; Shan, B.; Zhang, H.; Ren, J.; et al. Preliminary analysis on the source properties and seismogenic structure of the 2017 Ms7.0 Jiuzhaigou earthquake. *Sci. China Earth Sci.* **2018**, *61*, 339–352. [CrossRef]
49. Cui, P.; Liu, S.Q.; Tang, B.X.; Chen, X.; Zhang, X. *Debris Flow Study and Prevention in National Park*; Science Press: Beijing, China, 2005; pp. 90–92. (In Chinese)
50. Jiang, T.; Yang, J.C.; Peng, S.E.; Wang, S.F.; Zhou, Y.; Qi, Y.L.; Wei, J.G. *Construction Drawing Design Report for the Control Projects of Zechawa and Xiajijie Debris Flow Disasters in Jiuzhaigou Valley*; Nuclear Industry Southwest Survey and Design Institute Co., Ltd.: Chengdu, China, 2009.
51. Sun, H.; You, Y.; Liu, J.-F. Experimental study on characteristics of trapping and regulating sediment with an open-type check dam in debris flow hazard mitigation. *J. Mt. Sci.* **2018**, *15*, 2001–2012. [CrossRef]
52. You, Y.; Liu, J.F.; Chen, X.C. Debris flow and its characteristics of Subao River in Beichuan county after 5.12 Wenchuan earthquake. *J. Mt. Sci.* **2010**, *28*, 358–366. (In Chinese)
53. Liu, J.; You, Y.; Chen, X.; Liu, J.; Chen, X. Characteristics and hazard prediction of large-scale debris flow of Xiaojia Gully in Yingxiu Town, Sichuan Province, China. *Eng. Geol.* **2014**, *180*, 55–67. [CrossRef]
54. Zhou, B.F.; Li, D.J.; Luo, D.F.; Lv, R.R.; Yang, Q.X. *Guide to Prevention of Debris Flow*; Science Press: Beijing, China, 1991; pp. 80–95. (In Chinese)
55. Tang, C.; Rengers, N.V.; van Asch, T.W.; Yang, Y.; Wang, G. Triggering conditions and depositional characteristics of a disastrous debris flow event in Zhouqu city, Gansu Province, Northwestern China. *Nat. Hazards Earth Syst. Sci.* **2011**, *11*, 2903–2912. [CrossRef]
56. Xiong, M.Q.; Meng, X.M.; Wang, S.Y.; Guo, P.; Li, Y.J.; Chen, G.; Qing, F.; Cui, Z.J.; Zhao, Y. Effectiveness of debris flow mitigation strategies in mountainous regions. *Prog. Phys. Geogr.* **2016**, *40*, 768–793. [CrossRef]
57. Costa, J.E. Paleohydraulic reconstruction of flash-flood peaks from boulder deposits in the Colorado front range. *Geol. Soc. Am. Bull.* **1983**, *94*, 986–1004. [CrossRef]
58. Lanzoni, S.; Gregoretti, C.; Stancanelli, L.M. Coarse-grained debris flow dynamics on erodible beds. *J. Geophys. Res. Earth Surf.* **2017**, *122*, 592–614. [CrossRef]
59. Chanson, H. *The Hydraulics of Open Channel Flow: An Introduction*, 2rd ed.; Elsevier Butterworth-Heinemann: Oxford, UK, 2004.
60. Subramanya, K. *Flow in Open Channels*, 3rd ed.; Elsevier Butterworth-Heinemann: Oxford, UK, 2009.
61. Ministry of Water Resources of the People's Republic of China (MWR). *Technique Guideline for Emergency Disposal of Landslide Lake*; SL451-2009; Standards Press of China: Beijing, China, 2009; pp. 22–25. (In Chinese)
62. Dai, R.Y.; Wang, Q. Research on the maximum discharge of dam-breaking. *J. Hydraul. Eng.* **1983**, *2*, 15–23. (In Chinese)
63. Takahashi, T. *Debris Flow Mechanics, Prediction and Countermeasures*; Taylor & Francis Group: London, UK, 2007; pp. 45–47. [CrossRef]
64. Schoklitsch, A. Der geschiebetrieb und die geschiebefracht. *Wasserkraft Wasserwirtschaft* **1934**, *29*, 37–43.
65. Du, C.; Yao, L.-K.; Shakya, S.; Li, L.-G.; Sun, X.-D. Damming of large river by debris flow: Dynamic process and particle composition. *J. Mt. Sci.* **2014**, *11*, 634–643. [CrossRef]
66. Helley, E.J. *Field Measurement of the Initiation of Large Bed Particle Motion in Blue Creek Near Kalmath, California*; US Geological Survey Professional Paper; United States Government Printing Office: Washington, DC, USA, 1969.
67. Williams, G.P. Paleohydrological methods and some examples from Swedish fluvial environments: I cobble and boulder deposits. *Geogr. Ann. Ser. A Phys. Geogr.* **1983**, *65*, 227–243. [CrossRef]
68. Clarke, A.O. Estimating probable maximum floods in the Upper Santa Ana Basin, Southern California, from stream boulder size. *Environ. Eng. Geosci.* **1996**, *2*, 165–182. [CrossRef]

69. Zhao, Z.X.; He, J.J. *Hydraulics*; Tsinghua University Press: Beijing, China, 2010; pp. 176–177. (In Chinese)
70. China Association of Geological Hazard Prevention (CAGHP). *Specification of Design for Debris Flow Prevention*; T/CAGHP 021-2018; China University of Geosciences Press: Wuhan, China, 2018; pp. 3–6. (In Chinese)
71. Rickenmann, D. Hyperconcentrated Flow and Sediment Transport at Steep Slopes. *J. Hydraul. Eng.* **1991**, *117*, 1419–1439. [CrossRef]
72. You, Y.; Chen, X.; Liu, J. 8.13" extra large debris flow disaster in Wenjia gully of Qingping Township, Mianzhu, Sichuan Province. *J. Catastrophol.* **2011**, *26*, 68–72. (In Chinese)
73. Cui, P.; Zhou, G.G.D.; Zhu, X.H.; Zhang, J.Q. Scale amplification of natural debris flows caused by cascading landslide dam failures. *Geomorphology* **2013**, *182*, 173–189. [CrossRef]
74. Zhou, G.G.D.; Cui, P.; Chen, H.Y.; Zhu, X.H.; Tang, J.B.; Sun, Q.C. Experimental study on cascading landslide dam failures by upstream flows. *Landslides* **2012**, *10*, 633–643. [CrossRef]
75. Chen, X.Q.; Chen, X.Z.; Zhang, W.F.; Chen, K.T.; Yang, D.X.; Hu, K.; Gong, X.L.; Si, G.W.; Xiong, Z. *Investigation Report for the Control of Zechawa Debris Flow Disaster in Jiuzhaigou Valley, Jiuzhaigou County*; Institute of Mountain Hazards and Environment, Chinese Academy of Sciences: Chengdu, China, 2017. (In Chinese)
76. Cui, P.; Liu, S.; Tang, B.; Chen, X. Debris flow prevention pattern in national parks-taking the world natural heritage Jiuzhaigou as an example. *Sci. China* **2003**, *46*, 1–11. [CrossRef]
77. Vagnon, F. Design of active debris flow mitigation measures: A comprehensive analysis of existing impact models. *Landslides* **2019**. [CrossRef]
78. Zhang, S.A. Comprehensive Approach to the Observation and Prevention of Debris Flows in China. *Nat. Hazards* **1993**, *7*, 1–23. [CrossRef]
79. Cui, P.; Zeng, C.; Lei, Y. Experimental analysis on the impact force of viscous debris flow. *Earth Surf. Process. Landf.* **2015**, *40*, 1644–1655. [CrossRef]
80. Vagnon, F.; Segalini, A. Debris flow impact estimation on a rigid barrier. *Nat. Hazards Earth Syst. Sci.* **2016**, *16*, 1691–1697. [CrossRef]
81. Ministry of Land and Resources of the People's Republic of China (MLR). *Specification of Geological Investigation for Debris Flow Stabilization*; DZ/T0220-2006; Standards Press of China: Beijing, China, 2006; pp. 2–3. (In Chinese)

MDPI

Article

A Study on Interaction between Overfall Types and Scour at Bridge Piers with a Moving-Bed Experiment

Wei-Lin Lee [1], Chih-Wei Lu [2,*] and Chin-Kun Huang [1]

[1] Department of Hydraulic and Ocean Engineering, National Cheng Kung University, Tainan 701401, Taiwan; glaciallife@gmail.com (W.-L.L.); ckhuang@mail.ncku.edu.tw (C.-K.H.)

[2] Department of Civil and Construction Engineering, National Taiwan University of Science and Technology, Taipei 106335, Taiwan

* Correspondence: cwlu@mail.ntust.edu.tw; Tel.: +886-2-2737-6563

Abstract: River slopes can be changed due to an extreme event, e.g., a large-scale earthquake. This can uplift a riverbed greatly and thereby change the behavior of the river flow into a free or submerged overfall. Corresponding damage, including extreme erosion, on bridge piers located in the river can take place due to the aforementioned flow conditions. A reconstructed bridge pier in the same location would also experience a similar impact if the flow condition is not changed. It is important to identify these phenomena and research the mechanism in the interaction between overfall types and scour at bridge piers. Therefore, this paper is aimed at studying a mechanism of free and submerged overfall flow impacts on bridge piers with different distances by a series of moving-bed experiments. The experiment results showed clearly that bridge pier protection requires attention particularly when the pier is located in the maximum scour hole induced by the submerged overfall due to the z directional flow eddies. In many other cases, such as when the location of the bridge pier was at the upstream slope of a scour hole induced by a flow drop, a deposition mound could be observed at the back of the pier. This indicates that, while a pier is at this location, an additional protection takes place on the bridge pier.

Keywords: bridge pier; overfall; scour; landform change impact on pier

Citation: Lee, W.-L.; Lu, C.-W.; Huang, C.-K. A Study on Interaction between Overfall Types and Scour at Bridge Piers with a Moving-Bed Experiment. *Water* **2021**, *13*, 152. https://doi.org/10.3390/w13020152

Received: 15 December 2020
Accepted: 8 January 2021
Published: 11 January 2021

Publisher's Note: MDPI stays neutral with regard to jurisdictional claims in published maps and institutional affiliations.

1. Introduction

Unexpected free or submerged overfall conditions in a river flow can occur due to a force within the earth that causes the riverbed to uplift. The changed condition of river flow could have an impact on the safety of downstream river structures. A major earthquake that occurred in 21 September 1999 dramatically changed many landforms in central Taiwan, such as a local rise in the Da-Ja River inducing a flow drop in the riverbed. This flow drop is very close to the Pai-Furn bridge pier and could induce additional erosion. Figure 1a demonstrates a bridge that was damaged in the 921 Earthquake in 1999 in Taiwan and the surrounding river bed was significantly affected. The newly constructed bridge was completed in less than 2 years due to the importance for transportation (shown in Figure 1b). However, it can be seen in Figure 1c that the bridge pier was again exposed to river flow in six years due to significant nearby scour. Figure 1b,c display that the riverbed level had been apparently lowered down 4.5 m deep near the pier with a diameter of 3.6 m.

Figure 1. (**a**) Changed river bed and damaged bridge due to the 921 Earthquake in 1999; (**b**) Newly constructed bridge in the same location; (**c**) Water drop and local scour at the bridge piers.

Many researchers devoted themselves to studying local scour below drop structures and at bridge piers (Dey 2014) [1]. Some researchers focused on the local scour below drop structures; for example, Schoklitsch (1932) has been the pioneering researcher and proposed an empirical relationship to estimate the equilibrium scour depth for flow-over structures [2]. Moore (1943), Rand (1955), Akram (1979), and Little and Murphey (1982) studied the energy change due to the drop [3–6]. Smith and Strang (1967) found that the profile change of a riverbed was strongly affected by the size of the river bed materials [7]. Mason and Arumugam (1985) reviewed the empirical formulas of equilibrium scour depth under a falling jet that started in 1932, and they proposed a modified formula that includes the effect of tailwater depth [8]. Hoffmans (1998) derived relations to predict the maximum scour depth in the equilibrium phase based on the Newton's second law of motion [9].

Hoffmans (2009) introduced an index to represent the strength of loose material and extended previous relations to predict the sum of the maximum scour depth and the tailwater depth [10]. D'Agostino and Ferro (2004) proposed an empirical formula to estimate the equilibrium scour depth of weir type drop structures based on the high crest of the weir and the flow depth over a weir [11]. Yager et al. (2012) extrapolated an approach to predict the scour depth and geometry of A-, U-, and W-shaped rock weirs from the case of two-dimensional flow [12]. Melville (2014) used a small-scale experiment to investigate the scour at a bridge foundation in the vicinity of a sluice gate and low wire [13].

With the aforementioned research, the effect of different types of drop structures, different conditions of approach flow, and different materials of sediment have been investigated, and varied empirical formulas for the characteristic of local scour due to the drop structure have been proposed. On the other hand, some researchers focused on the local scour at bridge piers; for example, Breusers et al. (1977) reviewed a series of literature regarding theory, model, and field data about the local scour around cylindrical piers and suggested a set of designs for protection against scour [14]. Ahmed and Rajaratnam (1998) reported that smooth, rough, and mobile beds impacted the flow features and pier scour [15]. Graf and Istiarto (2002) experimented with the equilibrium scour depth around a cylinder pier and investigated the vortex system around a cylinder pier based on measurement of the acoustic Doppler velocity profiler (ADVP) [16].

Dey and Raikar (2007) experimented with the developing scour depth around a cylinder pier and investigated the features of the vortex system in the intermediate and equilibrium stages [17]. Ataie-Ashtiani and Aslani-Kordkandi (2013) experimented with the developing scour depth around a single pier and two piers in tandem on a roughly flat bed and investigated the difference of flow features in the implemented experiments [18]. Euler et al. (2014) investigated the local scour in the vicinity of pillar-like objects through experimental studies and compared the results with field data [19].

These studies contribute to the flow feature impact on pier scour. Other studies focused on scoring features. For example, Baker (1980) derived a formula to estimate the equilibrium scour depth in front of a cylindrical bridge pier and compared the results with the results of Baker (1979), Breusers et al. (1977), and Chabert and Engeldinger (1956) [14,20–22]. Chiew and Melville (1987) proposed an empirical relationship that was related to the equilibrium depth scour, particle size of sediment, and flow condition, and compared their results with the findings of Chee (1982) and Melville (1984) [23–25]. Elliott and Baker (1985) investigated the feature of scour depth under the effect of lateral spacing between bridge piers [26]. Melville and Chiew (1999) indicated that the development of the equilibrium scour depth can be related with the size of the pier, size of the sediment, and approach flow velocity [27].

Sheppard et al. (2004) indicated that the wash load concentration impacts the scale of the equilibrium scour depth under clear-water conditions [28]. Ataie-Ashtiani and Beheshti (2006) derived an empirical relationship to estimate the maximum local scour depth for the pile group and compared their results with the reports of Melville and Coleman (2000) and Richardson and Davis (2001) [29–31]. Khosronejad et al. (2012) investigated the features of clear-water scour around the geometry of cylindrical, square, and diamond bridge piers through experiments and numerical simulation [32]. According to the aforementioned research, the mechanism of local scour at bridge piers has been investigated comprehensively by the theory, experiment, field data, and numerical model, and empirical formulas for the equilibrium depth scour at bridge piers have been proposed. However, there are few papers, to the authors' knowledge, focusing on the interaction between overfall types and scour at bridge piers.

This paper was focused on probing the mechanism of the scouring effect on piers considering the different bridge locations and a flow-drop-induced scour hole. Two types of overfall, which included the submerged type and free overfall type, were researched herein. In the submerged type, the velocity component in the vertical direction was relatively smaller and, therefore, gave a smaller effect on the riverbed scour. The free type of overfall, on the other hand, produced a strong velocity component in the vertical direction and induced a more dramatic riverbed change. It is necessary to discuss in detail the response of piers suffering these two types of overfall and to take these responses into consideration in engineering practice.

2. Procedure of the Experimental Work

The configuration for the experiment flume equipped with a circulation flow system is shown in Figure 2a. The total length of the flume was 15 m, the width was 1 m, and slope of the flume bed was 1/1000. The wall of the flume was 0.8 m in height composed of transparent tempered glass. A deeper part of the flume bed had 2 m length, 1 m width, and 0.4 m depth located at 7 m upstream from the scour development. A flat flume bed was provided with 4.4 m in length, which was 81 times the hydraulic radius of 5.4 cm for a fully developed flow. H represents the difference in height from the water level upstream of the lifted platform to the tailwater level. d_s indicates the depth of scouring.

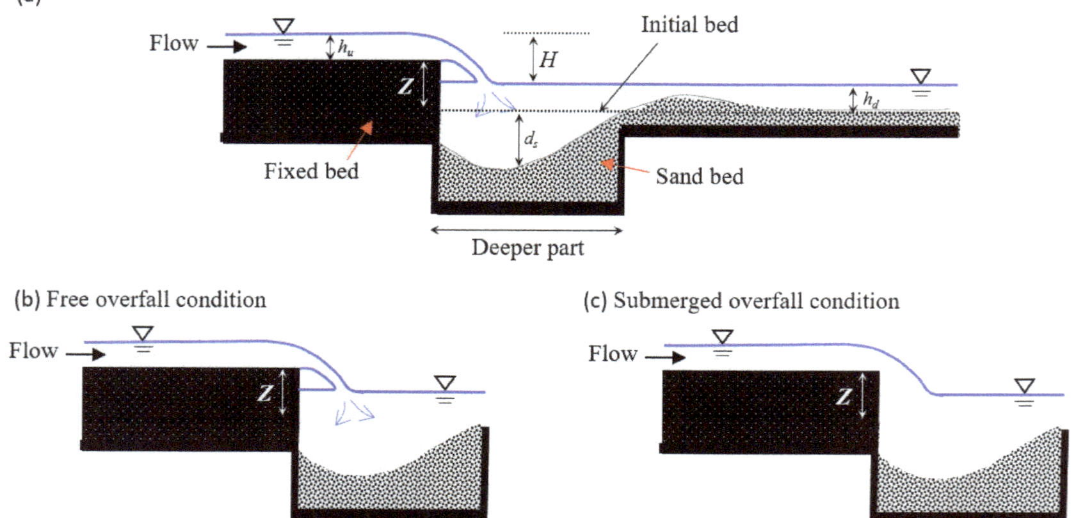

Figure 2. (**a**) Configuration of experiment flume. (**b**) Schematic of free overfall condition. (**c**) Schematic of submerged overfall condition.

For the sediment using in the experiment, we assumed that the river bed was composed of medium sand. The median diameter (D_{50}) of sediments in this experiment was 0.46 mm, and the standard deviation of the sediments (σ_g) was 1.69. Randkivi and Ettema (1977) suggested that the σ_g should be smaller than 1.3 to avoid the armor layer in the development of local scour [33]. The flume was paved using homogenous sediments at a 2.5 cm depth in the zone out of the scour area to provide a similar roughness in the alluvial bed.

This was a clear-water scour test. The experiment was designed so that the local scour occurred only due to the influence of the drop structure and pier. In other words, the clear-water flow could trigger scour in the moving-bed when there was no drop structure and pier in the experiment flume. Accordingly, the ratio of the designed velocity of approach flow (V) and the critical mean approach flow velocity of the using sediment (V_c) was given as 0.5. Melville and Sutherland (1988) suggested that the critical mean approach flow velocity (V_c) can be estimated using following equation:

$$\frac{V_c}{V_{*c}} = 5.75 \times \log\left[5.53 \times \frac{h_d}{D_{50}}\right] \quad (1)$$

where h_d is the depth of the downstream flow, and V_{*c} is the shear velocity of the using sediment [34]. Melville and Sutherland (1988) proposed a Shields chart for the threshold condition of uniform sediment in water, and the shear velocity was suggested as

0.018 m/sec for using sediment [34]. To satisfy the aforementioned conditions, in the experiment, the boundary conditions of flow were: critical flow depth h_c = 2.4 cm, upstream flow depth h_u= 5.4 cm, upstream velocity 27.6 cm/sec, upstream Froude number Fr = 0.38, downstream flow depth h_d = 9.5 cm, downstream velocity (V) 15.9 cm/sec, and downstream Froude Number Fr = 0.16.

The authors changed the difference in height between the river bed and the crest of flow-control structure (Z) to produce two different conditions: free overfall and submerged overfall, with the same boundary conditions upstream and downstream, to study the interacting behaviors between the pier and overfall. The schematics of the free overfall and submerged overfall are shown in Figure 2b,c. The values of Z were selected as 8 and 12 cm in which the submerged overfall took place at Z = 8 cm (H = 3.9 cm) with no air vent occurrence and the free overfall took place at Z = 12 cm (H = 7.9 cm) with air vent occurrence.

Melville and Chiew (1999) and Dey (2014) indicated that the approach flow can no longer move the sediment from the scour hole when the scour is at equilibrium in the clear-water condition [1,27]. Accordingly, the equilibrium time was selected based on the development process of the scour hole in the free overfall test, in which 83.5% of the 24 h erosion was reached in 5 h as shown in Figure 3. We also observed that the deposition of the dune downstream was segregated after 4 h of erosion due to the lack of sediment supplementation from upstream. This showed that the overfall energy was dispersed in the scour hole so that the erosion was reduced. Therefore, the scouring behavior in the fifth erosion hour was chosen for discussion in this paper.

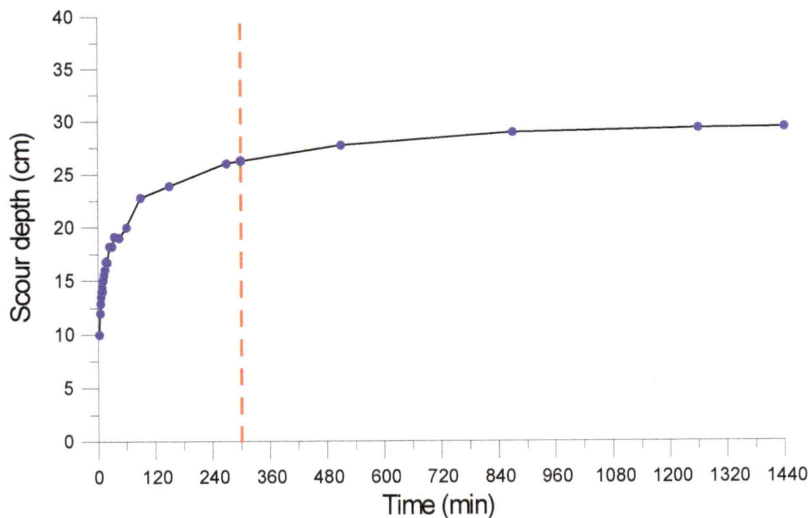

Figure 3. Development of the scouring process (the broken line was selected as the equilibrium hour).

A significant bed would be changed in the free overfall test; therefore, piers were installed at five locations:

1. At half distance between the maximum scour depth and upstream, which was 16 cm (L_a) to the flow-control structure, namely Case A.
2. At the maximum scour depth, which was 32 cm (L_b) to the flow-control structure, namely Case B.
3. At half of the inclined slope of the scouring hole, which was 58.7 cm (L_c) to the flow-control structure, namely Case C.
4. At the boundary between the scouring hole and original bed, which was 85.5 cm (L_d) to the flow-control structure, namely Case D.

5. At the deposition zone of the downstream, which was 130.8 cm (L_e) to the flow-control structure, namely Case E.

On the other hand, piers were installed at three different locations for the submerged overfall test:

1. At upstream, which was 17.3 cm (L_f) to the flow-control structure, namely Case F.
2. At the scouring hole, which was 46.4 cm (L_g) to the flow-control structure, namely Case G.
3. At the boundary between the scouring hole and original bed, which was 73.5 cm (L_h) to the flow-control structure, namely Case H.

The above-mentioned cases are summarized in Table 1, and the schematic of the pier locations for free overfall and submerged overfall is shown in Figure 4. In addition, the free type and submerged type of overfall where no pier was installed in the experiment flume were also carried out and named "free overfall w/o pier" and "submerged overfall w/o pier", respectively. Lastly, a pier in the experiment flume was implemented without any type of overfall condition (Z = 0 cm) and named "Pure Bridge".

Table 1. List of the locations of piers.

Type of Overfall	Free Overfall					Submerged Overfall		
Case [1]	A [2]	B [3]	C [4]	D [5]	E [6]	F [2]	G [3]	H [4]
Height of overfall (Z)	12	12	12	12	12	8	8	8
The distance from pier to flow-control structure (L_i)	16 (L_a)	32 (L_b)	58.7 (L_c)	85.5 (L_d)	130.8 (L_e)	17.3 (L_f)	46.4 (L_g)	73.5 (L_h)

[1] A–E: free overfall flow F–H: Submerged overfall flow. [2] A and F: Pier located at the upstream slope of the flow-drop-induced scour hole. [3] B and G: Pier located at the maximum scour point of the flow-drop-induced scour hole. [4] C and H: Pier located at the downstream slope of the flow-drop-induced scour hole. [5] D: Pier located at the edge of the flow-drop-induced scour hole. [6] E: Pier located far from the scour hole.

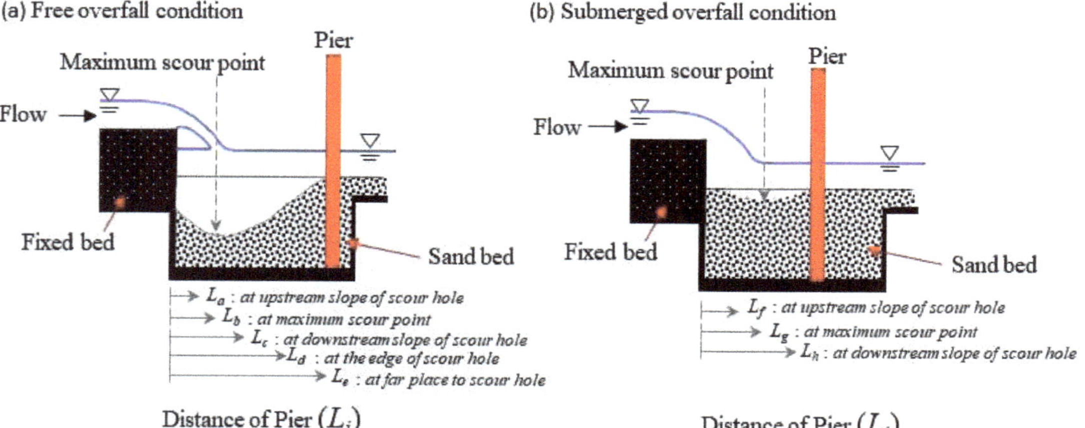

Figure 4. Schematic graphs of the pier locations: (**a**) Free overfall condition. (**b**) Submerged overfall condition.

3. Observations from the Experiment and Discussion

3.1. Profile of the Scouring Development of Free Overfall without the Effect of the Pier

In Figure 5, the time history of profile of scouring development in the case of free overfall w/o a pier is shown.

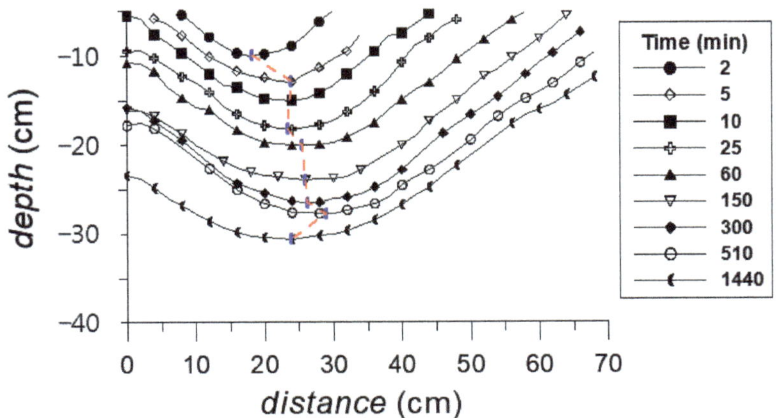

Figure 5. Development of the scouring process in the case of free overfall w/o a pier.

The horizontal axis represents the direction of flow, 0 is at the location of the flow-control structure, the vertical axis is the scouring depth, and the broken red line in the figure links the maximum scour depth at each observed time. We investigated that the maximum scour depth moved deeper and more downstream. During the scouring process, we also found that the slope of the scouring hole sometimes fell backward toward the scouring hole. Two counter-rotation eddies that were produced by the overfall affected the slope of the scouring hole where the slope at upstream was less steep than the slope at downstream because the counter-rotating eddies downstream were stronger than the ones that rotated upstream. The eddy provided the drag force along the slope surface, which increased the resistance of the sediment fall due to gravity.

The mechanism of development of the scouring hole was that the two counter-rotating eddies brought up the sediment to the slope at the downstream side, and gradually a small dune was formed. Euler et al. (2014) investigated the mechanism using a tracer, which allowed a visualization of the turbulent eddying and was similar to the observations in our experiments [19]. While a deeper and wider scouring hole was dug by the overfall, the dune was moved further downstream. On the other hand, the sediment of the slope of the scouring hole upstream occasionally slid into the hole while the hole was being dug wider and deeper. The sliding sediment was brought away downstream randomly. The slope at the downstream of the scouring hole was steeper than the original at-rest angle of the sediment deposits because the eddies provided a floating force along the slope surface that supported the sediments to stay at the same location until slope instability due to the occurrence of toe erosion induced by scour.

The maximum equilibrium scour depth was about 26.7 cm at 1440 min in the condition of free overfall w/o pier as shown in Figure 5. Many researchers proposed different empirical formulas for the maximum equilibrium scour depth under varied conditions of structure, sediment material, and approach flow [2,8–11,21,24,29]. For the condition of free overfall, Mason and Arumugam (1985) mentioned that the empirical formula for the maximum equilibrium scour depth has general form.

$$d_s = \alpha_1 \frac{V^{\alpha_2} H^{\alpha_3}}{D_{50}^{\alpha_3}} \quad (2)$$

in which α_1, α_2, α_3, and α_4 are all coefficients [8]. These coefficients were represented by different values in individual studies, and our study lists some suggested values from Mason and Arumugam (1985) in Table 2 [8]. In the procedure of the experiment work, the approach velocity (V) was 15.9 cm/sec, the value of H was 7.9 cm in the condition of free overfall, and the median diameter (D_{50}) of sediments was 0.46 mm. The comparison of the equilibrium scour depth in the experiment and with the empirical formulas of other authors can be obtained in Table 2. These results illustrated that the scour depth had close to an equilibrium state in the experiment.

Table 2. Coefficients for use in Equation (2) (Mason and Arumugam) [8].

Author (year)	α_1	α_1	α_1	α_1	d_s (m)	Error (m)
Hartung (1959)	1.4	0.64	0.36	0.32	0.222	−0.045
Chee and Kung (1974)	1.663	0.6	0.2	0.1	0.359	0.092
Machado (1980)	1.35	0.5	0.3145	0.0645	0.255	−0.012
INCYTH (1981)	1.413	0.5	0.25	0	0.299	−0.032

3.2. Interaction between Piers and Overfall-Induced Erosion in Plain View

3.2.1. Free Overfall Impact on Pier at Different Location

Regarding free overfall, there were five locations in the experiment as shown in Figure 6. Figure 6a shows the case where no pier was installed in the experiment, and we observed that the geographic changes in the flume were mostly two-dimensional except at the boundaries. The distance of the maximum scour depth was about 32 cm, and the distance to the original bed level was about 80 cm.

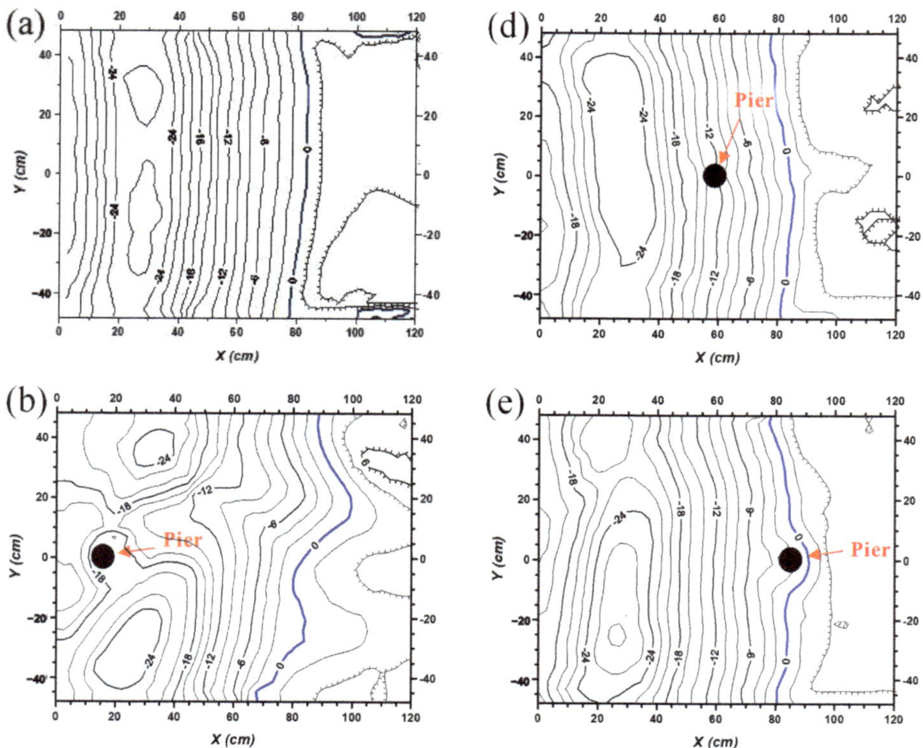

Figure 6. *Cont.*

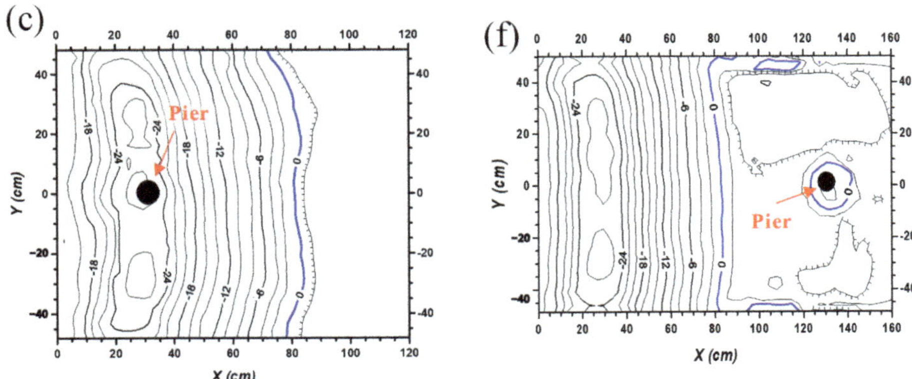

Figure 6. Contour lines of the river bed in the free overfall condition: (**a**) Free overfall w/o pier; (**b**) Case A; (**c**) Case B; (**d**) Case C; (**e**) Case D; (**f**) Case E.

Figure 6b shows the results of Case A where the location of the pier was upstream of the scouring hole. In this case, the nappe directly impacted onto the pier instead of the river bed. The neighborhood of the pier was influenced by the pier and, therefore, deformed largely. Two sides of the pier were further eroded than in the previous case because the circulating flow took place after the nappe hit on the pier and the flow increased the erosion. However, the energy of the nappe reduced after hitting the pier, and thus less erosion occurred to the downstream.

Figure 6c shows Case B where the pier was located at the location of the maximum scour depth. A similar geography to the case without piers was observed, and therefore we concluded that the erosion induced by the pier in this case was not influential.

Figure 6d or 6e demonstrate a slight change of the river bed near the pier (Case C and D). Figure 6f shows that the erosion took place only at the neighborhood of the pier while the pier was located at the deposition area (Case E).

Overall speaking, in the free overfall condition, when the pier location was upstream of the scouring hole (L_a), significant erosion was found in the front of the pier along with a significant deposition in the back. When the pier location was far from the local scour (L_e), some erosion and deposition took place in the front of the pier and in the back of the pier, respectively. The localized scour in the vicinity of the pier was induced by the approaching flow similar to classical local scour at the bridge (Dey 2014) [13]. When the pier location was at the edge of the local scour (L_d), the erosion depth in the front was lower than the original river bed, and the erosion took place in the back of the pier as well. However, the eroded river bed level was still higher than the original bed.

As to the above discussions, while the pier location was at the downstream slope of the scour hole (L_c), the erosion in the front of pier was similar to the case of free overfall w/o pier. This indicates that the pier did not affect the characteristics of erosion. However, significant deposition occurred in the back of the pier in this case, and this caused the total erosion to be reduced. When the pier location was at the maximum scour point (L_b), greater erosion took place compared with at the maximum scouring depth, and slight deposition occurred in the back of the pier.

3.2.2. Submerged Overfall Impact on Piers at Different Locations

Figure 7a shows that, in the case of submerged overfall without pier installation, the erosion was much less than in the case of free overfall, and the major scouring area was moved downstream. In Case F with the pier located upstream of the scour hole (L_f), Figure 7b shows that no significant geography changes of the river bed in the front of pier were found when comparing with the previous cases. However, significant deposition was observed at the back of the pier. When the pier location was at the maximum scouring depth (Case G, L_g), the erosion in the front of pier became much more significant compared with the previous case, and it decreased in the back of pier in Figure 7c. In this case, the maximum scour depth shows a significant increase. In Figure 7d, the overfall condition shows a limited impact on the pier in Case H (L_h).

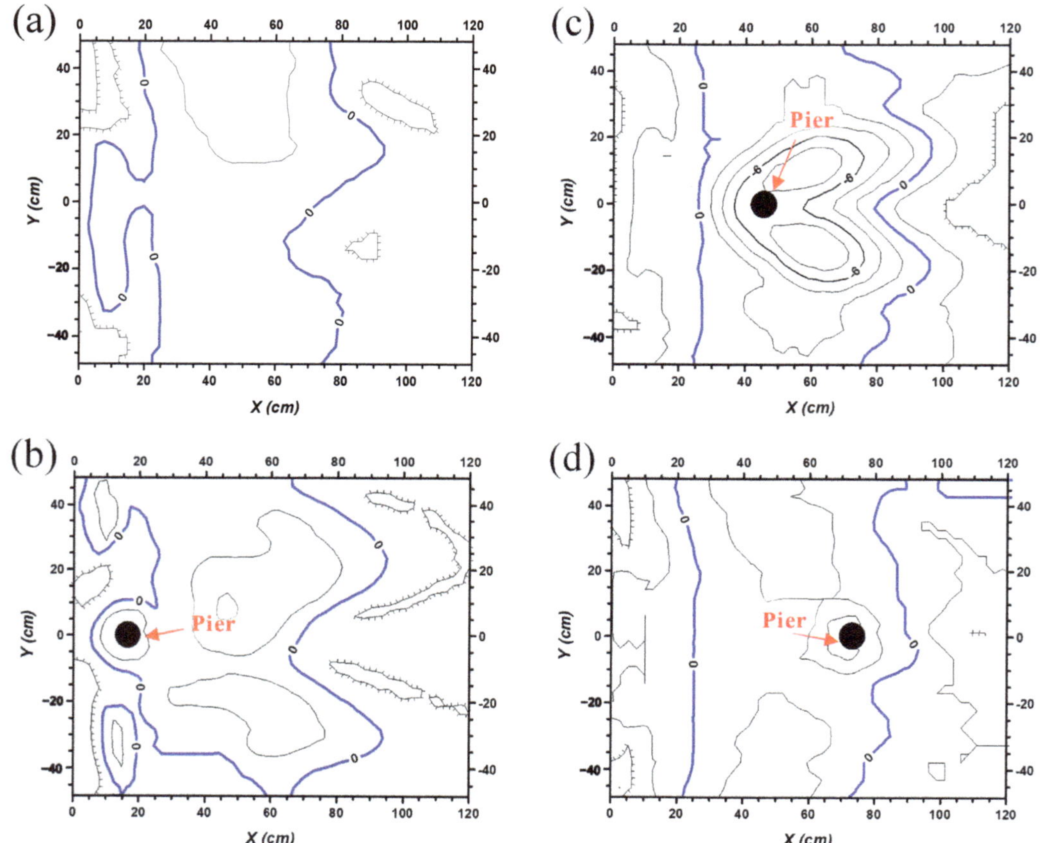

Figure 7. Contour lines of the river bed in the submerged overfall condition: (**a**) Submerged overfall w/o pier; (**b**) Case F; (**c**) Case G; (**d**) Case H.

Overall, in the submerged overfall condition, the scouring depth was clearly much smaller than in the free overfall cases. The most significant result was in Case G.

3.3. Interaction between Piers and Overfall-Induced Erosion in Side View

The profile change of the center line of the river bed can be seen in Figure 8. This demonstrates that there was no significant change of the landforms at the location in the front of pier when the water drop induced scour located upstream (Case A). The situation is similar to the case of the free overfall w/o pier. However, a relatively large deposition was observed in the back of pier for the pure water drop scour condition (Case A). This condition presented increased scour depth due to the pier when the pier was located at the maximum erosion depth of the flow drop hole (Case B). A relatively larger scour to pure water drop condition but no significant deposition was observed when the pier was located at the downstream slope of the scour hole (Case C). When the pier was located far from the scour hole, a localized scour was found in the front of the pier, and some deposition was also observed in the back of the pier (Case E). Case D presented the pier located at the edge of the scour hole scour hole, and the erosion occurrence in the front of pier became more significant and lower than the initial bed level. On the other hand, erosion in the back of pier took place as well. However, the river bed level was still higher than the initial river bed level.

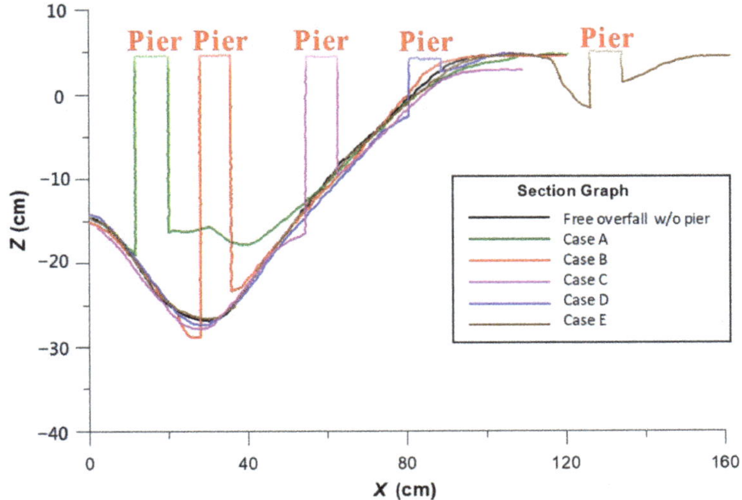

Figure 8. Comparisons of the center line of the vertical profile of the channel in the free type overfall drop.

We concluded that, while the pier located at the upstream slope of the maximum scour depth was induced by overfall, the scour that occurred in the front of pier was similar to the pure water drop inducing scour, which indicates that the scouring characteristic was not influenced by the pier. However, a significant deposition was observed in the back of pier. This revealed that the total scouring was reduced. This implies that better protection for the river bed can be found compared with the case of free overfall o/w pier when the pier is located at the upstream slope of the scour hole.

When the pier was located at the downstream slope of the scour hole, an increased scour depth was found in the front of pier when compared with the original scour hole, and some deposition was observed at the back of pier. The experiments demonstrated that the change in depth of the river bed was at the minimum when the pier was located at the edge of the scour hole. When the pier was located far from the scour hole, a localized scour in the vicinity of pier was induced by the approaching flow without the impact of free overfall. These results imply that that the bridge pier was more secure when it was located at the edge of the scour hole.

The scour depth in the submerged overfall condition was found to be smaller than in the free overfall condition as shown in Figure 9. When the pier was located upstream of the maximum scour point (Case F), a deeper scour was found in the front of pier compared with the initial river bed, and there was a deposition at the back of pier. When the pier was located at the point of maximum scour (Case G), a significant scour was observed in the front and the back of the pier, and this was also deeper than for the initial river bed. When the pier was located in the initial river bed (Case H), scour was found at the front and back of the pier and was smaller than in Case G.

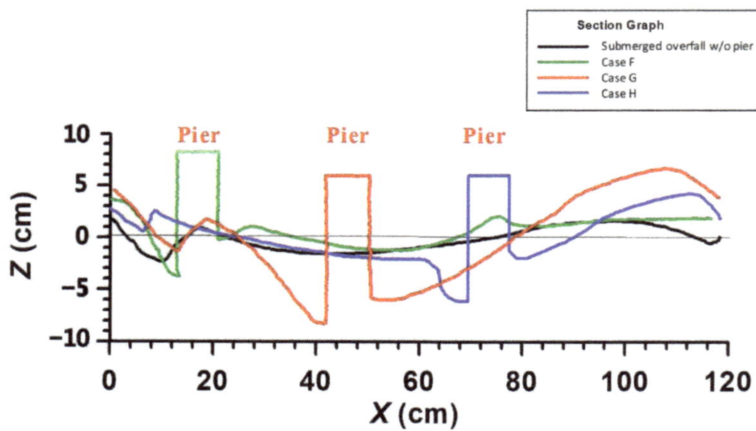

Figure 9. Comparisons of the center line of the vertical profile of the channel in the submerged type overfall drop.

We concluded that the scouring characteristics would be varied with the pier locations at the scour hole and that the most significant scour was found at the point of the maximum scour location induced by water drop. This implies that the bridge pier was a smaller influence of the local scour when it was closer to the location of the submerged overfall.

The maximum scour depth and its location change due to the interaction between overfall type and pier's location can be investigated based on our experiments. The pier's location (L_i), the maximum scour depth (d_s), and its location (L_{scour}) in each experiment were listed in Table 3. By comparing the conditions with and without pier, d_s/d_{o-s} and $L_{scour}/L_{o-scour}$, the effect of pier's location on the maximum scour depth and its location can be investigated. In the condition of free overfall, when $L_i > L_{o-scour}$, the maximum scour depth and its location due to drop structure were not affected by the pier. When $L_i < L_{o-scour}$, the location of maximum scour depth was changed according to L_i, and the maximum scour depth became smaller than in the case of w/o pier. In the condition of submerged overfall, when $L_i > L_{o-scour}$, the location of maximum scour depth was changed based on L_i, and the maximum scour depth was larger than in the case of w/o pier obviously. This result implied that the empirical formulas for the characteristic of local scour due to the drop structure, i.e., Mason and Arumugam (1985) [21], could be used when the overfall condition is free type and $L_i > L_{o-scour}$.

Table 3. Maximum scour depth and its location change due to the overfall type and pier's location.

Experiments	L_i (cm)	d_s (cm)	L_{scour} (cm)	d_s/d_{o-s}	$L_{scour}/L_{o-scour}$
Free overfall w/o pier	w/o pier	−26.7	29.2	-	-
Case A	16	−18.7	11.7	70.2%	40.0%
Case B	32	−28.8	28.2	107.9%	96.5%
Case C	58.7	−27.8	29.6	104.3%	101.2%
Case D	85.5	−27.2	28.2	101.8%	96.5%
Case E	130.8	−26.5	29.2	99.4%	100.0%
Submerged overfall w/o pier	w/o pier	−2.3	10.7	-	-
Case F	17.3	−3.7	13.1	158.9%	122.6%
Case G	46.4	−8.3	41.8	355.9%	390.8%
Case H	73.5	−6.2	69.8	265.1%	652.4%

3.4. Scour Conditions at Pier Surroundings Due to Overfall

In the condition that the flow drop depth (Z) was set at 12 cm, the model of the pier was positioned at five different locations in the experiment facility. The centerline of the vertical profile of the flow drop inducing scour was represented by Case A–E in this paper. We used a camera in the hollow pier model to record the process of the experiment tests over 5 h, and Figure 10 shows the scour depth of the surroundings of the pier in the condition of free overfall. In the same way, a model of pier was positioned at three different locations in the condition of submerged overfall and represented by Case F–H in the centerline of the vertical profile of the flow drop inducing scour. Figure 11 depicts the sour depth of the surroundings of the pier in the condition of submerged overfall. In Figures 10 and 11, the position of the pier at 0 degrees is the location where the approaching flow hits the pier.

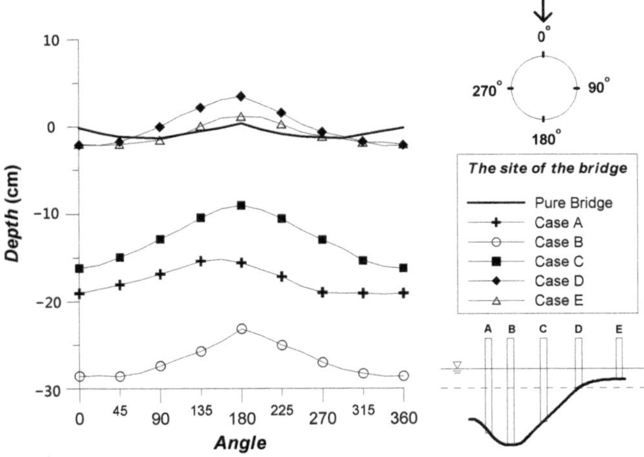

Figure 10. Scour depth distributions of the pier's surroundings in the free type of overfall drop.

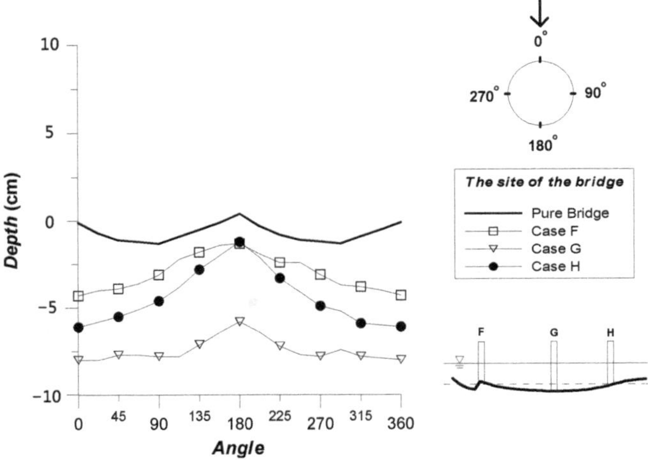

Figure 11. Scour depth distributions of the pier's surroundings in the submerged type of overfall drop.

3.4.1. Scour Conditions at the Pier Surroundings Due to Free Overfall

Figure 10 shows that, in Case A, although the location of the pier was at the upstream slope of the scour hole, the bottom of the pier was scoured due to a reversed flow induced by the flow drop. The maximum scour point of the surroundings of the pier was at the position of 0 degrees. This reveals that a water jet along the river bed from upstream dominated the scour characteristics. There was an unstable condition at 90 degrees and 270 degrees due to interactions from the reversed flows and water jet from upstream, which led to an asymmetric scour at the surroundings of the pier. In Case B, when the pier was located at the point of the maximum scour, the water jet lost most of its energy after hitting the river bed; therefore, the scour depth located from 0 degrees to 45 degrees on the upstream side was almost the same.

The deposition, found in the downstream, was out of 45 degrees, and it deposited greater at around 60 degrees and scoured the least at 180 degrees. In Case C, we found that the scour distribution curve of the pier's surroundings appeared to be greatly affected by the pier inducing scour. However, this was not true, in fact, as the occurrence of the scour mostly occurred upstream of the scour hole. Cases D and E presented cases with the location of the pier at the edge of the scour hole and far from the scour hole. We found that Case D was affected by the sediment loaded flow from the bottom of the scour hole, and therefore, the local scour in front of the pier was not apparent. The scour hole in front of the pier in Case E, on the other hand, was mostly dominated by the pier itself, as the location of the pier in this case was away from the flow drop induced scour hole.

3.4.2. Scour Conditions at the Pier Surroundings Due to the Submerged Overfall

Figure 11 shows that, in Case F, the pier was located in the deposited mound and was hit by a submerged flow jet directly. The results show a full scour hole developed right after the deposited mound was affected by the overfall surrounding the pier, and was not found to be strongly affected by the water jet. In Case G, no significant landform was found in the pier's surrounding, we found in Figure 7c that scour holes developed with a shape of mullet roe surrounding the pier. Both depths in the holes were found to be greater than the ones in front of the pier. This indicates that the water jet caused by the submerged overfall in the x direction was stronger than in the z direction. In Case H, the pier was located downstream of the scour hole, and therefore a greater range of landforms could be observed.

4. Conclusions

This paper focused on probing the mechanism of the scouring effect on piers considering different bridge locations and the flow drop induced scour hole through a series of experiments. Two types of overfall, submerged and free overfall, were applied in the experiment. This mechanism is expected to draw attention from both engineering and academic specialists regarding protecting bridges in newly changed landforms.

Our concluding remarks can be drawn as follows:

Location of the pier vs. the free overfall:

1. The scour surrounding bridge pier in the free overfall condition was mainly controlled by the overfall.
2. When the pier's location was at the upstream slope of the scour hole, better protection to the river bed was found compared with the case of the free overfall w/o pier.
3. When the pier's location was at the maximum scouring point in the scour hole, this deepened the scour depth in the front of pier in a limited manner. Reconstructed bridge piers should not be located here.
4. When the pier's location was at the downstream slope of the scour hole, the pier did not clearly change the impact of the free overfall on the river bed. However, the depth of scour at the vicinity of the bridge pier was still deep enough to expose the pillar in the approaching flow.

5. When the pier's location was at the edge of the scour hole, the scour depth that occurred in front of the pier was similar to the case of the submerged overfall w/o pier. The depth of the localized scour at the vicinity of the bridge pier was the minimum under the interaction between the free overfall and the scour at the bridge pier. This implies that a reconstructed bridge pier would be more secure here.
6. When the pier's location was far from the local hole, the localized scour surrounding the bridge pier was induced by the approaching flow only and without the influence of the free overfall.

Location of pier vs. submerged overfall:

1. Scour at the pier's surroundings was dominated by the flow drop inducing jet, and a relatively deeper scour was be developed due to water jets that were closer to the river bed. This implies that a reconstructed bridge pier should not be located in the area influenced by the submerged overfall.
2. The most significant scour depth at the front of pier was investigated in the condition where the pier location was at the maximum scour point induced by the submerged overfall. A reconstructed bridge pier should not be located here.
3. When the pier was at the maximum point of scour, it induced more scour hole development due to the disturbances caused by the water jet and pier. The depth of those scour holes would be even larger than the scour depth at the pier's surrounding.

Author Contributions: Conceptualization, C.-K.H. and C.-W.L.; methodology, C.-K.H. and C.-W.L.; formal analysis, C.-K.H., C.-W.L., and W.-L.L.; C.-W.L. and W.-L.L. wrote the manuscript, and all authors contributed to improving the paper. All authors have read and agreed to the published version of the manuscript.

Funding: This research received no external funding.

Institutional Review Board Statement: Not applicable.

Informed Consent Statement: Not applicable.

Data Availability Statement: Data sharing not applicable.

Acknowledgments: We thank the reviewers for their useful comments and suggestions.

Conflicts of Interest: The authors declare no conflict of interest.

References

1. Dey, S. Bedforms. In *Fluvial Hydrodynamics*; GeoPlanet: Earth and Planetary Sciences; Springer: Berlin/Heidelberg, Germany, 2014; Available online: https://doi.org/10.1007/978-3-642-19062-9_8 (accessed on 11 January 2021).
2. Schoklitsch, A. Kolkbildung unter Uberfallstrahlen. *Wasserwirtschaft* **1932**, *24*, 341–343.
3. Gill, M.A. Hydraulics of rectangular vertical drop structures. *J. Hydraul. Res.* **1979**, *17*, 289–302. [CrossRef]
4. Little, W.C.; Murphey, J.B. Model study of low drop grade control structures. *J. Hydraul. Div.* **1982**, *108*, 1132–1146.
5. Moore, W.L. Energy loss at the base of free overfall. *Trans. ASCE* **1943**, *108*, 1343–1360.
6. Rand, W. Flow geometry at straight drop spillways. *Proc. ASCE* **1955**, *81*, 1–13.
7. Smith, C.D.; Strang, D.K. Scour in stone beds. In Proceedings of the 12th IAHR-Congress, International Association for Hydraulic Research, Delft, The Netherlands, 11–14 September 1967.
8. Mason, P.J.; Arumugam, K. Free jet scour below dams and flip buckets. *J. Hydraul. Eng.* **1985**, *111*, 220–235. [CrossRef]
9. Hoffmans, G.J.C.M. Jet scour in equilibrium phase. *J. Hydraul. Eng.* **1998**, *124*, 430–437. [CrossRef]
10. Hoffmans, G.J. Closure problem to jet scour. *J. Hydraul. Res.* **2009**, *47*, 100–109. [CrossRef]
11. D'Agostino, V.; Ferro, V. Scour on alluvial bed downstream of grade-control structures. *J. Hydraul. Eng.* **2004**, *130*, 24–37. [CrossRef]
12. Yager, E.M.; Dietrich, W.E.; Kirchner, J.W.; McArdell, B.W. Prediction of sediment transport in step-pool channels. *Water Resour. Res.* **2012**, *48*. [CrossRef]
13. Melville, B.W. Scour at various hydraulic structures: Sluice gates, submerged bridges and low weirs. *Australas. J. Water Resour.* **2014**, *18*, 101–117. [CrossRef]
14. Breusers, H.N.C.; Nicollet, G.; Shen, H. Local scour around cylindrical piers. *J. Hydraul. Res.* **1977**, *15*, 211–252. [CrossRef]
15. Ahmed, F.; Rajaratnam, N. Flow around bridge piers. *J. Hydraul. Eng.* **1998**, *124*, 288–300. [CrossRef]
16. Graf, W.H.; Istiarto, I. Flow pattern in the scour hole around a cylinder. *J. Hydraul. Res.* **2002**, *40*, 13–20. [CrossRef]

17. Dey, S.; Raikar, R.V. Characteristics of horseshoe vortex in developing scour holes at piers. *J. Hydraul. Eng.* **2007**, *133*, 399–413. [CrossRef]
18. Ataie-Ashtiani, B.; Aslani-Kordkandi, A. Flow field around single and tandem piers. *Flow Turbul. Combust.* **2013**, *90*, 471–490. [CrossRef]
19. Euler, T.; Zemke, J.; Rodrigues, S.; Herget, J. Influence of inclination and permeability of solitary woody riparian plants on local hydraulic and sedimentary processes. *Hydrol. Process.* **2013**, *28*, 1358–1371. [CrossRef]
20. Baker, C.J. Vortex Flow around the Bases of Obstacles. Ph.D. Thesis, University of Cambridge, Cambridge, UK, 1979. Available online: https://doi.org/10.17863/CAM.14045 (accessed on 11 January 2021).
21. Baker, C. Theoretical approach to prediction of local scour around bridge piers. *J. Hydraul. Res.* **1980**, *18*, 1–12. [CrossRef]
22. Chabert, J.; Engeldinger, P. *Etude des Affouillements Autour des Piles des Ponts*; Laboratoire National D'hydraulique: Chatou, France, 1956.
23. Chee, R.K.W. Live-Bed Scour at Bridge Sites. Master's Thesis, Auckland University, Auckland, New Zealand, 1982.
24. Chiew, Y.M.; Melville, B.W. Local scour around bridge piers. *J. Hydraul. Res.* **1987**, *25*, 15–26. [CrossRef]
25. Melville, B.W. Live-bed scour at bridge piers. *J. Hydraul. Eng.* **1984**, *110*, 1234–1247. [CrossRef]
26. Elliott, K.R.; Baker, C. Effect of pier spacing on scour around bridge piers. *J. Hydraul. Eng.* **1985**, *111*, 1105–1109. [CrossRef]
27. Melville, B.W.; Chiew, Y.-M. Time scale for local scour at bridge piers. *J. Hydraul. Eng.* **1999**, *125*, 59–65. [CrossRef]
28. Sheppard, D.M.; Odeh, M.; Glasser, T. Large scale clear-water local pier scour experiments. *J. Hydraul. Eng.* **2004**, *130*, 957–963. [CrossRef]
29. Ataie-Ashtiani, B.; Beheshti, A.A. Experimental investigation of clear-water local scour at pile groups. *J. Hydraul. Eng.* **2006**, *132*, 1100–1104. [CrossRef]
30. Melville, B.W.; Coleman, S.E. *Bridge Scour*; Water Resources Publication: Highlands Ranch, CO, USA, 2000.
31. Richardson, E.V.; Davis, S.R. *Evaluating Scour at Bridges*; No. FHWA-NHI-01-001; Federal Highway Administration: Washington, DC, USA, 2001.
32. Khosronejad, A.; Kang, S.; Sotiropoulos, F. Experimental and computational investigation of local scour around bridge piers. *Adv. Water Resour.* **2012**, *37*, 73–85. [CrossRef]
33. Raudkivi, A.J.; Ettema, R. Effect of sediment gradation on clear water scour. *J. Hydraul. Div.* **1977**, *103*, 1209–1213.
34. Melville, B.W.; Sutherland, A.J. Design method for local scour at bridge piers. *J. Hydraul. Eng.* **1988**, *114*, 1210–1226. [CrossRef]

Article

Landslide Susceptibility Based on Extreme Rainfall-Induced Landslide Inventories and the Following Landslide Evolution

Chunhung Wu

Department of Water Resources Engineering and Conservation, Feng Chia University, Taichung 40724, Taiwan; chhuwu@fcu.edu.tw; Tel.: +886-4-2451-7250 (ext. 3223)

Received: 15 September 2019; Accepted: 6 December 2019; Published: 11 December 2019

Abstract: Landslide susceptibility assessment is crucial for mitigating and preventing landslide disasters. Most landslide susceptibility studies have focused on creating landslide susceptibility models for specific rainfall or earthquake events, but landslide susceptibility in the years after specific events are also valuable for further discussion, especially after extreme rainfall events. This research provides a new method to draw an annual landslide susceptibility map in the 5 years after Typhoon Morakot (2009) in the Chishan River watershed in Taiwan. This research establishes four landslide susceptibility models by using four methods and 12 landslide-related factors and selects the model with the optimum performance. This research analyzes landslide evolution in the 5 years after Typhoon Morakot and estimates the average landslide area different ratio (LAD) in upstream, midstream, and downstream of the Chishan River watershed. We combine landslide susceptibility with the model with the highest performance and average annual LAD to draw an annual landslide susceptibility map, and its mean correct ratio ranges from 62.5% to 73.8%.

Keywords: extreme rainfall-induced landslide susceptibility model; landslide ratio-based logistic regression; landslide evolution; Typhoon Morakot; Taiwan

---

1. Introduction

Deaths and economic losses due to natural disasters have drastically increased in Taiwan over the past two decades, especially after the 1999 Chichi earthquake [1]. In Taiwan, landslides and debris flows are the major causes of serious rainfall-induced disasters. The death toll due to Typhoon Morokot in 2009 was around 703, and the death toil due to the associated landslides and debris flow disasters was over 500, including 465 deaths caused by the Xiaolin deep landslide [2]. The number and intensity of the heavy rainfall events are expected to increase in the future in Taiwan [3], and the occurrences of landslides and debris flows over the next decade are expected to increase. Therefore, the assessment of landslide susceptibility is an important consideration for disaster prevention or mitigation in Taiwan.

Landslide susceptibility assessment models can be created based on heuristic, deterministic, and statistical approaches. Among these approaches, statistical methods are the most popular because of the development of geographic information systems and remote sensing techniques. The processes involved in evaluating landslide susceptibility by establishing a susceptibility model using statistical approaches include selecting landslide-related factors, creating a database, acquiring the most suitable fitting equations from the statistical model for landslide occurrence, and calibrating or validating models. The prediction accuracy of most statistical landslide susceptibility models can exceed 70% [4–7]. Technological statistical methods for predicting landslide susceptibility have improved to involve artificial neural networks [8], machine learning [9,10], empirical methods based on big data [11,12], and artificial intelligence [13]; the prediction accuracy of landslide susceptibility models using these technologies range from 80% to 90%.

Several rainfall events with a return period of more than 100 years have occurred in Taiwan, particularly in the Central and Southwestern regions, over the past two decades. Extreme rainfall events in Taiwan typically occur when daily rainfall >800.0 mm (such as that in Southern Taiwan during Typhoon Morakot in 2009 and Northern Taiwan during Typhoon Soudelor in 2015) or hourly rainfall intensity >80.0 mm/h (such as that in Northeastern Taiwan during Typhoon Megi in 2010). Extreme rainfall events also result in serious disasters. The accumulated rainfall and rainfall intensity during Typhoon Morakot in Southern Taiwan is representative of extreme rainfall events. The 48- and 72-h accumulated rainfall at most rainfall stations during Typhoon Morakot in Southwestern Taiwan exceeded the accumulated rainfall record of the 200-year return period [2], and the average rainfall intensity from 12:00 a.m. to 8:00 p.m. on 8 August 2009 in the midstream of the Chishan River watershed was more than 80.0 mm/h. Some studies have started to emphasize the seriousness of extreme rainfall-induced landslide or debris flow [8,9] or have created landslide susceptibility models for regions with high annual and daily rainfall [10] because of the increasing occurrence frequency of extreme rainfall events. Rainfall is widely used as a factor for building landslide susceptibility models, but the pattern and extent of rainfall should be emphasized [14,15]. Therefore, the assessment of landslide susceptibility is a crucial consideration for disaster prevention and mitigation in Taiwan [16].

The novelty of this research includes the applicability of landslide susceptibility models using statistical modeling in areas with dense landslide distribution and the process of drawing annual landslide susceptibility maps in the years after extreme rainfall events. The landslides induced by Typhoon Morakot in the Chishan River watershed were densely distributed, and the landslide types mainly included debris falls, translational landslides, riverbank landslides, and large-scale landslides. Areas with dense landslide distributions induced by an extreme rainfall event are uncommon globally, and discussing the performance of landslide susceptibility maps using statistical models in dense landslide distribution areas is valuable. Another novelty in this research is that a large amount of sediment was deposited downhill or transported into the river after numerous landslides occurred in the area. The large amount of sediment deposited randomly in the river resulted in sinuous rivers and subsequent riverbank landslides. Landslide susceptibility in some specific areas in the years after extreme rainfall events did not decrease; they instead increased, because riverbank landslides increased substantially. Numerous articles have focused on predicting landslide susceptibility for specific rainfall or earthquake events, and this research suggests that annual landslide susceptibility maps after specific rainfall or earthquake events should be emphasized as well.

This research compares and analyzes the applicability of landslide susceptibility assessment models based on four methods by using the extreme rainfall-induced landslide inventory and suggests a process of drawing landslide susceptibility maps in the years after extreme rainfall events. The four methods used in the study include landslide ratio-based logistic regression (LRBLR) [17], frequency ratio (FR), weights of evidence (WOE), and instability index (II) [18]. The landslide susceptibility model is combined with 12 factors. The validation of the landslide susceptibility model in this research adopts the area under receiver operating curves and confusion matrix methods. The research selected the landslide susceptibility model with the best performance of the four as the basis for drawing landslide susceptibility maps after Typhoon Morakot. Furthermore, this study analyzes the long-term landslide evolution from 2008 to 2014 in the Chishan River watershed. The landslide evolution analysis from 2008 to 2009 identifies geomorphic characteristics of extreme rainfall-induced landslide-prone locations, whereas the analysis from 2009 to 2014 analyzes the difference in landslide count and area induced by Typhoon Morakot from 2010 to 2014 to understand the long-term evolution of landslides in the Chishan River watershed. Finally, the study draws annual landslide susceptibility maps from 2010 to 2014 by combining the extreme rainfall-induced landslide susceptibility model and the statistical data from landslide evolution from 2010 to 2014 in the Chishan River watershed. Figure 1 presents the flowchart of this research.

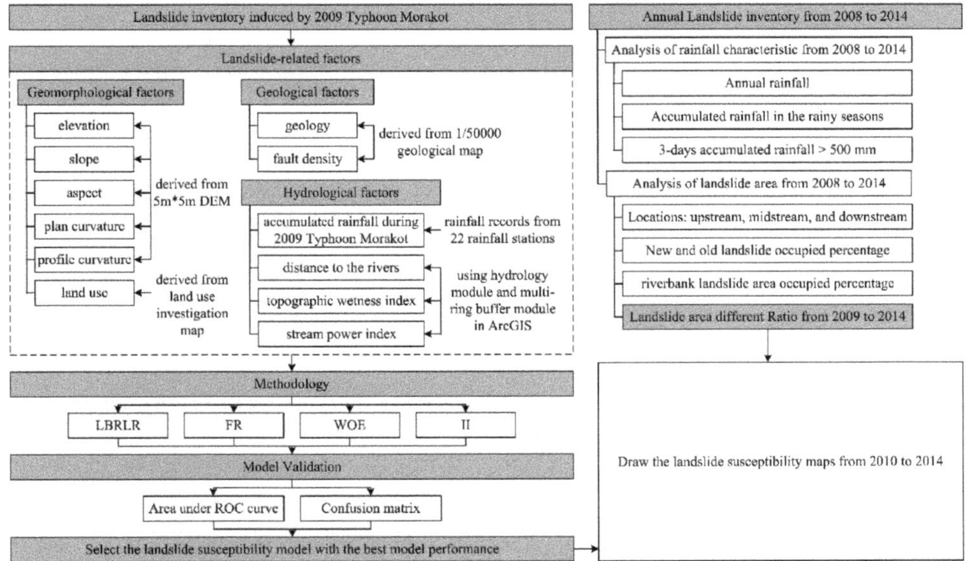

Figure 1. The flow chart of this research.

2. Materials and Methods

2.1. Research Area and Extreme Rainfall Event

2.1.1. Research Area: Chishan River Watershed in Southwestern Taiwan

The Chishan river watershed (Figure 2) is a tributary watershed of Kaoping river watershed in Southwestern Taiwan. Kaoping river watershed ranks 11th in terms of suspended load in the world [19,20]. The mean sediment yield (5.9 kg/m^2/year) and physical denudation rate (655.8 g/m^2/year) of the Kaoping watershed are 1.96 and 4.37 times larger than that of mountainous rivers throughout the world [21,22]. The high suspended sediment quantity show that Kaoping river watershed is a soil erosion-, landslide-, and debris flow-prone watershed due to the fragile geology, steep terrain, and heavy rainfall.

The area of the Chishan river watershed is around 819 km^2 with the mean elevation and slope of 838 m (Figure 2) and 22.4° (Table 1). The average annual precipitation is 4468 mm. The mean precipitation in the rainy season, i.e., from May to October, occupies 83% to 89% of the mean annual precipitation, and that in the dry seasons, from November to April, only occupies 11% to 17%. The land use in the research area consists of forest (65.0%), agriculture (23.2%), development (4.2%), river (2.7%), and bare land (5.0%) based on the land use investigation maps produced in 2008 by National Land Surveying and Mapping Center in Taiwan. The main strata (Figure 3 and Table 2) in the research area includes the Miocene Changchihkeng formation (26.2% of the watershed), the Holocene alluvium (17.8% of the watershed), middle Miocene Nankang formation and equivalents (10.8% of the watershed), and the Miocene Tangenshan sandstone (10.5% of the watershed) based on the 1/5000 basin geological map in Taiwan [23].

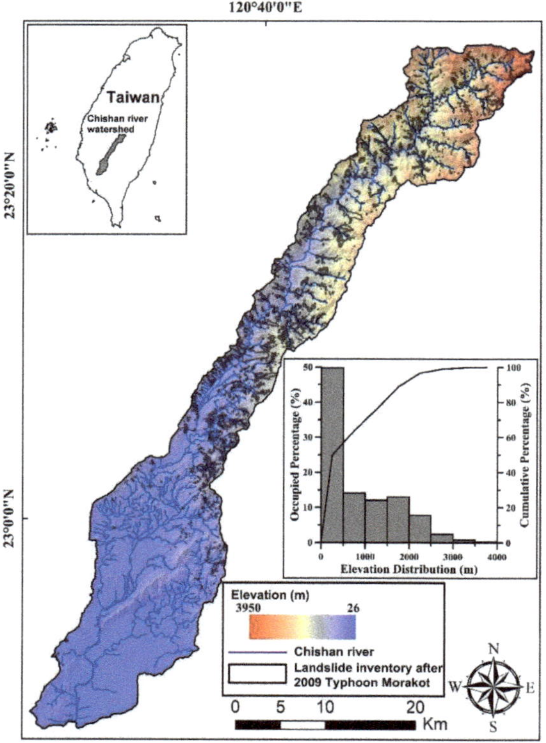

Figure 2. The distribution of elevation, river, and landslide inventory induced by 2009 Typhoon Morakot in the Chishan river watershed.

Table 1. The statistical data of twelve landslide-related factors in this research in the Chishan river watershed.

Variable (Unit)	Max	Min	Mean	Median	S.D. *	Skewness	Kurtosis
Geomorphological factors							
Elevation (m)	3979.4	26.3	830.61	505.5	806.32	0.881	2.762
Slope (degrees)	79.2	0.0	22.32	25.5	15.34	0.065	2.111
Aspect (degrees)	360.0	−1.0	188.37	193.3	102.50	−0.188	1.934
Land use	categorical variable						
Plan curvature	200.4	−200.4	−0.77	0	3.01	−0.376	63.282
Profile curvature	201.0	−271.4	−0.89	0	4.63	−0.719	127.152
Geological factors							
Geology	categorical variable						
Fault density (10^{-3} m^{-1})	2.4	0.0	0.07	0	0.28	4.213	21.591
Hydrological factors							
Accumulated rainfall (mm)	2174.1	1083.3	1671.0	1705.4	263.15	−0.565	2.746
Proximity to the rivers (m)	5641.0	0	352.1	221.4	178.9	4.579	13.212
Topographic wetness index	41.2	6.9	12.6	12.3	2.31	1.549	7.361
Stream power index	24.2	−9.0	2.4	2.8	3.73	−0.409	4.343

* Note: The S.D. indicates the standard deviation.

Table 2. The geological settings in the Chishan river watershed.

Ab.	Times	Strata	Lithology	Oc. (%)
a	Holocene	alluvium	gravel, sand and clay	17.8
Al	Pliocene	Ailiaochiao formation	thin alternation of siltstone and shale	7.6
Cc	Miocene	Changchihkeng formation	alternations of sandstone and shale	26.2
Gt	Pliocene-Pleistocene	Gutingken formation	mudstone with intercalated sandstone	1.3
Hh	Miocene	Hunghuatzu formation	thick-bedded siltstone, thick alteration of siltstone and sandstone	2.9
ig	none	igneous rock	igneous rock	0.0
Kz	Miocene-Pliocene	Kaitzuliao shale	shale	0.6
Le	Pleistocene	Liukuei formation	conglomerate, sandstone, sandy shale and mudstone	0.4
Lo	Pleistocene	Linkou conglomerate	conglomerate with mudstone interbeds, intercalated with sheet or lenticular sandstone	0.9
M2	Middle Miocene	Nankang formation and equivalents	augillite or slate	10.8
Nc	Miocene	Nanchuang formation and equivalents	sandstone and shale interbeded with igneous rock	3.6
Nl	Pliocene	Nanshihlun sandstone	thick sandstone, mudstone, alternations of sandstone and shale, thick carbonaceous shale with intercalated sandstone	0.5
Si	Miocene	Sanming shale	shale intercalated with thin-bedded siltstone	2.3
Sp	Eocene	Shihpachungchi formation	slate with mate-sandstone	1.1
t Tc	Pleistocene-Holocene Eocene	Terrace gravel Tachien sandstone	mud, sand, and gravel meta-sandstone with slate	5.9 1.3
Tn	Miocene	Tangenshan sandstone	sandstone intercalated with shale	10.5
Wa	Miocene-Pliocene	Wushan formation	thin alternation of sandstone and shale	0.5
Ya	Eocene	Yushanchushan formation	meta-sandstone and slate interbeded	0.7
Ys	Pliocene	Yenshuikeng shale	massive shale	4.9

Note: Ab. means abbreviation and Oc. refers to the occupied percentage of the strata in the Chishan river watershed.

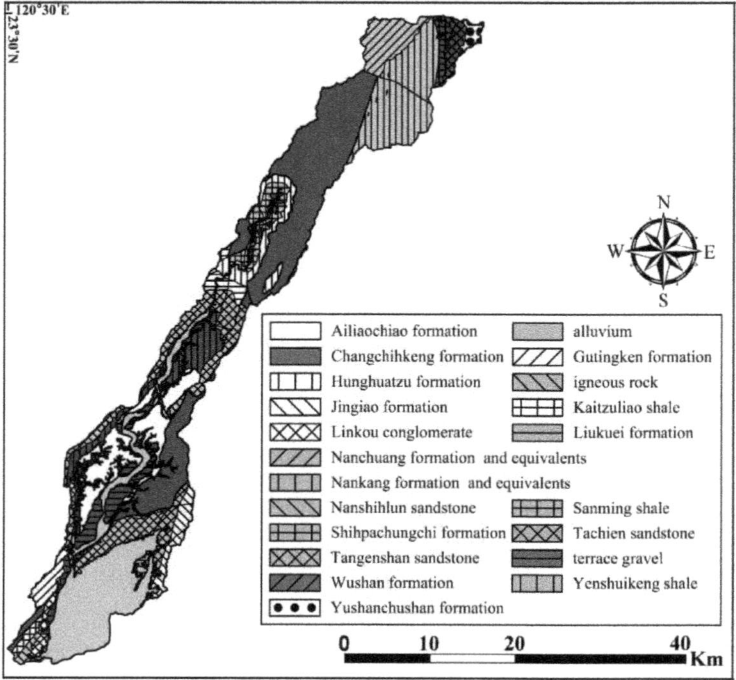

Figure 3. The geological settings of the Chishan river watershed.

2.1.2. Extreme Rainfall Event: 2009 Typhoon Morakot

Typhoon Morakot struck Southern Taiwan between 6 and 10 August 2009. The rainfall distribution in the Chishan river watershed based on the rainfall records from 22 rainfall stations is shown in Figure 4. The rainfall ranges from 1083 to 1990 mm with an average of 1528.0 mm. The 24-h, 48-h, and 72-h accumulated rainfall exceeded the 200-year return-period accumulated rainfall [2]. The accumulated rainfall during the most intense rainfall period, i.e., 1 pm to 12 pm on 8 August 2009, was 577.0 mm to 786.5 mm, equal to a mean rainfall intensity of 48.1 mm/h to 65.5 mm/h in this period. The 2389 landslide cases (Figure 2) induced by 2009 Typhoon Morakot in the Chishan river watershed were extracted from high resolution SPOT 5 images [24,25]. The area of each identified landslide polygons ranges from 264 m^2 to 3.5 km^2. The total landslide area in the Chishan river watershed is around 33.5 km^2 with the landslide ratio (*LR*) of 4.1%.

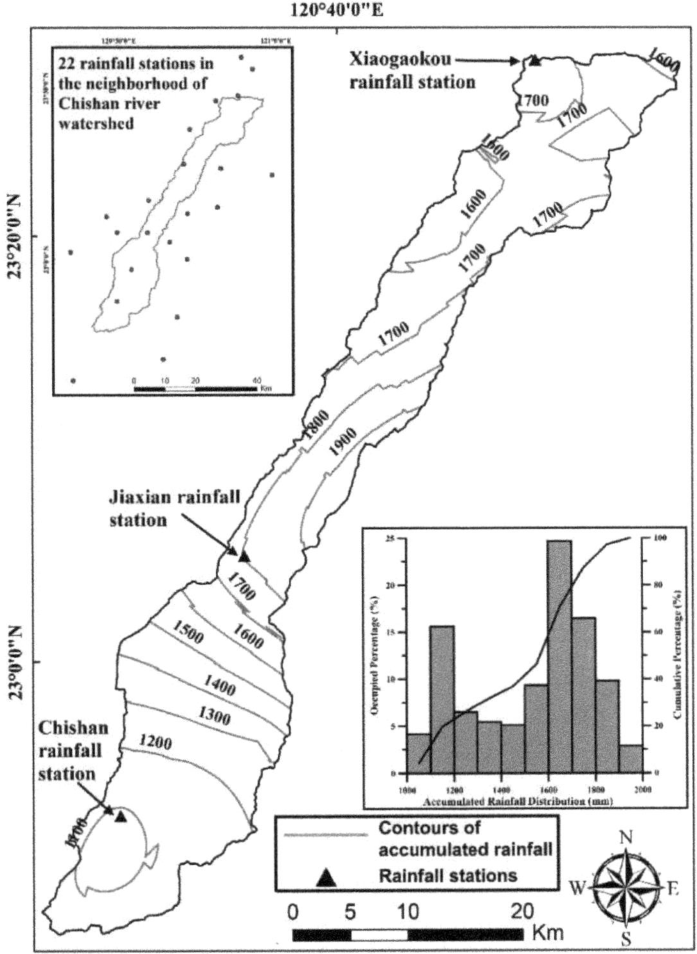

Figure 4. The distribution of accumulated rainfall during the 2009 Typhoon Morakot and rainfall stations in the Chishan river watershed.

2.2. Research Methodology for Landslide Susceptibility Mapping and Long-Term Landslide Evolution

2.2.1. Landslide-Related Factors

Based on the literature [18] and data availability, this research selects a total of 12 factors as the basis for establishing the landslide susceptibility model and can be classified into geomorphological, geological, and hydrological. Geomorphological factors include elevation, slope, aspect, land use, plan curvature, and profile curvature. The elevation (Figure 2), slope, aspect, plan curvature, and profile curvature factors are derived from a 5-m digital elevation model (DEM), whereas the land use factor adopts the land use investigation map in Taiwan, which was produced in 2008 by the National Land Surveying and Mapping Center. Because the Chishan River watershed is an erosion- and landslide-prone watershed, we adopt plan and profile curvature factors to describe divergence and convergence of water flow and runoff and infiltration mechanisms.

Geological factors include geology and fault density, and this study adopts a 1/50,000 geological map of the Chishan River watershed [23] to draw the geological setting map and estimate fault density.

Geological formations in the Chishan River watershed are fragile and landslide-prone, and six fault lineaments pass through the Chishan River watershed, particularly in midstream. Hydrological factors include the accumulated rainfall during Typhoon Morakot, proximity to the rivers, topographic wetness index (TWI), and stream power index (SPI). This research uses the accumulated rainfall during Typhoon Morakot (from 20:30 on 5 August to 05:30 on 10 August 2009) to describe the influence of heavy rainfall on landslides. Rainfall records from 22 rainfall stations within or near the Chishan River watershed were collected to draw the distribution of accumulated rainfall in the watershed (Figure 3). Furthermore, headward erosion- and bank erosion-induced landslide cases occupy a considerable portion of the landslide inventory. The area within 300 m of the rivers occupies approximately 43.9% of the Chishan River watershed area, but the landslide area within 300 m of the rivers after Typhoon Morakot occupies approximately 52.8% of the total landslide area. This research adopts the hydrology module in ArcGIS to draw the river distribution and estimate the TWI and SPI. The TWI is defined as the natural logarithm ratio of the local upslope area drainage per contour length to the local slope angle and describes the water saturation in the surface soil layer. The SPI is defined as the product of the natural logarithm of both slope and flow accumulation. The SPI describes the erosion strength of river flow and is suitable for determining riverbank landslide locations.

2.2.2. Landslide Susceptibility Methodology: Landslide Ratio-Based Logistic Regression Method (LRBLR)

The purpose of logistic regression analysis is to find the best fitting equation (Equation (1)) to describe the dependent variable (landslide or not landslide, Y in Equation (1)) and the independent parameters (landslide-related factors, X_n in Equation (1)):

$$\mathrm{logit}(Y) = \beta_0 + \beta_1 X_1 + \beta_2 X_2 + \ldots \quad (1)$$

$$\ln \frac{p}{1-p} = \mathrm{logit}\ (Y) = \beta_0 + \beta_1 X_1 + \beta_2 X_2 + \ldots + \beta_n X_n \quad (2)$$

where β_0 is a constant and β_n is the nth regression coefficient. The landslide susceptibility P can be written as Equation (2). Wu [17] suggested that the performance of landslide susceptibility model using the logistic regression method with the *LR* index is better than that using the original logistic regression method. Landslide ratio (*LR*) refers to the ratio of the landslide area in a specific area to that in the total watershed area. This research follows the suggestions from Wu [17] and reclassifies the categories of all variables according to *LR*. The number of landslide ratio classifications (*LRC* number) in a specific category is marked as 1, 2, 3, 4, 5, 6, 7, 8, 9, 10, and 11 as the *LR* in a specific category is <1.0%, 1.0–2.0%, 2.0–3.0%, 3.0–4.0%, 4.0–5.0%, 5.0–6.0%, 6.0–7.0%, 7.0–8.0%, 8.0–9.0%, 9.0–10.0%, and >10.0%, respectively. All variables in the *LRBLR* analysis are categorical variables.

The total grid count in the Chishan river watershed is approximately 3.17×10^7 grids, including 3.14×10^7 non-landslide grids and 1,356,104 landslide grids. The non-landslide grid count is approximately 23.2 times greater than the landslide grid count. In this study, all grid counts were attempted to be placed into the statistical software for logistic regression analysis, which was difficult to be analyzed in the statistical software, and the result was dominated by the non-landslide grid. This research was based on the study conducted by Yesilnacar and Topcal [26], who performed a random sampling analysis. Twenty random sampling datasets were picked, and each dataset included 1,356,104 and 1,356,104 landslide and non-landslide grids, respectively. The 20 random sampling datasets were analyzed using the SPSS software to obtain the Cox-Snell R^2 value and Nagelkerke R^2 value. Only when the Cox-Snell R^2 and Nagelkerke R^2 values from the logistic regression analysis were greater than 0.15, the dataset was considered as useful and valid [26] in the research. In this research, datasets with the highest Cox-Snell R^2 and Nagelkerke R^2 values were picked from 20 random sampling datasets, and coefficients from logistic regression were used to develop the landslide susceptibility model.

2.2.3. Landslide Susceptibility Methodology: Frequency Ratio Method (FR)

Lee and Talib [27] suggested that landslide susceptibility should be directly proportional to *LR* in a specific area, i.e., the landslide susceptibility in an area with a dense landslide distribution should be high. Frequency ratio value (*FR*) can be a useful index when establishing a landslide susceptibility map. The *FR* value can be calculated as the ratio of the occupied percentage of landslide area in specific category in specific landslide-related factor to the occupied percentage of area in specific category in specific landslide-related factor. The *FR* value in each category of each factor is estimated in Table 3. The *FR* value in each category of every landslide-related factor can be calculated, and the sum of *FRs* can be used as the landslide susceptibility index (*LSI*, Equation (3)). FR_n represents the frequency ratio value of the *n*th landslide-related factor. If the *FR* value in a specific category of the specific landslide-related factor >1, this means the landslide in the specific category of the specific landslide-related factor has a high correlation to the landslide distribution, while a value of <1 indicates a lower correlation.

$$LSI = \sum_{1}^{n} FR_n \tag{3}$$

Table 3. Coefficient values of landslide-related factors based on four methods.

Factors	Area (Km^2)	Landslide Area (Km^2)	*LR* (%)	*LRC* Number	*FR*	*C*	*Di*
			Elevation (m)				
<250	285.2	0.3	0.1	1	0.03	−4.16	1.00
250–500	122.2	3.8	3.1	4	0.75	−0.35	3.26
500–750	64.3	6.7	10.4	11	2.51	1.13	8.74
750–1000	51.2	5.7	11.0	11	2.68	1.18	9.27
1000–1250	45.5	5.5	12.2	11	2.91	1.28	10.00
1250–1500	52.8	5.0	9.4	10	2.28	0.98	8.03
1500–1750	53.0	2.9	5.5	6	1.32	0.31	5.03
1750–2000	53.4	1.8	3.4	4	0.81	−0.23	3.45
2000–2250	39.7	0.9	2.4	3	0.55	−0.65	2.63
2250–2500	23.6	0.3	1.4	2	0.31	−1.24	1.88
2500–2750	14.2	0.2	1.1	2	0.34	−1.12	1.98
2750–3000	5.2	0.1	1.8	2	0.46	−0.80	2.37
>3000	7.9	0.8	10.2	11	2.44	0.97	8.52
			Slope (°)				
<10	219.0	1.1	0.5	1	0.12	−2.44	1.00
10–20	102.5	2.8	2.8	3	0.66	−0.48	4.14
20–30	172.8	8.0	4.7	5	1.12	0.15	6.81
30–40	190.2	13.2	7.0	8	1.68	0.79	10.00
40–50	95.5	6.5	6.8	7	1.65	0.62	9.87
50–60	30.9	1.7	5.4	6	1.33	0.31	8.03
>60	7.4	0.5	6.7	7	1.64	0.53	9.80
			Aspect				
flat	20.7	0.0	0.0	1	0.00	−4.52	1.00
North	82.9	3.6	4.4	5	1.05	0.05	8.11
Northeast	73.0	4.0	5.5	6	1.32	0.33	10.00
East	86.0	4.1	4.7	5	1.15	0.17	8.80
Southeast	102.4	4.4	4.3	5	1.04	0.04	8.03
South	107.7	4.7	4.4	5	1.05	0.06	8.14
Southwest	116.0	4.7	4.1	5	0.98	−0.03	7.63
West	120.4	4.6	3.8	4	0.92	−0.10	7.25
Northwest	108.9	3.8	3.5	4	0.84	−0.20	6.71

Table 3. *Cont.*

Factors	Area (Km^2)	Landslide Area (Km^2)	*LR* (%)	*LRC* Number	*FR*	*C*	*Di*
			Land use				
agriculture	189.5	1.1	0.6	1	0.14	−2.24	1.00
forest	547.7	28.5	5.2	6	1.26	0.99	2.86
development	43.1	0.4	1.0	2	0.22	−1.57	1.13
neighborhood of river	22.3	0.3	1.1	2	0.32	−1.17	1.30
bare land	15.7	3.6	22.8	11	5.53	2.03	10.00
			Accumulated rainfall (mm)				
<1200	161.9	0.1	0.0	1	0.01	−4.48	1.00
1200–1400	97.9	0.1	0.1	1	0.02	−3.88	1.00
1400–1600	118.3	4.1	3.4	4	0.83	−0.22	3.88
1600–1800	336.5	18.7	5.5	6	1.34	0.58	5.68
>1800	103.6	11.0	10.6	11	2.56	1.27	10.00
			Geology				
a	145.8	0.2	0.1	1	0.03	−3.65	1.09
Al	61.8	0.9	1.5	2	0.35	−1.13	2.01
Cc	214.4	14.1	6.6	7	1.59	0.73	5.55
Gt	10.9	0.1	0.6	1	0.22	−1.55	1.64
Hh	24.1	2.6	10.6	11	2.60	1.08	8.47
ig	0.3	0.0	4.9	5	0.00	−0.26	1.00
Kz	5.0	0.0	0.0	1	0.00	−3.08	1.00
Le	3.3	0.0	0.0	1	0.00	−2.66	1.00
Lo	7.7	0.0	0.0	1	0.00	−3.51	1.00
M2	88.4	3.7	4.2	5	1.01	0.01	3.90
Nc	29.4	0.7	2.3	3	0.57	−0.59	2.65
Nl	3.9	0.0	0.0	1	0.00	−2.83	1.00
Si	18.8	1.0	5.3	6	1.28	0.27	4.68
Sp	8.7	0.4	4.2	5	1.11	0.11	4.18
t	48.4	0.4	0.9	1	0.20	−1.70	1.57
Tc	10.8	0.3	3.1	4	0.67	−0.42	2.92
Tn	86.3	7.0	8.1	9	1.96	0.84	6.62
Wa	4.3	0.0	0.0	1	0.00	−2.93	1.00
Ya	5.5	0.7	13.1	11	3.07	1.23	9.81
Ys	40.3	1.8	4.5	5	1.08	0.08	4.09
			Fault density (10^{-3} m/km^2)				
<5	764.6	28.2	3.7	4	0.89	−1.11	1.00
10	22.8	2.1	9.4	10	2.23	0.89	4.57
15	23.8	2.6	11.1	11	2.64	1.10	5.68
20	5.3	0.6	11.2	11	2.74	1.10	5.93
>20	1.7	0.3	16.0	11	4.27	1.61	10.00
			proximity to the river (m)				
<250	320.9	15.1	4.7	5	1.14	0.23	10.00
250–500	192.0	8.1	4.2	5	1.02	0.02	7.83
500–750	123.4	4.9	4.0	4	0.96	−0.05	6.72
750–1000	78.4	3.0	3.8	4	0.92	−0.09	6.07
>1000	103.5	2.8	2.7	3	0.65	−0.49	1.00

Table 3. *Cont.*

Factors	Area (Km^2)	Landslide Area (Km^2)	*LR* (%)	*LRC* Number	*FR*	*C*	*Di*
			Plan curvature				
<−0.6	182.3	11.3	6.2	7	1.50	0.58	10.00
−0.6 − −0.3	91.5	3.6	3.9	4	0.95	−0.06	4.84
−0.3 − 0	173.1	3.9	2.2	3	0.54	−0.75	1.00
0 − 0.3	81.9	3.5	4.3	5	1.03	0.04	5.61
0.3 − 0.6	93.9	3.2	3.5	4	0.82	−0.23	3.64
>0.6	195.5	8.4	4.3	5	1.04	0.05	5.66
			Profile curvature				
<−5	28.5	1.1	4.0	4	0.93	−0.07	1.63
−5–−3	22.4	0.9	4.0	4	0.97	−0.03	2.29
−3–−1	82.6	3.8	4.6	5	1.11	0.13	4.73
−1–1	552.0	20.5	3.7	4	0.90	−0.31	1.00
1–3	80.2	4.7	5.8	6	1.42	0.42	10.00
3–5	22.1	1.2	5.5	6	1.31	0.30	8.20
>5	30.4	1.6	5.3	6	1.27	0.27	7.50
			Topographic Wetness index				
<10	44.9	2.2	5.0	5	1.08	0.08	8.69
10–12	258.2	12.0	4.7	5	1.02	0.04	8.03
12–14	239.1	12.9	5.4	5	1.19	0.29	10.00
14–16	84.3	2.7	3.2	4	0.70	−0.41	4.25
16–18	40.9	0.8	2.0	2	0.43	−0.91	1.00
>18	21.0	0.7	3.2	4	0.73	−0.33	4.59
			Stream Power Index				
<−2	106.6	2.4	2.3	3	0.54	−0.70	2.08
−2–1	66.9	3.0	4.4	5	1.08	0.09	8.56
1–4	415.5	20.7	5.0	5	1.20	0.44	10.00
4–7	159.1	6.4	4.0	4	0.97	−0.04	7.22
7–10	53.1	1.0	1.8	2	0.45	−0.85	1.00
>10	16.9	0.4	2.4	3	0.57	−0.59	2.41

2.2.4. Landslide Susceptibility Methodology: Weight of Evidence Method (WOE)

The *WOE* method was proposed by Bonham-Carter (1994) [28], and the assessment equations in *WOE* method can be written as Equations (4)–(6) [5]:

$$W^{+} = ln\left[\frac{A_1/(A_1 + A_2)}{A_3/(A_3 + A_4)}\right] \tag{4}$$

$$W^{-} = ln\left[\frac{A_2/(A_1 + A_2)}{A_4/(A_3 + A_4)}\right] \tag{5}$$

$$C = W^{+} - W^{-} \tag{6}$$

where A_1 (A_3) is the landslide area (not-landslide) in a specific category of specific landslide-related factor and A_2 (A_4) is the total landslide (not-landslide) area not in the specific category of specific landslide-related factor. The W^+ (W^-) value represents the landslide-induced positive (negative) weight of the specific category in the landslide-related factor. The weights contrast value (C) is the difference between W^+ and W^- and represents the spatial association between the specific category in the landslide-related factor and landslide occurrence [5]. The landslide susceptibility in a specific grid can be calculated as the summation of C values in each landslide-related factor.

2.2.5. Landslide Susceptibility Methodology: Instability Index Method (II)

The instability index (*II*) method was proposed by Jian [29] to assess the slope stability. The calculation process for assessing the landslide susceptibility using the *II* method can be divided into two parts: the normalized grades (*D*) of each category in each landslide-related factor and the weighting value (W_e) of each landslide-related factor. The *D* and *We* values can be written as Equations (7) and (8) [30]:

$$D_i = \frac{9(X_i - X_{min})}{X_{max} - X_{min}} + 1 \tag{7}$$

$$We_i = \frac{V_i}{V_1 + V_2 + \ldots + V_n} \tag{8}$$

$$\text{landslide susceptibilty} = D_1^{We_1} \times D_2^{We_2} \times \ldots D_n^{We_n} \tag{9}$$

where X_i can be calculated as the ratio of *LR* in the *i*th category to the total *LR* in all categories in a specific landslide-related factor, while X_{min} (X_{max}) represents the minimum (maximum) ratio value in all categories of the landslide-related factor. V_n in Equation (9) is the coefficient of variation of the X_i values in all categories for the nth landslide-related factor. The *D* value is a normalization value to show the landslide-induced influence of a specific category in all categories.

2.2.6. Validation and Similarity of Landslide Susceptibility Models

The area under the receiver operating characteristic curve (*AUC*) and the confusion matrix are the two methods to assess the model performance of landslide susceptibility models in this research. The receiver operating characteristic curve is obtained by plotting the sensitivity value on the vertical axis and the 1-specificity value on the horizontal axis, and the *AUC* value is adopted as an index to assess the model performance. The model performance can be considered as failed, poor, fair, good, and excellent with *AUC* values (the area under the receiver operating characteristic curve) ranges of 0.5–0.6, 0.6–0.7, 0.7–0.8, 0.8–0.9, and 0.9–1.0, respectively.

This research uses the confusion matrix [18] concept to set four indexes. The *PLCR* (*PLWR*) is the ratio of the predicted-landslide area within (not within) the range of landslide inventory to the total landslide area, and the *PNLCR* (*PNLWR*) is the ratio of the predicted-non-landslide area outside (not outside) the range of landslide inventory to the total non-landslide area. The mean correct ratio (*MCR*) is the mean of *PLCR* and *PNLCR*, while the mean wrong ratio (*MWR*) is the mean of *PLWR* and *PNLWR*.

This research adopts the correlation analysis to assess the similarities of four landslide susceptibility models. The similarities between the two models is very weak, weak, moderate, strong, and very strong when the correlation coefficient from the correlation analysis is 0.0–0.2, 0.2–0.4, 0.4–0.7, 0.7–0.9, and 0.9–1.0, respectively.

2.2.7. Long-Term Landslide Evolution Analyses

The analysis of long-term landslide evolution includes the analysis of long-term rainfall records and the difference analysis of the annual landslide distribution from 2008 to 2014 in the Chishan river watershed. We collect the rainfall record in the Chishan river watershed from 2008 to 2014 to analyze the long-term rainfall distribution. The Chishan river watershed can be classified into three sun-watersheds, including upstream, midstream, and downstream watersheds (Figure 3). This research selects a representative rainfall station in each watershed, including the Xingaokou station in the upstream watershed, the Jiaxian station in the midstream watershed, and the Chishan station in the downstream watershed (Figure 3), based on the rainfall station location and rainfall record data availability. The research estimates the annual rainfall, the accumulated rainfall in the rainy seasons, i.e., from May to October, and also estimates the counts of the accumulated rainfall of three days in a row over 500 mm to understand the inducing strength from heavy rainfall in a specific year.

The long-term landslide evolution analyses in this research means that we adopt the annual landslide inventories from 2008 to 2014, and analyze the difference of annual landslide distribution and expanding or contracting of the total landslide area in every year. The annual landslide inventory used in this research was produced by the Forestry Bureau in Taiwan and the landslide inventory was identified from the Formosat-2 images with the spatial resolution of 2 m shot during January to July every year.

The landslide distribution of the Chishan river watershed after 2009 Typhoon Morakot was strongly related to the landslide location and proximity to the river [27]. In this study, the landslide distribution of the upstream, midstream, and downstream of the Chishan river watershed in 2008 to 2014 was analyzed. The areas of the upstream, midstream, and downstream river watershed were 210.0, 250.3, and 357.9 km^2, respectively. Furthermore, areas within 300 m of the rivers were defined as riverbank areas and areas 300 m outside the rivers were identified as non-riverbank areas. A landslide located on the riverbank area was recognized as a riverbank landslide and a landslide that was not located on the riverbank area was recognized as a non-riverbank landslide.

In this study, the count and area of the landslide and the new and old landslide ratio of each year of the Chishan river watershed were estimated. A new landslide grid refers to when a landslide was identified this year, but was not identified as a landslide in the previous year, whereas the old landslide grid refers to when a landslide was identified both in this year and the previous year. The new and old landslide ratio is the ratio of the sum of the new and old landslide area to the total landslide area in a specific year. The purpose of the new or old landslide comparison from two annual landslide inventories from 2008 to 2009 is different to that from 2009 to 2014. In this study, the definition of new or old landslide from 2008 to 2009 is according to the aforementioned definition but that from 2009 to 2014 is the comparison of the landslide inventory in 2009 and the following year. For example, the old landslide ratio of 62.9% in 2014 suggests that 62.9% of the landslide grid in 2014 was also identified as a landslide grid in 2009.

3. Results

3.1. Extreme Rainfall-Induced Landslide Characteristics

The extreme rainfall-induced landslide characteristics in the Chishan river watershed can be explained based on the statistical data in Table 3. If we consider the *LR* of >5.0% as an obvious landslide-prone area, the top three obvious landslide-prone areas in all categories of 12 landslide-related factors are the bare land category in land use factor (*LR* = 22.8%), the area with fault density $>20 \times 10^{-3}$ m/km^2 category in fault density factor (*LR* = 16.0%), and the Yushanchushan formation (*Yn*) category in geology factor (*LR* = 13.1%). The three factors with the largest variance of *LR* are the land use factor (154.71), accumulated rainfall factor (112.18), and the geology factor (108.52), while those with the smallest variance of *LR* are the profile curvature factor (17.80), the proximity to the river factor (19.06), and the plan curvature (32.05). The geological setting and rainfall distribution are key factors for landslide distribution in the Chishan river watershed. The area of the strata with the *LR* > 5.0 occupies 42.6% of the total area in the Chishan river watershed, but the landslide area in the same area occupies 75.0% of the total landside area. Lithology in the categories with the *LR* > 5.0 in the geology factor is all about sandstone, shale, siltstone, and slate. The total area with accumulated rainfall >1400 mm during the 2009 Typhoon Morakot occupies 68.2% of the watershed area, while the landslide area with accumulated rainfall >1400 mm occupies 99.4% of the total landslide area in the watershed.

3.2. Landslide Susceptibility Models Usinf Four Methods

The landslide susceptibility mapping is followed by using four methods, and the research selects the landslide susceptibility model with the best performance from four models for the following research. In the process of establishing the landslide susceptibility model using the *LRBLR* method, every category in each landslide-related factor is marked a *LRC* number based on *LR* (Table 3).

The highest mean *LRC* number is 9.4 in the fault density factor, while the highest variation of *LRC* number is 94.5% in the land use factor. The research picks 20 random sampling datasets for the logistic regression analyses and selects the result for the random sampling dataset with the largest Cox & Snell R^2 value and Nagelkerke R^2 value. Only if the two indexes, including Cox & Snell R^2 value and Nagelkerke R^2 value, from the logistic regression analysis result using the random sampling datasets are greater than 0.15, the dataset is useful and valid [7] in the research. This research picks the dataset with the highest Cox & Snell R^2 value (0.346) and Nagelkerke R^2 value (0.461) from 20 random sampling datasets, and uses the coefficients resulted from the logistic regression for developing the landslide susceptibility model. The coefficient of each landslide-related factor from logistic regression analysis is listed in Table 4.

Table 4. Coefficients of landslide-related factors in the landslide ratio-based logistic regression analyses.

LRC Number	Coe*	*LRC* Number	Coe*	LRC Number	Coe*
elevation		geology		Plan curvature	
1	—v	1	—	3	—
2	0.510	2	0.026	4	0.487
3	1.123	3	0.281	5	0.662
4	1.487	4	0.784	7	0.841
6	1.889	5	0.821	Profile curvature	
10	2.327	6	0.892	4	—
11	2.483	7	0.978	5	0.065
slope		9	1.148	6	0.119
1	—	11	1.124	Topographic wetness index	
3	0.632	Land use		2	—
5	0.725	1	—	4	0.981
6	1.142	2	0.657	5	1.123
7	1.168	6	0.742	Stream power index	
8	1.392	11	1.183	2	—
aspect		accumulated rainfall		3	0.221
1	—	1	—	4	1.103
4	16.623	4	2.032	5	1.123
5	17.145	6	2.685	Constant	−21.652
6	17.862	11	3.112		
fault density		proximity to the river			
4	—	3	—		
10	−0.174	4	0.124		
11	0.235	5	0.521		

Note: Coe* means the coefficient of category in the twelve landslide-related factors from landslide ratio-based logistic regression analysis.

The *FR* and *C* values in the process of establishing the landslide susceptibility map using *FR* and *WOE* methods are listed in Table 3, while the *D* and *W* values using *II* method are also listed in Tables 3 and 5. The landslide susceptibility maps using four methods are shown in Figure 5. The mean landslide susceptibility, standard deviation and variance of landslide susceptibility values using the *LRBLR* method are 0.533%, 0.280%, and 52.5%, while those using the *FR* method are 0.387%, 0.185%, and 47.8%. The mean landslide susceptibility, standard deviation and variance of landslide susceptibility values by the *WOE* method are 0.549%, 0.222%, and 40.4%, while those using the *II* method are 0.325%, 0.199%, and 61.2%. The accumulated percentages from 0 to 0.5 of landslide susceptibility using *FR* and *II* methods are 71.1% and 79.2%, while those from 0.5 to 1.0 of landslide susceptibility using *LRBLR* and *WOE* methods are 65.6% and 65.8%.

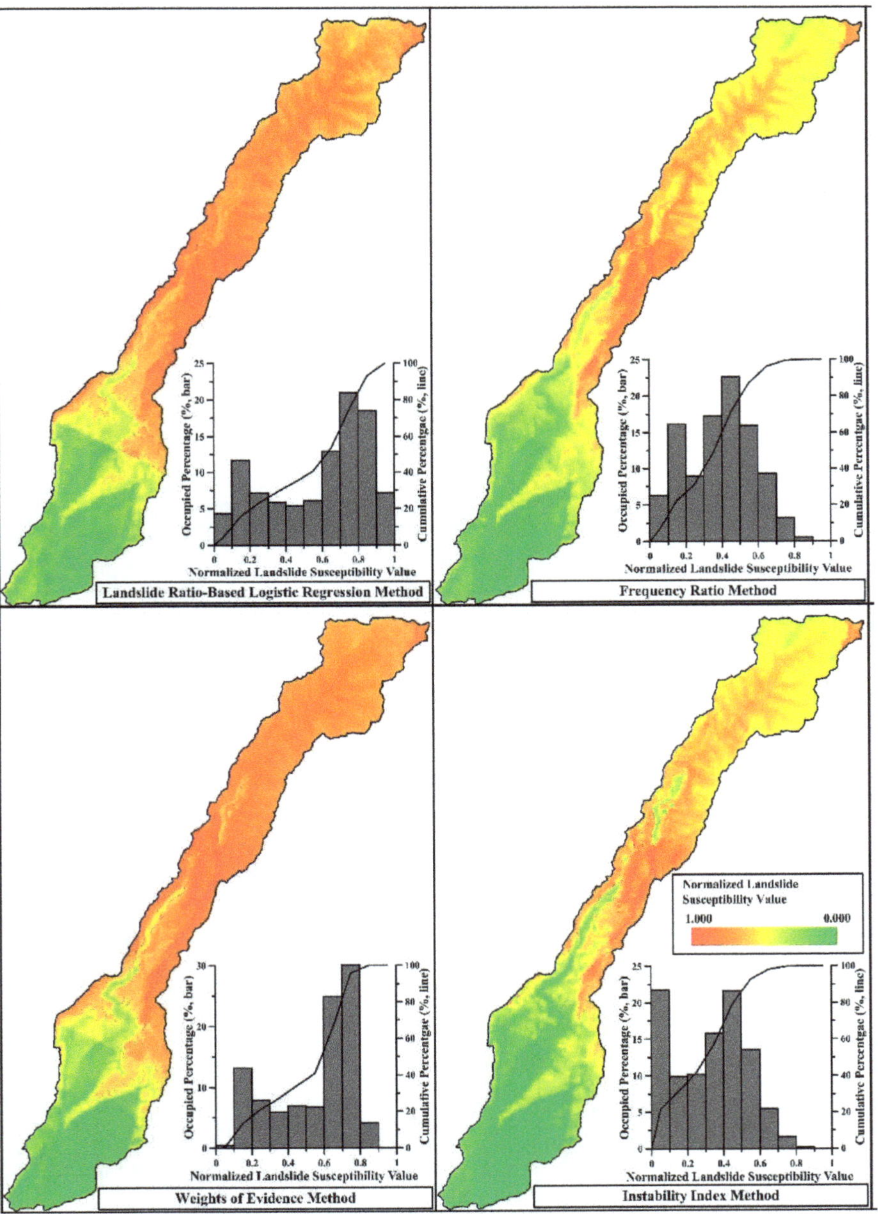

Figure 5. The landslide susceptibility maps using four methods based on the landslide inventory after 2009 Typhoon Morakot in the Chishan river watershed.

Table 5. Weights of landslide-related factors based on the *II* method.

Variables	Elevation	Slope	Aspect	Land Use	Accumulated Rainfall	Geology
S.D.	5.538	4.843	3.856	6.140	3.920	3.500
mean	4.430	2.430	1.552	9.499	4.397	3.798
variance	79.941	50.083	40.246	154.713	112.178	108.520
W.V.	0.109	0.069	0.055	0.212	0.153	0.148
Variables	**Fault Density**	**Proximity to the Rivers**	**Plan Curvature**	**Profile Curvature**	**TWI**	**SPI**
S.D.	10.280	3.880	4.067	4.700	3.917	3.317
mean	4.425	0.740	1.303	0.837	1.318	1.315
variance	43.041	19.062	32.049	17.801	33.656	39.653
W.V.	0.059	0.026	0.044	0.024	0.046	0.054

Note: S.D. means the standard deviation, W.V. means the weighting value, TWI refers to the Topographic wetness index, and SPI refers to stream power index.

The performance of landslide susceptibility models based on four methods is considered from good to fair [28], because the *AUC* value of each method is *LRBLR* (0.803) > *WOE* (0.789) > *FR* (0.762) > *II* (0.721). The confusion matrix of four landslide susceptibility models is shown in Table 6. This research only explains the *PLCR*, *PNLCR*, and *MCR* data of landslide susceptibility models using four methods, because the summation of *MCR* and *MWR* is 1.0. The *PLCR* value of the landslide susceptibility model is 96.2% for *LRBLR*, 75.3% for *FR*, 92.3% for *WOE*, and 62.7% for *II*, while the *PNLCR* is 45.6% for *LRBLR*, 60.3% for *FR*, 46.3% for *WOE*, and 62.2% for *II*. The *MCR* of landslide susceptibility model is 70.9% for *LRBLR*, 67.8% for *FR*, 69.3% for *WOE*, and 64.0% for *II*. Based on the performance of landslide susceptibility models, including the *AUC* and *MCR* values, this research considers that the landslide susceptibility model using the *LRBLR* method is the most suitable model in four landslide susceptibility models in the Chishan river watershed.

Table 6. Confusion matrix of landslide susceptibility models using the four methods and in 2010 to 2014.

Statistical Data between 4 Methods					**Statistical Data from 2010 to 2014**				
	LRBLR	***FR***	***WOE***	***II***	**2010**	**2011**	**2012**	**2013**	**2014**
PLCR	96.2	75.3	92.3	62.7	94.6	95.2	96.4	91.8	62.6
PNLCR	45.6	60.3	46.3	65.2	30.4	34.6	36.0	55.7	75.5
MCR	70.9	67.8	69.3	64.0	62.5	64.9	66.2	73.8	69.1
PLWR	3.8	24.7	7.7	37.3	5.4	4.8	3.6	8.2	37.4
PNLWR	54.4	39.7	53.7	34.8	69.6	65.4	64.0	44.3	24.5
MWR	29.1	32.2	30.7	36.1	37.5	35.1	33.8	26.3	31.0

Note: *PLCR* and *PNLCR* refer to the predicted landslide correct ratio and the predicted non-landslide correct ratio, respectively, *PLWR* and *PNLWR* refer to the predicted landslide wrong ratio and predicted non-landslide wrong ratio, respectively, and *MCR* and *MWR* refer to the mean correct ratio and mean wrong ratio, respectively.

3.3. Rainfall Records from 2008 to 2014 in the Chishan River Watershed

The rainfall records in the Chishan river watershed from 2008 to 2014 are shown in Figure 6 and Table 7. The mean accumulated rainfall during the rainy seasons and annual rainfall from 2008 to 2014 are 2784 mm and 3488 mm in the upstream watershed, 2478 mm and 3168 mm in the midstream watershed, and 2408 mm and 2595 mm in the downstream watershed. The accumulated rainfall during the rainy seasons in the upstream watershed occupied 72.5% to 85.6% of the annual rainfall from 2008 to 2014, while those in the midstream and downstream watershed occupied over 87.0%.

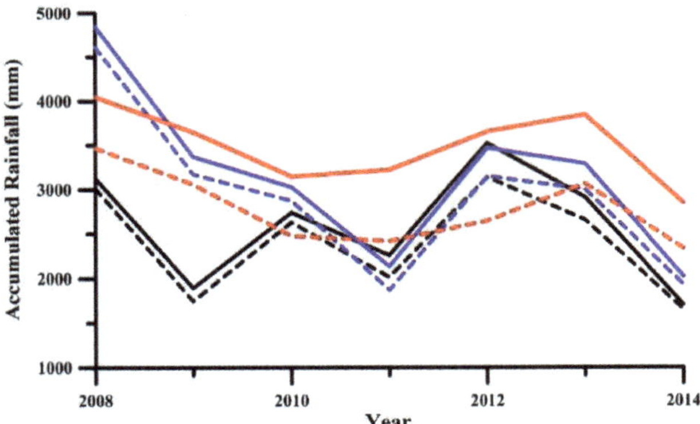

Figure 6. The annual rainfall (solid lines) and accumulated rainfall during the rainy season (dash lines) from three representative rainfall stations, including Xingaokou station (red line), Jiaxian station (blue line), and Chishan station (black line), in the Chishan river watershed from 2008 to 2014.

Table 7. The 3-day accumulated rainfall >500 mm records from 2008 to 2014 in the Chishan river watershed.

Year	The Rainfall Events with the 3-Day Accumulated Rainfall > 500 mm
	Chishan station in the downstream watershed
2008	572 mm from 17 to 19 July
2009	900 mm from 6 to 10 August
2010	No event
2011	No event
2012	577 mm from 10 to 12 June
2013	820 mm from 29 to 31 August
2014	No event
	Jiaxian station in the midstream watershed
2008	1018 mm from 17 to 19 July and 618 mm from 13 to 15 September
2009	2142 mm from 6 to 10 August
2010	677 mm from 18 to 20 September
2011	No event
2012	706 mm from 10 to 12 June
2013	567 mm from 21 to 23 August and 852 mm from 29 to 31 August
2014	No event
	Xingaokou station in the upstream watershed
2008	566 mm from 17 to 19 July, 883 mm from 13 to 15 September, and 601 mm from 28 to 30 September
2009	2076 mm from 6 to 10 August
2010	No event
2011	No event
2012	784.5 mm from 10 to 12 June
2013	No event
2014	501.5 mm on 23 July

The research collects the heavy rainfall or typhoon events with the 3-day accumulated rainfall over 500 mm from 2008 to 2014 in the Chishan river watershed and lists in Table 7. The most accumulated rainfall in 3 days were 1018 mm in 2008 and 2142 mm in 5 days in 2009 in the midstream watershed, and 604 mm in 2008 and 2076 mm in 2009 in the upstream watershed. Furthermore, the most accumulated rainfall was 572 mm in 3 days in 2008 and 900 mm in 5 days in 2009 in the downstream watershed.

The comparison of the rainfall concentration in 2008 and 2009 can explain why the landside ratio in 2009 is larger than that in 2008. The accumulated rainfall during the rainy season in 2008 ranges from 3012 mm to 4615 mm, and that in 2009 ranges from 1747 mm to 3173 mm. The concentrated rainfall during 2009 Typhoon Morakot is the key factor for the dense landslide distribution in the Chishan river watershed. The 3-day accumulated rainfall in 2008 in the Chishan river watershed ranges from 566 mm to 1018 mm, while that in 2009 ranges from 900 mm to 2142 mm. The rainfall concentration during specific heavy rainfall or typhoon events is a key factor for inducing landslides in the Chishan river watershed.

3.4. Landslide Distribution from 2008 to 2014 in the Chishan River Watershed

The annual landslide distributions and statistical data from 2008 to 2014 are shown in Figure 7 and Table 8. The landslide distributions from 2008 to 2014 in the Chishan river watershed are concentrated in the midstream and upstream watersheds. The landslide counts and area in 2009 are 3.4 times and 7.4 times larger than those in 2008 due to 2009 Typhoon Morakot. The landslide area lowers gradually from 2009 to 2012, and raises slight from 2012 to 2013, and lower again from 2013 to 2014. The landslide counts and area in 2014 are only 69.8% and 53.4% of those in 2009. The landslide area from 2010 to 2014 shows that the landslide area in the following years after 2009 Typhoon Morakot gradually decreases if without any heavy rainfall event with more accumulated rainfall than that during 2009 Typhoon Morakot.

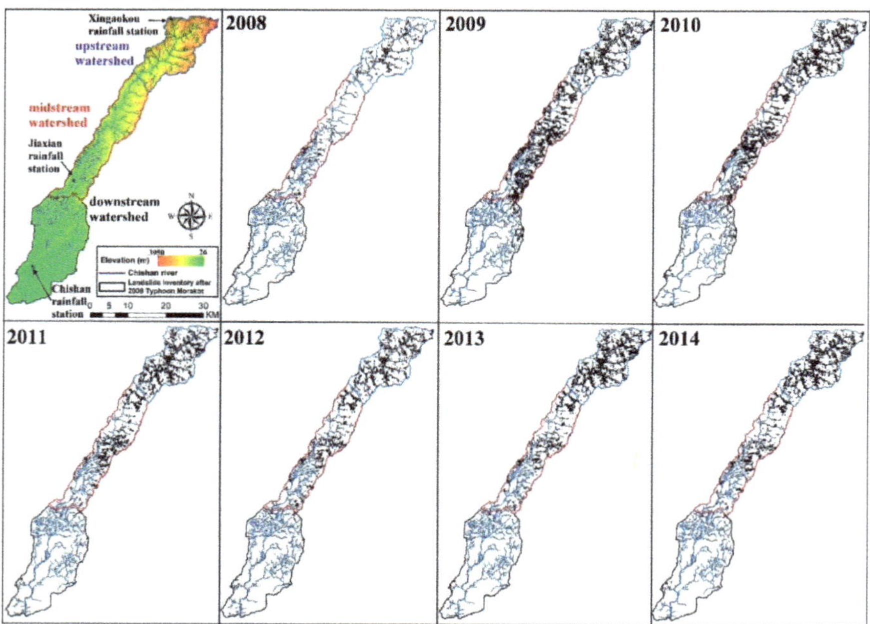

Figure 7. The landslide distribution from 2008 to 2014.

Table 8. The statistical data of landslide distribution from 2008 to 2014.

Year	2008	2009	2010	2011	2012	2013	2014
N	710	2389	2750	1551	1501	1957	1667
A (km^2)	4.6	33.9	26.2	19.0	14.8	18.3	18.1
OP (%)	—	8.7	66.2	68.7	71.4	61.7	62.9
NP (%)	—	91.3	33.8	31.3	28.6	38.3	37.1

Notes: N and A refer to the landslide count and landslide area (km^2), respectively, and OP and NP refer to the old and new landslide percentage (%), respectively.

The new landslide ratio in 2009 rises to 91.3% due to the concentrated rainfall during 2009 Typhoon Morakot. The new landslide occupied percentage from 2009 to 2014 is <38.3%. This means that the landslide induced by 2009 Typhoon Morakot still plays an important role in the annual landslide inventory from 2010 to 2014.

Table 9 lists the statistical data of landslide ratio, new and old landslide percentage in the upstream, midstream, and downstream watershed of the Chishan river watershed from 2008 to 2014. Most of landslide distribution from 2010 to 2014 still overlaps the landslide distribution induced by 2009 Typhoon Morakot. The mean old landslide percentage from 2010 to 2014 in the upstream, midstream, and downstream watersheds are 60.1%, 76.1%, and 49.7%, respectively. The old landslide percentage in the upstream, midstream, and downstream watersheds in 2014 are 56.7% and 76.0%, and 45.8%, respectively, and this means that near or over 50% of landslide induced by Typhoon Morakot in 2009 is still hard to recover in 2014.

Table 9. The landslide ratio, new and old landslide percentages in the upstream, midstream, and downstream of the Chishan river watershed from 2008 to 2014.

Watershed	2008	2009	2010	2011	2012	2013	2014
Old and New landslide percentage in the upstream watershed (%)							
Old	—	18.4	60.6	61.4	67.4	54.5	56.7
New	—	81.6	39.4	38.6	32.6	45.5	43.3
Old and New landslide percentage in the midstream watershed (%)							
Old	—	4.5	72.6	77.1	76.6	78.0	76.0
New	—	95.5	27.4	22.9	23.4	22.0	24.0
Old and New landslide percentage in the downstream watershed (%)							
Old	—	8.8	39.8	62.0	52.8	47.9	45.8
New	—	91.2	60.2	38.0	47.2	52.1	54.2

The statistical data of riverbank-landslide and non-riverbank-landslide from 2008 to 2014 in the Chishan river watershed is shown in Figure 8. The area of the riverbank-landslide and non-riverbank-landslide in the downstream of the Chishan river watershed from 2008 to 2014 are still smaller than 1.0 km^2, and the downstream watershed can be considered as a non-landslide-prone area. The area of riverbank-landslide and non-riverbank-landslide in the upstream watershed in 2009 are 3.3 and 3.5 times larger than those in 2008, while those in the midstream watershed are 13.5 and 17.9 times larger than those in 2008. The area of riverbank-landslide and non-riverbank-landslide in the midstream watershed in 2014 are only 25.3% and 30.5%, respectively, of those in 2009, while those in the upstream watershed are 122% and 112%, respectively, of those in 2009. This shows that most of landslide induced by 2009 Typhoon Morakot in the midstream watershed has been gradually recovery in 2014, but that in the upstream watershed was still hard to recover in 2014.

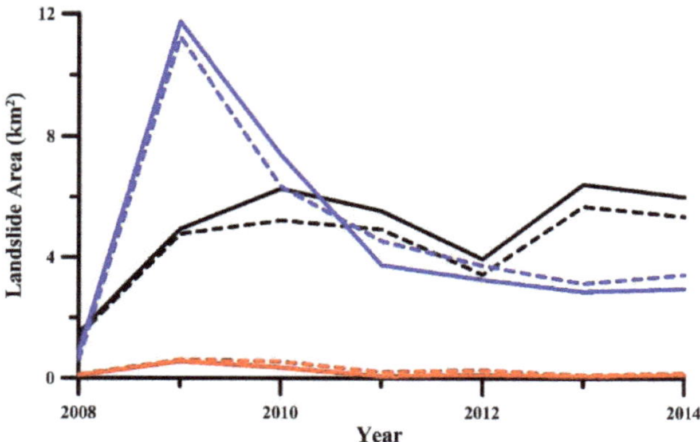

Figure 8. Area of the riverbank-landslide (solid lines) and non-riverbank-landslide (dash lines) in the upstream (black line), midstream (blue line), and downstream (red line) from 2008 to 2014.

3.5. Landslide Susceptibility in the Following 5 Years after the 2009 Typhoon Morakot in the Chishan River Watershed

The long-term landslide evolution analysis in this research has proved that the landslide distribution in 2010 to 2014 has a high correlation to the landslide distribution induced by the 2009 Typhoon Morakot. We suggest that the annual landslide susceptibility maps of the Chishan river watershed from 2010 to 2014 can be the combination of the landslide susceptibility map after the 2009 Typhoon Morakot and the average landslide area different ratio (*LAD*) to the power of the year interval number between 2009 to the specific year in 2010–2014. The *LAD* ratio in this study can be defined as the ratio of the total landslide area in a specific year from 2010 to 2014 to the total landslide area in 2009 of the watershed. For example, the annual landslide susceptibility map in 2012 is the production of the landslide susceptibility model after the 2009 Typhoon Morakot and the *LAD* value to the power of 3.

The annual landslide susceptibility map in 2010 to 2014 of the Chishan river watershed was drawn based on two assumptions. The first assumption was that no previous heavy rainfall event occurred with more accumulated rainfall than that of the 2009 Typhoon Morakot. This assumption is valid for the Chishan river watershed based on data shown in Figure 5 and Table 7. Second, the *LAD* ratio in a specific area was considered to be constant in the 5 years following the 2009 Typhoon Morakot.

Given the difference in the landslide evolution in the upstream, midstream, and downstream areas of the Chishan river watershed and for the riverbank and non-riverbank areas, the Chishan river watershed was classified into six subareas, including the riverbank and non-riverbank areas in the upstream, midstream, and downstream watersheds. The research uses the landslide area in 2009 and 2014 in the same subareas to calculate the *LAD* value. The average *LAD* values in the riverbank and non-riverbank areas in the midstream watershed from 2010 to 2014 were 0.760 and 0.788, respectively, whereas those in the downstream watershed were 0.732 and 0.789, respectively. The average *LAD* values of the riverbank and non-riverbank areas in the upstream watershed from 2010 to 2014 were 1.04 and 1.02, respectively.

The annual landslide susceptibility of each subarea of the river watershed in a specific year from 2010 to 2014 is the production of landslide susceptibility in 2009 and the *LAD* ratio to the power of the year interval. The annual landslide susceptibility distributions of the Chishan river watershed from 2010 to 2014 are shown in Figure 9, and the statistical data of the annual landslide susceptibility from 2010 to 2014 are shown in Table 6. The *MCR* value of the landslide susceptibility model using the landslide ratio-based logistic regression (*LRBLR*) method in 2009 was 70.9%, and the *MCR* values of the annual landslide susceptibility models from 2010 to 2014 ranged from 62.5% to 73.8%. The *MCR* values of the annual landslide susceptibility maps from 2010 to 2014 are still acceptable.

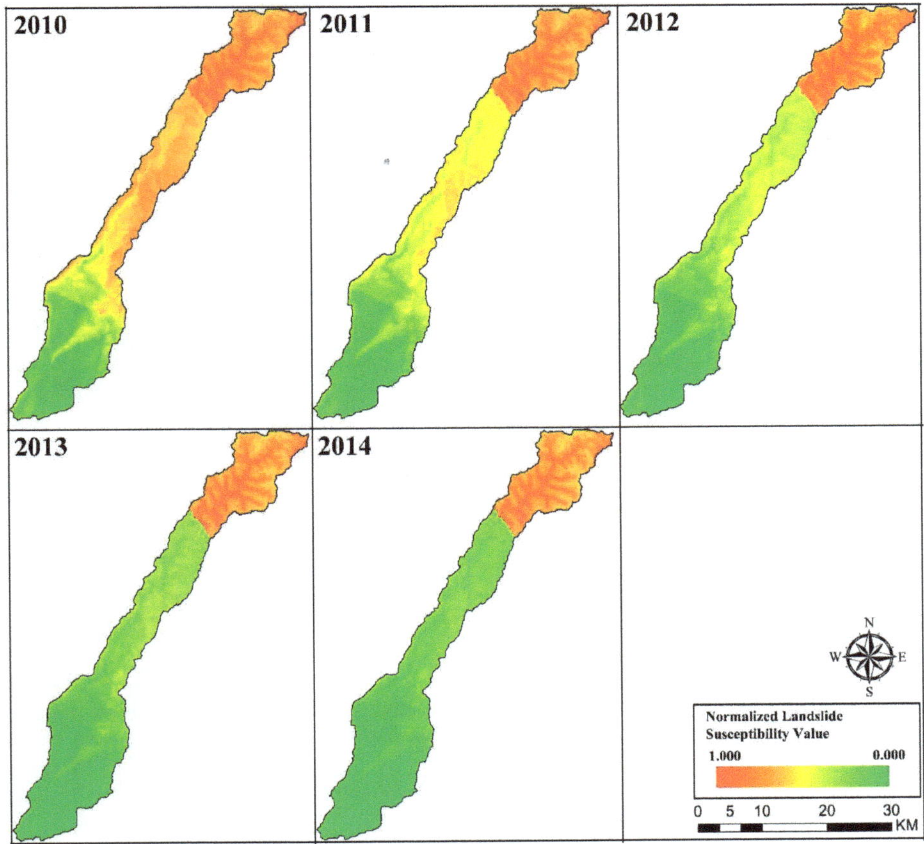

Figure 9. The annual landslide susceptibility from 2010 to 2014 in the Chishan river watershed.

4. Discussion

4.1. Applicability of Landslide Susceptibility Models to the Areas with Dense Landslide Distribution

The similarity and difference of landslide susceptibility models using four methods can help to understand the applicability of each landslide susceptibility model to the areas with dense landslide distribution induced by extreme rainfall events. The correlation analysis result of the four landslide susceptibility models is listed in Table 10.

Table 10. The correlation coefficients of landslide susceptibility models based on four methods.

Landslide Susceptibility Models	*LRBLR*	*FR*	*WOE*	*II*
LRBLR	1.000	0.838	0.924	0.782
FR	0.838	1.000	0.807	0.886
WOE	0.924	0.807	1.000	0.763
II	0.782	0.886	0.763	1.000

The similarity between the landside susceptibility models based on *LRBLR*, *FR*, *WOE*, and *II* methods is strong to very strong. The four landslide susceptibility models can be classified into two groups based on the similarity, including the first group with the landslide susceptibility maps based on *LRBLR* and *WOE* methods and the second group with the landslide susceptibility maps based on

FR and *II* methods. The distribution of landslide susceptibility based on the methods in the first group is somewhat different to that in the second group.

The difference of landslide susceptibility models based on four methods is the process how the assessing grade of each category and weighting value of each landslide-related factor are decided in the specific method. The concept of landslide ratio is used in the four methods, such as *LRC* classification in *LRBLR*, *FR* value in *FR*, W^+ value in *WOE*, and the *D* value in *II*. The landslide susceptibility value by using *FR* or *WOE* methods is with equal weighting value, while using *LBRLR* or *II* methods gives a different weighting value. The accumulated rainfall, geology distribution, and land use factors should be the top three key factors for building the landslide susceptibility models based on the variance of landslide ratio in each factor in the Chishan river watershed based on the above-mentioned analysis. The factors used in building the extreme rainfall-induced landslide susceptibility model should be with different weighting values, so the *FR* and *WOE* methods are not suitable methodologies to build the landslide susceptibility model in the Chishan river watershed after the 2009 Typhoon Morakot.

The process of building the landslide susceptibility models by using *LRBLR* and *II* methods are with *LR* ratio and weighting values, but the difference between the two methodologies is the method by which landslide susceptibility values can be estimated. The landslide susceptibility value by using the *II* method is the product of the landslide susceptibility value of each factor, and the landslide susceptibility value in each factor was <1.0. The mean landslide susceptibility value by using the *II* method was 0.325, which is only 60.9% of the mean landslide susceptibility value obtained using the *LBRLR* method. Using the product to combine each landslide susceptibility value of each factor by using the *II* method underestimates the landslide susceptibility. In this study, the *II* method is considered suitable to develop the landslide susceptibility in the area with mild landslide distribution. The landslide susceptibility value using the *LRBLR* method is the summation of the assessment value of each category of each factor, and the assessment value of each category of each factor is determined by the SPSS software. The *LRBLR* method was considered suitable to develop the landslide susceptibility model for the extreme rainfall-induced landslide susceptibility model.

4.2. Evolution of Landslide Distribution in the Following 5 Years after the 2009 Typhoon Morakot

The landslide evolution in 2010 to 2014 is different in the upstream, midstream, and downstream of the Chishan river watershed and must be discussed in detail. The landslide ratio in the upstream watershed was 1.37% in 2008, 4.62% in 2009, and 5.40% in 2014. The landslide ratio in the upstream watershed from 2010 to 2014 was larger than that in 2009, except 2012. On average, the landslide inventory from 2010 to 2014 in the upstream watershed was composed of 60.1% old landslide that had originated from the 2009 typhoon Morakot and 39.9% new landslide. This means that the landslide in the upstream watershed following the 2009 typhoon Morakot is difficult to recover and easily induced by the mild heavy rainfall events. The landslide distribution in the upstream watershed in 2009, 2010, 2012, and 2014 is shown in Figure 10 for detailed discussion. The river intersection area (red rectangles in Figure 10) and the river source area (red circles in Figure 10) are the two main areas where the landslide is problematic to recover and easily induced from 2010 to 2014.

The midstream watershed has the most landslide area after 2009 Typhoon Morakot in the Chishan river watershed. The landslide ratio in the midstream reaches peak (9.19%) in 2009 and decreases gradually to 2.56% in 2014. On average, the landslide inventory from 2010 to 2014 in the midstream watershed is composed of 76.1% old landslide originating from 2009 Typhoon Morakot and 23.9% new landslide. This means that the landside in the midstream watershed is easily induced only by extreme rainfall events and recovers quickly in 5 years after extreme rainfall events. The composition of strata should be among important factors for the different landslide recovery in the upstream and midstream watersheds. The main strata in the upstream watershed include the Nankang formation, Changchihkeng Formation, and Nanchuang formation. The accumulated occupied percentage of the three strata in the upstream is around 87.9%. The composition of the three strata is slate, sandstone and shale, i.e., three landslide-prone lithologies. The landslide occurred in the three strata in the upstream watershed in 2009 occupied 85.3% of the total landslide in the upstream, while that in 2014 still occupied 87.8% of the total landslide in the upstream.

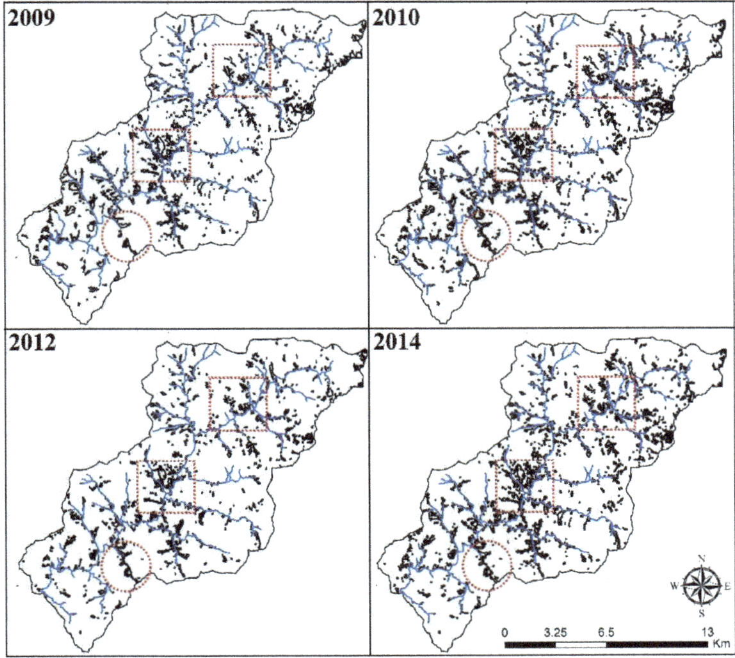

Figure 10. The landslide distributions in 2009, 2010, 2012, and 2014 in the upstream of the Chishan river watershed.

The recovery of riverbank landslide from 2010 to 2014 was different in the upstream and midstream of the Chishan river watershed. Wu [31] mentioned that the riverbank landslide after the 2009 Typhoon Morakot in Taiwan was difficult to recover because of the excessive sediment yield from numerous landslides and debris flow that deposited randomly in the river and resulted in serious riverbank landslide. The area of the riverbank landslide induced by the 2009 Typhoon Morakot in the midstream watershed was recovered in 2014, whereas that in the upstream watershed increased in 2014. This is a notable and valuable observation for further discussion.

The riverbank landslide areas from 2012 to 2013 in the upstream and midstream watersheds are obvious comparisons for the difference in landslide recovery. The midstream watershed had suffered two heavy rainfall events with a 3-day accumulated rainfall of over 500.0 mm in August 2013 (Table 7), but the riverbank landslide area still decreased from 2012 to 2013. Additionally, the upstream watershed had suffered two heavy rainfall events on the same date, but the 3-day accumulated rainfall of Xingaokou station on 21–23 August and 29–31 August were only 449.5 mm and 353.5 mm, respectively. The riverbank landslide area in the upstream watershed in 2013 increased by 1.63 times of that in 2012, and the non-riverbank landslide area also increased by 1.66 times. In this study, the statistical data of landslide and rainfall in 2008 in the upstream watershed were adopted for obvious comparison. The annual rainfall in 2008 and in 2013 in the upstream watershed were 4049 and 3846 mm, respectively, and there were three heavy rainfall events with accumulated rainfall over 500.0 mm in 2008 and zero event in 2013 in the upstream watershed. The riverbank landslide ratio in the upstream in 2008 was 1.5%, and that in 2013 was 6.5%. These data demonstrate that the landslide proneness in the upstream watershed increased significantly after the 2009 Typhoon Morakot, whereas that in the midstream and downstream watershed decreased gradually. The upstream watershed should be considered the most important area in the Chishan river watershed to implement further engineering and disaster prevention based on the long-term landslide evolution analysis.

Another key consideration for future studies can be that the landslide distribution in a river watershed in the following years after extreme rainfall events is mostly overlapped with that induced by extreme rainfall events. The old landslide percentages in the upstream and midstream of the Chishan river watershed were still over 60.0% from 2010 to 2014. The landslide susceptibility maps after extreme rainfall events can be the basis for the annual landslide susceptibility in the years following extreme rainfalls. We suggest that the landslide susceptibility model should be developed after extreme rainfall or earthquake events, and the annual landslide maps in the years following extreme rainfall or earthquake events can be the combination of the landslide susceptibility model and LAD values to the power of the year interval. The LAD values should be estimated carefully in each of the subareas. The annual landslide susceptibility maps from 2010 to 2014 in the Chishan river watershed in this research also proves the aforementioned concept and can be used with acceptable accuracy.

5. Conclusions

This research draws annual landslide susceptibility maps in the years after specific extreme rainfall events. Numerous landslides were induced by Typhoon Morakot in the Chishan River watershed. Based on our analysis result, 61.7% of the landslide area from 2010 to 2014 upstream and midstream of the Chishan River watershed overlapped with that induced by Typhoon Morakot in 2009. This indicates that the landslide distribution following specific extreme rainfall events are strongly related to that induced by the events. We suggest that annual landslide susceptibility maps in the years after specific extreme rainfall events can be drawn on the basis of the landslide susceptibility maps induced by specific extreme rainfall events. Most landslides in the years after specific extreme rainfall events were riverbank landslides induced by sinuous rivers that resulted from the large amount of sediment deposited in the river from the dense landslide after Typhoon Morakot. We emphasize the importance of riverbank landslides and explain how to assess susceptibility to them in the 5 years after Typhoon Morakot.

The research selects 12 landslide-related factors as the basis for establishing landslide susceptibility models using four methods, and the highest-performing landslide susceptibility model of the four methods is the *LRBLR* method. Accumulated rainfall, geology distribution, and land use are the top three key factors for establishing the landslide susceptibility model based on the variance of landslide ratios in each factor. Furthermore, we adopt the annual landslide inventories from 2008 to 2014 in the Chishan River watershed to analyze the long-term landslide evolution. The mean old landslide percentages from 2010 to 2014 upstream, midstream, and downstream of the Chishan River watershed are 60.1%, 76.1%, and 49.7%, respectively. The study calculates the mean LAD in the riverbank and non-riverbank areas upstream, midstream, and downstream of the Chishan River watershed. We suggest that the annual landslide susceptibility maps of the Chishan River watershed from 2010 to 2014 can be the combination of the landslide susceptibility map after Typhoon Morakot and the average LAD to the power of the year interval number between 2009 to the specific year from 2010 to 2014. We can roughly draw the annual landslide susceptibility map in the Chishan River watershed from 2010 to 2014. We compare the annual landslide inventories and susceptibility map from 2010 to 2014 in the Chishan River watershed, and the mean correct ratios from 2010 to 2014 range from 62.5% to 73.8%.

Funding: This research was funded by National Science Council in Taiwan (MOST 107-2313-B-035-001). And The APC was funded by National Science Council in Taiwan.

Conflicts of Interest: The authors declare no conflict of interest.

References

1. National Fire Agency. *Loses Caused by Natural Disasters*; Ministry of Interior: Taipei, Taiwan, 2019. Available online: https://www.nfa.gov.tw/cht/index.php?code=list&ids=233 (accessed on 30 November 2019).
2. Wu, C.H.; Chen, S.C.; Chou, H.T. Geomorphologic characteristics of catastrophic landslides during Typhoon Morakot in the Kaoping watershed, Taiwan. *Eng. Geol.* **2011**, *123*, 13–21. [CrossRef]
3. Shiu, C.J.; Liu, S.C.; Chen, J.P. Diurnally asymmetric trends of temperature, humidity, and precipitation in Taiwan. *J. Clim.* **2009**, *22*, 5635–5649. [CrossRef]
4. Conforti, M.; Pascale, S.; Robustelli, G.; Sdao, F. Evaluation of prediction capability of the artificial neural networks for mapping landslide susceptibility in the Turbolo River catchment (northern Calabria, Italy). *CATENA* **2014**, *113*, 236–250. [CrossRef]
5. Regmi, N.; Giardino, J.R.; Vitek, J.D. Modeling susceptibility to landslides using the weight of evidence approach: Western Colorado, USA. *Geomorphology* **2010**, *115*, 172–187. [CrossRef]
6. Wang, L.J.; Guo, M.; Sawada, K.; Lina, J.; Zhang, J.C. Landslide susceptibility mapping in Mizunami City, Japan: A comparison between logistic regression, bivariate statistical analysis and multivariate adaptive regression spline models. *CATENA* **2015**, *135*, 271–282. [CrossRef]
7. Yilmaz, I. Landslide susceptibility mapping using frequency ratio, logistic regression, artificial neural networks and their comparison: A case study from Kat landslides(Tokat-Turkey). *Comput. Geosci.* **2009**, *35*, 1125–1138. [CrossRef]
8. Cama, M.; Lombardo, L.; Conoscenti, V.; Agnesi, V.; Rotigliano, E. Predicting storm-triggered debris flow events: Application to the 2009 Ionian Peloritan disaster (Sicily, Italy). *Nat. Hazards Earth Syst. Sci.* **2015**, *15*, 1785–1806. [CrossRef]
9. Goetz, J.N.; Brenning, A.; Petschko, H.; Leopold, P. Evaluating machine learning and statistical prediction techniques for landslide susceptibility modeling. *Comput. Geosci.* **2015**, *81*, 1–11. [CrossRef]
10. Pham, B.T.; Shirzadi, A.; Shahabi, H.; Omidvar, E.; Singh, S.K.; Sahana, M.; Asl, D.T.; Ahmad, B.B.; Quoc, N.K.; Lee, S. Landslide Susceptibility Assessment by Novel Hybrid Machine Learning Algorithms. *Sustainability* **2019**, *11*, 4386. [CrossRef]
11. Sameen, M.I.; Sarkar, R.; Pradhan, B.; Drukpa, D.; Alamri, A.M.; Park, H.J. Landslide spatial modelling using unsupervised factor optimisation and regularised greedy forests. *Comput. Geosci.* **2020**, *134*, 104336. [CrossRef]
12. Shafizadeh-Moghadam, H.; Minaei, M.; Shahabi, H.; Hagenauer, J. Big data in Geohazard; pattern mining and large scale analysis of landslides in Iran. *Earth Sci. Inform.* **2018**. [CrossRef]
13. Lombardo, L.; Cama, M.; Conoscenti, C.; Maerker, M.; Rotigliano, E. Binary logistic regression versus stochastic gradient boosted decision trees in assessing landslide susceptibility for multiple-occurring landslide events: Application to the 2009 storm event in Messina (Sicily, southern Italy). *Nat. Hazards* **2015**, *79*, 1621–1648. [CrossRef]
14. Bragagnolo, L.; da Silva, R.V.; Grzybowski, M.V. Landslide susceptibility mapping with r. landslide: A free open-source GIS-integrated tool based on Artificial Neural Networks. *Environ. Model. Softw.* **2020**, *103*, 104565. [CrossRef]
15. Hong, H.Y.; Liu, J.Z.; Zhu, A.X. Landslide susceptibility evaluating using artificial intelligence method in the Youfang district (China). *Environ. Earth Sci.* **2019**, *78*, 488. [CrossRef]
16. Lee, C.T.; Fei, L.Y. Nationwide Landslide Hazard Analysis and Mapping in Taiwan. *Eng. Geol. Soc. Territory* **2015**, *2*, 971–974.
17. Wu, C.H. Landslide susceptibility mapping by using landslide ratio-based logistic regression: A case study in the Southern Taiwan. *J. Mt. Sci.* **2015**, *12*, 721–736. [CrossRef]
18. Su, M.B.; Chen, Y.H.; Fang, C.J. Analysis of landslide susceptibility by using instability index method. *J. Soil Water Conserv. Technol.* **2009**, *4*, 9–23.
19. Water Resources Agency. *The Analysis of the Rainfall and River Discharge During Typhoon Morakot*; Water Resources Agency, Ministry of Economic Affairs: Taipei, Taiwan, 2009; pp. 24–25.
20. Liu, J.T.; Liu, K.J.; Huang, J.C. The effect of a submarine canyon on the river sediment dispersal and inner shelf sediment movements in southern Taiwan. *Mar. Geol.* **2002**, *181*, 357–386. [CrossRef]
21. Milliman, J.D.; Syyitski, J.P.M. Geomorphic/tectonic control of sediment discharge to the ocean: The importance of small mountainous rivers. *J. Geol.* **1992**, *100*, 525–544. [CrossRef]

22. Hung, J.J.; Hung, P.Y. Carbon and nutrient dynamics in a hypertrophic lagoon in southwestern Taiwan. *J. Mar. Syst.* **2003**, *42*, 97–114. [CrossRef]
23. Sinotech Consultants. *Geological Investigation and Database Construction for the Upstream Watershed of Flood-Prone*; Central Geological Survey, Ministry of Economic Affairs, R.O.C: Taipei, Taiwan, 2007.
24. Wu, C.H.; Chen, S.C.; Feng, Z.Y. Flooding process of Xiaolin landslide dam failure triggered by extreme rainfall of Typhoon Morakot in Taiwan. *Landslides* **2014**, *11*, 357–367. [CrossRef]
25. Central Geological Survey. *The Topographic and Geological Database*; Central Geological Survey: Taipei, Taiwan, 2009. Available online: http://gwh.moeacgs.gov.tw/mp/Portal/index.cfm (accessed on 30 November 2019).
26. Yesilnacar, E.; Topal, T. Landslide susceptibility mapping: A comparison of logistic regression and neural networks methods in a medium scale study, Hendek region (Turkey). *Eng. Geol.* **2005**, *79*, 251–266. [CrossRef]
27. Lee, S.; Talib, J.A. Probabilistic landslide susceptibility and factor effect analysis. *Environ. Geol.* **2005**, *47*, 982–990. [CrossRef]
28. Bonham-Carter, G.F. *Geographic Information Systems for Geoscientists: Modeling with GIS*; Pergamon Press: Oxford, UK, 1994.
29. Jian, L.B. Application of Geographic Information System in the Quantitative Assessment of Slope Stability. Master's Thesis, National Chung Hsing University, Taichung, Taiwan, 1992.
30. Shou, K.J.; Wu, C.C.; Lin, J.F. Predictive analysis of landslide susceptibility in the Kao-ping watershed, Taiwan under climate change conditions. *NHESS* **2015**, *3*, 575–606. [CrossRef]
31. Wu, C.H. Comparison and Evolution of Extreme Rainfall-Induced Landslides in Taiwan. *Int. Soc. Photogramm. Remote Sens.* **2017**, *6*, 367. [CrossRef]

Article

Tree-Ring Based Chronology of Landslides in the Shirakami Mountains, Japan

Kinuko Noguchi [1,2], Ching-Ying Tsou [1,*], Yukio Ishikawa [3], Daisuke Higaki [4] and Chun-Yi Wu [5]

1 Faculty of Agriculture and Life Science, Hirosaki University, Hirosaki 036-8561, Japan; noguchi.kinuko.k0@elms.hokudai.ac.jp
2 Graduate School of Agriculture, Hokkaido University, Sapporo 060-8589, Japan
3 Shirakami Research Center for Environmental Sciences, Faculty of Agriculture and Life Science, Hirosaki University, Hirosaki 036-8561, Japan; yishi@hirosaki-u.ac.jp
4 Nippon Koei Co., Ltd., Tokyo 102-8539, Japan; a9024@n-koei.co.jp
5 Department of Soil and Water Conservation, National Chung Hsing University, Taichung 40227, Taiwan; cywu@nchu.edu.tw
* Correspondence: tsou.chingying@hirosaki-u.ac.jp

Abstract: The N-Ohkawa landslide, and the southern section of the Ohkawa landslide, occurred during the snow-melt seasons of 1999 and 2006, respectively, in the Shirakami Mountains, Japan. This paper examines the response of trees in the Shirakami Mountains to landslides, and also investigates the spatio-temporal occurrence patterns of landslide events in the area. Dendrogeomorphological analysis was used to identify growth suppression and growth increase (GD) markers in tilted deciduous broadleaved trees and also to reveal the timing of the establishment of shade-intolerant tree species. Analysis of the GD markers detected in tree-ring width series revealed confirmatory evidence of landslide events that occurred in 1999 and 2006 and were observed by eyewitnesses, as well as signals from eight additional (previously unrecorded) landslide events during 1986–2005. Furthermore, shade-intolerant species were found to have become established on the N-Ohkawa and southern Ohkawa landslides, but with a lag of up to seven years following the landslide events causing the canopy opening.

Keywords: tree ring; dendrogeomorphology; landslide; landslide activity; deciduous broadleaved tree; Shirakami Mountains

Citation: Noguchi, K.; Tsou, C.-Y.; Ishikawa, Y.; Higaki, D.; Wu, C.-Y. Tree-Ring Based Chronology of Landslides in the Shirakami Mountains, Japan. *Water* **2021**, *13*, 1185. https://doi.org/10.3390/w13091185

Academic Editor: Matthew Therrell

Received: 22 March 2021
Accepted: 23 April 2021
Published: 25 April 2021

1. Introduction

Landslides are common in mountainous regions, and can be driven by tectonic, climatic, and/or human activities [1,2]. Landslides can create permanently unstable sites, and as a result, can drastically alter landscape morphology, damage forest environments, and even endanger life. Identifying the spatial and temporal patterns of landslide occurrence is vital for environmental management and minimizing the losses associated with landslides. However, information regarding past landslide events is scarce and almost always incomplete.

Dendrogeomorphology can be used as a proxy indicator of past landslide activity at the scale of years [3–5]. This dating technique is based on the analysis of annual growth rings in trees, with the mixed signals being filtered to isolate the signal indicative of landslide events from non-landslide disturbances, such as climate variations, insect epidemics, and human activity, encoded within the tree-ring chronologies [6,7]. Landslides cause disturbances in tree growth that are preserved as variations within the tree-ring width series. These growth disturbances (hereafter GD) can take several forms, namely, abrupt growth release (wider annual rings), suppression (narrower annual rings), and the formation of compression wood that results from the elimination of neighboring trees, damage to the root, crown or stems, and stem tilting [5,8]. Dating of landslide reactivation by interpretation of these GD markers preserved within annual-ring-width series has been performed using a moving-window approach to smooth out non-landslide fluctuations [9] or evaluating the change rate of the annual ring width if it exceeds a certain threshold value [10]. Additionally, other studies have dated landslides using different thresholds (e.g., the event-response (I_t) index and number of GD markers) [10,11]. Although the amount of research has increased in recent years, no systematic standard approach has yet been proposed and the choice of an appropriate definition and threshold appears to be site-specific.

Dendrogeomorphological studies of landslides have been performed using conifers in the European Alps and Americas [5,8,12]. In North America, Carrara [13] identified synchronous abrupt reductions in annual ring width in tree samples. He suggested that these tree responses were the result of damage during a landslide and was thus able to date the landslide event to 1693 or 1694 and infer that the trigger was an earthquake. With a focus on abrupt reductions in annual ring width and the formation of compression wood on the tilted side stem in the French Alps, Lopez-Saez et al. [10] assessed eight different stages of landslide reactivation over the past 130 years and found that landslide reactivation was associated with seasonal rainstorms. Recently, Lopez-Saez et al. [14] added abrupt increases in annual ring width as another type of growth disturbance, and this enabled reconstruction of 26 reactivation phases of landslides between 1859 and 2010 in the Swiss Alps. In the Orlické hory Mountains (Czech Republic), Šilhán [11] found that landslide activity is particularly associated with slide and creep effects, and the consequent growth disturbance can be identified in trees growing on the scarp and the landslide block. In contrast, there have been few such studies in Asia [3,12,15]. Recent studies have demonstrated that broadleaved trees are also useful for dating landslides and shown the need for additional case studies that consider, for example, an adequate variety of species and age classes [5,12].

Coherent landslides, which often move slowly (*Jisuberi* in Japanese), dominate in the Shirakami Mountains [16], but historical records relevant to landslide activity are scarce. In this study, we investigate the spatio-temporal patterns of landslide occurrence through analysis of the dendrogeomorphological record of 90 deciduous broadleaved trees from 12 species growing on landslide scarps and landslide moving bodies, which we refer to as the displaced blocks, on the right flank of the Ohakawa River, a tributary of the Iwaki River, within the Shirakami Mountains. Our main aims are: (i) to identify and interpret the GD markers (i.e., abrupt growth increase and growth suppression) preserved in the tree-ring series of trees growing on the landslide slopes; (ii) to investigate how these trees responded to landslides known to have occurred in the area; and (iii) to reconstruct the spatial and temporal patterns of landslide occurrence over the past 70 years using our GD data, as well as the timing of the establishment of shade-intolerant trees, and compare this with the limited eyewitness reports of landslides.

2. Study Area

The coherent landslides studied here were located on the right bank, and on an outside bend, of the meandering Ohkawa River, which originates from the eastern side of the Shirakami Mountains, northern Honshu Island, Japan (Figure 1). These landslides are covered by deciduous broadleaved trees dominated by Siebold's beech (*Fagus crenata*). The forest is a naturally regenerated, unmanaged secondary forest that developed after the original forest was selectively felled until 1967 [17]. The study area has a cool-temperate climate, with an average temperature of 8.1 °C and average annual rainfall of 2589 mm [18]. Each year, from November to the following April, the area is covered by snow to a maximum depth of about 2.2 m [18].

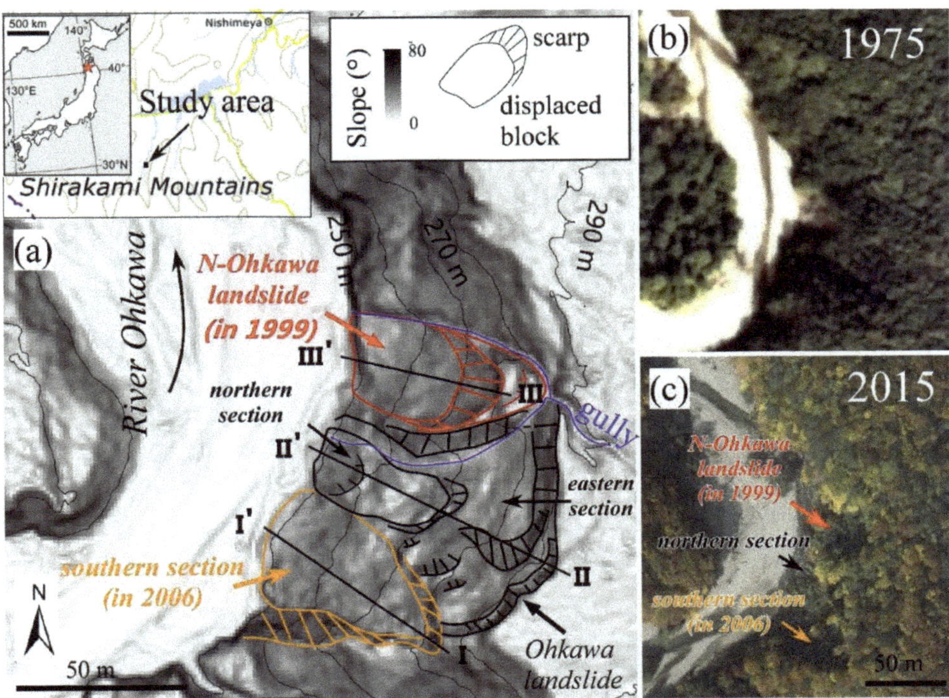

Figure 1. Coherent landslides, topographic map, and aerial photographs from the study area. (**a**) Landslide topography. Aerial photographs are from (**b**) 1975 and (**c**) 2015. The topographic map was constructed from a 1-m digital elevation model (DEM) based on LiDAR data provided by the Geospatial Information Authority of Japan (https://www.gsi.go.jp/, accessed on 12 April 2020). The landslide topography was interpreted using the slope image and the results were checked in the field. Topographic cross-sections (I–I′, II–II′, and III–III′) are shown in Figure 2.

Our study area contains two neighboring landslide slopes: the N-Ohkawa and Ohkawa landslides, that are located along a 40-m-high terraced scarp, with the river terrace top at elevations of 285 to 295 m (Figures 1a and 2). Terrace gravels were exposed at the edge of the terrace after the landslides. The bedrock is formed from the mid-Miocene Hayaguchigawa Formation, which consists primarily of acidic pyroclastic deposits, but also contains andesitic pyroclastic deposits, sandstones, and conglomerates [19] (Figure 2). The N-Ohkawa landslide comprises a single displaced block. In contrast, distinctive stair-like features are evident on the displaced block of the Ohkawa landslide, which also comprises two secondary scarps that separate the individual blocks within the larger block at its northern and southern ends (Figure 2). Minor gully features are present in the landslide slope. Based on its slope geometry, we divided the Ohkawa landslide into three sections; i.e., the eastern, northern, and southern sections, for the following discussion. The timing of these movements is not well constrained. However, limited information obtained from several eyewitness accounts recorded during site visits suggests that the major movements of the N-Ohkawa landslide and the southern section of the Ohkawa landslide occurred in April 1999 and May 2006, respectively [20]; other slope movements of the Ohkawa landslide occurred recently, as described in Section 4.3. In addition, the lower slope of the N-Ohkawa landslide seems to have failed beforehand, as indicated by the bare area seen on the aerial photograph from 1975 (Figure 1b). The N-Ohkawa landslide and the northern and southern sections of the Ohkawa landslide are visible on the aerial photograph from 2015 (Figure 1c).

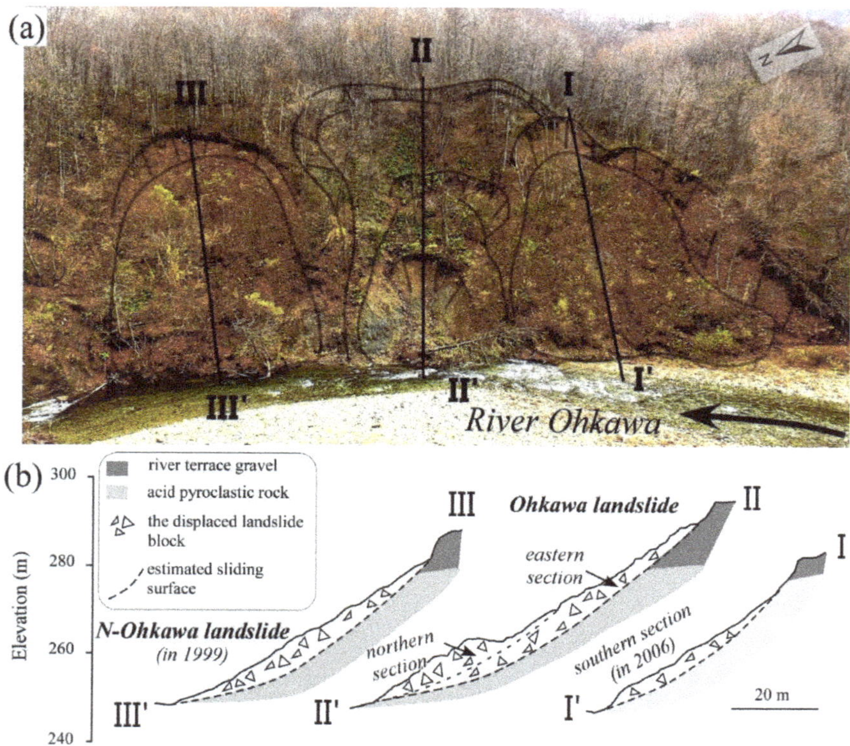

Figure 2. Topography and geological cross-sections of the studied landslide slopes. (**a**) A photograph of the study area. (**b**) Geological cross-sections of the N-Ohkawa landslide (III–III′), the eastern and northern sections of the Ohkawa landslide (II–II′), and the southern section of the Ohkawa landslide (I–I′). The photograph was taken in 2017. The cross-sections are based on the LiDAR DEM.

3. Methods

3.1. Sampling and Cross-Matching of Ring-Width Series

Increment cores were extracted from the upper side of the tilted stems of 90 living broadleaved trees using a Pressler increment borer (maximum length of 40 cm and diameter of 5.15 mm) between June and November 2019, on the main and secondary landslide scarps and on landslide-displaced blocks (Figure 3). The trees were sampled at trunk heights of 20–120 cm. According to the standard methods of dendrochronological research, increment core should be taken parallel to contour to avoid the development of reaction wood in tilted trees [21]. Tension wood develops on the upper side of leaning hardwood trees and typically has wider annual rings than on the lower side [3]. However, in the present study, we obtained cores oriented in the slope direction, because the formation of tension wood is, in itself, a good indicator of landslide movement [3]. Indeed, tension wood may not form in all tilted trees; therefore, wider annual rings may also be the result of growth release owing to, for example, the formation of canopy opening after landslide [5,8]. As such, responses resulting from both tilting and gap formation after landslides are included in our results.

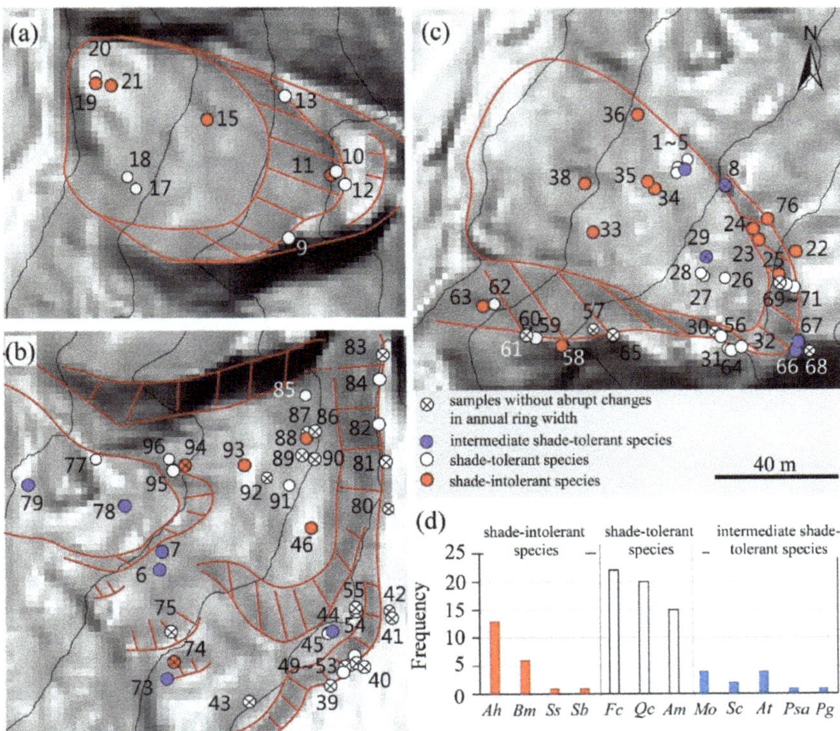

Figure 3. Locations of sampled trees and frequency distribution of tree species. (**a**) Locations of sampled trees at the N-Ohkawa landslide. (**b**) Locations of sampled trees in the eastern and northern sections of the Ohkawa landslide. (**c**) Locations of sampled trees in the southern section of the Ohkawa landslide. (**d**) Frequency distribution of tree species. The numbers on the maps are sample ID numbers.

We selected 21 samples from four shade-intolerant species (*Alnus hirsuta* (*Ah*), *Betula maximowicziana* (*Bm*), *Salix bakko* (*Sb*), and *Salix sachalinensis* (*Ss*)), 57 samples from three shade-tolerant species (*Acer pictum subsp. mono* (*Am*), *Fagus crenata* (*Fc*), and *Quercus crispula* (*Qc*)), and 12 samples from five intermediate shade-tolerant species (*Aesculus turbinata* (*At*), *Magnolia obovata* (*Mo*), *Prunus grayana* (*Pg*), *Prunus sargentii* (*Psa*), and *Sorbus commixta* (*Sc*); Figure 3). We collected 11 samples from the N-Ohkawa landslide and 79 samples (including 39 from the southern (2006) section) from the Ohkawa landslide. The cores were prepared and analyzed using standard procedures following Stokes and Smiley [22] and Speer [21]. The sample cores were prepared using a razor blade to maximize the visual resolution of the ring widths and were measured to the nearest 0.01 mm under a binocular zoom microscope (Olympus SZ61) using a precision measurement stage (Chuo Seiki LTD. LS-252D) attached to a digital output unit (Mitsutoyo Digimatic). After measurement, all cores were visually cross-dated by matching well-defined wide or narrow rings. In addition, longer chronologies (>60 years) of shade-tolerant species and several intermediate species were cross-dated by using a simple list method [23].

3.2. Identification of Growth Disturbance by Landslides in Tree-Ring Width Series and Age Determination of Shade-Intolerant Species

In this study, we considered two types of GD markers in the tree-ring width series: abrupt growth increase and abrupt growth suppression. GD markers were identified using the method described by Ishikawa et al. [7], in which a five year moving average of ring width is used to identify periods of abrupt growth increase or suppression as follows. A growth increase is defined as a doubling of the five year moving average of the ring width when compared with that of the previous five year period and a defined growth rate that fluctuates continuously above 1 for at least 10 consecutive years. Conversely, a growth suppression is defined as a halving of the five year average ring-width and a growth rate fluctuating continuously below 1 for at least 10 consecutive years. In addition, because of spatial irregularities in tree growth, the duration of the GD also depends on the sampling position [4,13]. Therefore, we took into account moderate levels of GD in which the defined growth rate persisted for less than 10, but more than five, consecutive years. In some cases, there is a slight time lag from the casual disturbance event in the GD markers extracted using the moving average method because of growth variation prior to and/or after the event. To avoid this inaccuracy, we carefully checked the ring-width pattern around the timing of the GD markers, and used the information to decide on the GD marker years in the tree-ring width series. Figure 4a–d shows representative examples of how the GD marker years were identified in the tree-ring series using the above method. No significant changes in annual ring width were found in 30 of the cores from the sampled trees (33%) and these cores were not considered for further analysis.

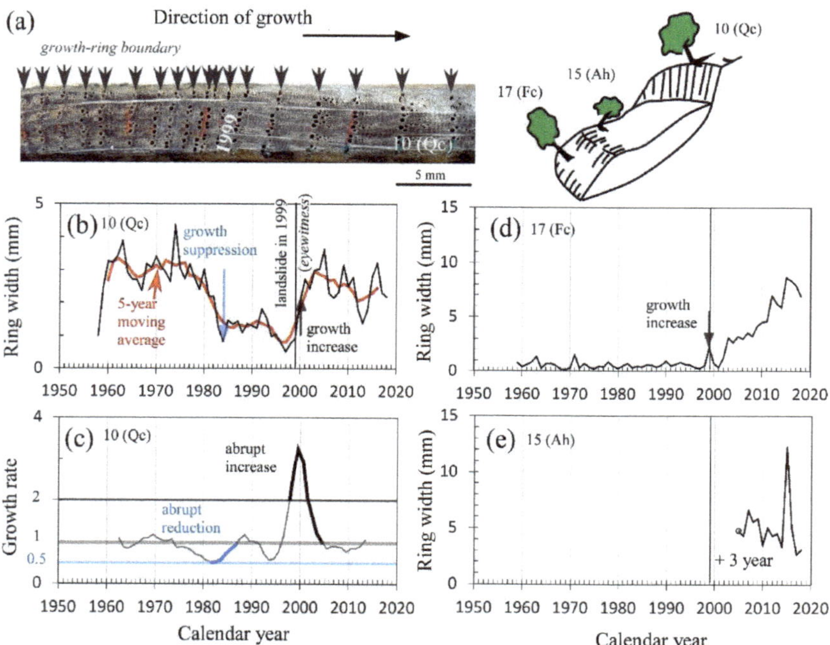

Figure 4. Representative cases from the N-Ohkawa landslide. (**a**) A micro-section of *Nu.* 10 (*Quercus crispula, Qc*) on the landslide scarp showing an abrupt increase in annual ring width in 1999. (**b**) A tree-ring width series and the 5 year moving average of the tree-ring width series of *Nu.* 10. Light blue and black arrows indicate identified response years of GD, growth suppression, and growth increase, respectively. (**c**) GD (i.e., growth suppression and growth increase) defined using growth rate of the tree *Nu.* 10. Note that the defined growth rate of growth suppression was continuously below 1 for 6 consecutive years, and that of growth increase was continuously above 1 for 7 consecutive years. (**d**) The tree-ring width series and identified GD from *Nu.* 17 (*Fagus crenata, Fc*) on the displaced landslide block. The annual ring width increased abruptly in 1999, followed by successive decreases in 2000 and 2001, and then by an increasing trend. (**e**) The tree-ring width series from *Nu.* 15 (*Alnus hirsuta, Ah*, the dominant shade-intolerant species in the study area). Open circle indicates the number of years for the tree to grow to the sample height.

The chronology of each of the previous landslides was expressed using the event-response (I_t) index, following Shroder [24], as follows:

$$I_t(\%) = \frac{\sum GD_t}{\sum N_t} \times 100 \quad (1)$$

where GD_t is the number of trees showing GD in their tree-ring record in year t, and N_t is the number of sampled trees for each landslide alive in year t. Due to the limited number of samples and detected GD markers available to identify landslide reactivation years, thresholds of $GD_t \geq 2$ and $I_t \geq 15\%$ were used. Additionally, the reported year of landslide events and year of establishment of shade-intolerant tree species were also used to assist our interpretation of the dendrochronological effects of landslide activity.

The establishment of shade-intolerant species is indicative of the development of large gaps in the canopy at some point in the past, and these gaps were most probably caused by landslides [25,26]. Consequently, the ages of individual younger trees from shade-intolerant species were determined (Figure 4e) based on the number of rings counted in the cores and the number of years required for seedlings to reach coring height estimated using an age–height regression relationship. The age–height regression (age (years) = 0.025 × height (cm), R^2 = 0.27) was established from 15 specimens of *Ah* (<2.5 m in height) sampled at the Shirakami Natural Science Park of Hirosaki University, 4 km from the study area, where the growth conditions are similar to those in our study area because of their similar elevations. However, the ages for trees of shade-tolerant species and intermediate shade-tolerant species to reach coring height were not estimated, and these trees were used only to identify GD markers on tree-ring width series, as described above. The tree-ring record of these samples were inspected between 1950 and 2019.

4. Results and Discussion

4.1. Spatial Distribution of Tree Ages and GD in Tree-Ring Width Series

The age of the trees sampled around the N-Ohkawa and Ohkawa landslides was 48.2 ± 22.4 years (average ± 1 SD), with a median of 56 years. The youngest tree was 6 years old and the oldest was 101 years old. Figure 5a shows the spatial distribution of the tree ages of 60 trees used for landslide dating. Older ages tend to be concentrated near the scarps, where the majority of trees were 51–70 years old. Trees of 51–90 years in age were also sparsely distributed on the displaced block and many of these were back-tilted, which suggests that the trees were moved down hillsides during landslide transport by rotation along a circular feature of the sliding surface [11,27]. The younger trees (<20 years old) on the scarps and displaced blocks were the shade-intolerant species *Ah*, *Sb*, and *Ss*. GD markers were not detected in these younger trees (Figure 5).

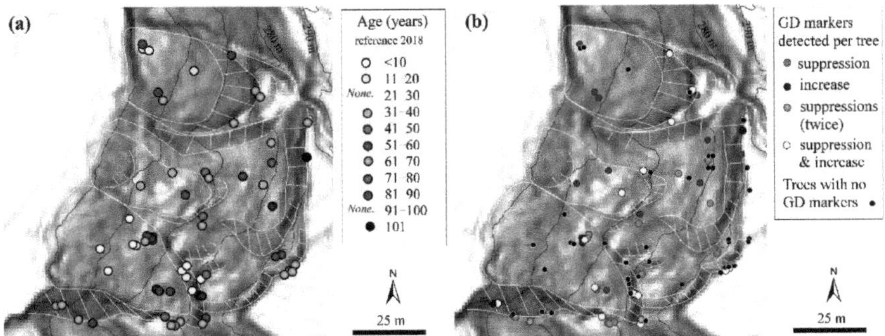

Figure 5. Spatial distribution of tree ages and detected GD markers and trees with no GD markers. (**a**) Spatial distribution of the ages of 60 trees sampled for dendrogeomorphological analysis. (**b**) Spatial distribution of detected GD markers for individual trees.

In total, 64 GD events (including 39 moderate GDs) were identified from 47 trees (Figure 5b). Growth suppression (34 GDs, 53%) occurred in slightly more trees than growth increase (30 GDs, 47%). This higher frequency of growth suppression has also been reported in other similar works [14,28]. The highest frequency (45%) of first-detected GD within the tree-ring width series occurred for trees aged between 16 and 30 years. Individual trees with two GDs (e.g., growth suppression and increase or multiple growth-suppression events) were detected mainly on the landslide scarps. Nevertheless, in a few cases, two GD markers were also detected in trees on the landslide blocks.

4.2. Summary of GD in Tree-Ring Width Series

The GD markers associated with the N-Ohkawa landslide occurred mainly between 1998 and 2001 (Figure 6a). Samples *Nu.* 9 and 20 on the landslide scarp and the displaced landslide block, respectively, showed wider annual rings in 1998, one year before the landslide event that eyewitnesses reported as occurring in 1999. Two samples (*Nu.* 17 and 18) on the displaced landslide block showed wider annual rings in 1999 in response to the landslide occurrence. Following the event in 1999, trees on the landslide scarp presented wider annual rings in 2000 and 2001. Furthermore, narrow annual rings in 1970, 1982, and 1983 were detected in trees on the landslide scarp. In addition, shade-intolerant trees on the displaced block appeared in the early 2000s, which is a lag of 3–5 years after the growing season following the landslide in 1999 (Figure 6a).

Figure 6b–d summarizes the GD for the Ohkawa landslide. In the eastern area of the landslide, a sample (*Nu.* 82) from the landslide scarp showed wider annual rings in 1956 as the earliest GD marker in the study area (Figure 6b). The majority of GD events appear to be clustered after the 1980s, with six GDs on the landslide scarp and eight GDs on the displaced block. In the northern section of the landslide, samples Nu. 95 and 96 on the landslide scarp showed wider annual rings in 1973 and 1995, respectively. In addition, three trees on the landslide scarp recorded GD markers in 2000 and narrow annual rings were also identified for the same year in a sample (*Nu.* 78) from the displaced block (Figure 6c). Furthermore, wider annual rings were detected in 2005 and 2006 in samples from the displaced blocks. In the southern section of the landslide, 29 GDs were identified between the 1960s and the 2010s (Figure 6d). GDs appear to be concentrated in the 1980s and the late 2000s. In particular, GDs detected between 2006 and 2009 are considered to be the consequence of the landslide event that occurred (based on eye-witness reports) in 2006. Notably, shade-intolerant species appeared for the first time on the scarp and displaced block between the late 2000s and the 2010s, a lag of 2–7 years behind the growing season following the landslide event in 2006 (Figure 6d). The two major eye-witnessed landslide events, the N-Ohkawa landslide in 1999 and the southern section of the Ohkawa landslide in 2006, can be identified in the tree-ring records, which suggests that other GDs detected in the study area may also be indicative of historical landslides that were large enough to remove and damage the trees.

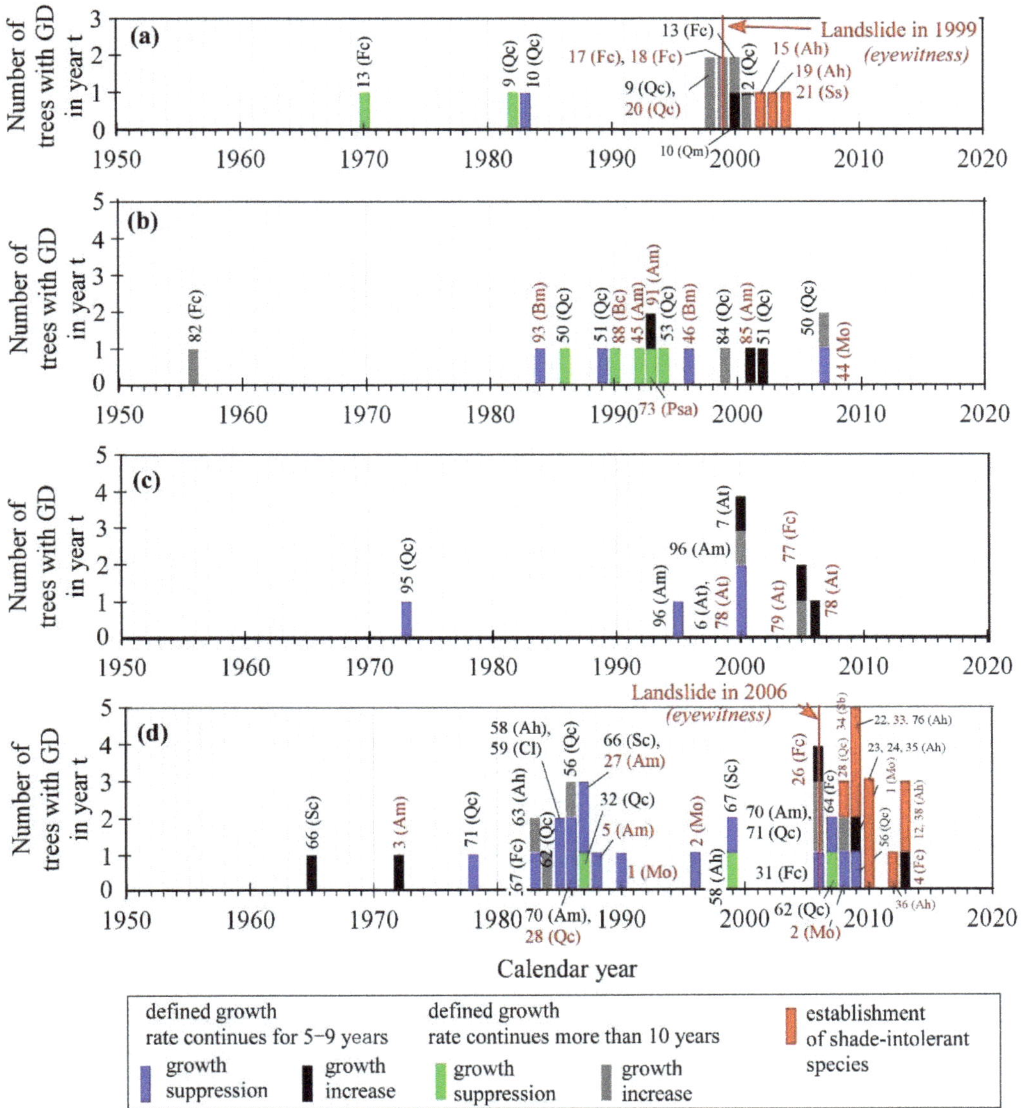

Figure 6. Summary of GD markers identified in the trees and year of establishment of shade-intolerant species for the: (**a**) N-Ohkawa landslide, (**b**) eastern section of the Ohkawa landslide, (**c**) northern section of the Ohkawa landslide, and (**d**) southern section of the Ohkawa landslide. Black and red text indicates samples from the landslide scarp and the displaced landslide block, respectively. Sample locations are shown in Figure 3.

4.3. Dendrochronological Investigations of Spatial and Temporal Patterns of Landslide Reactivation

The analysis of GD markers enabled the identification of landslide events on the studied slopes (Figure 7). These previous slides are summarized in Figure 8 with reference to the locations of trees with GD markers and field observations. For the N-Ohkawa landslide, apart from the landslide reported in 1999, additional landslide activity was detected in 1998 (Figure 7a). In addition, an event took place in 2000 that was detected using samples from the landslide scarp, implying an enlargement of the scarp (Figures 7a and 8). In the eastern section of the Ohkawa landslide, for which there are no reported landslides, we detected two landslide events that took place in 1993 and 2007 (Figure 7b). The landslide events in 1993 and 2007 may suggest an episode of regressive enlargement of the landslide scarp along the terrace scarp (Figure 8). This is supported by field observations showing terrain below the landslide scarp with collapsed debris deposited on a pre-existing landslide mass. The observations suggest that the present landslide unit may have grown from gradual accumulation of landslide debris from repeated landslides, in combination with retrogressive enlargement. For the northern section of the Ohkawa landslide, two landslide events were identified in 2000 and 2005 (Figure 7c). These landslide events have not been previously reported; however, the landslide aftermath can be observed on the aerial photograph from 2015 (Figure 1c). Our analysis suggests that a large landslide might have been initiated in 2000 (as three of the four GDs were identified on the scarp; Figure 6c) and experienced further downward movement in 2005 (as GDs were detected on the landslide block; Figures 6c and 8). Furthermore, ongoing movement is evident on the downslope section, which is bounded by a secondary scarp up to 7 m in height in the lower section. This section is cut by a minor gully, in which surface water is concentrated, and which affected the area before and after a local failure in 2017 (Figure 9a). For the southern section of the Ohkawa landslide, apart from the landslide in 2006, two additional previously unknown events were dated to 1986 and 1987 (Figure 7d). Tension cracks were observed on the crown of the Ohkawa landslide along a ridgeline (Figure 9b,c), from which a crack developed into a lateral scarp of the southern section of the Ohkawa landslide in 2006 (Figure 9c).These observations suggest progressive movement prior to the catastrophic failure in 2006 (Figure 8). In addition, in the middle portion of the southern section of the Ohkawa landslide, a disrupted slide (5 m wide, 25 m long, and 20 m travel distance) was also observed in 2009 [29] (Figure 9d). At the northeastern end of the disrupted slide, within the landslide block of the southern section of the Ohkawa landslide, downward slope movement of about 8 m occurred between 2009 and 2014 [29]. The foot of the downslope section is undergoing river toe erosion, this may steepen the slope and facilitate further movement [29,30]. The landslide activity in 1998, as indicated by the GD markers, might also have progressed to become the major event in 1999 on the N-Ohkawa landslide block.

Our dendrochronological study using 60 deciduous broadleaved trees from 12 species for landslide analysis is unique on global scale [12]. In this contribution, we illustrate that the obtained chronology of landslide activity is in agreement with eyewitness reports of the major landslide events in 1999 and 2006, which suggests the GD markers and index values (where $GD_t \geq 2$ and $I_t \geq 15\%$ are adjusted based on the number of disturbed trees available for analysis) employed in this study may provide a critical assessment of past landslide occurrence in the study area and in those areas with similar environmental conditions. Shade-intolerant tree species are typically established between 2–7 years after landslides. However, this lag may reflect the severe erosion that can continue for several years after a landslide, thus limiting tree establishment [26].

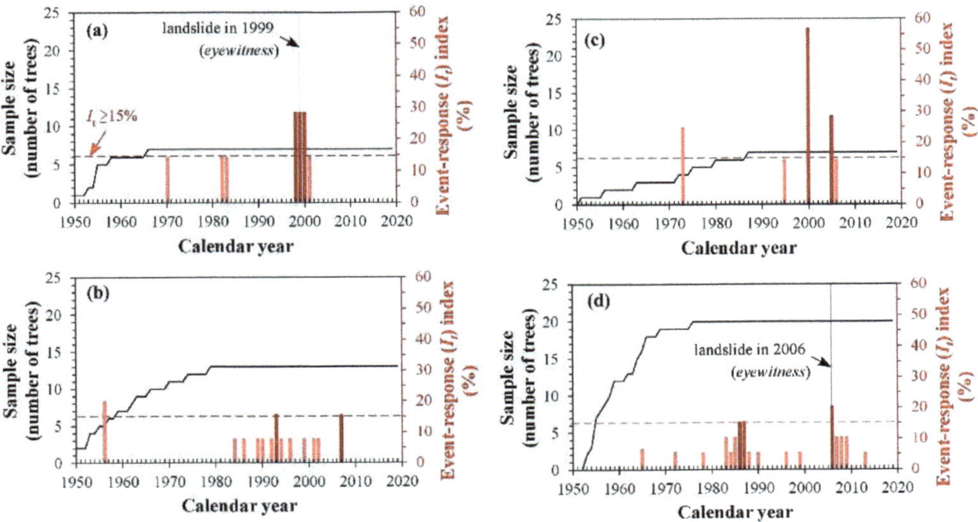

Figure 7. Dendrochronological investigations of past landslide events (dark red columns) expressed using the I_t index and number of disturbed trees. (**a**) Chronology of the N-Ohkawa landslide. (**b**) Chronology of the eastern section of the Ohkawa landslide. (**c**) Chronology of the northern section of the Ohkawa landslide. (**d**) Chronology of the southern section of the Ohkawa landslide.

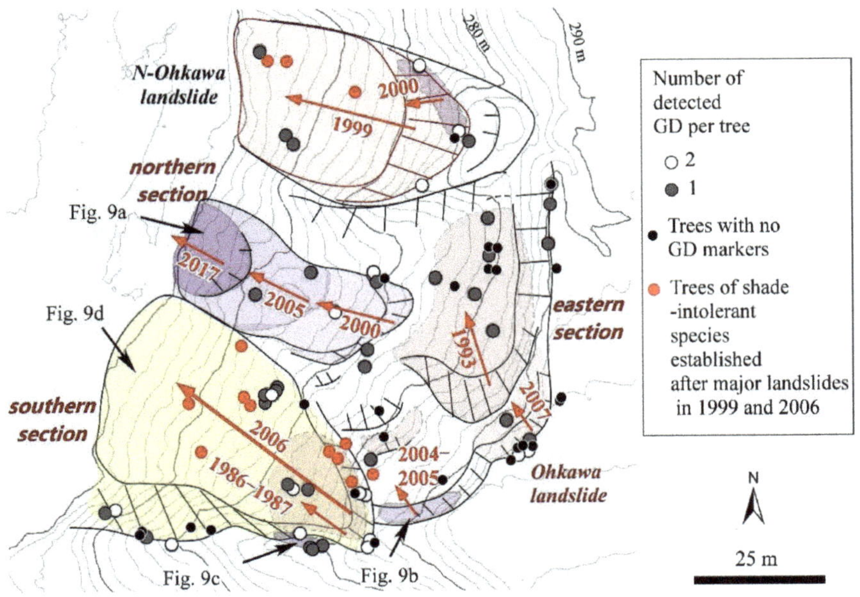

Figure 8. Summary of past landslide events in the study area and distribution of detected GDs and trees established after the landslide events.

Figure 9. Representative examples of slope movements in the study area. (**a**) Ongoing movement on the downslope part of the northern section of the Ohkawa landslide. (**b**) Enlargement of about 2 m tension crack identified in the landslide scarp of the eastern section of the Ohkawa landslide. (**c**) Close-up view of the tension crack. The crack became the lateral scarp of the southern section of the Ohkawa landslide and had enlarged to about 20 m by July 2006. (**d**) A disrupted slide in the downslope part of the southern section of the Ohkawa landslide. The month and year in which photographs were taken are indicated at bottom-right in each panel. Locations of the photographs are indicated in Figure 8.

5. Conclusions

The spatial and temporal development of the coherent N-Ohkawa and Ohkawa (consisting of the eastern, northern, and southern sections) landslides were investigated using tree-ring chronologies from tilted deciduous broadleaved trees in the Shirakami Mountains, northern Honshu Island, Japan. In total, we identified 64 GD markers (i.e., periods of growth suppression or growth increase) from 47 trees, as well as the year of establishment of 13 trees from shade-intolerant species over about 70 years.

Our dendrogeomorphological analysis allowed us to identify the GD markers related to two major eye-witnessed landslide events; i.e., the N-Ohkawa landslide in 1999 and the southern section of the Ohkawa landslide in 2006. Shade-intolerant tree species became established after a lag of 2–7 years after the events in response to canopy opening by the landslides. Other GDs were used to reconstruct previously unknown events within the local landslide chronology. The reconstruction of the N-Ohkawa landslide added precursory landslide activity in 1998 and a local enlargement of the landslide scarp in 2000. In addition, the reconstruction of the Ohkawa landslide indicated episodes of regressive enlargement of the landslide scarp from 1993 to 2007 in the eastern section. In the northern section of this landslide, the landslide slope might have been undergoing sliding to form the current landslide scarp observed in 2000. The slope may have moved progressively downwards in 2005 and its secondary scarp on the downslope locally expanded in 2017. In addition, the reconstruction of the southern section of the Ohkawa landslide suggested that progressive movements may have developed in 1986 and 1987; i.e., before the landslide event in 2006.

Author Contributions: K.N. conducted field investigations, collected tree samples, and analyzed the tree cores, and also collaborated with the corresponding author in the preparation of the manuscript. C.-Y.T. conducted field investigations, landslide interpretation, and the collection and interpretation of tree samples, and also drafted this manuscript. Y.I. conducted field investigations, collected tree samples, and analyzed and interpreted the tree cores. D.H. conducted field investigations and landslide interpretation. C.-Y.W. conducted landslide interpretation and performed calculations. All authors have read and agreed to the published version of the manuscript.

Funding: This work was supported by JSPS KAKENHI Grant Numbers 16K20893 and 19K15257.

Institutional Review Board Statement: Not applicable.

Informed Consent Statement: Not applicable.

Acknowledgments: The authors are grateful to Hajime Makita and Mitsuharu Kudo of Shirakami-Matagisha for providing useful and informative discussions regarding our study. We also thank Hisako Furukawa and Ryunosei Sato of Hirosaki University for their assistance in the field. We acknowledge the Tsugaru Forest Management Office, Tohoku Forest Management Bureau, and Ministry of Agriculture Forestry and Fisheries, Japan for permission to access mountain areas within their territories, and for generous logistical support. The Geospatial Information Authority of Japan provided the LiDAR DEM. We are grateful to three anonymous reviewers whose comments improved the paper.

Conflicts of Interest: The authors declare that they have no competing interests.

References

1. Tsou, C.Y.; Chigira, M.; Higaki, D.; Sato, G.; Yagi, H.; Sato, H.P.; Wakai, A.; Dangol, V.; Amatya, S.C.; Yatagai, A. Topographic and geologic controls on landslides induced by the 2015 Gorkha earthquake and its aftershocks: An example from the Trishuli Valley, central Nepal. *Landslides* **2018**, *15*, 953–965. [CrossRef]
2. Alexander, D. On the causes of landslides: Human activities, perception, and natural processes. *Evniron. Geol. Water Sci.* **1992**, *20*, 165–179. [CrossRef]
3. Higashi, S. Abnormal annual rings as a presage of landslides. *Bull. Hokkaido For. Exp. Stn.* **1968**, *6*, 19–39. (In Japanese with English Abstract)
4. Stoffel, M.; Bollschweiler, M. Tree-ring analysis in natural hazards research—An overview. *Nat. Hazards Earth Syst.* **2008**, *8*, 187–202. [CrossRef]
5. Stoffel, M.; Corona, C. Dendroecological Dating of Geomorphic Disturbance in Trees. *Tree-Ring Res.* **2014**, *70*, 3–20. [CrossRef]
6. Norton, D.A.; Ogden, J. Dendrochronology: A review with emphasis on New Zealand applications. *N. Z. J. Ecol.* **1987**, *10*, 77–95.
7. Ishikawa, Y.; Krestov, P.V.; Namikawa, K. Disturbance history and tree establishment in old-growth *Pinus koraiensis*-hardwood forests in the Russian Far East. *J. Veg. Sci.* **1999**, *10*, 439–448. [CrossRef]
8. Šilhán, K. Dendrogeomorphology of landslides: Principles, results and perspectives. *Landslides* **2020**, *17*, 2421–2441. [CrossRef]
9. Braam, R.R.; Weiss, E.E.J.; Burrough, P.A. Spatial and temporal analysis of mass movement using dendrochronology. *Catena* **1987**, *14*, 573–584. [CrossRef]
10. Lopez-Saez, J.; Corona, C.; Stoffel, M.; Astrade, L.; Berger, F.; Malet, J.P. Dendrogeomorphic reconstruction of past landslide reactivation with seasonal precision: The Bois Noir landslide, southeast French Alps. *Landslides* **2012**, *9*, 189–203. [CrossRef]

11. Šilhán, K. Dendrogeomorphic chronologies of landslides: Dating of true slide movements? *Earth Surf. Proc. Landf.* **2017**, *42*, 2109–2118. [CrossRef]
12. Tumajer, J.; Treml, V. Meta-Analysis of Dendrochronological Dating of Mass Movements. *Geochronometria* **2013**, *40*, 59–76. [CrossRef]
13. Carrara, P.E. Movement of a large landslide block dated by tree-ring analysis, Tower Falls Area, Yellowstone National Park, Wyoming. In *Integrated Geoscience Studies in the Greater Yellowstone Area—Volcanic, Tectonic, and Hydrothermal Processes in the Yellowstone Geoecosystem*; Morgan, L.A., Ed.; U.S. Geological Survey Professional Paper: Washington, DC, USA, 2007; pp. 43–49.
14. Lopez-Saez, J.; Morel, P.; Corona, C.; Bommer-Denns, B.; Schlunegger, F.; Berger, F.; Stoffel, M. Tree-ring reconstruction of reactivation phases of the Schimbrig landslide (Swiss Alps). *Geomorphol. Relief Process. Environ.* **2017**, *23*, 265–276. [CrossRef]
15. Keck, J.; Hsiao, C.-Y.; Lin, B.-S.; Chan, M.-H.; Wright, W. Spatiotemporal landslide activity derived from tree-rings: The Tieliku landslide, northern Taiwan. *J. Chin. Soil Water Conserv.* **2014**, *45*, 36–48.
16. Doshida, S. Relationship between landslide distribution and geological units. *J. Jpn. Landslide Soc.* **2015**, *52*, 271–281. (In Japanese with English Abstract) [CrossRef]
17. Kudo, M.; (Shirakami-Matagisha, Hirosaki, Aomori, Japan). Personal communcation, 2020.
18. Ishida, S. General meteorological conditions of the Shirakami Natural Science Park, 2015. *Shirakiami-Sanchi Bull. Shirakami Inst. Environ. Sci. Hirosaki Univ.* **2016**, *5*, 1–9.
19. Ozawa, A.; Tsuchiya, N.; Sumi, K. *Geology of the Nakahama District. Quadrangle Series Scale 1: 50,000*; Geological Survey of Japan: Tsukuba, Japan, 1983; p. 62, (In Japanese with English Abstract).
20. Makita, H.; (Shirakami-Matagisha, Hirosaki, Aomori, Japan). Personal communcation, 2018.
21. Speer, J.H. *Fundamentals of Tree-Ring Research*; The University of Arizona Press: Tucson, AZ, USA, 2010.
22. Stokes, M.A.; Smiley, T.L. *An Introduction to Tree-Ring Dating*; The University of Arizona Press: Tucson, AZ, USA, 1968.
23. Yamaguchi, D.K. A simple method for cross-dating increment cores from living tree. *Can. J. For. Res.* **1991**, *21*, 414–416. [CrossRef]
24. Shroder, J.F. Dendrogeomorphological analysis of mass movement on Table Cliffs Plateau, Utah. *Quat. Res.* **1978**, *9*, 168–185. [CrossRef]
25. Ito, S.; Nakamura, F. Forest disturbance and regeneration in relation to earth surface movement. *Jpn. Soc. For. Environ.* **1994**, *36*, 31–40, (in Japanese with English Abstract).
26. Grau, H.R.; Easdale, T.A.; Paolini, L. Subtropical dendroecology—Dating disturbances and forest dynamics in northwestern Argentina montane ecosystems. *Forest. Ecol. Manag.* **2003**, *177*, 131–143. [CrossRef]
27. Okamoto, Y.; Higaki, D. Tree lean indicating landslide activity in Shirakami Mountains, northeast Japan. *J. Jpn. Landslide Soc.* **2013**, *50*, 130–136. (In Japanese) [CrossRef]
28. Chalupova, O.; Silhan, K.; Kapustova, V.; Chalupa, V. Spatiotemporal distribution of growth releases and suppressions along a landslide body. *Dendrochronologia* **2020**, *60*, 125676. [CrossRef]
29. Higaki, D.; Yuguchi, K.; Kumagai, N.; Makita, H. The relationship between changes in riverbed morphology and frequent landslides in the Shirakami Mountains. *Shirakiami-Sanchi Bull. Shirakami Inst. Environ. Sci. Hirosaki Univ.* **2016**, *5*, 10–16.
30. Chigira, M.; Yagi, H. Geological and geomorphological characteristics of landslides triggered by the 2004 Mid Niigta prefecture earthquake in Japan. *Eng. Geol.* **2006**, *82*, 202–221. [CrossRef]

Article

Experimental Study on Landslides in Terraced Fields in the Chinese Loessial Region under Extreme Rainfall

Yongfu Wen [1], Peng Gao [2,3,*], Xingmin Mu [2,3], Mengzhen Li [1], Yongjun Su [1] and Haixing Wang [1]

1. Department of Hydraulic Engineering, Hebei University of Water Resources and Electric Engineering, Cangzhou 061001, China; 18706882796@163.com (Y.W.); monica_lmz@163.com (M.L.); 15130800077@163.com (Y.S.); hebeishuiyuanwanghaixing@163.com (H.W.)
2. State Key Laboratory of Soil Erosion and Dryland Farming on the Loess Plateau, Institute of Soil and Water Conservation, Northwest A&F University, Xianyang 712100, China; xmmu@ms.iswc.ac.cn
3. State Key Laboratory of Soil Erosion and Dryland Farming on the Loess Plateau, Institute of Soil and Water Conservation, Chinese Academy of Sciences and Ministry of Water Resources, Xianyang 712100, China

* Correspondence: gaopeng@ms.iswc.ac.cn; Tel.: +86-137-2045-8182

Abstract: Due to the development of the scale of tractor-ploughed terraces, terraces have been increasing in number, while global climate change is causing frequent extreme rainfall events in the Loess Plateau, resulting in many terrace landslides. To study the mechanism and process of shallow landslides and deep slip surface of terraces induced by extreme rainfall in loess hill and gully area, we conducted a laboratory model test of a terrace under artificial rainfall and used the Swedish arc strip method. The research results are as follows. The mechanism of shallow landslides in terraces is rill erosion accelerating rainfall infiltration, suspending the slope, and increasing its bulk density. The destruction process of shallow landslides can be roughly divided into six processes, and the earth volume of the landslide is 0.24 m^3. The mechanism of the deep sliding surface in terraces occurs under the combined action of water erosion and gravity erosion. The soil moisture content increases, which decreases the anti-sliding moment and increases the sliding moment, and the safety factor becomes less than the allowable limit for terraces. The deep sliding deformation area of the terrace was 0~1.0 m below the slope surface, slip surface radius was 1.43 m, the slip surface angle was 92°, and the deep sliding surface began to form earlier than terraced shallow landslides. The displacement of the characteristic points increased from the slope top, to the slope center, and to the slope foot, with maximum displacements of 40.3, 15.5, and 6.0 mm, respectively.

Keywords: laboratory model test; extreme rainfall; rill erosion; shallow landslides; deep lip surface; safety factor

Citation: Wen, Y.; Gao, P.; Mu, X.; Li, M.; Su, Y.; Wang, H. Experimental Study on Landslides in Terraced Fields in the Chinese Loessial Region under Extreme Rainfall. *Water* **2021**, *13*, 270. https://doi.org/10.3390/w13030270

Received: 17 December 2020
Accepted: 19 January 2021
Published: 22 January 2021

1. Introduction

With the implementation and promotion of slope-to-terrace projects, large areas of sloping fields have been built into terraced fields [1]. The construction of terraced field projects has changed the minor features of sloping fields, reducing surface runoff and increasing soil infiltration, thus effectively improving soil moisture content, which plays a crucial role in reducing soil erosion and increasing grain yield in the surrounding areas [2,3]. However, after the implementation of slope-to-terrace projects, back-slope terraces in particular experience significantly increased rainfall infiltration, which also leads to the reduction in soil shear strength, thus increasing the risk of landslides [4,5]. For example, an extreme rainstorm event occurred in Yan'an China, which caused a large area of terraces to collapse and landslide in July 2013, as shown in Figure 1.

In recent years, due to the large-scale development of tractor-ploughed terraces, terraces have been constantly increasing in number, while global climate change has led to frequent occurrences of extreme rainfall events in the Loess Plateau, resulting in many terrace landslides [6]. Long-term production practice has proved that the outer edges of exposed terraces experience shallow landslides over time, which leads to the loss of its water storage and soil conservation functions [7]. When extreme rainfall produces runoff on a field's surface, the terraces seriously erode. This kind of hillslope is featured by collapsibility, strong water permeability, vulnerability, and so on, so rainfall conditions can easily induce shallow landslides, which seriously affect agricultural production [8]. As a consequence, an in-depth study of the process and mechanisms of shallow landslides and deep slip surface of terraces under extreme rainfall is theoretically significant for disaster prevention and mitigation and has practical value for agricultural production in Northern Shaanxi [9–13].

Figure 1. Shallow landslides of terraces in Yan'an China, in July 2013.

Research on the mechanisms of the rainfall-induced general slope instability of soil, rock, and soil–rock mixture has mainly focused on field measurements, numerical simulations and model tests [14–17]. The model test has favorable intuition and can comprehensively consider various factors, simulate complex boundary conditions, and reflect the deep interaction of landslides under the condition of basically meeting the similarity principle [18]. Using the model test of soil slope instability induced by rainfall, Lin et al. [19] discussed the influence of the characteristics of precipitation on slope instability, and thereby selecting appropriate rainfall warning parameters. Zuo et al. [20] studied the laws of seepage, deformation, damage, and particle migration of accumulating soil slope under rainfall conditions through a rain-triggered landslide model test of accumulation bodies with different gradations; they also discussed the influence of particle size on the stability of accumulation soil slope. Li et al. [21] constructed an artificial rainfall simulation test of slopes with different angles and studied the changing laws of the front-end thrust, moisture content and deformation of the slope. Jeong et al. [22] comprehensively analyzed landslides caused by rainfall through laboratory tests, field tests and numerical analysis. Their results showed that landslide activity is closely related primarily to rainfall, soil properties, slope geometry, and vegetation. Numerical analysis was also performed to confirm the effect of these factors on landslide occurrence. Aleotti [23] identified the empirical triggering thresholds for Piedmont and proposed an NI-NCR (where NI is normalized intensity with respect to the annual precipitation, where NCR is the normalized cumulative critical rainfall) diagram. Xu et al. [24], Wang et al. [25], Tohari et al. [26], and Huang et al. [27] conducted rainfall landslide model tests and study the influence of compactness, silt particle content, water level, and other factors on pore water pressure, water content, landslide start-up and development, and failure mode.

The above studies considered the deformation of slope under the condition of rainfall infiltration–seepage interaction, which is mainly concentrated on engineered and natural slopes in China [28,29]. Current research on terraces has mostly focused on the benefits of water and sediment reduction and the rill erosion of terraces [30–33]. However, research on rainfall-induced shallow landslides of terraces is still lacking in China. A number of studies have performed some advantageous explorations. For instance, Jiang [34] discussed the design of terrace sections in the Loess Plateau by considering the requirements for small construction quantity, less land loss, good stability, convenience for cultivation, and conduciveness for crop growth of the terrace ridge. Given the problem of the steep or slow ridges in terrace construction in the sandy, mountainous area of Linqu County, Ge [35] analyzed the stability of the ridge shear test, and the results showed that the slope angle of fine gravel sand ought to be 37°, and that of fine sand should be 34°. Zhang et al. [36] used the Green–Ampt model to study the slope stability of terraced fields based on crop irrigation infiltration and discussed the position of the potential sliding surface. Yang et al. [37] selected the terraced ridge in Southern Shaanxi province as the research object and explored the failure forms and causes of the ridge through an indoor conventional triaxial shear test. Liu [38] studied the changes in the failure time and safety factor of a horizontal terrace, a separated slope terrace and an original slope (for contrast) under a rainfall infiltration intensity of 28~38 mm/h and a side slope gradient of 15–30° using ABAQUS software. Derbyshire [39] discusses how terraces in the Loess Plateau can maintain good stability under a natural state, but tend to be eroded and collapse under rainfall infiltration.

In this study, we selected terraces as the research object, and indoor model tests and the Swedish arc strip method were used to study the mechanism and mode of shallow landslides and deep sliding surfaces in terraces under extreme rainfall conditions. The following assumptions were made for the test. We ignore the influence of: (1) the model's side wall on the test results; (2) the internal sensors on the test results; and (3) the soil disturbance on the test results. The research is important for the agricultural development of the loess hilly and gully region as it provides: (1) a reliable theoretical basis and abundant experimental data for slope collapse and instability prevention, and disaster mitigation, monitoring, and forecasting of terraces; (2) parameters for the optimized design of terraces; and (3) a method for studying multi-level terraces and terraced landslides in the basin.

2. Materials and Methods

2.1. Test Soil Properties

The soil used in this model test was obtained from Zhifanggou, Ansai County, Shaanxi province, China; the depth of sediment deposition is 6~8 m, so it belongs to the category of loessal deposits [40]. The basic parameters of the test soil are shown in Table 1. According to the light compaction test, the maximum dry bulk was 1.703 g/cm^3 and the optimal moisture content was 19.3%. The particle size of the soil was measured by a Marven laser hondrometer, with a measured range of 0~2 mm, and the characteristic values of average grading in the Table 1 are as follows: clay particle ($\leq$0.002 mm) content was 12.1%, silt particle (0.05–0.002 mm) content was 52.6%, and sand (2–0.05 mm) content was 35.4%, which showed that the soil sample contained fewer particle size series and that the difference between coarse and fine particle sizes was small. The curvature coefficient (CC) of the particle grading curve was 1.79, which is well-graded.

Table 1. Basic property indexes of soil.

Great Group	Natural Density (g.cm^{-3})	Natural Water Content (%)	Dry Density (g.cm^{-3})	Cohesion (Kpa)	Internal Friction Angle (°)
Calcic Cambisols	1.32	7.86	1.21	13.85	20.10

2.2. Test Device

The test equipment was divided into four major systems: test object, rain, data monitoring, and image capture systems (Figure 2).

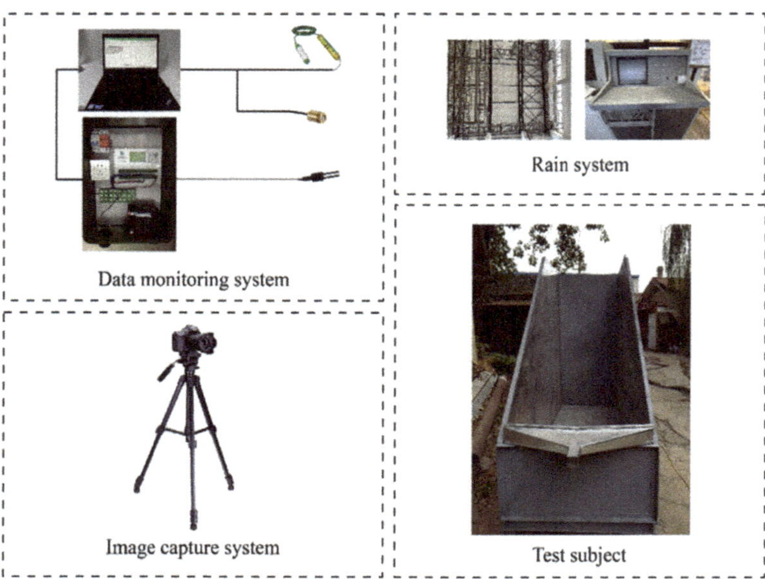

Figure 2. Schematic diagram of model test device.

The test object system involved a terrace groove with a length of 2.8 m, a width of 1 m and a height of 2.1 m. The height of the slope filled in the test was 1.2 m, with a gradient of 65°. The front and back edges and one side of the model were surrounded by steel plates, and the wall surface of which was smoothed by applying a layer of Vaseline to reduce the influence of the model boundary effect on the test. On the remaining side, transparent plexiglass with a thickness of 1 cm was used as a visual window to help observe the movement of soil at any time in the process of the test. To facilitate the observation of soil movement, a rectangular grid measuring 10 × 20 cm was drawn onto the transparent poly, and steps were placed close to the steel plate to facilitate the measurement of the channel shape parameters and flow velocity during the test. A catch basin was set up at the front edge to collect runoff sediment.

The rain system device was developed by the Institute of Soil and Water Conservation, Ministry of Water Resources, Chinese Academy of Sciences. The rainfall device's height is 16 m, which can measure the terminal speed of all raindrops. The range of rainfall intensity was 40–260 mm/h, the rainfall uniformity was more than 80% and the maximum duration of rainfall was 12 h. The rainfall area was composed of two independent rainfall experimental areas. The effective rainfall area of a single experimental area was 4 × 9 m, which can accurately simulate natural rainfall [41].

The data monitoring system was composed of a RR-7120 water content sensor, KPE-200 kPa pore water pressure sensor, Campbell 257 soil suction sensor, and an LDS-S-200 displacement monitoring sensor. Each sensor was connected to the corresponding collection system through a data line, and then the data in the system is exported and sorted through a computer. The data collection frequency of the water content sensor was 1 min, (unit: %); the data collection frequency of the pore water pressure sensor was 1 min (unit: Ka); the acquisition frequency of the suction sensor data acquisition system was 1 min (unit: Ka); the displacement monitoring sensor data acquisition system had an acquisition frequency of 1 min (unit: mm).

For the image capture system, a Canon EOS M50 camera was set up on the side facing the transparent plexiglass at a height of 0.85 m, to clearly capture the downward movement of the wet front on the side of the soil.

2.3. Soil-Filling and Sensor Embedding

A square sift iron was applied for screening to ensure the maximum particle size of the model's soil filling would be less than 1 cm. Then, the soil was evenly spread, sprayed with an appropriate amount of water and evenly mixed to make the density and moisture content of the soil close to that of the undisturbed soil. However, in the process of model filling, the structure of the soil, the particle gradation, the stratum structure, the soil cracks, and so on, will change to some extent, which is inevitable. For this test, we adopted the method of layered compaction and filling: the soil prepared before the test was evenly divided into 17 layers, each layer 10 cm thick, and the side wall of the terrace was compacted with a discus. After the compaction of each layer was completed, samples were taken from several different parts with a cutting ring. The wet density of each layer of soil was 1.32~1.40 g/cm^3, and the moisture content was around 7.5%. After the placement into layers was complete, the geometric dimensions of a 65° slope in the model was obtained by manual slope cutting after the stratified filling.

To ensure the accuracy of the monitoring data and minimize the impact of sensors on the test results, we arranged the fewest sensors possible. Figure 3a provides a cross-sectional view of terraced soil filling and sensor embedding, showing that two moisture content sensors were arranged on the side close to the slope every 20 cm. Due to the large soil width, one more was arranged on the bottom side, for a total of 13 sensors. To study the mechanical properties of the slope, three pore water pressure sensors and suction sensors were arranged on the top, middle and toe of the slope with a vertical distance of 20 cm from the slope surface. To accurately determine the shape of the slip surface, 5 to 7 displacement sensors were arranged every 20 cm, for a total of 36 displacement sensors. To reduce the influence of the sensor on the test results, the sensor was arranged in the middle of the whole soil, as shown in Figure 3b.

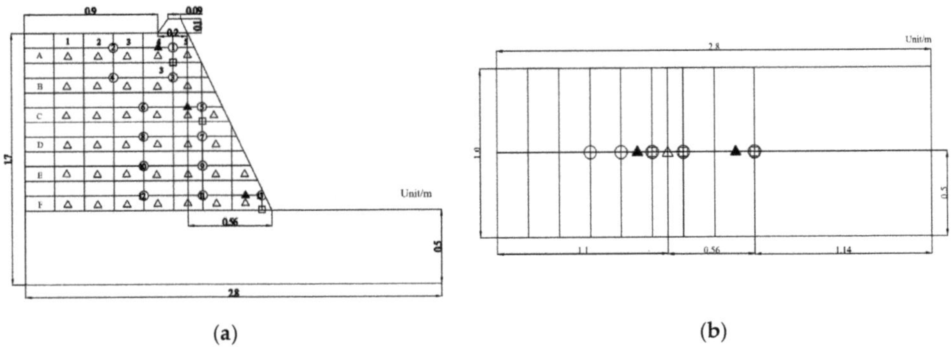

Figure 3. Layout drawing of test sensors. (**a**) Left view of device, (**b**) top view of device. Note: Different letters indicate soil layers; Numbers indicate soil columns; ○ indicates soil water content sensor; □ indicates pore water pressure sensor; △ indicates monitor point of displacement; ▲ indicates soil water suction sensor.

2.4. Test Method

The experiment started at 10:00 a.m. on 11 November 2017 and ended at 8:00 a.m. on the 12 November. It was carried out in Area II of the artificial rainfall hall of the Institute of Soil and Water Conservation, Ministry of Water Resources, Chinese Academy of Sciences. According to the hydrological data for Yan'an in July 2013, and the actual situation of slope movement, the data were divided into five periods of rainfall, each lasting for 1 hour, with rainfall intervals of 1 h, a rainfall intensity of 2.5 mm/min and a total rainfall of 750 mm. The test was repeated once on 20 November 2017, and the average of the data from the two tests was used as the test result for analysis, and the standard deviation of each dataset was found to be within 0.2, so the data were considered reasonable and reliable.

To ensure the uniformity and stability of the rain intensity, the slope of the terraces was covered with plastic sheeting before the test. The rain intensity was calibrated around the model trough and the top. When the rain intensity stabilized, we quickly uncovered the plastic sheet and started timing. When the water flow on the slope of the terraces was in a laminar flow state and flowed from top to bottom to the water outlet, it was regarded as the start of runoff. We recorded the runoff time, and then restarted the clock. After the trial runoff, runoff and sediment samples were collected every 1 min. After rills appeared on the slope, the time of rill appearance was recorded, and we measured the size with a measuring tape with an accuracy of 1 mm every 2 min. When the rill length exceeded 10 cm, we measured the width every 10 cm along the length of the rill. The average value was used as the width of the rill. The measurement density was increased in locations where the morphological mutation of the rill was obvious. Simultaneously the slope and wet front morphology were recorded every 20 min with a digital camera, and the camera shooting frequency was increased during the period of severe morphological changes. After the test was over, the sediment samples were allowed to stand for 6 h and the supernatant liquid was poured out, then the sediment samples were dried in an oven (105°) and weighed by electronic scale. The sensor data were imported into an Excel table for data preprocessing.

When calculating the moisture content data, the sensor data at the same time point of each layer were averaged, and this average value was regarded as the soil moisture content of this layer. Because the data collection frequency was very fast, a large amount of data were generated. Therefore, we selected the representative data to draw figures under the premise of not affecting the changing of the parameter curve.

3. Results and Discussion

3.1. Mechanism and Process of Shallow Landslides of Terraces

3.1.1. Mechanism of Shallow Landslides of Terraces

The results suggest that the shallow landslides of terraces were caused by rill erosion. Most of the rills in the loessial soil were developed by a single drop sill, which was mainly manifested by the headway erosion of the gully head and the collapse of the side wall [42,43] (Figure 4a). By measuring the traceable erosion pattern of terraces, we found that the maximum width of the gully head was 34.25 cm, the maximum depth was 21.32 cm, and the maximum length was 75.86 cm. The total erosion amount was about 270.96 kg. The sediment yield rate of five rainfalls was 181, 475.67, 1707.17, 1624.33, and 527.83 g/min. The sediment yield rate showed a trend of increasing first and then decreasing. Before the runoff, when the exposed slope surface was hit by large raindrops, the surface soil structure was destroyed, and the soil particles splashed up and fell back to the slope surface, forming raindrop splash erosion (Figure 4b). After the runoff, the erosion developed from raindrop splash erosion to layered surface erosion. The time from splash erosion to surface erosion was one hour (Figure 4c). The reason for this finding is that the runoff was low at the initial stage of runoff generation and the runoff eroding force was less than the anti-erosion ability of soil resistance. With the increasing runoff, runoff eroding force also increased. Isolated and sporadic falling ridges were generated in the terrace ridge and the vulnerable parts of the side slope's soil. When the terrace ridge was filled with water, the terrace ridge breached under the action of hydraulic erosion (Figure 4d) and the water cut down along the breach to form obvious gullies, with an average width of 17.8 cm (Figure 4e). Due to the erosion, transportation, and accumulation of overtopping flow, the erosion gully continuously eroded and undercut longitudinally, eroding the gully bank and widening the gravity collapse horizontally. The sediment carried by the slope flow was fan-shaped deposition around the gully mouth at the toe of the slope, forming an alluvial fan, covering an area of 176.64 cm^2 (Figure 4f). The erosion gully further developed and constantly degraded and widened. Due to the difference in the density of terraces and the non-uniformity of the filling materials, the velocities of anti-erosion of the terraces differed. The weak position started easily, and erosion occurred first, forming a scouring pit.

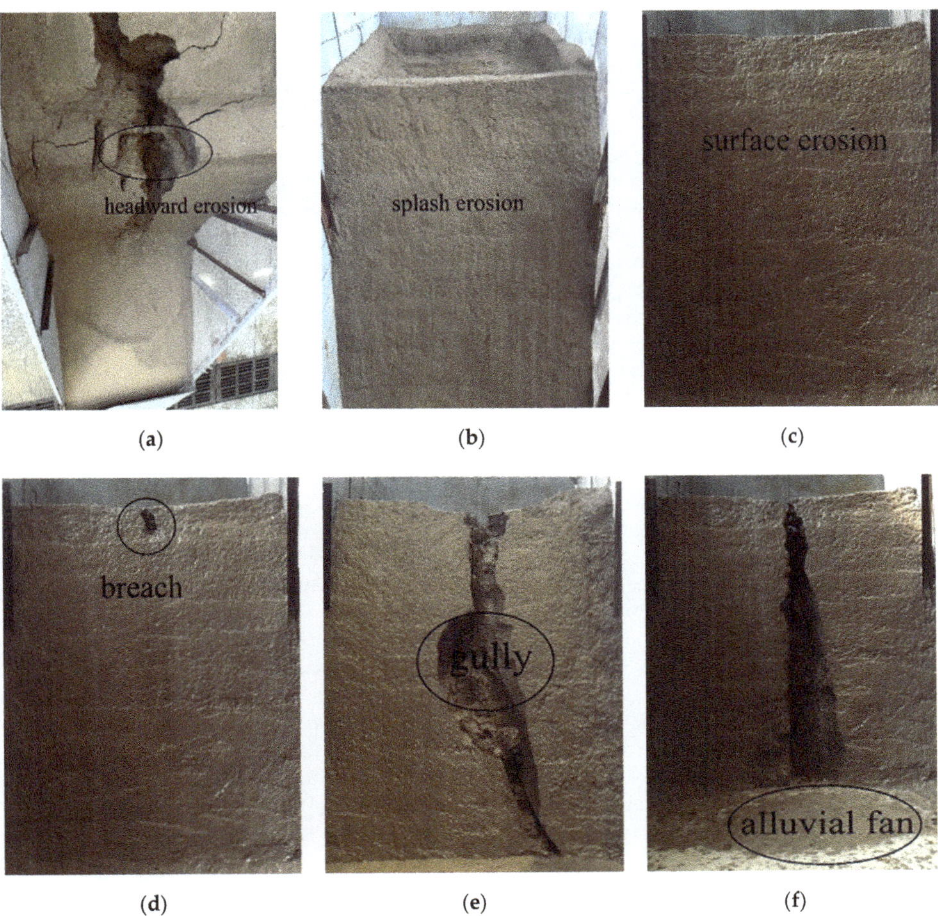

Figure 4. Morphological characteristics of terraces at different rainfall times; (**a**) Headward erosion, (**b**) Splash erosion, (**c**) Surface erosion, (**d**) Breach, (**e**) Gully, (**f**) Alluvial fan.

The above characteristics all indicate that many side wall collapses occur in the process of rill formation, which is consistent with the phenomenon observed in the test process. It is generally thought that the composition of soil particles is an important factor affecting soil erosion resistance. The finer the particle composition, the stronger the cohesion. To a certain extent, when the soil forms a mass structure, and its anti-erosion ability will be higher. In particular, the clay content in the soil significantly enhances the anti-dispersion ability of the wet soil layer. It can be seen from Table 1 shows that silt (0.002~0.2 mm) and clay (<0.002 mm) only accounted for around 35% of the total particles in the model. The overall stability of the soil was poor. In addition, the content of sand particles was high, and the soil was loose and porous. Therefore, the rill side wall of the loessal soil easily lost stability and collapsed under the action of runoff erosion and soil moisture. As such, the main forms of the rill development process of loessal soil are wall collapse and traceable erosion, and the randomness is significant [44]. Han et al. [45] and Acharya et al. [46] also showed that traceable erosion is the most active sediment yield factor in rill development, and the collapse of the ditch wall mainly occurred on steep slopes above 65°. Both the rill and cut trench in this experiment were similar in shape to the above research results, but the size was larger, because the rainfall intensity and total rainfall were higher than in the above experimental conditions.

3.1.2. Process and Mode of Shallow Landslides of the Terraces

According to a series of deformation characteristics of the side slope during the test, the mode of this kind of shallow landslide of terraces under rainfall conditions is summarized in this section. The deformation mode is shown in Figures 5 and 6 and can be described as follows: (i) The stage of water accumulation on terraced fields: As the rainfall continued, the soil of the terraces gradually became saturated, which reduced the infiltration capacity of the soil, while the rainfall gradually increased, resulting in water accumulated on the terraces. The height of stagnant water was 3.5 cm. (ii) The formation and development stage of the breach: After rainfall had been occurring for a period of time, the loess on the terrace surface and its slope surface reached saturation. The ponding on the field surface crossed the ridge, forming surface runoff and flowing down the slope. When the water flows through the ridge, the erosion of the ridge formed a breach with area of 706 cm^2. With the continuous erosion of water and rainfall, the erosion degree of the ridge increased and the breach expanded. (iii) The erosion of the waterfall nappe flow: The ponding flowed along the breach and formed plumes. Under the combined action of hydraulic forces and gravity, the discharge flow formed a multi-stage drop sill on the slope surface. The maximum discharge and maximum velocity of the breach occurred at this stage, and the ponding on the terrace surface dropped rapidly. (iv) The formation and development stage of the erosion gully: With continuous rainfall, the multi-level drop sill was connected by the water flow, forming an erosion ditch, and the width and depth of the erosion ditch gradually increased along the slope shoulder to the slope toe, finally forming the alluvial fan at the slope toe. The average erosion gully width was 17.8 cm. (v) The stage of superficial-layer shallow landslides of the terraces: With increases in the width and depth of the erosion gully, the soil on both sides of the slope was suspended. In addition, the soil is constantly saturated with water, the gravity increased and the cohesion decreased, resulting in superficial-layer shallow landslides of the terraces, and the soil volume of the landslide was about 0.24 m^3 (vi) The terraces tend to be stable: After the superficial-layer shallow landslides of the terraces, the side slope grade of the residual slope was very small; even if the soil was in a saturated state, it would not easily collapse.

The pattern of shallow landslides in terraces is similar to that in the earth dams. Zhong et al. [47] used experimental methods to simulate the mechanism and mode of earth dam failure, they proposed that the most important reasons for earth dam failure are the overflow of water, the crest of the dam breaking, and the huge instantaneous downflow washing the earth dam. The dam break test (dam height 6 m) funded by the EU IMPACT project [48] and the dam break test (dam height 1.5 to 2.3 m) carried out by Hanson et al. [49] of the United States Department of Agriculture both simulated the earth dam break mode, which is similar to the landslide pattern of the terraces, but the mechanism is different. The dam break mechanism of action of an earth dam is through the upstream water flow forming an infiltration line inside the dam body due to the seepage effect, or the upstream water flowing over the top to destroy the earth dam. The terrace landslides mechanism is through the soil losing cohesion due to rainfall infiltration and the erosion ditch making the terraces lose their integrity, leading to local shallow landslides. The results of this study are different from those of Sun et al. [50] because there is a border dike on the loess slope in the natural state. When the rainfall intensity is greater than the infiltration rate, the slope will produce runoff. For terraced fields, the border dike plays a role in water storage and soil conservation. Therefore, under the same rainfall conditions, terraces are more capable of resisting catastrophic rainfall than sloping fields.

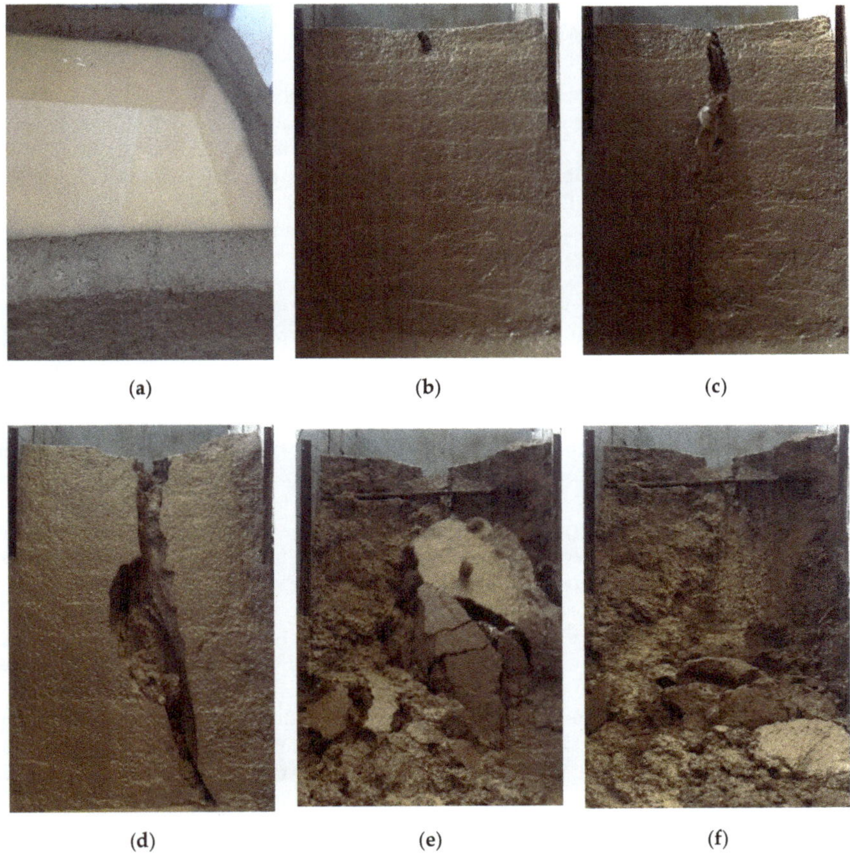

Figure 5. Shallow landslide process in terraces. (a) First stage, (b) Second stage, (c) Third stage, (d) Fourth stage, (e) Fifth stage, (f) Sixth stage.

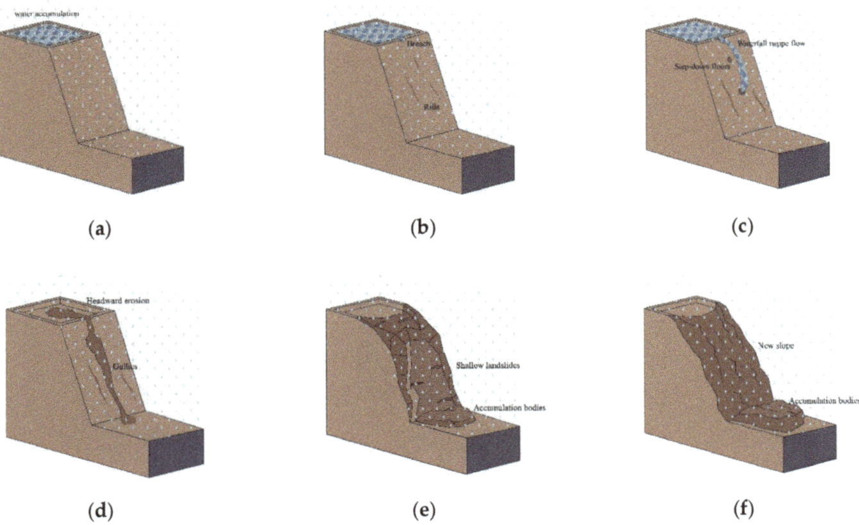

Figure 6. Three-dimensional diagram of shallow landslide process in terraces. (a) First stage, (b) Second stage, (c) Third stage, (d) Fourth stage, (e) Fifth stage, (f) Sixth stage.

3.2. The Factors of Deep Slip Surface of Terraces

3.2.1. Rainfall Infiltration

The curve in Figure 7 shows the relationship between average rainfall (infiltration), infiltration percentage, and rainfall frequency. The infiltration rate was 79%, and the infiltration rate gradually decreased with the increase in surface runoff. In the third rainfall event, the infiltration rate decreased to 49.3% and then remained steady at approximately 26.14%. More than half of the rainfall turned into surface runoff, causing soil erosion and rain erosion on the slope's surface and formed rills. The infiltration of rainwater from the slope surface to the slope was an unsaturated-to-saturated seepage process, and the change in infiltration rate with time was related not only to the original humidity and matrix suction of unsaturated soil, but also to the physical characteristics and structure of the soil from the side slope. Generally, at the early stage of infiltration, infiltration capacity is greater than rainfall intensity and the infiltration rate is higher, so infiltration is pressureless. After a period of time, the soil begins to saturate, the gradient of soil moisture content decreases, the matrix suction reduces, and the infiltration capacity lowers. When rainfall intensity is greater than the soil's infiltration capacity, slope runoff occurs, which is pressure infiltration. Finally, with rainfall, the infiltration rate gradually decreases until it tends to be constant, reaching the stable infiltration stage. Figure 8a shows the relationship curve between infiltration rate and rainfall time for five rainfall events, and Figure 8b provides partial enlarged view; the figure shows, the infiltration rate generally presents the same decreasing trend. The steady infiltration rate ranged from 0.74 mm/min to 0.77 mm/min, and the results are similar to those in the literature [51]. This is because with progressing rainfall, the infiltrating rainwater continuously increased the soil moisture content, which saturated the surface soil causing the infiltration rate to gradually decrease. However, the infiltration rate of the topsoil was relatively low and stable after saturation, and the infiltration rate of rainwater was further reduced due to the small amount of rainwater infiltration inside. For the third and fourth rainfall events, the infiltration rate first increased and then decreased. The reason for the increase in the third rainfall's infiltration rate was that cracks appeared on the terrace surface and the infiltrated rainwater could speedily travel deep through the cracks; on the other hand, due to the water-retaining effect of the ridge, ponding formed on the horizontal surface, which accelerated the infiltration rate. The reason for the increase in the infiltration rate in the fourth rainfall was that the superficial layer of the terrace collapsed and the rainwater rapidly entered into the soil along the gully.

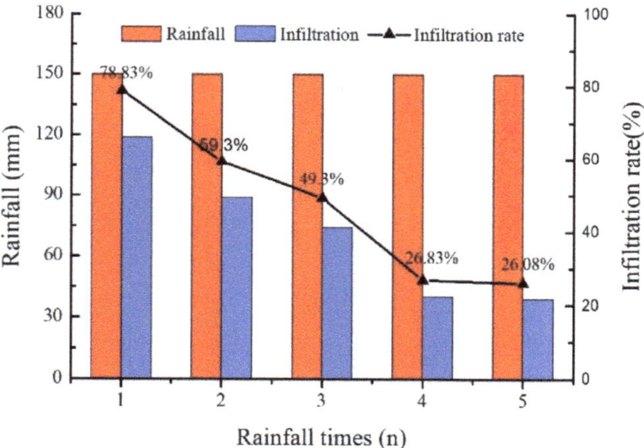

Figure 7. Rainfall amounts and percentage of infiltration.

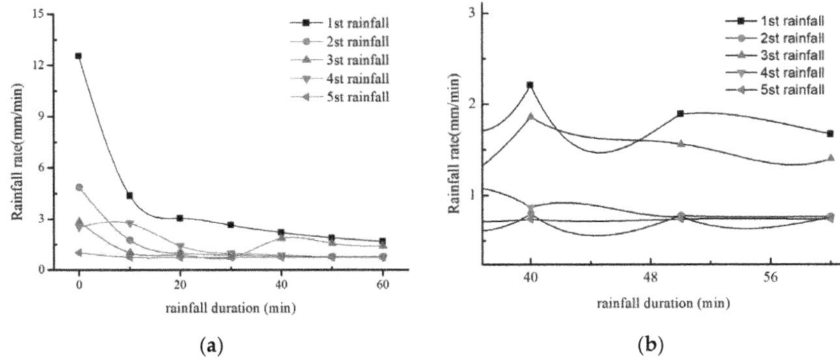

Figure 8. Relationship between infiltration rate and rainfall time in five rainfalls; (**a**) The relationship curve between infiltration rate and rainfall time for five rainfall events, (**b**) partial enlarged view of the Figure 8a.

In this study, the rainfall infiltration law of terraces is similar to that of Liu et al. [52], but the infiltration rate is lower than the latter, because Liu et al. considered the development characteristics of the root system in vegetation and soil, which has strong water storage and soil conservation capabilities and can intercept more rainfall and runoff. Compared with Huang et al. [53], the infiltration rate of this result is relatively high because of the different nature of the soil. The initial moisture content of the soil in Huang et al.'s research was high, and the initial moisture content will shorten the saturation time of the soil. The results of this test provide reference value for the study of terrace infiltration in loess hilly and gully areas, especially for mechanically-repaired horizontal terraces.

With the continuous infiltration of rainwater, the color of the soil from the model's side slope gradually darkened with the increase of in moisture content, and the infiltration peak appeared at the boundary between the dark- and light-colored soils. In this test, the change in infiltration peak was recorded by a camera set up on the side of the transparent poly to judge and calculate the infiltration depth and infiltration rate of rainwater.

Figure 9 depicts the wetting front from different rainfall time points: when the rainfall duration was 10 min, the downward depth of the wet front was 9.7 cm; when the rainfall duration was 60 min, the downward depth of the wet front was 28.8 cm; and when the rainfall duration was 180 min, the downward depth of the wet front was 35.5 cm. The wet front was basically linear with the rainfall time. The wetting front moved downward with increasing duration of rainfall and the migration rate of the horizontal wetting front was faster than that of the side slope surface. The reason for this finding is that the horizontal plane could effectively intercepted the rainwater, and the ponding accelerated the infiltration rate of the horizontal surface [54]. However, for the inclined slope, most of the rainwater formed runoff along the side slope surface, and only a small part of the rainwater infiltrated. Figure 10 shows that with continuous rainfall, a proportion of the rainwater on the terrace surface infiltrated, another part formed slope runoff, and some of it infiltrated along the back wall of the model box, where collapse deformation occurred. Due to the different materials of the back wall and the soil, the infiltration speed was faster than that of the soil; thus, the wetting front moved rapidly in the back wall.

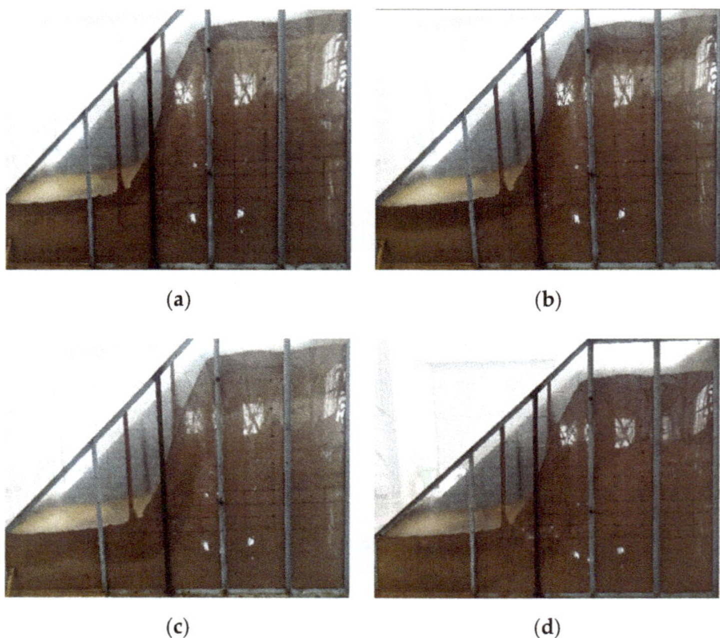

(**a**) (**b**)

(**c**) (**d**)

Figure 9. Infiltration peaks at different rainfall time; (**a**) T = 10 min, (**b**) T = 60 min, (**c**) T = 180 min, (**d**) T = 300 min.

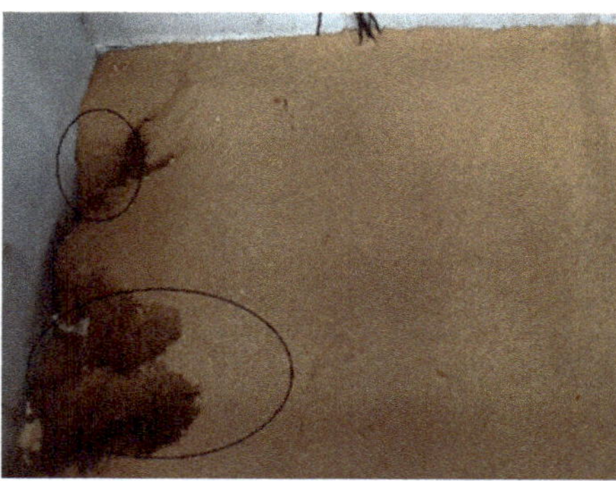

Figure 10. Collapsible deformation on the posterior wall.

The downward movement of the wet front was similar to the results reported by Tian et al. [55], but the downward movement rate of the wet front was larger than that recorded by them, because the terraced ridges have the capacity to store water, and the accumulated water increases the water infiltration gradient and accelerates the rate of water downward movement. This also proves that the measures of slope conversion can effectively intercept rainfall, and the benefits of water and sediment reduction are obvious in the Loess Plateau. This is also supported by the research results of Bai et al. [56].

3.2.2. Water Content

In accordance with the monitored data from the slope moisture sensor, the change in slope moisture content in the whole process can be understood. Figure 11a shows that the moisture content of each soil layer at different depths of slope varied with rainfall duration and post-rain duration. Within 0.5 h of the beginning of rainfall, the moisture content of layers A, B, and C of the slope increased to 0.83%/h, and the change was obvious, while the moisture content of soil in layers D, E, and F changed very little, due to the infiltration of rainwater going from shallow to deep. With the increase in rainfall time, rainwater infiltrated and the moisture content of the whole slope increased continuously. The increased rate of moisture content of the upper soil was about 2~2.83%/h, which is higher than the previous value of 0.83%/h, due to the continuous increase in soil moisture content, the decrease in matrix suction, and the increase in the permeability coefficient. For D, E, and F, there is an obvious break point in the curve, and the moisture content suddenly increased at about 4:00 p.m., which indicated that there were violent activities in the soil; deep cracks appeared in the soil, the soil began to lose stability and the sliding surface formed. Thus, the change in moisture content strongly influenced soil failure. First, the increase in moisture content led to an increase in pore water pressure and a decrease in effective stress, thus resulting in a decrease in soil shear strength; second, the increase in moisture content increased the permeability of the water, which led to a decrease in side slope stability. The dual effect of rainfall infiltration leading to water content variation may be an important reason for the side slope's rainfall-induced instability. Figure 11b shows that the moisture content at the top of the slope rose the fastest and had the largest change range. At around 1:00 p.m. on 11 November, due to the shallow landslides of the slope body, the moisture content of the slope rapidly dropped to 0%, followed by the toe of the slope, with the smallest change and the smallest range in the middle of the slope.

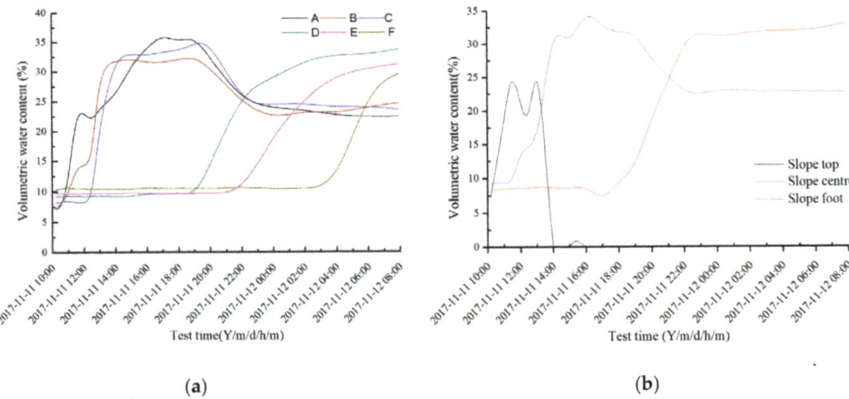

Figure 11. Variation of volumetric water content of soil layers and feature points with time; (**a**) The moisture content of each soil layer at different depths of slope varied with rainfall duration, (**b**) The moisture content at the top, center, and foot of the slope.

3.2.3. Pore Water Pressure and Suction

Figure 12 shows the variation in pore water pressure and suction with rainfall. Figure 12a shows that the pore water pressure first increased and then tended to be stable with the rainfall duration. The variation in pore water pressure at the top of the slope ranged most before the shallow landslides of the slope, and the maximum value was 2.4 kPa. Since the sensor was exposed outside the slope after shallow landslides, it rapidly dropped to 0 kPa. The change in pore water pressure at the measuring point at the toe of slope was slightly later than that at the top of slope, with a maximum value of 3.6 kPa. At the end of the rainfall event, the pore water pressure gradually decreased and finally tended to be stable around 2.5 kPa. The change in pore water pressure at the measuring point in the middle of the slope changed later than at the measuring points at the top and toe of the slope, with the water infiltration reaching the measuring point at around 4:00 p.m., then increasing gradually, and finally tending to be about 2.8 kPa. As Figure 12b shows, the suction (negative pore water pressure) varied from 7.4 to 14.6 kPa, showing a sharp decrease at first and then a stable trend. The reason for this finding is that with rainfall infiltration, the moisture content of each measuring point increased, and the suction decreased sharply. After the rainfall, the moisture content of each measuring point decreased slowly due to evaporation, so the suction increased slowly and finally tended to be stable. Notably, the Campbell 257 soil suction sensor uses an indirect-method suction sensor (with a measuring range of 200 kPa), but the air intake value of soil material is small, so when the suction of the soil sample is lower than 10 kPa, the measurement accuracy of the sensor is poor.

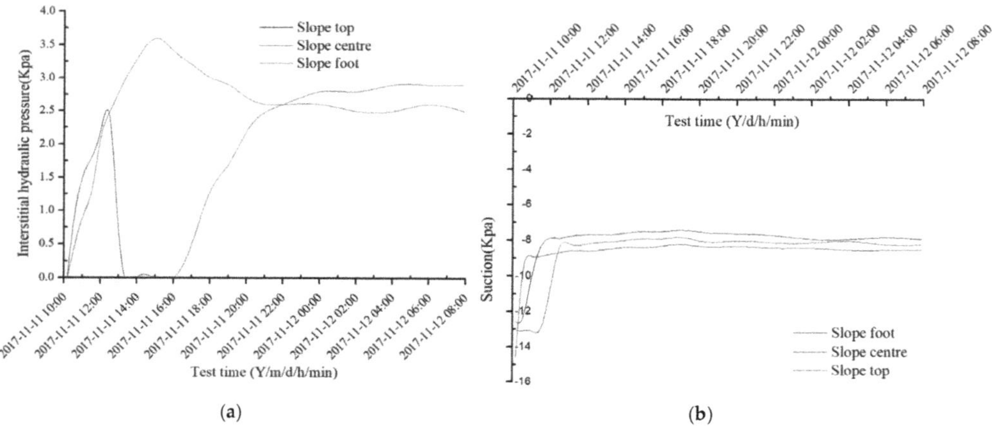

Figure 12. Variation of pore water pressure and soil water suction of feature points with time; (**a**) Variation of pore water pressure at top, center, and foot of slope, (**b**) soil water suction at top, center, and foot of slope.

3.3. *Shape Characteristics and Mechanism of Deep Slip Surface*

3.3.1. Shape Characteristics

The displacement of measuring points near the top of slope (A1, A2, A3, A4, and A5) was 0.5–40 mm and the direction was 40–50° to the horizontal; the displacement of measuring points near the slope surface (B5, C6, and D6) was 2.5–15.6 mm and the direction was 60–70° to the horizontal; the displacement of measuring points (E7 and F7) near the slope toe was 1.2–6.0 mm and the direction was 80–85° to the horizontal. The displacement of the deepest measuring points (C1, D1, D2, E1, E2, E3, F1, F2, F3, F4, and F5) was almost 0 mm. Through analysis of the above measurements, we found that the movement of the soil near the slope was the most intense; the displacement was the largest here, as was the displacement change rate, as it was the main active area of soil. With increasing depth of the measuring points, the displacement of the soil became increasingly small. When the depth reached the deepest points, there was no displacement. With the increase in depth, the angle between the displacement of each measuring point and the horizontal direction was increasingly large, and some points were close to 90°. This shows that both horizontal and vertical movements occurred in the soil landslides, but with the increase in depth, the horizontal movement transformed into vertical movement, and finally, at a certain depth, the soil movement was close to vertical movement. The approximate depth of the deep slip surface can be determined from the tracer point with no displacement. If the displacement of B1, D2, and F5 points is 0 mm, the depth of the deep slip surface is approximately 0.35, 0.75, and 1.15 m, respectively. Combining the depth of the deep slip surface obtained by each tracing point moving to 0 mm and the staggered fracture at the trailing edge of the landslide, the position of the deep slip surface can be determined; the slip surface radius was 1.43 m. The shape of the deep slip surface is shown in Figure 13.

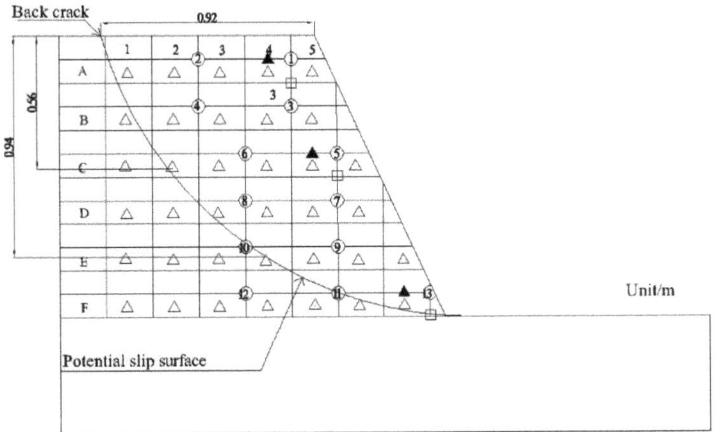

Figure 13. The deep shape of sliding surface. Different letters indicate soil layers; Numbers indicate soil columns; ○ indicates soil water content sensor; □ indicates pore water pressure sensor; △ indicates monitor point of displacement; ▲ indicates soil water suction sensor.

Rainfall is an important factor that causes landslides in terraces, it increases the sliding moment and reduces the anti-skid moment, finally forming a slip surface similar to a circular arc. This is similar to the results of studies by Zhang et al. [36] and Liu et al. [38]. However, the abovementioned studies used numerical simulation methods to study terraced landslides, which are less convincing. Our test explains the mechanism and process of terraced landslides and fully verifies the accuracy of the above-mentioned studies. However, this test only studied a single terrace, and we did not consider landslides on multi-level terraces, which will be the focus and direction of future research. Wu et al. [57] studied the causes of landslides in terraced fields in the loess area caused by over-irrigation, and the landslides at the Heifangtai can be classified into two different types based on their composition: loess landslides and loess-bedrock landslides, characterized by high-speed, long-distance sliding and low-speed, short-distance sliding respectively. Agnoletti et al. [58], taking terraced fields as the research object, explained that the terraced field can better reduce the possibility of shallow landslide disasters relative to slope field, and have less impact on deep landslides under extreme rainfall conditions. This is also consistent with the results of our experimental study, highlighting the guiding significance of this study for terraced landslides in the whole loess hilly and gully area.

3.3.2. Mechanical Mechanism

To study the mechanism and process of deep sliding surfaces in terraces under extreme rainfall, the safety factor of sliding surfaces in terraces was calculated using the Swedish slice method. This method divides the soil above the slip surface into several strips to analyze the force and moment equilibrium on each strip, and to obtain the safety factor of soil stability under the limit equilibrium state [59]. In this experiment, the soil strips above the sliding surface are divided into seven vertical strips according to the location of the sensors. Before solving the safety factor, the following assumptions are made: (i) the force between the strips has little effect on the overall stability, which can be ignored; (ii) the moisture content of each soil strip is the average value of all moisture content sensors on the soil strip; (iii) the cohesive force and internal friction angle of each soil block are used form [60], namely, $c_i = \alpha w_i^{-\beta}$. As shown in Figure 14, according to the equilibrium condition of radial force

$$N_i = W_i cos\alpha_i$$

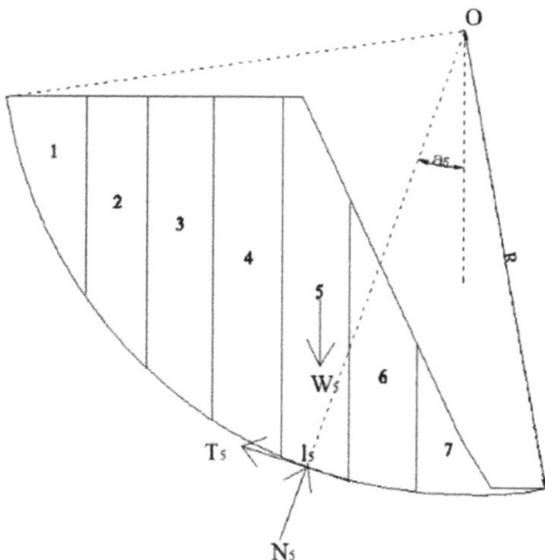

Figure 14. Analysis of the forces of the Swedish slice method.

According to the limit equilibrium condition on the arc surface,

$$T_i = \frac{c_i l_i + N_i tan\varphi_i}{F_s}$$

The anti-sliding moment generated on the sliding surface is,

$$\sum T_i R = \sum \frac{c_i l_i + N_i tan\varphi_i}{F_s} \times R$$

From the moment balance, we can finally obtain

$$F_s = \frac{\sum \alpha w_i^{-\beta} l_i + W_i cos\alpha_i tan\varphi_i}{\sum W_i sin\alpha_i}$$

here α_i is the bottom slope angle of the strip i; W_i is the sum of the self-weight of strip i and the upper load; N_i is the total normal force at the bottom of strip i; T_i is the total tangential resistance of strip i at the bottom; F_s is the safety factor of the sliding arc; c_i is the cohesion of block i; l_i is the bottom length of the block i; φ_i is the internal friction angle of block i; R is the arc radius of the sliding surface; w_i is the soil water content, ($\times$100) α *and* β can be obtained by linear interpolation in Tables 2 and 3.

Table 2. Selection of cohesion parameters of unsaturated loess [61].

Soil Dry Density (g/cm^{-3})	α	β
1.2	37.67	1.602
1.3	42.3	1.615
1.4	79.33	1.782
1.5	108.901	1.795
1.6	56.687	1.503

Table 3. Selection of internal friction angle of unsaturated loess [61].

Soil Dry Density (g/cm^{-3})		1.2	1.3	1.4	1.5	1.6
soil water content	7%	20.8	21	21.1	21.6	22.6
	9%	20.4	20.7	20.9	21.5	22.5
	11%	20.1	20.3	20.7	21.3	22.3
	13%	19.8	20	20.5	21.2	22.1
	15%	19.6	19.6	20.3	20.9	21.9
	17%	19.3	19.4	20.1	20.8	21.6
	19%	18.9	19.2	19.9	20.8	21.4
	21%	18.6	19	19.7	20.5	21.2
	23%	18.2	18.8	19.5	20.4	21.1
	25%	17.9	18.6	19.3	20.2	21
	27%	17.6	18.4	19.1	20.2	20.9
	29%	17.2	18.2	18.9	20	20.8
	31%	16.8	18	18.7	19.9	20.6
	33%	16.5	17.8	18.5	19.8	20.5

Figure 15 shows the variation law of terraces' safety factor with test time. The safety factor first decreased, then increased, and finally slowly increased and tended to be stable. From 10:00 a.m. to 3:00 p.m., with the continuous infiltration of rainfall, the self-weight of the upper soil on the sliding surface increased, which increased the sliding torque. The decrease in cohesion led to the decrease in anti-sliding torque, which was the reason for the decrease in the safety factor. From 3:00 p.m. to 4:00 p.m., due to the sliding torque being greater than the anti-sliding torque, the sliding surface gradually formed, and the time of deep sliding surface began to form earlier than the terraced shallow landslides; 4:00 p.m. to 5:00 p.m., due to the terraces forming a new stable state; the slope was slowed down and the safety factor was larger than the initial stage. From 5:00 p.m. to 9:00 p.m., the change law was similar to that of the initial rainfall, which also showed that rainfall infiltration was the dominant factor leading to safety factor; after 9:00 p.m., the safety factor increased slowly and tended to be stable, because after the rainfall, the soil inside the terraces slowly dried and evaporated naturally, so that the soil moisture content decreased slowly and tended to be stable. The mechanical mechanism of deep slip surfaces in terraces was that through the sliding moment of the sliding body being greater than the anti-sliding moment. The formation of a deep sliding surface in a terraced slope was mainly the result of the interaction of hydraulic erosion and gravity erosion, due to rainfall infiltration, soil moisture content increase, pore water pressure increase, and suction decrease. It increased the bulk density of the sliding body, thereby increasing the sliding torque; however, rainfall infiltration reducing the cohesion and internal friction angle of the sliding body, thereby reduced the anti-sliding torque. With the continuous rainfall, the anti-sliding torque was equal to the sliding torque at a certain time, and the terraced slope was in the limit equilibrium state. The sliding surface began to develop and form from this moment, and the development of the erosion gully accelerated the formation process of the sliding surface. In this study, the rainfall threshold for deep landslides in terraces was 500 mm, which is similar to the results of Zhuang et al. [3], and provides data support for landslides in loess hilly and gully areas.

3.3.3. Variation in Characteristic Points Displacement with Accumulated Rainfall

Figure 16 shows the relation curve between the displacement of three characteristic monitoring points (A5, C6, and F7) and the accumulated rainfall. With the increase in accumulated rainfall, the soil displacement gradually increased, with the largest displacement occurring at the top of the slope, the second largest at the slope center, and the smallest at the foot of the slope. At around 2:00 p.m. on 11 November, the displacement increased sharply. After the collapse, the displacement increased slowly and remained unchanged at 8:00 a.m. on 12 November. The displacements of the top, center and foot of the slope were 40.3 mm, 15.6 mm, and 6.0 mm, respectively.

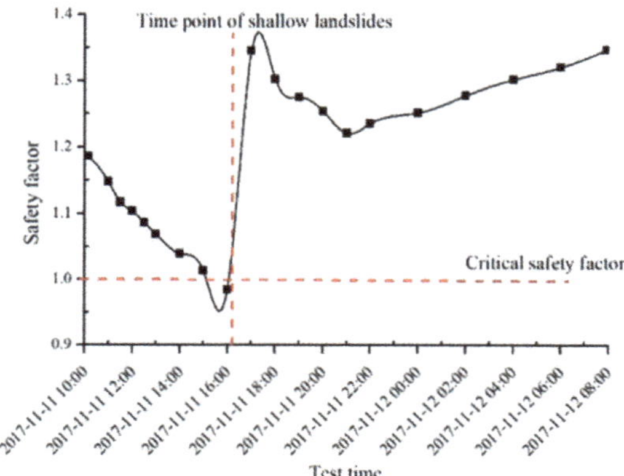

Figure 15. The changes of safety factor of terraces with test time.

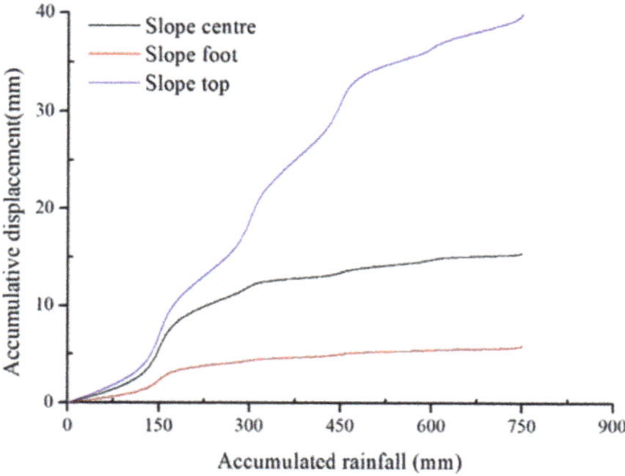

Figure 16. Displacement of characteristic points and its comparison with accumulative rainfall.

The variation law of the slope top, the middle of the slope, and the characteristic points of the slope foot was similar to that reported by Chen et al. [62], but the displacement was generally larger than the above research because of the high rainfall intensity and the long rainfall duration. In future studies, we must examine the impact of different rainfall intensities and total rainfall on terraced landslides.

4. Summary and Conclusions

In this paper, where we selected newly built bare-land terraces as the research object, the laboratory model test method was used to study the mechanism and process of shallow landslides and deep slip surface in terraces under extreme rainfall conditions. The main conclusions were as follows.

Shallow landslides in terraces are formed under the interaction of water erosion and gravity erosion, and the main driving force is from headward erosion and rill erosion. The superficial-layer shallow landslides of the terraces under the action of extreme rainfall can be divided into six stages. The width of the erosion ditch on the terraced slope was 17.8 cm and the volume of shallow landslides in terraces was 0.24 m^3.

The mechanism of the slip surface in the terraces is rainfall infiltration, which increases the water content of the soil, increases the pore water pressure, and decreases the suction force, which leads to the anti-slip torque being less than the sliding torque, causing a slip surface to appear inside the terrace. Under extreme rainfall conditions, terraces formed a circular sliding surface with a radius of 1.43 m and an angle of 92°. The appearance of this slip surface was earlier than the appearance of shallow landslides in terraces, and rill erosion accelerated the formation of deep slip surfaces. The threshold of rainfall that caused deep landslides in terraces was 500 mm.

Author Contributions: P.G. and X.M. conceived and designed the research theme. Y.W. and P.G. collected the data and designed methods. Y.W. analyzed the data and interpreted the results. Y.W. and P.G. wrote and edited the paper. M.L.; Y.S.; H.W. processed the images and forms. All authors have read and agreed to the published version of the manuscript.

Funding: This research was funded by the National Key Scientific Research Project (Grant No. 2016YFC0501707).

Institutional Review Board Statement: Not applicable.

Informed Consent Statement: Not applicable.

Data Availability Statement: Data was contained within the article.

Acknowledgments: We thank the editors of the journal and the reviewers for their useful comments and suggestions to improve the paper quality greatly. Special thanks to Guangju Zhao and Wenyi Sun from Northwest A&F University, State Key Lab Soil Eros & Dryland Farming Loess P, China and Chinese Acad Sci & Minist Water Resources, Inst Soil & Water Conservat provided valuable feedback on an earlier version of this manuscript.

Conflicts of Interest: The authors declare no conflict of interest.

References

1. Wei, W.; Chen, D.; Wang, L.; Daryanto, S.; Li, J.; Yu, Y.; Lu, Y.; Sun, G.; Feng, T. Global synthesis of the classifications, distributions, benefits and issues of terracing. *Earth Sci. Rev.* **2016**, *159*, 388–403. [CrossRef]
2. Li, Y.; Lindstrom, M.J. Evaluating Soil Quality-Soil Redistribution Relationship on Terraces and Steep Hillslope. *Soil Sci. Soc. Am. J.* **2001**, *65*, 1500–1508. [CrossRef]
3. Zhuang, J.; Peng, J.; Wang, G.; Iqbal, J.; Wang, Y.; Li, W.; Xu, Q.; Zhu, X. Prediction of rainfall-induced shallow landslides in the Loess Plateau, Yan'an, China, using the TRIGRS model. *Earth Surf. Process. Landf.* **2017**, *42*, 915–927. [CrossRef]
4. Van Dijk, A.I.J.M.; Bruijnzeel, L.A. Runoff and soil loss from bench terraces. 1. An event-based model of rainfall infiltration and surface runoff. *Eur. J. Soil Sci.* **2004**, *55*, 299–316. [CrossRef]
5. Fan, J.; Wang, Q.J.; Scott, B.J. Soil water depletion and recharge under different land cover in China's Loess Plateau. *Ecohydrology* **2015**, *9*, 396–406. [CrossRef]
6. Dijk, A.I.J.M.; Bruijnzeel, L.A. Terrace erosion and sediment transport model: A new tool for soil conservation planning in bench-terraced steep lands. *Software* **2003**, *18*, 839–850.
7. Rodrigo-Comino, J.; Seeger, M.; Iserloh, T.; González, J.M.S.; Ruiz-Sinoga, J.D.; Ries, J.B. Rainfall-simulated quantification of initial soil erosion processes in sloping and poorly maintained terraced vineyards—Key issues for sustainable management systems. *Sci. Total Environ.* **2019**, *660*, 1047–1057. [CrossRef] [PubMed]
8. Wen, Y.F.; Gao, P.; Mu, X.M. Response of soil erosion to rainfall intensity in terraced slope in the Loess Plateau. *J. Sediment Res.* **2017**, *6*, 46–51. (In Chinese)
9. Shi, Q.; Wang, W.; Guo, M.; Chen, Z.; Feng, L.; Zhao, M.; Xiao, H. The impact of flow discharge on the hydraulic characteristics of headcut erosion processes in the gully region of the Loess Plateau. *Hydrol. Process.* **2020**, *34*, 718–729. [CrossRef]
10. Tony, L.T.; Zhan, G.W.; Jia, Y.M. An analytical solution for rainfall infiltration into an unsaturated infinite slope and its application to slope stability analysis. *Int. J. Numer. Anal. Methods Geomech.* **2012**, *37*, 1737–1760.
11. Jiao, J.Y.; Wang, W.Z. Quality and soil-water conservation effectiveness of level terrace on the Loess Plateau. *Trans. CSAE* **1999**, *2*, 59–63. (In Chinese)
12. Bandara, S.; Ferrari, A.; Laloui, L. Modelling landslides in unsaturated slopes sub-jected to rainfall infiltration using ma-terial point method. *Int. J. Numer. Anal. Methods Geomech.* **2016**, *40*, 1358–1380. [CrossRef]
13. Ma, K.-C.; Tan, Y.-C.; Chen, C.-H. The influence of water retention curve hysteresis on the stability of unsaturated soil slopes. *Hydrol. Process.* **2011**, *25*, 3563–3574. [CrossRef]
14. Wang, J.D.; Gu, T.F.; Zhang, M.S. Experimental study of loess disintegration charac-teristics. *Earth Surf. Process. Landf.* **2019**, *44*, 1317–1329. [CrossRef]
15. Garcia, E.; Oka, F.; Kimoto, S. Numerical analysis of a one-dimensional infiltration problem in unsaturated soil by a seepage-deformation coupled method. *Int. J. Numer. Anal. Methods Geomech.* **2011**, *35*, 544–568. [CrossRef]

16. Ivanov, V.I.; Arosio, D.; Tresoldi, G.; Hojat, A.; Zanzi, L.; Papini, M.; Longoni, L. Investigation on the Role of Water for the Stability of Shallow Landslides—Insights from Experimental Tests. *Water* **2020**, *12*, 1203. [CrossRef]
17. Bui, D.T.; Pradhan, B.; Lofman, O.; Revhaug, I.; Dick, Ø.B. Regional prediction of landslide hazard using probability analysis of intense rainfall in the Hoa Binh province, Vietnam. *Nat. Hazards* **2012**, *66*, 707–730. [CrossRef]
18. Vonstorch, H.; Zorita, E.; Cubasch, U. Downscaling of global cli-mate-change estimates to regional scales—An application to iberian rainfall in wintertime. *J. Clim.* **1993**, *6*, 1161–1171. [CrossRef]
19. Lin, H.Z.; Yu, Y.Z.; Li, G.X. Influence of rainfall characteristics on soil slope failure. *J. Rock Mech. Eng.* **2009**, *1*, 198–294. (In Chinese)
20. Zuo, Z.B.; Zhang, L.L.; Wang, J.H. Model tests on rainfall-induced colluvium land-slides: Effects of particle-size distribution. *J. Geotech. Eng.* **2015**, *7*, 1319–1327. (In Chinese)
21. Li, H.Q.; Sun, H.Y.; Sun, X.M. Influence of rainfall infiltration on slopes by physical model test. *J. Geotech. Eng.* **2009**, *3*, 589–594. (In Chinese)
22. Jeong, S.; Lee, K.; Kim, J.; Kim, Y. Analysis of Rainfall-Induced Landslide on Unsaturated Soil Slopes. *Sustainability* **2017**, *9*, 1280. [CrossRef]
23. Aleotti, P. A warning system for rainfall-induced shallow failures. *Eng. Geol.* **2004**, *73*, 247–265. [CrossRef]
24. Xu, G.M.; Wang, G.L.; Gu, X.W. Centrifuge modeling for instability of excavated slope inexpansive soil due to water in-filtration. *J. Geotech. Eng.* **2006**, *2*, 270–273. (In Chinese)
25. Wang, G.; Sassa, K. Factors affecting rainfall-induced flowslides in laboratory flume tests. *Géotechnique* **2001**, *7*, 587–599. [CrossRef]
26. Tohari, A.; Nishigaki, M.; Komatsu, M. Laboratory Rainfall-Induced Slope Failure with Moisture Content Measurement. *J. Geotech. Geoenvironmental Eng.* **2007**, *133*, 575–587. [CrossRef]
27. Huang, C.-C.; Lo, C.-L.; Jang, J.-S.; Hwu, L.-K. Internal soil moisture response to rainfall-induced slope failures and debris discharge. *Eng. Geol.* **2008**, *101*, 134–145. [CrossRef]
28. Yilmaz, I.; Karacan, E. A Landslide in Clayey Soils: An Example from the Kızıldag Region of the Sivas-Erzincan Highway (Sivas-Turkey). *Environ. Geosci.* **2008**, *9*, 35–42. [CrossRef]
29. Benn, J.L. Landslide events on the West Coast, South Island, 1867–2002. *N. Z. Geogr.* **2005**, *61*, 3–13. [CrossRef]
30. Yang, Q.; Zhao, Z.; Benoy, G.; Chow, T.L.; Rees, H.W.; Bourque, C.P.-A.; Meng, F.-R. A Watershed-scale Assessment of Cost-Effectiveness of Sediment Abatement with Flow Diversion Terraces. *J. Environ. Qual.* **2010**, *39*, 220–227. [CrossRef]
31. Li, X.H.; Yang, J.; Zhao, C.Y. Runoff and sediment from orchard terraces in southeastern china. *Land Degrad. Dev.* **2012**, *25*, 184–192. [CrossRef]
32. Li, H.J.; Gao, J.E.; Zhang, Y.X. Analysis of Yan'an extreme rainfall characteristics and impacts of erosion disasters on terraces. *J. Soil Water Conserv.* **2016**, *6*, 79–84. (In Chinese)
33. Stavi, I.; Rozenberg, T.; Al Ashhab, A.; Argaman, E.; Groner, E. Failure and Collapse of Ancient Agricultural Stone Terraces: On-Site Effects on Soil and Vegetation. *Water* **2018**, *10*, 1400. [CrossRef]
34. Jiang, D.S. The design of the cross-sections of terrace in the loess plateau of china. *J. Soil Water Conserv.* **1987**, *02*, 28–36. (In Chinese)
35. Ge, Y.Q. Analysis of the stability of terrace soil edge in sandy mountains. *Soil Water Conserv.* **1999**, *7*, 32–33. (In Chinese)
36. Zhang, J.; Han, T.C.; Dou, H.Q.; Ma, S.G. Research on slope stability of terrace based on crop irrigation infiltration. *J. Sichuan Univ.* **2014**, *S1*, 79–85. (In Chinese)
37. Yang, J.; Li, G.L.; Wei, Z. Teat on expansion ratio and mechanical property of terrace soil ridges of southern shaanxi. *Soil Water Conserv.* **2015**, *5*, 40–43. (In Chinese)
38. Liu, J.L.; Tian, J.; Zheng, T.T. Optimized design of loess terrace based on slope stability. *Sci. Soil Water Conserv.* **2020**, *4*, 21–28. (In Chinese)
39. Derbyshire, E. Geological hazards in loess terrain, with particular reference to the loess regions of China. *Earth Sci. Rev.* **2001**, *1–3*, 253–266. [CrossRef]
40. Chen, J.; Wang, Y.j.; Chen, Y. Rb and Sr Geochemical Characterization of the Chinese Loess Stratigraphy and Its Implications for Palaeomonsoon Climate. *Acta Geol. Sin. Engl. Ed.* **2010**, *74*, 279–288.
41. Zheng, F.-L.; Huang, C.-H.; Norton, L.D. Vertical Hydraulic Gradient and Run-On Water and Sediment Effects on Erosion Processes and Sediment Regimes. *Soil Sci. Soc. Am. J.* **2000**, *64*, 4–11. [CrossRef]
42. Li, X.; Wang, L.; Hong, B.; Li, L.; Liu, J.; Lei, H. Erosion characteristics of loess tunnels on the Loess Plateau: A field investigation and experimental study. *Earth Surf. Process. Landf.* **2020**, *45*, 1945–1958. [CrossRef]
43. Wu, H.; Xu, X.; Zheng, F.; Qin, C.; He, X. Gully morphological characteristics in the loess hilly-gully region based on 3D laser scanning technique. *Earth Surf. Process. Landf.* **2018**, *43*, 1701–1710. [CrossRef]
44. Shen, N.; Wang, Z.-L.; Zhang, Q.; Wu, B.; Liu, J.; Nan, S. Modelling the process of soil detachment by rill flow on steep loessial hillslopes. *Earth Surf. Process. Landf.* **2020**, *45*, 1240–1247. [CrossRef]
45. Han, P.; Ni, J.R.; Li, T.H. Headcut and bank landslip in rill evolution. *J. Basic Sci. Eng.* **2002**, *2*, 115–125. (In Chinese)
46. Acharya, G.; Cochrane, T.A. Rainfall induced shallow landslides on sandy soil and impacts on sediment discharge: A flume based investigation. In Proceedings of the 12th Conference of International Association for Computer Methods and Advances in Geomechanics, Goa, India, 1–6 October 2008.
47. Zhong, Q.M.; Chen, S.S.; Mei, S.A. Breach Mechanism and Breach Process Simulation of Homogeneous Cohesive Earthen Dam Due to Overtopping. *Adv. Eng. Sci.* **2019**, *51*, 25–32. (In Chinese)

48. Morris, M.W.; Hassan, M.A.A.M.; Vaskinn, K.A. Breach formation: Field test and la-boratory experiments. *J. Hydraul. Res.* **2007**, *45*, 9–17. [CrossRef]
49. Hanson, G.J.; Cook, K.R.; Hunt, S.L. Physical modeling of overtopping erosion and breach formation of cohesive embankments. *Trans. ASAE* **2005**, *48*, 1783–1794. [CrossRef]
50. Sun, P.; Wang, G.; Wu, L.Z. Physical model experiments for shallow failure in rain-fall-triggered loess slope, Northwest China. *Bull. Eng. Geol. Environ.* **2019**, *78*, 4363–4382. [CrossRef]
51. Yuan, J.P.; Lei, T.W.; Guo, S.Y. Study on spatial variation of infiltration rates for small watershed in loess plateau. *J. Hydraul. Eng.* **2001**, *10*, 88–92. (In Chinese)
52. Liu, X.Y.; Wang, F.G.; Yang, S.T. Study on harmony equilibrium between water re-sources and economic society development. *J. Hydraul. Eng.* **2014**, *45*, 793–800. (In Chinese)
53. Huang, J.; Wu, P.; Zhao, X. Effects of rainfall intensity, underlying surface and slope gradient on soil infiltration under simulated rainfall experiments. *Catena* **2013**, *104*, 93–102. [CrossRef]
54. Cui, G.; Zhu, J. Infiltration Model Based on Traveling Characteristics of Wetting Front. *Soil Sci. Soc. Am. J.* **2018**, *82*, 45–55. [CrossRef]
55. Tian, S. Indoor Model Test Study on Fill Slope under Earthquake and Rainfull. *Pearl River* **2019**, *40*, 61–70. (In Chinese)
56. Bai, J.; Yang, S.; Zhang, Y.; Liu, X.; Guan, Y. Assessing the Impact of Terraces and Vegetation on Runoff and Sediment Routing Using the Time-Area Method in the Chinese Loess Plateau. *Water* **2019**, *11*, 803. [CrossRef]
57. Wu, W.J.; Su, X.; Meng, X.M. Characteristics and Origin of Loess Landslides on Loess Terraces at Heifangtai, Gansu Province, China. *Appl. Mech. Mater.* **2014**, *694*, 455–461. [CrossRef]
58. Agnoletti, M.; Errico, A.; Santoro, A.; Dani, A.; Preti, F. Terraced Landscapes and Hydrogeological Risk. Effects of Land Abandonment in Cinque Terre (Italy) during Severe Rainfall Events. *Sustainability* **2019**, *11*, 235. [CrossRef]
59. Baker, R. Determination of the critical slip surface in slope stability computations. *Int. J. Numer. Anal. Methods Geomech.* **1980**, *4*, 333–359. [CrossRef]
60. Dang, J.Q.; Li, J. The Influence of Water Content on the Strength of Unsaturated Loess. *J. Northwest Agric. Univ.* **1996**, *1*, 57–60.
61. Dang, J.Q.; Li, J. The structural strength and shear strength of unsaturated loess. *Shuili Xuebo* **2001**, *7*, 0079-05. (In Chinese)
62. Chen, W.; Luo, Y.S.; Wu, C.P. The Laboratory Model Test Sdudy of Loess Slope uder the Artificial Rainfall. *China Rural Water Hydropower* **2013**, *5*, 100–104. (In Chinese)

Article

Spatiotemporal Hotspots and Decadal Evolution of Extreme Rainfall-Induced Landslides: Case Studies in Southern Taiwan

Chunhung Wu * and Chengyi Lin

Department of Water Resources Engineering and Conservation, Feng Chia University, Taichung 40724, Taiwan; jknokiajk88@yahoo.com.tw
* Correspondence: chhuwu@fcu.edu.tw; Tel.: +886-424517250-3223

Abstract: The 2009 Typhoon Morakot triggered numerous landslides in southern Taiwan, and the landslide ratios in the Ailiao and Tamali river watershed were 7.6% and 10.7%, respectively. The sediment yields from the numerous landslides that were deposited in the gullies and narrow reaches upstream of Ailiao and Tamali river watersheds dominated the landslide recovery and evolution from 2010 to 2015. Rainfall records and annual landslide inventories from 2005 to 2015 were used to analyze the landslide evolution and identify the landslide hotspots. The landslide recovery time in the Ailiao and Tamali river watershed after 2009 Typhoon Morakot was estimated as 5 years after 2009 Typhoon Morakot. The landslide was easily induced, enlarged, or difficult to recover during the oscillating period, particularly in the sub-watersheds, with a landslide ratio > 4.4%. The return period threshold of rainfall-induced landslides during the landslide recovery period was <2 years, and the landslide types of the new or enlarged landslide were the bank-erosion landslide, headwater landslide, and the reoccurrence of old landslide. The landslide hotspot areas in the Ailiao and Tamali river watershed were 2.67–2.88 times larger after the 2009 Typhoon Morakot using the emerging hot spot analysis, and most of the new or enlarged landslide cases were identified into the oscillating or sporadic or consecutive landslide hotspots. The results can contribute to developing strategies of watershed management in watersheds with a dense landslide.

Keywords: landslide evolution; spatiotemporal cluster analysis; landslide hotspots

Citation: Wu, C.; Lin, C. Spatiotemporal Hotspots and Decadal Evolution of Extreme Rainfall-Induced Landslides: Case Studies in Southern Taiwan. *Water* **2021**, *13*, 2090. https://doi.org/10.3390/w13152090

Academic Editor: Su-Chin Chen

Received: 29 June 2021
Accepted: 27 July 2021
Published: 30 July 2021

1. Introduction

Landslides induced by large earthquakes or extreme rainfall events have been the main reason for disasters in the past two decades in Taiwan. Typhoon Morakot in 2009 dumped around 2000 mm of rainfall in 3 days in southern Taiwan [1], resulting in severe landslide-related disasters, including the catastrophic deep-seated Xiaolin landslide [2] and the following dam failure [3]. Over a decade since the 2009 Typhoon Morakot, sediment-related disaster events still occurred in the Kaoping River watershed in southern Taiwan. Although most landslides in southern Taiwan had been gradually recovered, the hillslope was still under high landslide susceptibility.

The rate and location of landslide recovery after the large earthquake or extreme rainfall events play essential roles in developing the watershed management strategies for watersheds with a dense landslide. The landslide recovery in the watersheds with dense landslides after large earthquake events is related to the earthquake magnitude, geological settings, and fault distribution and characteristics [4–6], while recovery after extreme rainfall events were mostly related to the distribution of drainage network [7]. The sediment yield from landslides or debris flow in the watersheds with dense landslides is usually the dominant factor behind the geomorphologic evolution, particularly in the upstream watershed. The randomly deposited sediment in narrow upstream reaches usually results in rivers gradually becoming sinuous and inducing bank-erosion landslides. Sediment from bank-erosion landslides usually increases the sinuosity of narrow reaches and changes the geomorphology of the river in the upstream watershed.

Long-term geomorphologic landslide evolution in watersheds is strongly related to spatiotemporal landslide distribution [8], which can be observed using the spatiotemporal cluster analysis with the high-resolution digital elevation model (DEM) and multi-annual landslide inventories [4,6]. Several researchers have discussed the changes in the distribution and activeness of landslides after extreme rainfall-induced [7] or earthquake-induced [5,8–11] events and found that the spatiotemporal distribution and activeness of landslides were key factors behind the geomorphologic evolution of watersheds. Identifying landslide hotspots and cold spots using multi-annual landslide inventories can help researchers analyze landslide activeness and recovery after large earthquake-induced landslide disasters [12].

The space-time cluster analysis (abbreviated as spatiotemporal cluster analysis) in ArcGIS Pro software [13] is a useful analysis tool that can describe data's spatial and temporal distribution patterns. This tool had been used to analyze the spread of the COVID-19 virus [14], road traffic accident occurrences [15], and the spread of air pollution [4,16] in recent years. Landslide disaster studies using the spatiotemporal cluster analysis have focused on discussing the long-term spatiotemporal distribution of disasters [5,8] and analyzing the relationship between disaster occurrence and related factors [6,9,17,18]. Spatiotemporal cluster analysis with multi-annual landslide inventories after extreme rainfall events can contribute to determining landslide hotspots and cold spots, identify locations where the landslide recovery was difficult, and analyze the reasons behind these factors. The use of spatiotemporal cluster analysis to observe landslide evolution trends and identify landslide clustering locations is more effective than only the spatial or temporal analysis of landslides.

The 2009 Typhoon Morakot (from 6–10 August 2009) caused the most severe rainfall-induced disaster event in the past two decades in Taiwan, and the return period accumulated 24 and 48 h of rainfall during the 2009 Typhoon Morakot in southern Taiwan exceeded 200 y [1]. The extreme rainfall event also caused numerous landslides and severe debris flow in southern Taiwan, and the landslide ratio (i.e., the ratio of the landslide area to watershed area) in the four sub-watersheds of the Kaoping River watershed after the typhoon exceeded 6.5% [1]. The geomorphologic evolution and developing trends of watersheds with dense landslides after 2009 Typhoon Morakot (abbreviated as after 2009) in southern Taiwan are worthy of discussion. The Ailiao river watershed (abbreviated as *ARW*) and Tamali river watershed (abbreviated as *TRW*) were the watersheds with the highest landslide ratio in southeastern and southwestern Taiwan after 2009. The *ARW* and *TRW* were selected to observe the landslide evolution from 2005 to 2015 and identify the landslide hotspots and cold spots using the spatiotemporal cluster analysis. The evolution characteristic of extreme rainfall-induced landslide events in Taiwan was also compared with that of large earthquake-induced landslide events in the world, and the cluster location and reason of new or enlarged landslides in the following years after 2009 were analyzed in the study.

2. Research Areas

2.1. Ailiao River Watershed (ARW)

The Ailiao river watershed (abbreviated as *ARW*) is located in southwestern Taiwan (Figures 1 and 2), and the area is 623.3 km^2. The average elevation and average slope in the *ARW* are 1006 m and 30.5°. The average annual precipitation is 3716 mm based on the records of six rainfall stations from 2005 to 2015 in the neighborhood of *ARW* (Figure 2a). The average precipitation in the rainy seasons, i.e., from May to October, occupies > 90% of the average annual precipitation. The land use distribution in the *ARW* is dominated by forest, which occupies 80.8% of the total watershed area. The main geological settings in the *ARW* consist of the Chaochou Formation, the Pilushan Formation, the Alluvium, and the Kaoling Schist (Figure 2b). The total precipitation during the 2009 Typhoon Morakot in the *ARW* was 2977 mm, i.e., around 80% of the average annual precipitation. The 2995 landslide cases (Figure 2a) were induced by the 2009

Typhoon Morakot in the *ARW*, and the landslide ratio, i.e., the ratio of the landslide area to the watershed area, was estimated as 7.6%. The landslides after 2009 centralized in the northeast *ARW*, especially in the A01 (8.2 km^2), A02 (6.7 km^2), A03 (2.5 km^2), A07 (3.2 km^2), and A11 (9.5 km^2) sub-watersheds (Figure 2b). The occupied percentage of the landslide cases with area > 100,000 m^2, 1000–100,000 m^2, and <1000 m^2 to all landslide cases in 2009 in the *ARW* were 3.5%, 73.0%, and 23.0%, respectively. The relation between the landslide length to width ratio and the mean slope in the *ARW* is shown in Figure 3; 93.1% and 57.6% of the landslide cases in 2009 in the *ARW* were of the landslide length to width ratio > 1.0 and ranged from 1.0 to 5.0. The rainfall-triggered slides, including the rotational and translational slides and flows on the hillslope with the slope > 30 degree, were the main landslide types in the *ARW*.

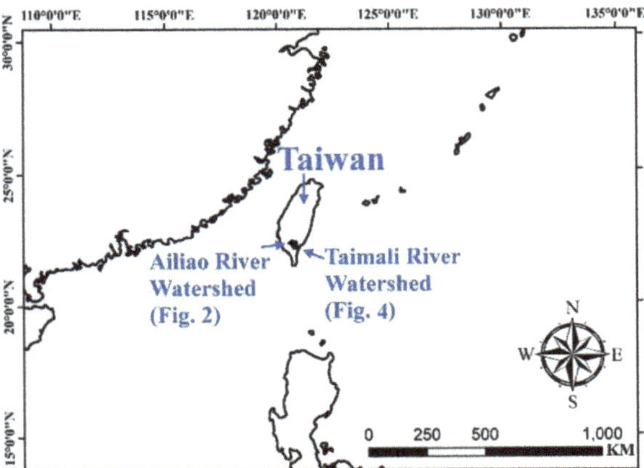

Figure 1. Location of Taiwan, Ailiao river watershed (abbreviated as ARW), and Taimali river watershed (abbreviated as TRW).

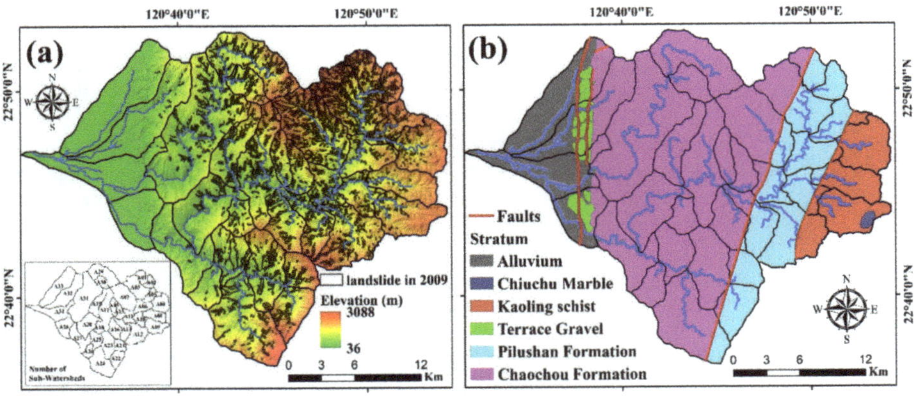

Figure 2. The distribution of elevation, landslide after 2009 Typhoon Morakot, and sub-watersheds (**a**), geological settings (**b**), in the *ARW*.

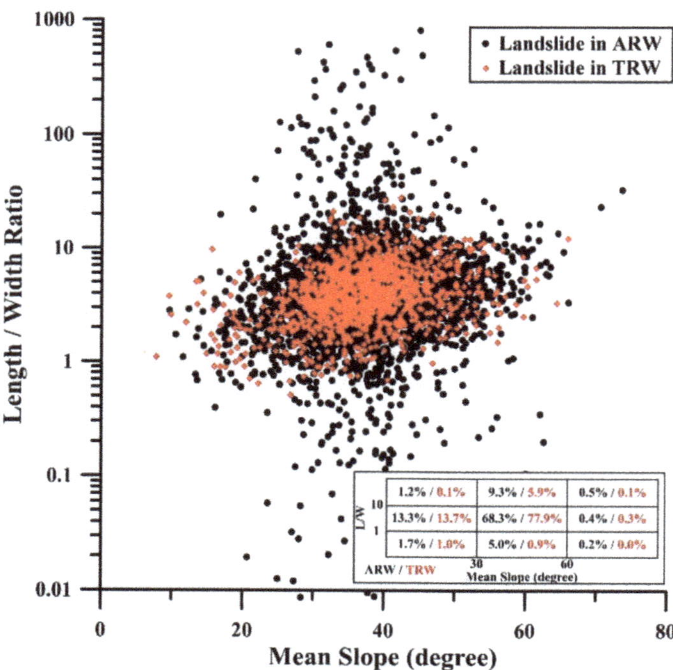

Figure 3. Relationship between the ratio of landslide length to width and mean slope of the landslide cases induced by 2009 Typhoon Morakot in the *ARW* and *TRW*.

2.2. *Taimali River Watershed (TRW)*

The Taimali River Watershed (abbreviated as *TRW*) is located in southeastern Taiwan (Figures 1 and 4) and the area is 264.9 km^2. The average elevation and slope in the *TRW* are 789.4 m and 30.4°, respectively. The average annual precipitation is 2185 mm based on the records of five rainfall stations from 2005 to 2015 in the neighborhood of *TRW* (Figure 4a). The average precipitation in the rainy seasons, i.e., from May to October, occupies 76% of the average annual precipitation. The land use distribution in the *TRW* consists of forest (81.59%), agricultural land (9.12%), water conservancy (4.32%), and others. The main geological settings in the *TRW* consist of three main strata, including the Chaochou Formation, the Pilushan Formation, and the Alluvium (Figure 4b). The total precipitation during the 2009 Typhoon Morakot in the *TRW* was 932.5 mm, i.e., around 42.7% of the average annual precipitation. The 1283 landslide cases (Figure 4a) were induced by 2009.

Typhoon Morakot in the *TRW*, and the landslide ratio was estimated as 10.7%. The landslide after 2009 centralized in the upstream *TRW*, especially in the T01 sub-watershed (121.6 km^2). The occupied percentage of the landslide cases with area > 100,000 m^2, 1000–100,000 m^2, and <1000 m^2 to all landslide cases in 2009 in the TRW were 4.2%, 71.2%, and 24.6%, respectively. The relation between the landslide length to width ratio and the mean slope in the TRW is shown in Figure 3; 98.1% and 64.3% of the landslide cases in 2009 in the TRW were of the landslide length to width ratio > 1.0 and ranged from 1.0 to 5.0. These data show that the majority landslide type of the landslide cases induced by the 2009 Typhoon Morakot in the TRW were rainfall-triggered slides on the steep slope.

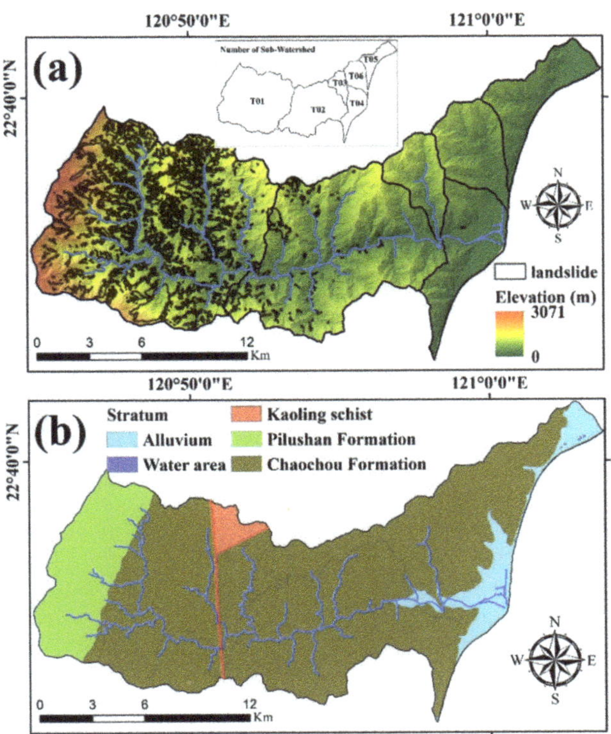

Figure 4. The distribution of elevation, landslide after 2009 Typhoon Morakot, and sub-watersheds (**a**), geological settings (**b**) in the TRW.

3. Data and Methods

3.1. Annual Landslide Inventories

The annual landslide inventories from 2005 to 2015 produced by the Forestry Bureau in Taiwan were used in this study, and the minimum landslide area in the annual landslide inventories was 100 m^2. Based on Varnes' classification [19], the majority of the landslide cases induced by 2009 Typhoon Morakot in southern Taiwan were rotational slides, translational slides, and flows [20–22].

3.2. Effective Accumulated Rainfall Index (EAR)

The *EAR* index (unit: mm) was used to assess the landslide-induced strength of rainfall events. The *EAR* index, defined in Equation (1), is the summation of daily rainfall on the assessment day (R_t) and the 7-day antecedent rainfall before the assessment day. The *K* coefficient, representing the decay constant, was set to 0.7 based on Taiwanese landslide research [23]. Equation (1) is calculated as follows:

$$EAR_t = \sum_{i=0}^{7} R_t \times K^i \quad (1)$$

The rainfall records used to estimate the *EAR* index value were collected from the representative rainfall stations at watersheds. For inclusion, the representative rainfall stations had to be located within the watershed, and the rainfall records from 2005 to 2015 had to be available without any missing data. The representative rainfall stations in the two watersheds are Ali station in the *ARW* and Jinfong station in the *TRW*. The annual landslide inventories were used in this study. It is challenging to find data on the time and date of landslide occurrences and estimate the rainfall threshold to induce the landslide. Rather than determining the precise time and date of landslide occurrences, the *EAR* values used in this study serve as reference coefficients to understand the intensity of landslides induced by typhoons and other heavy rainfall events each year from 2005 to 2015.

3.3. Landslide Topographic Position Analysis

The topographic position analysis method can be used to explain the main inducing factors of landslides [24]. Three parameters of the landslide on the hillslope, including the distance between the ridge and the crown of the landslide (D_P), the distance between the stream and the toe of the landslide (D_B), and the distance between the ridge of the hillslope and river (D_H), are used to explain the relative location of the landslide in the hillslope. The bubble plot is frequently used to draw the result of the topographic position analysis using the normalized distance from a landslide to the ridge (D_P/D_H) as the X-axis, the normalized distance from a landslide to the stream (D_B/D_H) as Y-axis, and the size of the bubble as the landslide area. If the bubbles are located in the upper-left portion of the bubble plot ($D_P/D_H < 0.5$ and $D_B/D_H > 0.5$), the landslide cases are located near the ridge and possibly induced by earthquake events [24]. If the bubbles are located in the lower-left portion of the bubble plot ($D_P/D_H > 0.5$ and $D_B/D_H < 0.5$), the landslide cases are located near the stream and possibly induced by rainfall or flooding events [1,21].

3.4. Spatiotemporal Cluster Analysis Method

We used the emerging hot spot analysis in the space-time cluster analysis tool in the ArcGIS Pro software to analyze the landslide evolution and identify the landslide hotspots and cold spots from 2005 to 2015. The emerging hot spot analysis tool can detect eight hotspot or cold spot trends, and the definition of the eight hot spot or cold spot trends had been described in Table 1 (revised from [14]). The emerging hot spot analysis was widely used in observing the evolution of the natural or artificial phenomenon but has still rarely been used to analyze the landslide evolution. The analysis unit in the study is a 5 m × 5 m grid, and the time step is a year. The clustering intensity of landslide in each analysis unit was estimated using the Getis-Ord Gi statistic [25], which considered the clustering intensity value for each analysis unit within the context of the values for the neighboring analysis unit. In the study, the neighborhood distance of the analysis unit was set as 25 m.

Table 1. The classifications and definition of emerging landslide hot spot and cold spot in the study.

Classification	Definition
Consecutive (CHS or CCS)	A landslide location with a single uninterrupted run of statistically significant hot spot or cold spot areas in the final year during the research time period. The landslide location has never been a statistically significant hot spot or cold spot before the final hot spot or cold spot run.
Diminishing (DHS or DCS)	A landslide location that has been a statistically significant hot spot or cold spot for 90% of the research time period, including the final year. In addition, the clustering intensity of landslide in each year is decreasing (increasing) overall and that decrease (increase) is statistically significant.
Historical (HHS or HCS)	The most recent year is not hot spot or cold spot, but at least 90% of the research time period has been a statistically significant hot spot or cold spot.
Intensifying (IHS or ICS)	A landslide location that has been a statistically significant hot spot or cold spot for 90% of the research time period. In addition, the clustering intensity of landslide for each year increased (decreased) overall and that increase (decrease) was statistically significant.
New (NHS or NCS)	A landslide location identified as a statistically significant hot spot or cold spot since the first year of the research time period but was not previously identified as a statistically significant hot spot or cold spot.
Oscillating (OHS or OCS)	A statistically significant hot spot or cold spot for the final year that has a history of also being a statistically significant cold spot or hot spot during a prior year. Less than 90% of the research time period have been statistically significant hot spot or cold spot.

Table 1. *Cont.*

Classification	Definition
Persistent (PHS or PCS)	A landslide location that has been a statistically significant hot spot or cold spot for 90% of the research time period with no discernible trend indicating an increase or decrease in the clustering intensity of landslide over time.
Sporadic (SHS or SCS)	A landslide location that is an on-again then off-again hotspot or cold spot. Less than 90% of the research time period have been statistically significant hot spot or cold spot, and none of the time-step intervals have been statistically significant colds pot or hot spot.
No pattern detected (No)	The analysis area does not fit any definition of hot spot or cold spot classifications

Note: The CHS and CCS are the abbreviations of consecutive hot spot and consecutive cold spot. The regulation of abbreviation is applied to each hot spot and cold spot in the study.

4. Decadal Analyses Results

4.1. Rainfall Distribution and Landslide Ratio

The *EAR* distributions from 2005 to 2015 in the two watersheds are shown in Figure 5 and Table 2. The average *EAR* value from 2005 to 2015 was 39.7 in the *ARW* and 29.7 in the *TRW*. The highest *EAR* values (EAR_h) in the *ARW* and *TRW* were 1926.9 and 1123.5 on 8 August 2009. The EAR_h and EAR_a (the average of the three highest *EAR* values in each year) from 2005 to 2008 in the two watersheds were larger than those from 2010 to 2015. The return periods of the top ten daily rainfall events from 2005 to 2008 in the two watersheds were estimated to be 10–50 years, and those from 2010 to 2015 were estimated to only be <2 year. The data demonstrated that the *EAR* value and the return periods of rainfall events from 2005 to 2008 in the two watersheds were larger than those from 2010 to 2015.

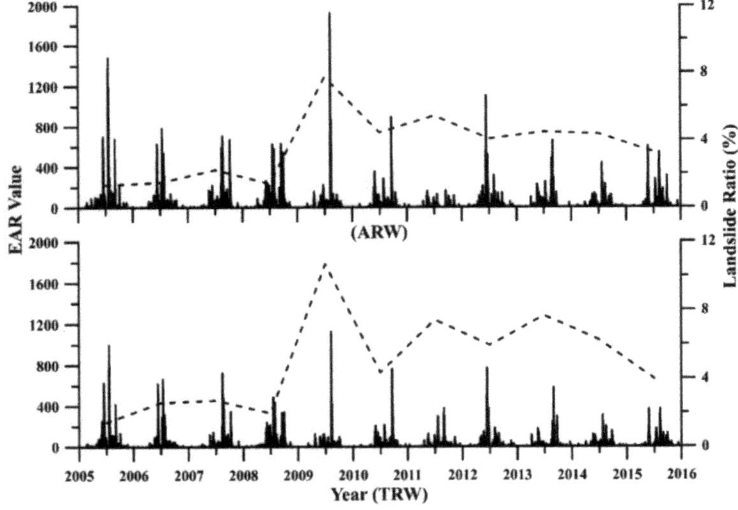

Figure 5. The distribution of effective accumulated rainfall index (abbreviated as EAR) value (black bar) and landslide ratio (dash line) from 2005 to 2015 in the ARW (up figure) and TRW (down figure).

The landslide ratios in 2009 in the *ARW* and *TRW* (Figure 5) were their historical peaks. The average landslide ratios in the *ARW* and *TRW* from 2005 to 2008 were 1.6% and 2.2%, respectively, and those from 2010 to 2015 were 4.3% and 5.9%, respectively. The trends of the landslide ratios in the two watersheds after 2009 were oscillating instead of stably decaying. The EAR_h and EAR_a in 2011 and 2013 in the *ARW* were smaller than those from 2005 to 2007, but the landslide ratio increased in 2011 and 2013. Other similar examples are shown in comparing the EAR_h, EAR_a, and landslide ratios in 2011 and 2013 in the

TRW. This data implied that landslides were more easily induced after the 2009 Typhoon Morakot. The rainfall factor was possibly not the only landslide-inducing factor in the two watersheds after 2009.

Table 2. The statistical data of the *EAR* values from 2005 to 2015 in the two watersheds.

Year	2005	2006	2007	2008	2009	2010	2011	2012	2013	2014	2015
					in the *ARW*						
EAR_h	1481.7	784.8	707.2	631.9	1926.9	894.5	163.5	1104.4	658.5	438.5	609.9
EAR_a	1353.7	700.6	680.9	616.5	1755.8	697.8	150.1	943.6	587.4	340.7	551.1
					in the *TRW*						
EAR_h	997.0	663.4	723.1	485.2	1123.5	759.3	374.6	766.2	576.2	304.8	367.3
EAR_a	856.3	620.6	593.6	449.6	944.0	588.9	332.0	670.3	494.2	250.8	351.0

Note: The EAR_h means the highest *EAR* value, and the EAR_a means the average of the three highest *EAR* values in each year.

4.2. Landslide Statistical Data

The research period was divided into three periods (i.e., 2005–2008, 2009, and 2010–2015) to analyze the changes in landslide distribution before and after 2009. The landslides' statistical data from 2005 to 2015 in the two watersheds are shown in Figure 6 and Table 3. The area and number of landslides from 2010 to 2015 in the two watersheds were larger than those from 2005 to 2008. From 2005 to 2015, the year with the most landslides was 2009, but the year with the most landslide numbers was 2013. In the *ARW*, for example, the landslide area in 2013 was 42% smaller than that in 2009, but the landslide number in 2013 was 31% higher than that in 2009. The same trend was observed in 2013 in the *TRW*. This data implies that most of the landslides induced by 2009 Typhoon Morakot gradually recovered, but some new landslides occurred in the two watersheds in 2013.

Figure 6. The area (solid line) and number (dash line) of landslide in the ARW (black) and TRW (red) from 2005 to 2015.

This study analyzed the landslide distribution at the sub-watershed scale to find the sub-watersheds in which landslides were induced in the years following the 2009 Typhoon Morakot. The landslide evolution trend index (abbreviated as *LET*) in this study was defined as the average change ratio of the landslide area from 2010 to 2015, and the *LET* was estimated in each sub-watershed of the two watersheds (Figure 7). A negative *LET* value indicates that the total landslide area in this sub-watershed gradually decreases, while a positive *LET* value indicates that the total landslide area gradually increases. The average *LET* value in the sub-watersheds was −0.022 and −0.072 km^2/year in the *ARW* and *TRW*.

Table 3. The average area and number of landslides in the two watersheds.

	Year	2005 to 2015	2005 to 2008	2009	2010 to 2015
ARW	Average landslide area (km^2)	22.8	10.1	48.4	27.1
	Average landslide number	1902.8	1132.3	2355	2341.2
TRW	Average landslide area (km^2)	13.3	5.8	28.4	15.8
	Average landslide number	766.3	482.0	1100	900.2

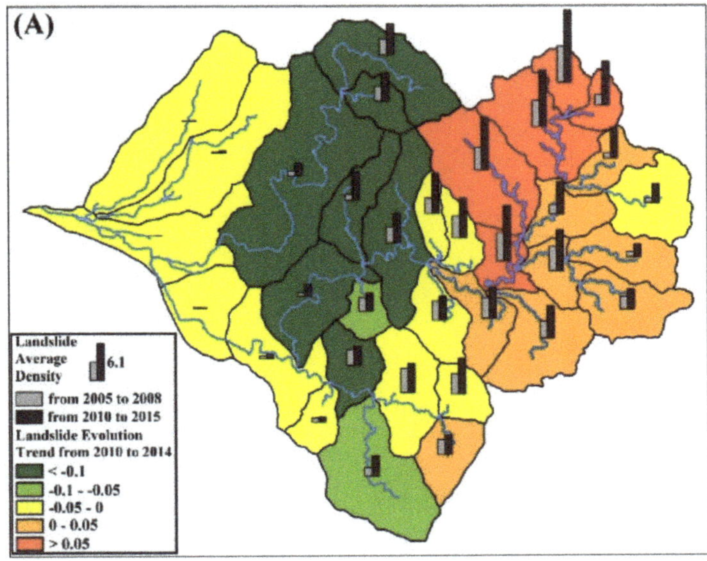

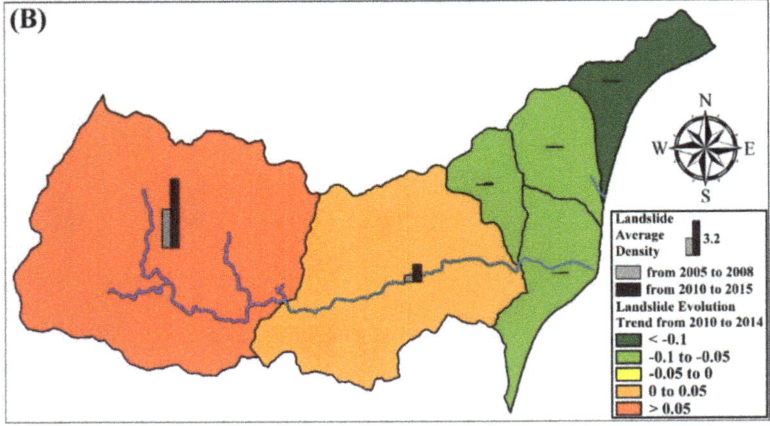

Figure 7. The average landslide density and landslide evolution trend index value from 2005 to 2015 in the *ARW* (**A**) and *TRW* (**B**).

The sub-watersheds with positive *LET* values were located upstream of *ARW* and *TRW*. There were 13 and 2 sub-watersheds with the positive *LET* values in the *ARW* and *TRW*, respectively, and the landslide ratio of the 15 sub-watersheds after 2009 was larger than 4.4%. There were six sub-watersheds with the LET values > 0.05, including A01, A02, A03, A07, and A11 in the *ARW* and T01 in the *TRW*, and the landslide ratio of the six sub-watersheds after 2009 was greater than 12.1%. The watershed areas in the A01, A02, A03, A07, A11, and T01 sub-watersheds were 8.2, 6.7, 24.7, 31.7, 9.5, and 121.6 km^2, respectively, and the landslide ratios after the 2009 Typhoon Morakot were 27.8%, 21.2%, 26.2%, 21.5%, 12.1%, and 20.7%, respectively. These results imply that the landslides in the sub-watersheds with a landslide ratio of >4.4% after 2009 in the *ARW* and *TRW* were difficult to recover and were easily induced or re-induced from 2010 to 2015.

4.3. Landslide Topographic Position Analysis

The study used the landslide topographic position analysis to examine the landslide evolution before and after 2009 in the *ARW* and *TRW*. The A03 (LET = 0.32 km^2/y), A31 (LET = −0.31 km^2/y), and T01 (LET = 0.43 km^2/y) sub-watersheds were selected for comparison of landslide evolution before and after 2009 (Table 4 and Figures 8 and 9). The area in the A31 sub-watershed was 33.9 km^2, and the landslide area and landslide ratio in 2009 in the A31 sub-watershed were 2.8 km^2 and 8.3%. The ratio of landslide area from 2009 to 2015 in the upslope, mid-slope, and downslope were 19.4%, 25.5%, and 38.2%, respectively, in the *ARW* and 27.6%, 29.8%, and 31.1% in the *TRW*, respectively. The landslide located in the downslope was the most difficult to recover from 2009 to 2015 in the slope.

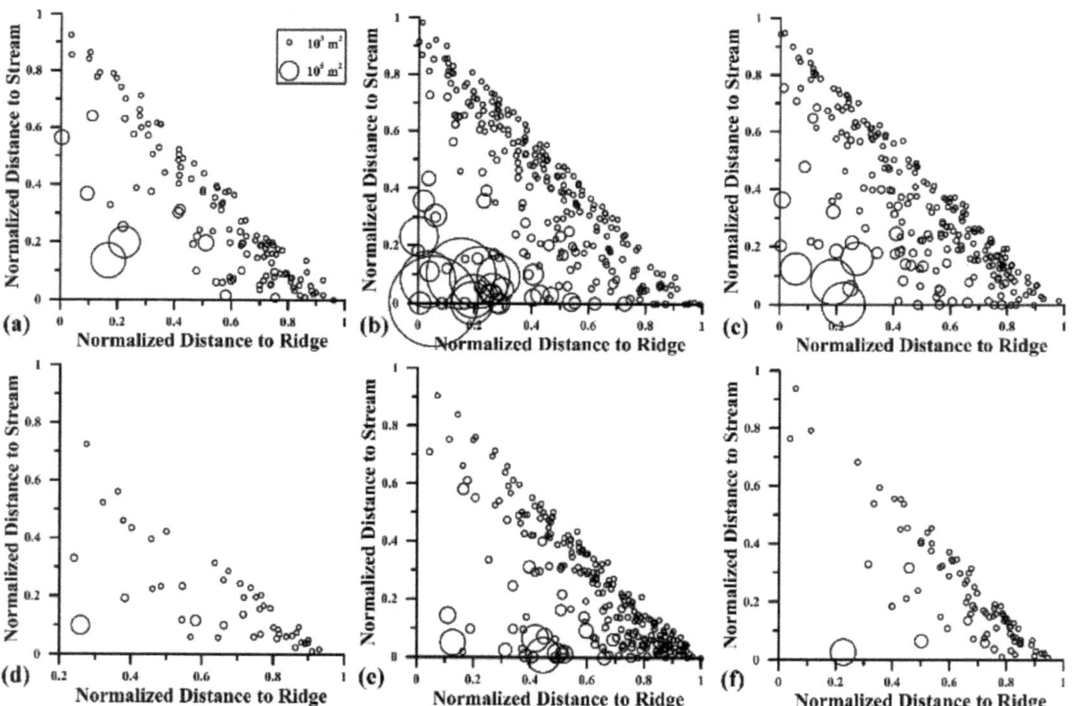

Figure 8. The topographic position analysis of landslide in 2008 (**a–d**), 2009 (**b–e**), and 2015 (**c–f**) in the A03 (up figures) and A31 (down figures) subwatersheds in the *ARW*.

Table 4. The topographic position analysis results in the *ARW* and *TRW*.

Time	B	2009	A	2015	B	2009	A	2015	B	2009	A	2015
	ARW				A03 sub-watershed				A31 sub-watershed			
UA	4.64	13.56	3.69	2.63	0.13	0.29	0.28	0.21	0.01	0.13	0.04	0.02
MA	0.81	2.39	2.20	0.61	0.76	5.27	2.88	1.92	0.15	1.31	0.48	0.21
DA	2.49	12.16	5.37	4.65	0.36	0.74	0.64	0.59	0.24	1.22	0.5	0.28
	TRW				T01 sub-watershed							
UA	2.05	4.78	3.37	1.32	0.09	0.23	0.2	0.15				
MA	0.38	3.19	1.09	0.95	0.45	4.76	0.03	0.01				
DA	1.59	5.92	2.56	1.84	0.37	1.11	0.9	0.76				

Note: "B" and "A" mean that the average landslide area before 2009, i.e., from 2005 to 2008 and after 2009, i.e., from 2010 to 2015. UA, MA, and DA mean the upslope, mid-slope, and downslope landslide area (km^2).

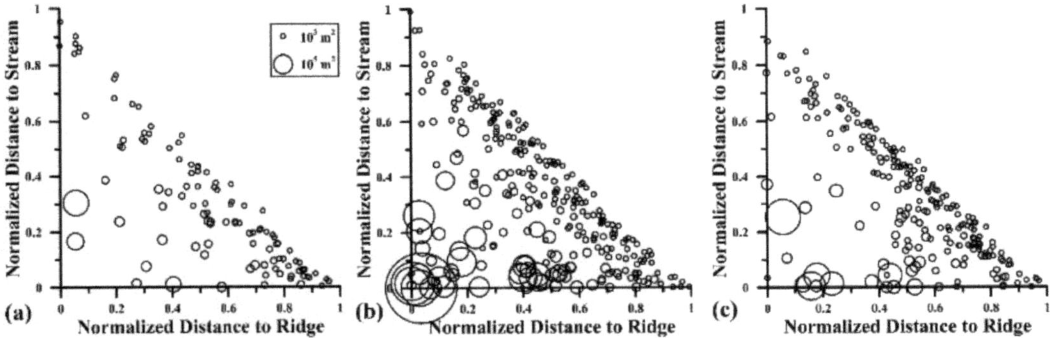

Figure 9. The topographic position analysis of landslide in 2008 (**a**), 2009 (**b**), and 2015 (**c**) in the T01 sub-watershed in the *TRW*.

A similar trend was also found in the A03, A31, and T01 sub-watersheds. The ratio of landslide area from 2009 to 2015 in the downslope was 79.7%, 23.0%, and 68.5% in the A03, A31, and T01 sub-watersheds, respectively. Figures 8 and 9 show that a reduction was observed in the number of upslope, mid-slope, and downslope landslides in the sub-watersheds, but the landslides in 2015 were concentrated in the downslope area. From 2009 to 2015, a large cluster of small-area landslides occurred downslope in the sub-watersheds, with poor recovery. Most of the landslides in the A03 and T01 sub-watersheds in 2015 were centered in areas with a normalized distance to a ridge of >0.7, meaning that the inducing factors should be related to the bank-erosion landslide, which was possibly induced by the sinuous rivers with huge amounts of deposited sediment.

4.4. Spatiotemporal Landslide Hotspot Analysis

The landslide ratios in the *ARW* and *TRW* after 2009 were 7.6% and 10.7%, and those were the top two highest landslide ratios in the watershed scale in Taiwan. It is interesting to understand the evolution of numerous landslides and compare the characteristic of landslide distribution before and after 2009 in the two watersheds. The evolutions of the landslide from 2005 to 2015 in the *ARW* and *TRW* were observed from the spatiotemporal landslide hotspot analyses (Table 5 and Figure 10) in the study.

The total areas from 2010 to 2015 in the two watersheds are 1.15–2.23 times larger than those from 2005 to 2008, and the increases in the landslide hot spot areas from before to after 2009 in the two watersheds were evident. The landslide hot spot areas from 2010 to 2015 in the two watersheds are 2.67–2.88 times larger than those from 2005 to 2008, and the landslide cold spot area is 1.73–1.93 times larger. This result means that the total time of areas identified as landslides from 2010 to 2015 is longer than that from 2005 to 2008. The landslide recovery was more difficult, and the landslide was easier to be clustered after than before 2009 Typhoon Morakot.

Table 5. The statistical data of spatiotemporal landslide hot spots and cold spots in the *ARW* and *TRW*.

Watershed	ARW			TRW		
year	05–15	05–08	10–15	05–15	05–08	10–15
HS (km^2)	17.6	7.8	22.5	8.8	4.8	12.8
CS (km^2)	13.5	10.3	20.3	6.8	5.5	9.5
No (km^2)	35.1	4.6	8.1	20.3	2.3	4.9
Total (km^2)	66.2	22.8	50.9	35.9	12.6	27.1

Note: The 05–15, 05–08, and 10–15 mean from 2005 to 2015, from 2005 to 2008, and from 2010 to 2015. The HS and CS mean the total area of all hot spots and cold spots, and the NO means the no pattern detected area.

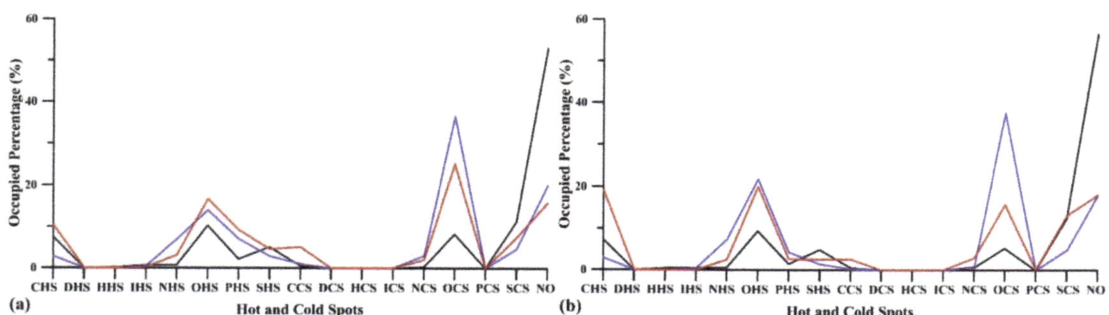

Figure 10. The occupied percentage of landslide hot spots and cold spots from 2005 to 2015 (black line), from 2005 to 2008 (blue line), from 2010 to 2015 (red line), and in the *ARW* (**a**) and *TRW* (**b**).

The no pattern detected area means that the time of area identified as a landslide is shorter than 90% of the research period (Table 1). The occupied percentages of the no pattern detected areas from 2005 to 2015 in the two watersheds are 53.0–56.5%, but those from 2005 to 2008 and from 2010 to 2015 are only 15.9% to 20.2%. This data means that 36.3–37.1% of landslide areas in the two watersheds recovered to the non-landslide areas in 4 to 9 years.

The landslide hot spots are centralized in OHS, SHS, and CHS in each research period, while the landslide cold spots are centralized in OCS, CCS, and SCS. The landslide hot spots and cold spots were reclassified into the main hot spots, the main cold spots, no pattern detected, and others. The main hot spots included OHS, SHS, and CHS, the main cold spots included OCS, CCS, and SCS, and the others included all the other hot spots and cold spots except the main hot spots and cold spots. After 2009, the main hot spots constituted 34.0–41.9% of all hot spots, whereas the main cold spots accounted for 31.6–37.8% of all cold spots.

Figures 11 and 12 present the main hot spots and cold spots from 2005 to 2015 in the two watersheds. The upstream sub-watersheds with dense landslide distributions were the main hot spot cluster areas in the two watersheds.

The main hot spots from 2005 to 2008 were discretely distributed in the upstream sub-watersheds of *ARW* and *TRW*, and those from 2010 to 2015 were densely clustered in the upstream of *ARW* and *TRW*, particularly in the A01 and T01 sub-watersheds.

Obvious increases in the average landslide ratios from after to before the 2009 Typhoon Morakot in the two watersheds were noted. The CHS were the hot spots that exhibited the largest area expansion from after to before the 2009 Typhoon Morakot, and the OCS were the cold spots that exhibited the largest area reduction. The CHS percentage increased by 7.5% to 16.3% from after to before the 2009 Typhoon Morakot, and the OCS percentage decreased from 11.4% to 21.8%. This means that the recovery of landslides induced by 2009 Typhoon Morakot was slower than that before 2009.

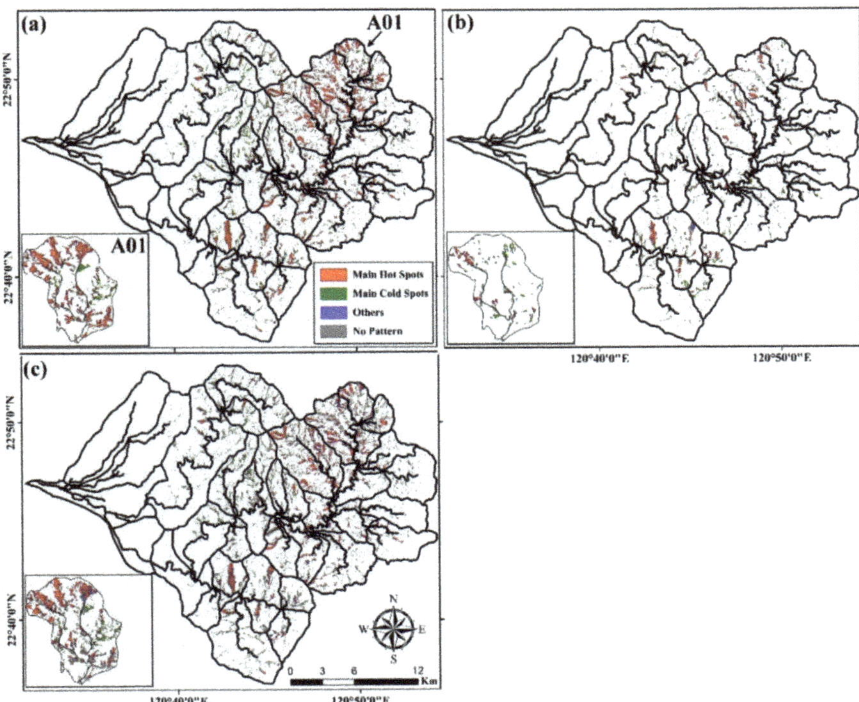

Figure 11. The main landslide hot spot and cold spot from 2005 to 2015 (**a**), from 2005 to 2008 (**b**), and from 2010 to 2015 (**c**) in the *ARW*.

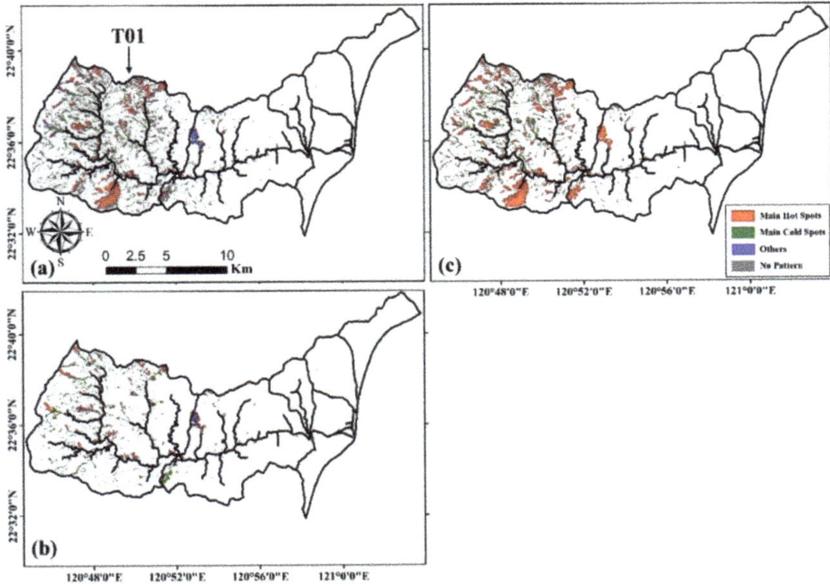

Figure 12. The main landslide hot spot and cold spot from 2005 to 2015 (**a**), from 2005 to 2008 (**b**), and from 2010 to 2015 (**c**) in the *TRW*.

The A01 sub-watershed was selected as the representative sub-watershed to explain the distribution of the main hot spots and cold spots in the study. The strata in A01 comprise the Pilushan and Chaochou formations (62.6% and 37.4%, respectively) from the Eocene epoch and Middle Miocene sub-epoch, respectively. The lithology of the Pilushan formation comprises slate with metasandstone and igneous rock, whereas that of the Chaochou formation is argillite and slate with an alternation of metasandstone or argillite. The main hot spots and main cold spots in the A01 sub-watershed increased substantially after Typhoon Morakot. From 2005 to 2008, 2010 to 2015, and 2005 to 2015, the main hot spots in the A01 sub-watershed constituted 3.0%, 17.0%, and 12.5%, respectively, and the main cold spots constituted 3.8%, 5.9%, and 4.2%, respectively. The main hotspots from 2010 to 2015 in the A01 sub-watershed were concentrated in the headwater landslides, bank-erosion landslides in sinuous reaches, and reoccurrence of older (from 2005 to 2008) landslides.

Mechanisms and triggering factors of landslide events, landslide areas with poor recovery, and geomorphological evolution trends can be explained, located, and predicted using the distributions of landslide hot spots and cold spots that were constructed through spatiotemporal analysis. The results of the spatiotemporal analysis over the various periods have different implications. Specifically, the hot spot and cold spot distributions from 2005 to 2015, from 2005 to 2008, and from 2010 to 2015 in the ARW and TRW represent the long-term landslide evolution.

5. Discussion

The prediction of landslide recovery in watersheds with dense landslides could be the key factor for watershed management. The characteristic of landslide recovery in the following years after the large earthquake or extreme rainfall events are worth comparing and discussing. We explained the recovery characteristic of extreme rainfall-induced landslides by comparing the landslide recovery conditions after the 2005 Kashmir earthquake [5], the 2008 Wenchuan earthquake [4,9], and the 2009 Typhoon Morakot in this study. The time, location, and rate of landslide recovery after the large earthquake or extreme rainfall events are the key discussion topics in this study.

The oscillating period was observed after the large earthquake or extreme rainfall events based on the annual landslide area data. The oscillating period can be defined as that the annual landslide area and landslide number in this period is an oscillating trend instead of a stable decline trend. The oscillating period for the serious earthquake-induced landslide events ranged from 3 to 5 years. The extreme rainfall-induced landslide events in the study were estimated as 5 years (Figure 4 and Table 6, from 2010 to 2014). The landslide in the watersheds in the oscillating period was active and easily induced, re-induced, or enlarged. The average annual landslide area decline rates (abbreviated as *LAD*) after 2014 were larger than that during the oscillating period (from 2010 to 2014), and the average *LAD* during or after the oscillating period in this study was also larger than those from the large earthquake-induced landslide events. This means that the recovery rate of the extreme rainfall-induced landslide was faster than that of large earthquake-induced landslide.

Table 6. Comparison of the average annual landslide area decline rate from the large earthquake events and the extreme rainfall events.

Events	Oscillating Period	LAD_O (km^2/Year)	LAD_A (km^2/Year)
2005 Kashimir Earthquake	2005–2010	0.32	0.99
2008 Wenchuan Earthauke	2008–2011	0.41	0.96
2009 Typhoon Morakot (ARW)	2009–2014	4.24	6.78
2009 Typhoon Morakot (TRW)	2009–2014	2.37	6.04

Note: LAD means the average annual landslide area decline rate (km^2/year), and LAD_O and LAD_A mean the LAD during the oscillating period and after the oscillating period.

The location and reason of new or enlarged landslides after the large earthquake or extreme rainfall events are worth discussing and comparing. The new or enlarged landslides in the following years after the 2005 Kashimir earthquake (including the active, very active, and extremely active landslides in [5]) were mostly located along the Muzaffarabad fault or in the high fractured and jointed rocks areas, or along with the drainage network, or in the source of the river and large landslide. Moreover, the new or enlarged landslides in the following years after the 2008 Wenchuan earthquake (the active landslides in [4]) were located in deep gullies, the source of debris flow and large landslides. Three factors, including the geological setting, the drainage network, and the landslide area, dominate the rate of landslide recovery after 2009 in the *ARW* and *TRW* in the study.

The statistical data and distribution of landslide evolution in the *ARW* and *TRW* are shown in Table 7 and Figure 13. The new or enlarged landslide in the following years after 2009 centralized in the northeast *ARW* and upstream *TRW*. The strata in the northeast *ARW* comprise 62.6% Pilushan formation (metasandstone and igneous rock) and 37.4% Chaochou formations (argillite and slate with an alternation of metasandstone or argillite), and three faults and anticlines also pass through the northeast ARW. The strata in the upstream *TRW* comprise the Chaochou formations (sandstone), kaolinite schist, and Pilushan formations (metasandstone and igneous rock), and three faults and anticlines also pass through the northeast *TRW*. Fractured slate, sandstone, or argillite are the main geological composition in the northeast *ARW* and upstream *TRW*, and also explain the reasons for the centralization of new or enlarged landslides in this area.

Table 7. Statistical data of landslide evolution from 2009 to 2010, 2013, and 2015 in the *ARW*.

Landslide Types	R Area	NR Area	NE			
			Area	Gully-Related	River-Related	Large-Related
2009–2010	24.8	23.5	4.3	2.41	0.02	1.34
2009–2013	26.1	22.2	5.7	3.07	0.05	1.84
2009–2015	28.1	20.2	7.2	4.04	0.08	2.54

Note: The unit of area in this table is km^2. R, NR, and NE mean the recovered, not recovered, and new and enlarged landslide, and the gully-related, river-related, and large-related mean the NE landslide located in the neighborhood of gully, river, and large landslide.

The landslide evolution results from the comparison of landslide inventories in two different years can be classified into three types, including recovered landslides, not recovered landslides, and new or enlarged landslides (Figure 13). The recovered landslide area from the comparison between 2009 and 2010 (Table 7) was the area identified as landslide in 2009 but not in 2010, and the not recovered landslide area were the areas identified as landslide in 2009 and 2010. The new or enlarged landslide area was the area identified as landslide in 2010 but not in 2009. The new or enlarged landslide in the *ARW* and *TRW* also centralized along with the drainage network, particularly in the upstream watersheds. Hugh sediment yield from the landslide in the upstream watershed with dense landslide should be the main reason. The landslide volume was estimated the empirical equations from Taiwan [26] for the landslide area < 10^6 m^2 and Italy [27] for the landslide area ≧ 10^6 m^2]. The landslide volume induced by 2009 Typhoon Morakot was estimated as 65.0×10^6 m^3 and 224.5×10^6 m^3 in the *ARW* and *TRW*. There were 848 landslide cases after 2009 in the northeast upstream *ARW*, including A01, A02, A03, A07, and A11 sub-watersheds, and 1138 landslide cases in the upstream *TRW*, i.e., the T01 sub-watershed. The landslide volume was estimated as 5.2×10^6 m^3 in the northeast upstream of the *ARW* and 223.9×10^6 m^3 in the T01 sub-watershed. Huge sediment was yielded, deposited in the narrow reaches, and dominated the evolution of landslide and river geomorphology in the northeast *ARW* and upstream *TRW*.

Huge sediment in the upstream watershed was continuously transported into the gullies and rivers and also resulted in the frequent occurrence of new or enlarged landslides in the neighborhood of gullies and rivers from 2010 to 2015 in the *ARW*. Moreover, 51.3%, 54.0%, and 58.2% of the landslide areas after 2009 in the *ARW* had been recovered in 2010, 2013, and 2015, respectively. The new or enlarged landslide area from 2010 to 2015 in the *ARW* showed a continuously increasing trend. The occupied percentage of a new or enlarged landslides located in the neighborhood of gullies from 2010 to 2015 was 53.9–56.1%, and the area of a new or enlarged landslide located in the neighborhood of the river from 2010 to 2015 also showed an increasing trend.

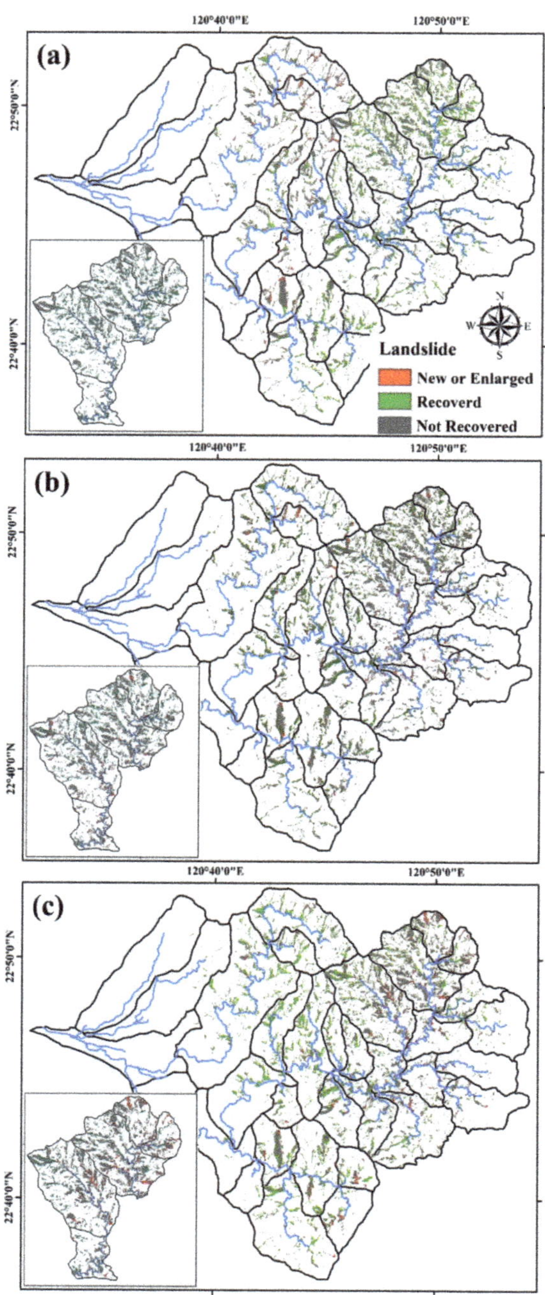

Figure 13. The landslide evolution from 2009 to 2010 (**a**), to 2013 (**b**), and to 2015 (**c**) in the ARW. The left down plot in each figure is the landslide evolution in the northeast *ARW*, including the A01, A02, A03, A07, and A11 sub-watersheds.

The centralization of new or enlarged landslides in the neighborhood of large landslides was mentioned [4,5], and it was also observed in the study. The occupied percentage of a new or enlarged landslide located in the neighborhood of large landslide cases from 2010 to 2015 in the *ARW* was 31.2–35.3%, particularly in the northeast *ARW*.

The dentification of landslide hot spots using the emerging hot spot analysis in this study can show the clustering strength of old, new, and enlarged landslides in space and time and provide a potential landslide location. The advantage of the identification of landslide hot spots using the emerging hot spot analysis is that we can estimate the maintenance time of landslides from the classification of landslide hot spots, and this information also contributes to making the priority of watershed management measures in the watersheds with dense landslides. The management strategy for the watersheds with huge sediment yield should be implemented considering the landslide evolution trend. The landslide evolution cases in the *ARW* and *TRW* in Taiwan demonstrated that controlling the sediment in the drainage network and the landslide boundary should be the priority after the extreme rainfall-induced landslide events.

6. Conclusions

This study used the rainfall analysis, spatiotemporal landslide hotspot analyses, and comparison analysis of large earthquake- and extreme rainfall-induced landslide evolution to understand the characteristic of rainfall-induced landslide evolution, which was useful in assessing the landslide activeness after an extreme rainfall event. We used the *EAR* to assess the landslide-induced strength of rainfall events from 2005–2015, and the *EAR* values in the *ARW* and *TRW* were larger than before after the 2009 Typhoon Morakot. The landslide evolution trend index (*LET*) was used to assess the recovery ratio of landslide area after 2009, and the *LET* value in most of the sub-watersheds in the *ARW* and *TRW* were ranged 0.022–0.072 km^2/year. However, some sub-watersheds in the *ARW* and *TRW*, particularly in the upstream watershed with the landslide ratio > 4.4%, were still of *LET* value > 0.05 km^2/year after 2009. The landslides downslope of sub-watersheds with positive *LET* values in the *ARW* and *TRW* after 2009 were easily induced, re-induced, or enlarged and difficult to recover based on the landslide topographic position analysis. Most of the new or enlarged landslides in the *ARW* and *TRW* after 2009 were classified into oscillating or sporadic or consecutive landslide hotspots and centralized along with the drainage network or large landslide boundary. The watersheds with dense landslides needed to spend 3–5 years, i.e., the oscillating period in the study, to achieve the stable landslide recovery based on the comparison of landslide recovery after the large earthquake or extreme rainfall events. The landslide area decline rates in the *ARW* and *TRW* after 2009 were 1.6–2.5 times larger after than during the oscillating period. The new or enlarged landslides after 2009 in the *ARW* and *TRW* was centralized in the huge sediment-deposited, narrow, and sinuous reaches or the boundary of a large landslide in the upstream watersheds with a geological composition of fractured slate, sandstone, or argillite. The findings from the study point out that the watershed management strategies in the watershed with dense landslides after the extreme rainfall-induced landslide events should be emphasized to control the huge sediment yield from the numerous landslides, particularly in the upstream watersheds.

Author Contributions: Conceptualization and methodology, C.W. and C.L.; software, C.L.; writing, C.W. and C.L.; supervision and funding acquisition, C.W. Both authors have read and agreed to the published version of the manuscript.

Funding: This work was supported by Ministry of Science and Technology, Taiwan; Grant number: MOST 108-2625-M-035-003-(Taiwan).

Institutional Review Board Statement: Not applicable.

Informed Consent Statement: Not applicable.

Data Availability Statement: Not applicable.

Acknowledgments: Financial supports from the Ministry of Science and Technology of Taiwan (R.O.C.) under contract MOST 108-2625-M-035-003- are appreciated.

Conflicts of Interest: The author declares no conflict of interest.

References

1. Wu, C.-H.; Chen, S.-C.; Chou, H.-T. Geomorphologic characteristics of catastrophic landslides during typhoon Morakot in the Kaoping Watershed, Taiwan. *Eng. Geol.* **2011**, *123*, 13–21. [CrossRef]
2. Wu, C.-H.; Chen, S.-C.; Feng, Z.-Y. Formation, failure, and consequences of the Xiaolin landslide dam, triggered by extreme rainfall from Typhoon Morakot, Taiwan. *Landslides* **2013**, *11*, 357–367. [CrossRef]
3. Feng, Z.-Y. The seismic signatures of the surge wave from the 2009 Xiaolin landslide-dam breach in Taiwan. *Hydrol. Process.* **2011**, *26*, 1342–1351. [CrossRef]
4. Li, J.; Wang, N.; Wang, J.; Li, H. Spatiotemporal evolution of the remotely sensed global continental PM2.5 concentration from 2000–2014 based on Bayesian statistics. *Environ. Pollut.* **2018**, *238*, 471–481. [CrossRef] [PubMed]
5. Shafique, M. Spatial and temporal evolution of co-seismic landslides after the 2005 Kashmir earthquake. *Geomorphology* **2020**, *362*, 107228. [CrossRef]
6. Yunus, A.P.; Fan, X.; Tang, X.; Jie, D.; Xu, Q.; Huang, R. Decadal Vegetation Succession from MODIS Reveals the Spatio-Temporal Evolution of Post-Seismic Landsliding after the 2008 Wenchuan Earthquake. *Remote Sens. Environ.* **2020**, *236*, 111476. [CrossRef]
7. Wu, C. Landslide Susceptibility Based on Extreme Rainfall-Induced Landslide Inventories and the Following Landslide Evolution. *Water* **2019**, *11*, 2609. [CrossRef]
8. Valenzuela, P.; Domínguez-Cuesta, M.J.; Mora García, M.A.; Jiménez-Sánchez, M. A Spatial-Temporal Landslide Inventory for the NW of Spain: BAPA Database. *Geomorphology* **2017**, *293*, 11–23. [CrossRef]
9. Liu, S.H.; Lin, C.W.; Tseng, C.M. A Statisticalmodel for the Impact of the 1999 Chi-Chi Earthquake on the Subsequent Rain-fall-Induced Landslides. *Eng. Geol.* **2013**, *156*, 11–19. [CrossRef]
10. Tang, C.; Jiang, Z.; Li, W. Seismic Landslide Evolution and Debris Flow Development: A Case Study in the Hongchun Catchment, Wenchuan Area of China. *Eng. Geol. Soc. Territ.* **2015**, *2*, 445–449. [CrossRef]
11. Zhang, S.; Zhang, L. Impact of the 2008 Wenchuan earthquake in China on subsequent long-term debris flow activities in the epicentral area. *Geomorphology* **2017**, *276*, 86–103. [CrossRef]
12. Yang, W.; Qi, W.; Wang, M.; Zhang, J.; Zhang, Y. Spatial and temporal analyses of post-seismic landslide changes near the epicentre of the Wenchuan earthquake. *Geomorphology* **2017**, *276*, 8–15. [CrossRef]
13. ESRI. [WWW Document]. Doc. ArcGIS. 2018. Available online: https://pro.arcgis.com/en/pro-app/tool-reference/analysis/enrich-layer.htm (accessed on 15 May 2021).
14. Barboza, G.E.; Schiamberg, L.B.; Pachl, L. A spatiotemporal analysis of the impact of COVID-19 on child abuse and neglect in the city of Los Angeles, California. *Child. Abus. Negl.* **2021**, *116*, 104740. [CrossRef]
15. Cheng, Z.; Zu, Z.; Lu, J. Traffic Crash Evolution Characteristic Analysis and Spatiotemporal Hotspot Identification of Urban Road Intersections. *Sustainability* **2018**, *11*, 160. [CrossRef]
16. Nielsen, C.; Amrhein, C.G.; Shah, P.S.; Stieb, D.M.; Osornio-Vargas, A.R. Space-time hot spots of critically ill small for gestational age newborns and industrial air pollutants in major metropolitan areas of Canada. *Environ. Res.* **2020**, *186*, 109472. [CrossRef]
17. Lin, S.C.; Ke, M.C.; Lo, C.M. Evolution of landslide hotspots in Taiwan. *Landslides* **2017**, *14*, 1491–1501. [CrossRef]
18. Shou, K.-J.; Lin, J.-F. Evaluation of the extreme rainfall predictions and their impact on landslide susceptibility in a sub-catchment scale. *Eng. Geol.* **2020**, *265*, 105434. [CrossRef]
19. Varnes, D.J. *Introduction to Landslides: Analysis and Control*; Transportation and Road Research Board; National Academy of Science: Washington, DC, USA, 1978; pp. 11–33.
20. Lu, S.Y.; Lin, C.Y.; Hwang, L.S. Spatial Relationships between Landslides and Topographical Factors at the Liukuei Experi-mental Forest, Southwestern Taiwan after Typhoon Morakot. *Taiwan J. For. Sci.* **2011**, *26*, 399–408.
21. Wu, C.Y.; Tsai, C.W.; Chen, S.C. Topographic Characteristic Analysis of Landslides in Kaoping River Watershed. *J. Chin. Soil Water Conserv.* **2016**, *47*, 156–164. (In Chinese)
22. Lin, Y.T.; Chang, K.C.; Yang, C.J. Object-Based Classification for Detecting Landslides and Vegetation Recovery—A Case at Baolai, Kaohsiung. *J. Chin. Soil Water Conserv.* **2018**, *49*, 98–109. (In Chinese)
23. Chen, C.-W.; Saito, H.; Oguchi, T. Analyzing rainfall-induced mass movements in Taiwan using the soil water index. *Landslides* **2016**, *14*, 1031–1041. [CrossRef]
24. Meunier, P.; Hovius, N.; Haines, J.A. Topographic site effects and the location of earthquake induced landslides. *Earth Planet. Sci. Lett.* **2008**, *275*, 221–232. [CrossRef]
25. Getis, A.; Ord, J.K. The Analysis of Spatial Association by Use of Distance Statistics. *Geogr. Anal.* **2010**, *24*, 189–206. [CrossRef]
26. Chan, H.C.; Chang, C.C.; Chen, S.C.; Wei, Y.S.; Wang, Z.B.; Lee, T.S. Investigation and Analysis of the Characteristics of Shallow Landslides in Mountainous Areas of Taiwan. *J. Chin. Soil Water Conserv.* **2015**, *46*, 19–28. (In Chinese)
27. Guzzetti, F.; Ardizzone, F.; Cardinali, M.; Galli, M.; Reichenbach, P.; Rossi, M. Distribution of landslides in the Upper Tiber River basin, central Italy. *Geomorphology* **2008**, *96*, 105–122. [CrossRef]

Article

The Analysis on Similarity of Spectrum Analysis of Landslide and Bareland through Hyper-Spectrum Image Bands

Shiuan Wan [1], Tsu Chiang Lei [2,*], Hong Lin Ma [1] and Ru Wen Cheng [1]

1 Department of Information Technology, Ling Tung University, Taichung 40851, Taiwan; shiuan123@teamail.ltu.edu.tw (S.W.); df916118@gmail.com (H.L.M.); 1989r67@gmail.com (R.W.C.)
2 Department of Urban Planning and Spatial Information, Feng Chia University, Taichung 40724, Taiwan
* Correspondence: tclei@fcu.edu.tw; Tel.: +886-0424517250

Received: 24 September 2019; Accepted: 12 November 2019; Published: 17 November 2019

Abstract: Landslides of Taiwan occur frequently in high mountain areas. Soil disturbance causes by the earthquake and heavy rainfall of the typhoon seasons often produced the earth and rock to landslide in the upper reaches of the catchment area. Therefore, the landslide near the hillside has an influence on the catchment area. The hyperspectral images are effectively used to monitor the landslide area with the spectral analysis. However, it is rarely studied how to interpret it in the image of the landslide. If there are no elevation data on the slope disaster, it is quite difficult to identify the landslide zone and the bareland area. More specifically, this study used a series of spectrum analysis to identify the difference between them. Therefore, this study conducted a spectrum analysis for the classification of the landslide, bareland, and vegetation area in the mountain area of NanXi District, Tainan City. On the other hand, this study used the following parallel study on Support Vector Machine (SVM) for error matrix and thematic map for comparison. The study simultaneously compared the differences between them. The spectral similarity analysis reaches 85% for testing data, and the SVM approach has 98.3%.

Keywords: landslide; image classification; spectrum similarity analysis

---

1. Introduction

Landslides cause a great loss of human lives and properties. Landslides are frequent phenomena in Taiwan in which a more effective solution to estimate landslide area is desired through considering the remote sensing data [1–3]. Conventionally, monitoring of landslides for their locations and distributions are generally used in situ or field geotechnical techniques through aerial photos by human-power or unmanned aerial devices [4–6]. In the past, the investigation of landslide areas requires much manpower, material resources, and funding, and is very time-consuming. Various modeling approaches have been taken in the form of multivariate statistical analyses or Data Mining techniques of landslide characteristics corresponding to past landslide records. Many researchers studied the landslide through various evaluation/estimation through a Geographic Information System [7,8] with different techniques. The usage of aerial images in large-scale land cover surveys is of great help to the problem [9–11]. Nowadays, spatial information technology is the most proper solution for spatial analysis, which is to effectively and accurately judge the landslide through remotely-sensed images [12]. Hyperspectral image data have been developed for more than 20 years. Hyperspectral image data combine the spectrum shape and image data. In general, the wavelengths of spectrum are divided into visible, near-infrared, and part-short-wave infrared—three different parts. Those instruments recorded the spectral reflection information of the material to obtain complete geospatial information quickly and extensively [13]. Due to the high spectral resolution of hyperspectral images, it can provide rich material details in landslide analysis [14,15].

Landslides cause lots of human life and economic losses every year. With the progressing techniques of spatial data survey in geosciences, large amounts of data for observing the change in the landslide area can easily be collected. Accordingly, the advancement and development of science and technology have enabled remotely to obtain large-scale and high-resolution quantitative information in a short period of time. To find the most valuable knowledge from the target, statistical classification and data mining techniques are usually used to predict the results of the analysis [2,4,8]. The aim of this research is to produce landslide susceptibility mapping by remote sensing data processing and GIS spatial analysis. To identify the unknown species, the spectral reflection diagram of the ground object could be used [16,17]. This action is like to discover the identification code of the ground object which can help us identify different features of land cover. Hyperspectral Imaging has a large number of bands and is almost continuous, which displays a relatively narrow on the spectral range of each band is relatively narrow. The amount of data obtained is huge and it can completely show slight differences in the spectrum of different features.

Due to the lack of accurate DEM (Digital Elevation Modeling) map/data in this study, only the hyperspectral with multi-band data is used to identify landslide and bareland based on a series of spectral intensities of the band reflection (see Figure 1). Therefore, the study aims to answer the question on whether the hyperspectral data can substitute for DEM data or not. On the other hand, landslide and bareland differentiation have drawn more attention to scientists and researchers. Landslide and bareland both have the same ingredient of soil but are usually at different locations on the hill. If the spectrum similarity analysis can be done to determine these two different categories, it could reduce a great amount of time in generating the DEM data/Map.

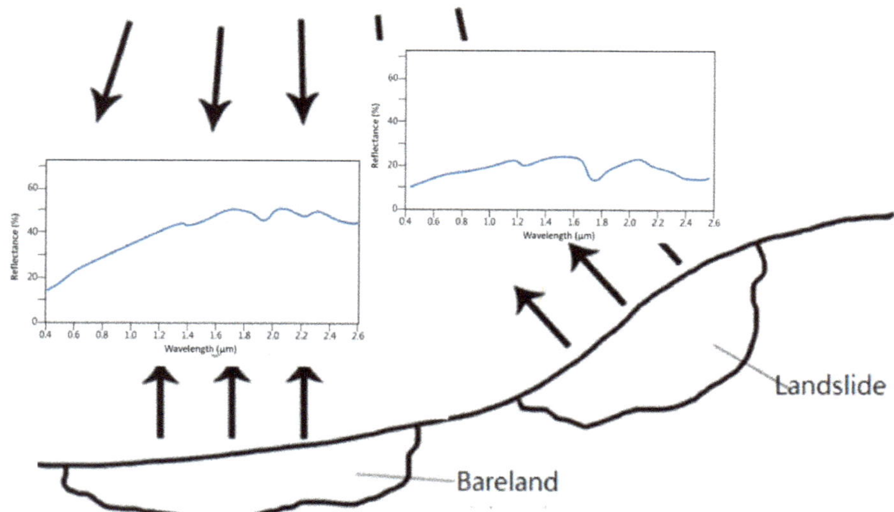

Figure 1. Theory for different reflection of target category (bareland vs. landslide).

In parallel studies, this study intends to use data mining methods: Support Vector Machine (SVM). A total of 72 spectral data of hyper-spectrum remote sensing images are distinguished from the traditional high-resolution data of traditional R, G, B and IR images which can clearly resolve the topography of the surface. If each category is carefully determined, it will be beneficial to compare them by similarity analysis. Between the classification of landslide and bareland, various machine learning classifiers may have different characteristics and solutions. The classification of the image can be conducted either in stage or in combination with each other. Therefore, the study intends to adopt the following two approaches: (a) Spectrum analysis and (b) Support Vector Machine (SVM). These two approaches are used to compare the outcomes in advantages and disadvantages, respectively.

2. Data Collection for Study Plan and Area

The study area is located in Zhuzizao Mountain, Nanxi District, Tainan City, Taiwan. It is located at the northeastern end of Tainan City, north of Dongshan District and Taipu Township of Chiayi, adjacent to Nanhua District in the east, Liujia District and Dazhong District in the west and the south of Yujing District. Nanxi District is located at the tail edge of the Alishan Mountain. The central part is the Dawu Ridge Basin. The hyperspectral image telemetry can reach a large area of the empirical area. To achieve the control and prediction of the collapse disaster, this study used the image data from the Chung-Hsing measurement Company in 2016. They purchased the UAV (Unmanned Aerial Vehicle), which is used to capture the hyperspectral image of Zhuzishan Mountain in Nanxi District, Tainan City for the study material.

2.1. Geomorphology

According to the plan of the Tainan City Landslide and Geostrophic Geological Sensitive Area (2014), the Nanxi District belongs to the river valley zone. Owing to the river originating from the Eastern Mountain, it shows a remarkable stream of excavation. At the same time, the cliff end with erosion is produced. A series of river bank terraces are formed under the action of undercut and side erosion; therefore, a small-scale vertical valley development is formed. The study area is located in the east of Meiling Scenic Area with an elevation of 1110 m. To the west, overlooking the Jianan Plain, the northwest side overlooks the Zengwen Reservoir, and the southeast is the Nanhua Reservoir. The terrain of this area has a large height difference in elevation, which is mainly composed of hilly terrain and plain terrain. The average elevation is between 800 and 1300 m. The geology is mainly composed of accumulated soil, and there are faults on both sides—the east and west. The earthquake-induced landslide caused this area soil condition to be very fragile.

2.2. Hydrographic System

The rivers in Tainan City include Bazhangxi, Jiushuixi, Zengwenxi, Yanxi, and Errenxi. The Ziwen River Basin originates from the Alishan Mountains. The drainage area is 1176.6 square kilometers and the longest is 138.5 km. The average slope of the riverbed is 1/200. The main tributaries are Houtunxi, Caixixi, and Guantianxi. It flows through Dongshan, Liujia, Annan, Yujing, Nanhua, Zuozhen, Shanshang, Dain, Guantian, Shanhua Madou, Anding, Xigang, and Qiqi on the Nanxi District of the study area, respectively. The study area is located near Tainan County in which there is Zengwen Reservoir (the largest reservoir in southern Taiwan). The mainstream originates from the Alishan Mountains, flows south to Zengwenxi, and flows southwest through the mountainous area to the Zengwen Reservoir. The strip has a total length of 138.5 km, an average slope of 1/57, and an average annual rainfall of about 2726 mm.

2.3. Geological Structure

The Tainan City Regional Disaster Prevention Plan (2016) is based on the data released by the Central Geological Survey of the Ministry of Economic Affairs in December 2016. It attributes to the historical landslide and ground slide area of about 69.11 square kilometers with landslide or ground slip conditions (with a sloping slope). The area is about 50.4 square kilometers with the buffer zone of 5 m is about 21.99 square kilometers, and the demarcation range is about 0.62 square kilometers. The total area is about 135.45 square kilometers (about 6.18% the total of area city).

In Figure 2, the location map of the landslide and geostrophic geological sensitive area in Tainan City is a plan for the Tainan landslide and geostrophic geological sensitive area (2016). The figure shows that, to increase the terrain steepness and aspect, the base map is overlaid with topographic shadow maps and the adjacent administrative boundaries.

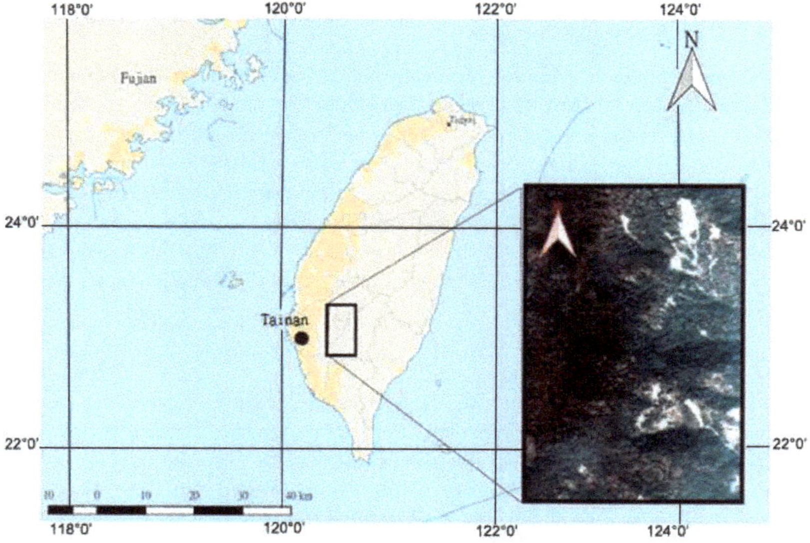

Figure 2. Study area.

2.4. Research Material

The spectral application image used in this study is the hyperspectral image of the Compact Airborne Spectrographic Imager (CASI) of the Bamboo-Waste Mountain in Nanxi District, Tainan City, which was provided by Taiwan Chung-Hsing Measurement in January and April 2016 as shown in Figure 3. The image scanning system CASI-1500 is manufactured by ITRES of Calgary, AB, Canada. The CASI-1500 instrument has a series of spectral wavelengths between 365 nm and 1050 nm, which is equivalent to the visible of near-infrared range. It can acquire 72 bands for this study with a spectral resolution of 3 nm and a spatial resolution of 1 m. Each band has its range and attribute of color, which is presented in Figure 3. Thus, the corresponding number of bands in the latter parts of this study is the same number presented here.

Band	1	2	3	4	5	6	7	8	9	10	11	12	13	14	15	16	17	18	19	20	21	22	23	24	25	26	27	28	29	30	31	32	33	34	35	36
	Blue															Green											Red									
nm	366	375	385	394	404	413	423	432	442	452	461	471	480	490	499	509	518	528	537	547	556	566	575	585	594	604	613	623	632	642	651	661	671	680	690	699

Band	37	38	39	40	41	42	43	44	45	46	47	48	49	50	51	52	53	54	55	56	57	58	59	60	61	62	63	64	65	66	67	68	69	70	71	72
	Near infrared																																			
nm	709	718	728	737	747	756	766	775	785	794	804	813	823	832	842	851	861	870	880	889	899	908	918	927	937	946	956	965	975	984	994	1003	1013	1022	1032	1041

Figure 3. The hyperspectral corresponding image band of range.

3. Research Method

3.1. Spectrum Similarity Analysis

We carefully selected 240 sampling data for vegetation areas (trees, grass, etc), bareland area, and landslide area, respectively. Spectrum similarity analysis becomes a well-accepted approach to reduce the data dimensionality of hyperspectral imagery. It retrieves several bands of important patterns in some sense by taking advantage of the all high spectral correlation. Verified by classification accuracy, it was expected that, just using a part of original bands, the accuracy is obtained rationally, whereas computational work is significantly reduced [18,19].

Figure 4a shows the entire research step. It includes two parallel approaches. One of the approaches is considering finding the similarity of the image bands width to attain the classification outcomes [20]. Figure 4b shows the similarity of image bands. The vegetation index threshold is found based on clustering analysis. The non-vegetation of the image is attained, which includes the bareland and landslide. All this is part of data normalization. Then, the progress of the similarity spectrum analysis is carried out. Two of the image layers are obtained (D_1 and D_2). The latter part of this paper will introduce the details on how the similarity classification of each pixel is identified.

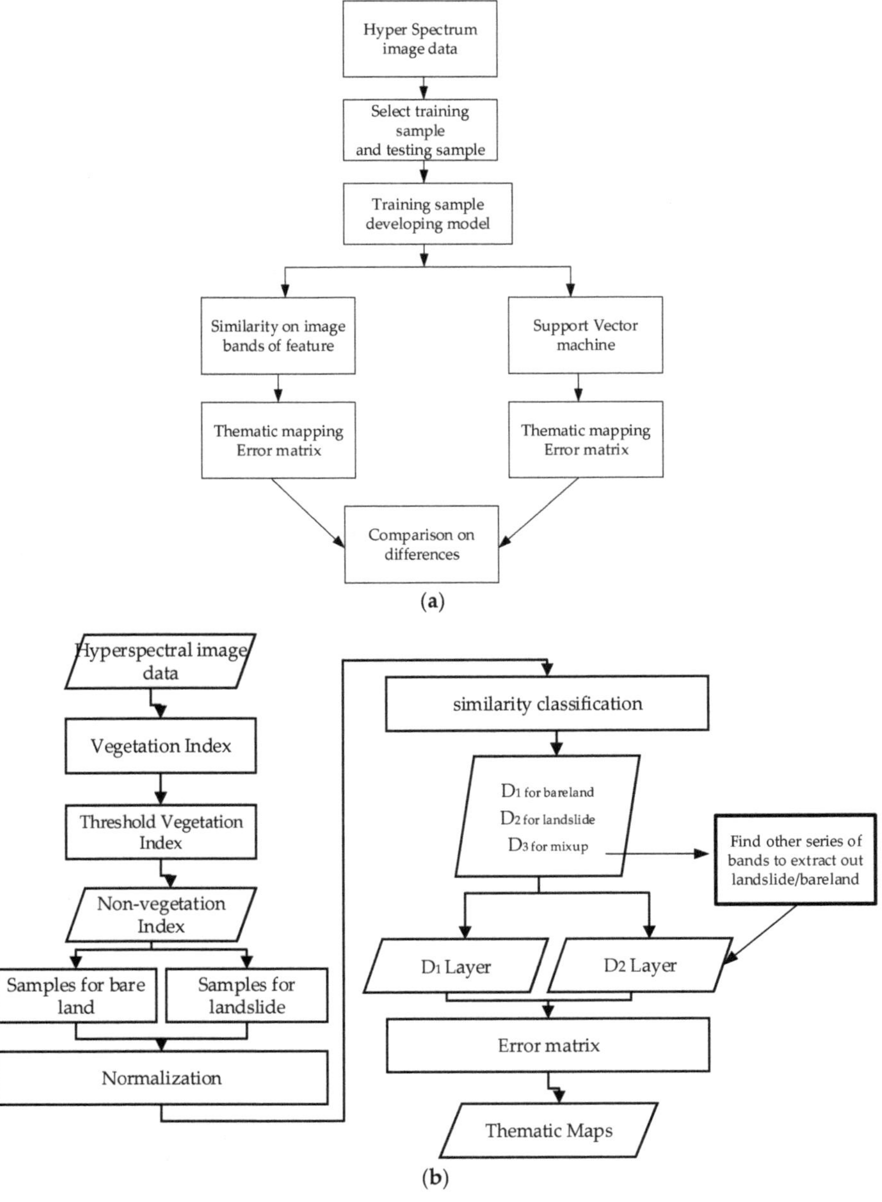

Figure 4. (**a**) research steps; (**b**) the steps for similarity classification.

Figure 5a presents the original investigation on the site for observing the location of landslide. All the image data for similarity were carefully checked by in situ investigation and compared to remote sensing data [21]. It was decided to extract serval samples as mentioned above for landslide and bareland. Thus, Figure 5b shows the longitude and latitude of the position of study and the accurate place of landslides. This area landslide belongs to block-slide. Bock slide is a kind of translational slide. The moving mass of soil and rocks has serval related units that move downslope as a relatively coherent mass. The largest size of the landslide is about 8×12 m^2, which is roughly measured by image data.

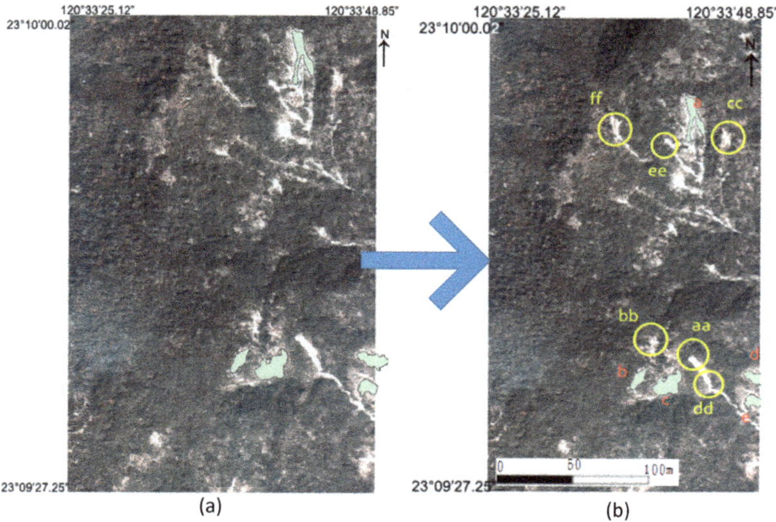

Figure 5. (**a**) original inventory map: landslide location of in situ investigation; (**b**) location of training sample, landslide: a, b, c, d, and e; bareland: aa, bb, cc, dd, ee, and ff.

The parallel study was used the SVM (Support Vector Machine) to access the classification on bareland and landslide. The thematic map is compared and the error matrix is also calculated.

3.2. Brief on Support Vector Machine

Support vector machines (SVMs) are well-accepted supervised learning methods used for classification [22]. The study considers the concept of improving statistical learning theory, generally applied as an effective classifier to solve many practical problems [23]. A special feature of this classifiers is to minimize the empirical classification error and maximize the geometric margin, simultaneously. Therefore, it is also known as a maximum margin classifier [24,25].

Linearly separable classes are the simplest cases for the analysis of three various classes (vegetation, landslide, and bareland). Assume the training data with k number of samples are presented as $\{x_i, y_i\}$, where $x \in R^N$ with an n-dimensional space, and $y \in \{+1, -1\}$ is the class label. These training patterns are linearly separable if there exists a vector w (determining the orientation of a discriminating plane) and a scalar b (determine the offset of the discriminating plane from the origin) such that

$$y_i(w \cdot x_i + b) - 1 \geq 0. \tag{1}$$

The hypothesis space is defined by the set of functions given by:

$$f_{w,b} = sign(w \cdot x + b). \tag{2}$$

If the set of examples is linearly separable, the goal of the SVMs is to minimize the value $||w_i||$. It is equivalent to finding the separating hyperplanes for which the distance between the classes of training data. It also measured along a line perpendicular to the hyperplane.

This distance is called the margin. The data points that are closest to the hyperplane are used to measure the margin. Thus, these data points are also called support vectors. Consequently, the number of support vectors should be small.

The problem of minimizing $||w_i||$ is solved by applying standard quadratic programming (QP) optimization techniques. It also trasforms the problem to a dual space by using Lagrangian multipliers. The Lagrangian is presented by introducing positive Lagrange multipliers λ_i, $i = 1, \ldots k$. The solution of the optimization problem is attained by considering the saddle point of the Lagrange function

$$L(w, b, \lambda) = \frac{1}{2}||w||^2 - \sum_{i=1}^{k} \lambda_i y_i (w \cdot x_i + b) + \sum_{i=1}^{k} \lambda_i. \quad (3)$$

The solution in Equation (5) needs $L(w,b,\lambda)$ to be minimized with respect to w and b and maximized with respect to $\lambda_i \geq 0$. Therefore, for a two-class problem, the decision rule separates the two classes that can be written as:

$$f(x) = sign\left(\sum_{i=1}^{k} \lambda_i y_i (x \cdot x_i) + b\right). \quad (4)$$

A soft margin problem for the case of SVMs is to handle the linearly non-separable data by Vapnik [22]. They concluded that the restriction of each training vector of a given class on the same side of the optimal hyperplane that applies the value. In $\xi_i \geq 0$, the SVM algorithm for the hyperplane maximizes the margin. At the same time, it minimizes a quantity proportional to the number of misclassification errors. This trade-off function between margin and misclassification error is also governed by a positive constant C such that $\infty > C > 0$. Thus, for non-separable data, Label (6) can be written as

$$L(w, b, \lambda, \xi, \mu) = \frac{1}{2}||w||^2 + C\sum_i \xi_i - \sum_i \lambda_i \{y_i (w \cdot x_i + b) - 1 + \xi_i\} - \sum_i \mu_i \xi_i, \quad (5)$$

where the μ_i are the Lagrange multipliers introduced to force the ξ_i to be positive. The solution of (7) is determined by the saddle points of the Lagrangian, by minimizing with respect to w, x, and b, and maximizing with respect to $\xi_i \geq 0$ and $\mu_i \geq 0$.

4. Results

As aforementioned, we select 120 of sampling data for training the model of vegetation, bareland, and landslide, respectively. The 40 pieces of data of each (vegetation, bareland, and landslide) categories to build the model. The study also randomly selects 40 pieces of data to verify the model as testing data. The study has been broken into two parts: spectral similarity analysis and support vector machine. As previously mentioned in Figure 1, the landslide mostly occurred on the slope that has the different responses of reflection on hyper-spectrum image data. Compared to the bareland, the ingredient of soil is the same as the landslide; however, most of them are located in the flat area. Thus, the reflection on hyper-spectrum image data must be different to a landslide. This is the objective to classify them by applying the similarity of a spectrum [26].

To introduce the overall accuracy, it can be formulated as

$$\mathrm{Ac} = \frac{TP + TN}{\textit{All the samples}}, \tag{6}$$

where *TP* is the true positive and *TN* is the true negative.

4.1. Spectral Similarity Analysis

The spectrum analysis entire study is divided into the following two steps:

1. Classify the vegetated areas and non-vegetated areas (similar to [18]).
2. Use the clustering analysis to separate bareland areas and landslide areas from non-vegetated areas (similar to [19]).

To achieve this task, the developed program scans all the bands to find the largest discrepancy of vegetation and non-vegetation for discriminating between these two categories. The program calculates and finds that the 34th band has the largest difference. The green lines in Figure 6a are rationally extracted. Figure 6b is generated to extract out vegetation parts (green line). The r-value on the y-axis is the response value of the reflection for various categories (vegetation, bareland, and landslide).

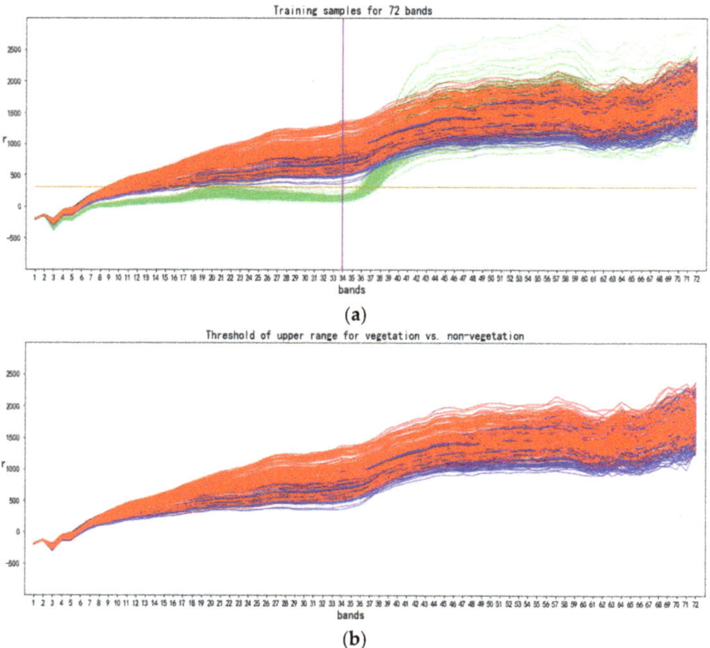

Figure 6. *Cont.*

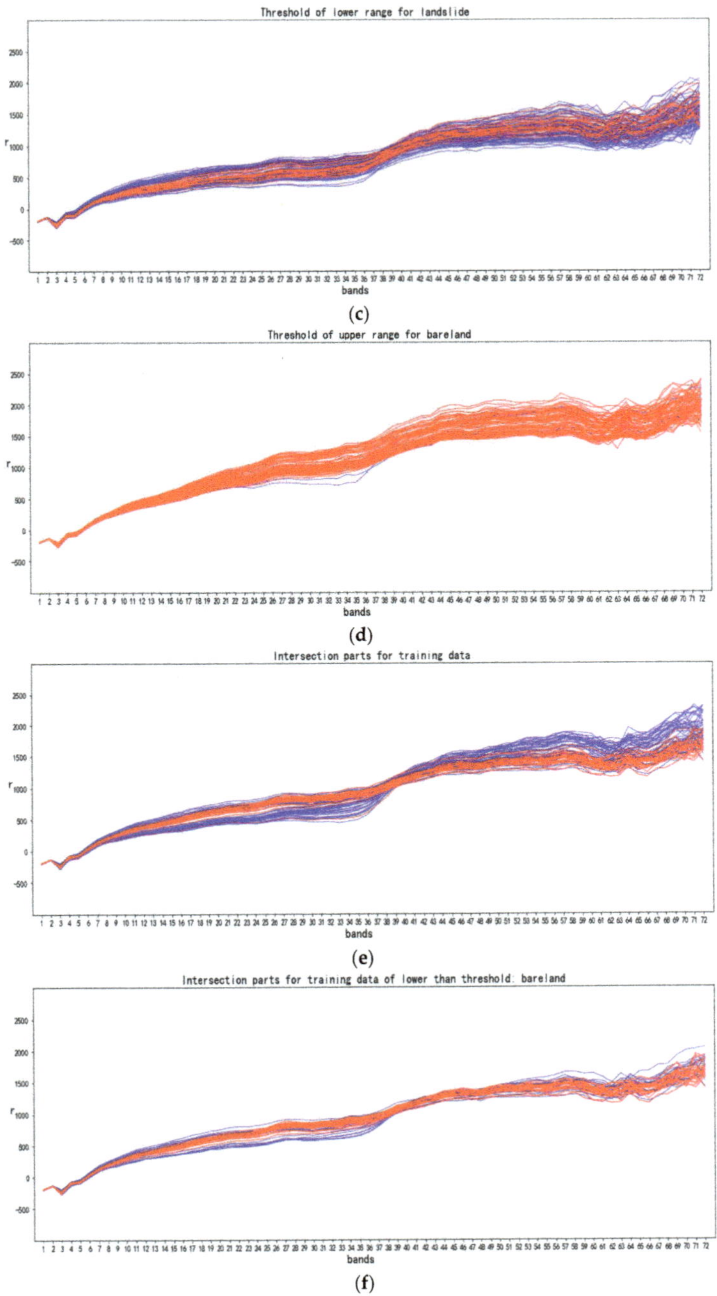

Figure 6. (**a**) three different categories for classification; (**b**) threshold of upper range for vegetation and non-vegetation; (**c**) the lower range on the threshold for landslide; (**d**) the upper range on the threshold for bareland; (**e**) intersection parts for training data of larger than threshold: landslide; (**f**) intersection parts for training data of lower than threshold: bareland (green: vegetation; blue: landside; red: bareland).

To obtain the best classification outcomes, the 1–72 bands are scanned to find the best part to distinguish the landslide and bareland. The developed program scans the data in Figure 6c to find the maximum discrepancy for landslide and bareland. A single cannot clarify the mixup data for landslide and bareland. Thus, a combination set of bands are requested to approach the goal. It is found that 38 to 42 bands are the best part in the spectrum analysis to attain the classification outcomes. First, the program adds the 38–42 bands of each data as a single band data (transfer the five-dimensional data to the one-dimensional data).

Then, the program generates a parametric r-value as an interval to attain three parts of the data: (a) lower the threshold(landslide), (b) upper the threshold(bareland), and (c) intersection part (mix-up part). Please refer to Figure 6b; Figure 6c; Figure 6d; and Figure 6e. The program gradually increases the r value to approach the optimal classification outcomes for landslide and bareland. For example, the program starts $r = 80$ and $\Delta r = 5$ and finds the error rate between classification on landslide and bareland. That is, the program gains $r = 95$, which is the best allowable value to cut the data into these three parts. The strategy is to approach the largest number of lower the threshold and upper the threshold. The minimal lowest number of intersection part is also requested. The program calculates each data after the summation and sets them as less than 95 as one group and greater than 95 as another. The strategy is the number of data of the largest group to the total number of data must be greater than 40%. The number of data of the smallest group to the total number of data must be smaller than 40%. Because in Data Mining, the portion of the number of data for each decision should be as close as possible. Applying these sets of data can be fairly and uniformly to develop the model.

Then, the program three parts for summing up band 38 to 42 is

$$\begin{cases} \textit{Lower than intersection} \leq 4105 \ldots \textit{landlside} \\ \textit{Greater than intersection} \geq 4828 \ldots \textit{bareland} \\ \textit{intersection} \ > 4105 \textit{ and } < 4828 \ldots \textit{mixup parts} \end{cases} . \tag{7}$$

After screening the band data, it is found that the data density variety is not uniform. Hence, different stepwise of a grouping data strategy is needed. In the mix-up parts, the program restarts to find the discrepancy between landslide and bareland. The solution takes a set of band values and uses the clustering technique to search the optimal set of possible outcomes. For instance, we found that the band numbers from 45 to 52 has the largest discrepancy. Then, the program sieves out the 45, 46, 48, 50, and 51 bands are the most useful information. That is, 47, 49, and 52 bands are eliminated from the data set. The program found that the band values in 45, 46, 48, 50, and 51 have the largest discrepancy between landslide and bareland. Then, the intersection ranges of bands of each piece of data are summed into a single value (five multi-band data into one-dimensional data). The summed maximum and maximum values are calculated, and the binary classification is executed. It is found that a finer value of 30 can be gradually increased as a stepwise to each line attribute for each categories (landslide and bareland). Then, the accuracy of each segmentation value is step-by-step calculated as the classification accuracy until the highest accuracy is approached. Each band of the data in the intersection range is clustered based on the rule of a finer interval being less than 30; the other group is greater than 30.

Determination value of 45, 46, 48, 50 and 51

$$\begin{cases} \textit{sum } r : \geq 6995 \ldots \textit{landslide} \\ \textit{sum } r : < 6995 \ldots \textit{bareland} \end{cases} . \tag{8}$$

The training data for generating this similarity model have 100% accuracy.

The thematic map (see Figure 7) is generated by inputting are the band data into the program. Green presents the vegetation, red for landslide and white for bareland. The major landslide areas (comparing in Figure 5) are almost found, and bareland is clearly found. The computational time is fast and a rough result is qualified. We also randomly picked 40 testing data to verify our spectrum analysis model. The error matrix is presented in Table 1. The overall accuracy is about 85%.

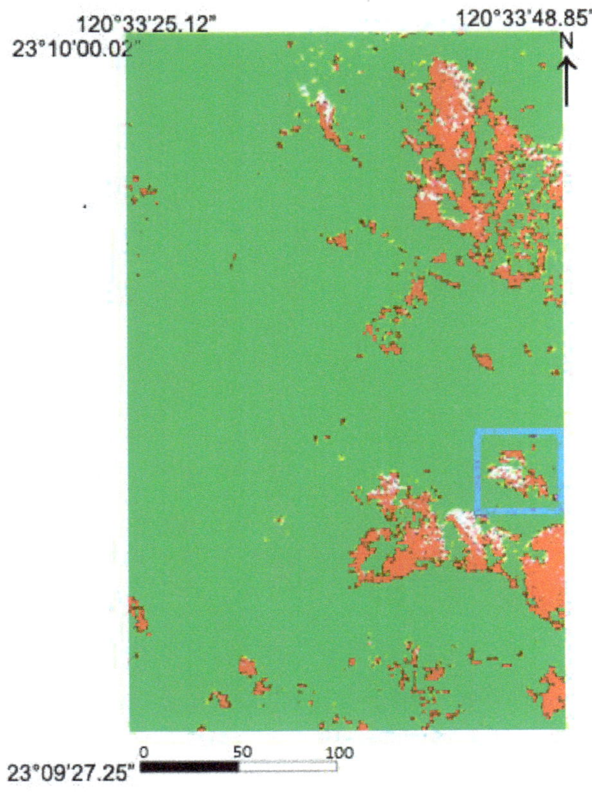

Figure 7. The thematic of spectrum analysis (green: vegetation, red: landslide and white: bareland).

Table 1. Error matrix for spectrum analysis.

Categories	Vegetation	Bareland	Landslide	User Accuracy	Commission Error
Vegetation	40	0	0	100%	0%
Bareland	0	32	8	80%	20%
Landslide	0	10	30	75%	25%
Producer Accuracy	100%	76.19%	78.95%	Overall accuracy = 85%	
Omission error	0%	23.81%	21.05%		

4.2. SVM

As part of the study, the Support Vector Machine is used as a parallel approach to examine the spectrum analysis. The objective of the support vector machine algorithm is to generate a hyperplane in n-dimensional space (n is the number of features) that accurately classifies the data points. The following steps are:

4.2.1. Step1: Normalization

The original data of the collected data sets (such as hyperspectral, multi-spectral, etc.) are normalized, and the values of the attribute data are standardized within the same range. This study converts all attribute values between −1 and 1, using the formula:

$$\text{d} = \frac{d - min_d}{((max_d - min_d) - 0.5) * 2}. \tag{9}$$

4.2.2. Step2: Cross-Validation

This study uses K-Fold Cross-Validation to first split the initial sample into K sub-samples (each sub-sample is independent from each other). A single sub-sample is the data for the validation model with the remaining K−1 samples. One of those sets of sub-samples is used for training. After repeating the above procedure K times, the K group classification correct rate will be obtained. In final, the data of the K group for the correct rate average value are estimated.

4.2.3. Step3: Model Selection for a Core Function

The functions of the support vector machine can be divided into four types: linear functions, polynomial functions, radial basis functions, and S functions. The user should select the core function based on different conditions. The parameters are adjusted for different kernel functions that are also different. The user has to adjust the kernel function and parameters according to the situation, which will have a significant impact on the prediction accuracy rate. In this study, the Radial Basis Function kernel (RBF) is taken for consideration. To obtain better model parameters, the Grid Search method repeats the test parameters C = 4.2 (penalty parameter) and g = 0.32 (gamma function) for possible combination and calculate the correct rate of its parameters (C, g). If it meets its condition, end the repeated test and output its best C and g parameters; otherwise, re-substitute with the new parameters until the combination is found.

This step is to optimize the optimal classification model obtained in the previous step. The testing data of the unknown result are substituted into the classification model construct by the previous step, and the obtained result will be aggregated in which the overall classification accuracy rate is calculated for performing the evaluation. It explores the effectiveness of machine learning under its selection points and different attribute data. The accuracy assessment of this study is divided into two parts: (1) the thematic map and (2) the error matrix. Figure 8 presents the thematic map for the overall condition in three categories. Green presents the vegetation, red for landslide, and white for bareland. Comparing Figure 8 to Figure 7, based on the image data in Figure 5, it presents a clearer and better accurate rate for the thematic map. The error matrix is also calculated in Table 2. The overall accuracy is 98.3%.

Comparing Figures 7 and 8 for the thematic map, the difference is clear. For instance, the blue rectangle part in Figure 7 renders a better interpretation of detecting the bareland. In the inventory map (Figure 5), this part presents as a bareland. The similarity analysis spectrum seems to provide a better prediction. However, SVM has a better interpretation of the integrity on landslide and bareland. The spatial information is a fundamental multi-temporal approach. The method can successfully be applied to serval periods of this area or another area. Thus, if the based rule of similarity spectrum can be developed successfully, the approximated location of landslide mapping can be rapidly generated.

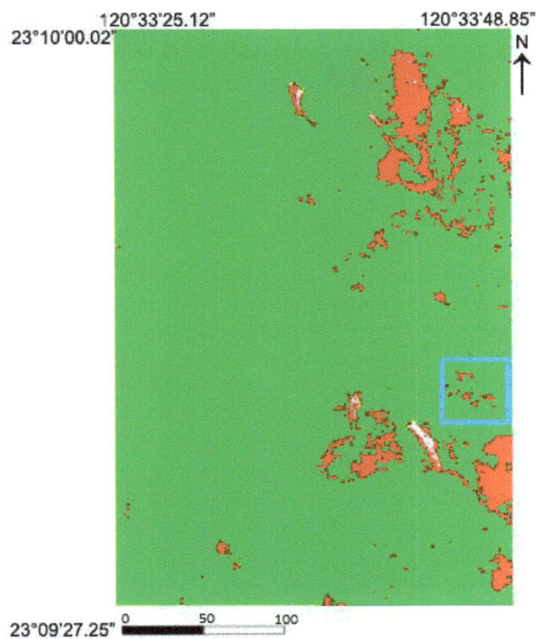

Figure 8. The thematic of support vector machine (green: vegetation, red: landslide and white: bareland).

Table 2. Error matrix for SVM.

Categories	Vegetation	Bareland	Bareland	User Accuracy	Commission
Vegetation	40	0	0	100%	0%
Bareland	0	38	0	100%	0%
Landslide	0	2	40	95.2%	4.8%
Producer	100%	95%	100%	Overall accuracy = 98.3%	
Omission error	0%	5%	0%		

5. Conclusions

The landslide and bareland are the most interesting topics that draw great attention to scientists and researchers. They both have the same soil ingredient but different locations on the hill. Landslides mostly displayed on the hill, which may produce destructive disasters for human beings. Owing to the lack of accurate DEM (Digital Elevation Modeling) map/data in this study, the hyperspectral data have been proved to identify landslide and bareland according to spectral intensities of reflection. A parallel study is designed to compare the spectral analysis approaches.

The study has three major contributions:

(a) The study proved that the hyperspectral image data can replace the DEM data by considering different land cover categories.

(b) The spectral similarity analysis can classify 100% of the vegetation area. Most of the landslide and bareland area is also being detected with a satisfactory level of the overall accuracy of 85%.

(c) The support vector machine is a superior classifier. However, the problem is that each of the training sample data must be supervised data. That is, each piece of in situ sampling data must be carefully labeled. The overall accuracy is 98.3%.

Author Contributions: S.W.: He was responsible for plan and design of this study. He also helped the student write the computer program. T.C.L.: He analyzed the data and discussion. H.L.M.: He wrote the computer program. R.W.C.: She used the program to plot the thematic map.

Funding: This research was funded by grant number 106-2119-M-275-002.

Acknowledgments: The authors expressed their gratitude for the National Science Council (106-2119-M-275-002)-sponsoring this work.

Conflicts of Interest: The authors declare no conflict of interest.

References

1. Van Westen, C.J.; Rengers, N.; Soeters, R. Use of Geomorphological Information in Indirect Landslide Susceptibility Assessment. *Nat. Hazards* **2003**, *30*, 399–419. [CrossRef]
2. Wan, S.; Lei, T.C.; Huang, P.C.; Chou, T.Y. Knowledge Rules of Debris Flow Event: A Case Study for Investigation ChenYu Lan River, Taiwan. *Eng. Geol.* **2008**, *98*, 102–114. [CrossRef]
3. Wan, S.; Lei, T.C.; Chou, T.Y. A novel data mining technique of analysis and classification for landslide problems. *Nat. Hazards* **2009**, *52*, 211–230. [CrossRef]
4. Wan, S. A spatial decision support system for extracting the core factors and thresholds for landslide susceptibility map. *Eng. Geol.* **2009**, *108*, 237–251. [CrossRef]
5. Lazzari, M.; Gioia, D. UAV images and historical aerial-photos for geomorphological analysis and hillslope evolution of the Uggiano medieval archaeological site (Basilicata, southern Italy). *Geomat. Nat. Hazards Risk GNHR* **2017**, *8*, 104–119. [CrossRef]
6. Lazzari, M.; Piccarreta, M. Landslide disasters triggered by extreme rainfall events: The case of Montescaglioso (Basilicata, southern Italy). *Geosciences* **2018**, *8*, 377. [CrossRef]
7. Lin, C.Y.; Chuang, C.W.; Lin, W.T.; Chou, W.C. Vegetation recovery and landscape change assessment at Chiufenershan landslide area caused by Chichi earthquake in central Taiwan. *Nat. Hazards* **2010**, *53*, 175–194. [CrossRef]
8. Pradhatsp, B.; Mansor, S.; Lee, S.; Buchroithner, M.F. Application of a data mining model for landslide hazard mapping. *ISPRS* **2008**, *37*, 187–196.
9. Wan, S.; Lei, T.C.; Chou, T.Y. A landslide expert system: Image classification through integration of data mining approaches for multi-category analysis. *Int. J. Geogr. Inf. Sci.* **2012**, *26*, 747–770. [CrossRef]
10. Kirschbaum, D.B.; Adler, R.; Hong, Y.; Lerner-Lam, A. Evaluation of a preliminary satellite-based landslide hazard algorithm using global landslide inventories. *Nat. Hazards Earth Syst. Sci.* **2009**, *9*, 673–686. [CrossRef]
11. Pradhan, B.; Lee, S.; Buchroithner, M.F. Remote sensing and GIS-based landslide susceptibility analysis and its cross-validation in three test areas using a frequency ratio model. *Photogramm. Fernerkund.-Geoinform.* **2010**, *1*, 17–32. [CrossRef]
12. Carrara, A.; Guzzetti, F.; Cardinali, M.; Reichenbach, P. Use of GIS technology in the prediction and monitoring of landslide hazard. *Nat. Hazards* **1999**, *20*, 117–135. [CrossRef]
13. Chang, C.I. Spectral Information Divergence for Hyperspectral Image Analysis. In Proceedings of the International Geoscience and Remote Sensing Symposium (IGARSS), Hamburg, Germany, 28 June–2 July 1999; Volume 1, pp. 509–511.
14. Aspinall, R.J.; Marcus, W.A.; Boardman, J.W. Considerations in Collecting, Processing, and Analysing High Spatial Resolution Hyperspectral Data for Environmental Investigations. *J. Geogr. Syst.* **2002**, *4*, 15–29. [CrossRef]
15. Borgogno Mondino, E.; Giardino, M.; Perotti, L. A neural network method for analysis of hyperspectral imagery with application to the Cassas landslide (Susa Valley, NW-Italy). *Geomorphology* **2009**, *110*, 20–27. [CrossRef]
16. Aguilera, C.A.; Aguilera, F.J.; Sappa, A.D.; Toledo, R. Learning Cross-Spectral Similarity Measures with Deep Convolutional Neural Networks. In Proceedings of the 2016 IEEE Conference on Computer Vision and Pattern Recognition Workshops, Las Vegas, NV, USA, 26 June–1 July 2016; pp. 267–275.
17. Iscen, A.; Avrithis, Y.; Tolias, G.; Furon, T.; Chum, O. Fast Spectral Ranking for Similarity Search. In Proceedings of the IEEE Computer Society Conference on Computer Vision and Pattern Recognition, Salt Lake City, UT, USA, 18–23 June 2018; pp. 7632–7641.

18. Hamad, D.; Biela, P. Introduction to Spectral Clustering. In Proceedings of the 2008 3rd International Conference on Information and Communication Technologies: From Theory to Applications, Damascus, Syria, 7–11 April 2008.
19. Nadler, B.; Galun, M. Fundamental limitations of spectral clustering. In *Advances in Neural Information Processing Systems*; Neural Information Processing Systems Foundation, Inc.: Vancouver, BC, Canada, 2007; pp. 1017–1024.
20. Fan, C.Y.; Fan, P.S.; Chan, T.Y.; Chang, S.H. Using hybrid data mining and machine learning clustering analysis to predict the turnover rate for technology professionals. *Expert Syst. Appl.* **2012**, *39*, 8844–8851. [CrossRef]
21. Guzzetti, F.; Mondini, A.C.; Cardinali, M.; Fiorucci, F.; Santangelo, M.; Chang, K.T. Landslide inventory maps: New tools for an old problem. *Earth-Sci. Rev.* **2012**, *112*, 42–86. [CrossRef]
22. Vapnik, V.N. *The Nature of Statistical Learning Theory*; Springer: New York, NY, USA, 1995; pp. 11–24.
23. Kecman, V. *Learning and Soft Computing, Support Vector machines, Neural Networks and Fuzzy Logic Models*; The MIT Press: Cambridge, MA, USA, 2001.
24. Khandoker, A.H.; Palaniswami, M.; Karmakar, C.K. Support Vector Machines for Automated Recognition of Obstructive Sleep Apnea Syndrome from ECG Recordings. *IEEE Trans. Inf. Technol. Biomed.* **2009**, *13*, 37–48. [CrossRef] [PubMed]
25. Pal, M. Support vector machines-based modelling of seismic liquefaction potential. *Int. J. Numer. Anal. Methods Geomech.* **2006**, *30*, 983–996. [CrossRef]
26. Van der Meer, F. The effectiveness of spectral similarity measures for the analysis of hyperspectral imagery. *Int. J. Appl. Earth Obs. Geoinf.* **2006**, *8*, 3–17. [CrossRef]

Article

A Landslide Probability Model Based on a Long-Term Landslide Inventory and Rainfall Factors

Chun-Yi Wu * and Yen-Chu Yeh

Department of Soil and Water Conservation, National Chung Hsing University, Taichung 402, Taiwan; cxz7.997.9@gmail.com

* Correspondence: cywu@nchu.edu.tw; Tel.: +886-4-2284-0381 (ext. 605)

Received: 11 February 2020; Accepted: 24 March 2020; Published: 26 March 2020

Abstract: The prediction and advanced warning of landslide hazards in large-scale areas must deal with a large amount of uncertainty, therefore a growing number of studies are using stochastic models to analyze the probability of landslide occurrences. In this study, we used a modified Thiessen's polygon method to divide the research area into several rain gauge control areas, and divided the control areas into slope units reflecting the topographic characteristics to enhance the spatial resolution of a landslide probability model. We used a 2000–2015 long-term landslide inventory, daily rainfall, and effective accumulated rainfall to estimate the rainfall threshold that can trigger landslides. We then employed a Poisson probability model and historical rainfall data from 1987 to 2016 to calculate the exceedance probability that rainfall events will exceed the threshold value. We calculated the number of landslides occurring from the events when rainfall exceeds the threshold value in the slope units to estimate the probability that a landslide will occur in this situation. Lastly, we employed the concept of conditional probability by multiplying this probability with the exceedance probability of rainfall events exceeding the threshold value, which yielded the probability that a landslide will occur in each slope unit for one year. The results indicated the slope units with high probability that at least one rainfall event will exceed the threshold value at the same time that one landslide will occur within any one year are largely located in the southwestern part of the Taipei Water Source Domain, and the highest probability is 0.26. These slope units are located in parts of the study area with relatively weak lithology, high elevations, and steep slopes. Compared with probability models based solely on landslide inventories, our proposed landslide probability model, combined with a long-term landslide inventory and rainfall factors, can avoid problems resulting from an incomplete landslide inventory, and can also be used to estimate landslide occurrence probability based on future potential changes in rainfall.

Keywords: landslide; rainfall threshold; landslide probability model; Taiwan

---

1. Introduction

Taiwan is a relatively new island formed by plate movements. Due to its high mountains, steep slopes, and relatively unstable geological conditions, as well as frequent typhoons and torrential rains, slopeland disasters are common in mountainous areas. Thus, slopeland hazard prevention and mitigation projects are necessary. In slopeland hazard prevention work, landslides have a high level of unpredictability. In particular, estimating the likelihood of landslides in large watersheds using deterministic models is difficult when no detailed geomorphological and hydrological data have been collected for the whole area. Therefore, the use of a stochastic model to assess landslide probability is more feasible. According to the definition, landslide hazard involves both spatial and temporal probability [1]. The analysis of landslide spatial probability is generally seen as a landslide susceptibility analysis in the research [2–12]. The landslide temporal probability, normally expressed in terms of frequency, return period, or exceedance probability [13], was analyzed in the research [14–21].

Methods of performing landslide temporal probability analysis can be classified as hydrological models and approaches based on exceedance probability [22]. The hydrological models employ infiltration models to determine the critical rainfall triggering landslides, which requires the estimation and validation of soil parameters over large areas, and therefore makes these models impractical for regional-scale applications. The approaches based on exceedance probability can be further subdivided into two types, where the first type employs a landslide inventory induced by a single rainfall event and rainfall data for that event to analyze the return period of the landslide event [8,23,24], and the second type employs a long-term landslide inventory to calculate the exceedance probability for the occurrence of landslides. Concerning the latter type, the Poisson probability model [17,25–29], binomial probability model, and empirical model [20] are commonly used to analyze the recurrence probability of landslide events. As a consequence, when a research area has a long-term landslide inventory, the Poisson probability model can be employed to estimate the temporal probability of landslides under the assumption that the frequency of future landslides occurring is the same as in the past. However, due to the constraints of this assumption, the Poisson probability model cannot separate the effect of geomorphological and hydrological factors on landslides, and therefore cannot be used to infer how landslide probability will change when climate change causes changes in the frequency of torrential rain and in the rainfall patterns. If the effects of geomorphological and hydrological factors can be considered separately and the occurrence probability of torrential rain events can be estimated independently, then the landslide temporal probability can be estimated correctly based on the change trends of the estimated torrential rain occurrence probability [30]. One approach to separate the effects of geomorphological and hydrological factors in landslide probability models is to employ the concept of conditional probability to separately estimate rainfall probability and landslide probability under these rainfall conditions. In this approach, a Poisson probability model is first used to calculate the exceedance probability of rainfall events that may trigger landslides, the landslide probability under these rainfall conditions is then calculated, and the two are multiplied to obtain the temporal probability of landslides [26,28,29].

Before calculating the probability of rainfall events that may trigger landslides, the scale of rainfall events that trigger landslides or the threshold rainfall events that must be exceeded to trigger landslides must first be understood. The minimum amount of rainfall needed to trigger landslides was first considered by Endo [31], and the rainfall thresholds triggering landslide events were quantified by Onodera et al. [32]. Campbell [33] suggested that the combined effect of both antecedent rainfall and rainfall intensity on the landslides needed to be considered, and a warning system could be based on the relationship between antecedent rainfall and critical rainfall [34]. Caine [35] used rainfall intensity and rainfall duration to establish global shallow landslide rainfall thresholds. Methods of establishing rainfall thresholds were classified as physical models and empirical models [36], where physical models employ detailed spatial information on hydrological, lithological, morphological, and soil characteristics as a basis for modeling the relationship between rainfall, infiltration, and landslide events. However, the information is hard to collect accurately over large areas. Empirical models can be grouped as thresholds combining rainfall duration, total event rainfall, or rainfall intensity parameters, thresholds considering antecedent rainfall, and thresholds combining other parameters, where the first two groups can be further subdivided into the following three categories based on the parameters used for determining rainfall thresholds [21]: the first category consists of intensity and duration parameters [18,20,29,34,35,37–39], the second category consists of antecedent rainfall conditions [26,28,29,40], and the third category consists of accumulative event rainfall and duration parameters [41]. Although rainfall intensity–duration models have been most commonly used in recent years [21], thresholds for rainfall-induced landslides may define the rainfall, soil moisture, or hydrological conditions that, when reached or exceeded, are likely to trigger landslides [36]. Some research has also suggested that groundwater and soil moisture are factors influencing the initiation of landslides [42,43], and antecedent rainfall can affect both of these factors. Accordingly, antecedent rainfall can be used to determine when landslides may occur [36]. In research on rainfall thresholds incorporating the antecedent rainfall conditions, large differences exist in the number of days of antecedent rainfall that were employed in each study. For example, daily rainfall was employed in conjunction with 15-day antecedent rainfall [39], both daily and 3-day cumulative rainfall were used [44], and 3-day and 30-day antecedent rainfall were employed [45]. Guzzetti et al. [36] suggested that the large variability in the antecedent rainfall may be attributed to three types of factors concerning the research area: diversity in lithological, morphological, vegetation, and soil conditions; differences in climatic regimes and meteorological circumstances; and the heterogeneity and incompleteness in the rainfall and landslide data used to establish the rainfall thresholds. As a consequence, the local conditions and availability of data in the research area must be assessed when choosing the number of days of antecedent rainfall.

Since the rainfall threshold determined using a single rain gauge for a large area constitutes one value for the entire area, as soon as rainfall reaches or exceeds the threshold, landslides may occur anywhere in that area, and knowing their precise locations is impossible. As a consequence, a denser array of rain gauges can be employed to acquire rainfall data with finer spatial resolution [21], and the research areas can be subdivided into analytical units with a smaller area, which can better account for the spatial variability of rainfall patterns in the analytical units and the spatial resolution of landslide prediction. However, 19.1% of recent studies on this subject failed to subdivide their research areas, and those studies that did subdivide their research areas had resulting analytical units with an average area of 302.0 km^2 [21]. For instance, a research area of 4660 km^2 was subdivided into 12 analytical units with an average area of 388.3 km^2 [39], but excessively large analytical units make it impossible to identify the precise possible locations of landslides. In addition, the subdivision approaches employed in some studies run into the problem of incomplete coverage. For instance, although a 25 km^2 research area was subdivided into eight analytical units, the landslide prediction results only represented the paths of roads in the subdivisions and not the entire subdivisions because most landslides (94%) in the study occurred on roadside slopes [26]. Althuwaynee et al. [28] divided the research area into six circular analytical units with their centers at rain gauges, but the analytical units did not cover the entire research area and also overlapped. Although these studies subdivided their research areas into different analytical units, the units could not provide a landslide probability distribution with a finer spatial resolution because they were excessively large, or experienced problems such as incomplete coverage and overlap. If the method of subdividing a research area into analytical units is improved so that the units are smaller in area, the spatial resolution of the landslide probability estimation results could be improved. There are seven types of analytical units subdivided in research areas: grid cell, terrain unit, unique condition unit, slope unit, geo-hydrological unit, topographical unit, and administrative unit [46,47]. The slope units are suitable for landslide probability analysis because they express topographic features and slope characteristics. In this study, we consequently selected slope units as our analytical unit.

2. Research Area and Materials

2.1. Environmental Setting of Taipei Water Source Domain

Taipei Water Source Domain is located in the northeast part of Taiwan and supplies tap water for five million people in the greater Taipei area. The area is characterized by hilly and mountainous topography, as well as the Xueshan Range extending to the northeast and a subrange of Mt. Qilan extending to the northwest, both of which account for the area's high terrain in the south and low terrain in the north. Elevations in the area range from 12 to 2130 m (Figure 1).

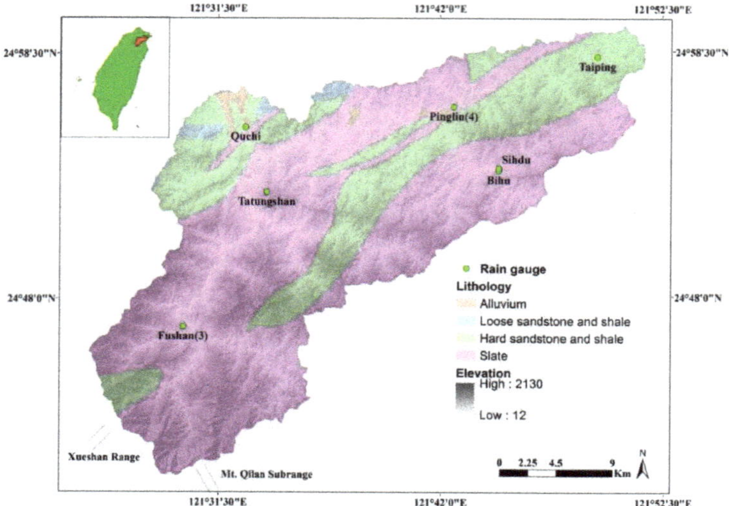

Figure 1. Distribution of elevation, lithology, and rain gauges in the Taipei Water Source Domain.

Concerning the distribution of lithology, we followed the classification approach proposed by Lin et al. [48] by dividing the Taipei Water Source Domain into areas underlain by alluvium, loose sandstone and shale, hard sandstone and shale, and slate. Whereas alluvium found at the confluence of rivers and in downstream areas covers only a small part of the research area, hard sandstone and shale as well as slate underlay most of the research area. Of these types of lithology, areas underlain by slate had the highest number of landslides and the greatest landslide area. Wu et al. [49] indicated that the areas underlain by hard sandstone and shale as well as slate in the Kaoping River Watershed had the highest landslide ratios in 2008 and 2009. This indicates that the lithology condition of most areas is fragile. Typhoons and torrential rain events can readily wash away unconsolidated sand and gravel as well as trigger landslides, which deposit large loads of sediment in rivers and reservoirs.

2.2. Rainfall Data

The rain gauges employed in this study were located as shown in Figure 1, and rainfall data between 1987 to 2016 from these rain gauges were used. Average daily rainfall for the entire area during the same period as the 2000–2015 landslide inventory is shown in Figure 2. Figure 2 shows that apart from the eight typhoon events causing the corresponding landslide inventory, other events of high daily rainfall occurred without a significant increase in landslides. As a consequence, apart from calculating the exceedance probability that rainfall events will exceed the rainfall threshold, we also calculated the probability of landslides when rainfall events exceed the threshold. In addition, Figure 3 shows the average daily rainfall and standard deviation of the eight typhoon events during the 2000–2015 period in each control area of a rain gauge divided by a modified Thiessen polygon method, considering the morphology of the area, proposed by Salvaticic et al. [19].

2.3. Landslide Inventory

After selecting eight major typhoon events occurring in the research area during the 2000–2015 period—typhoons Xangsane (2000), Nari (2001), Aere (2004), Sinlaku (2008), Morakot (2009), Parma (2009), Megi (2010), and Soudelor (2015)—we collected satellite images before and after each typhoon event, calculated and classified the normalized difference vegetation index (NDVI) to find the possible locations of landslide sites, and eliminated and revised unlikely landslide sites according to the slope, drainage, and land use maps in the study area. In the process of mapping the source areas of landslides from the satellite images, we often found that human mapping errors affected interpretation quality. We followed the recommended procedures proposed by Liu et al. [50] to map landslides in the research area. Table 1 shows the dates of the eight landslide events and landslide statistical data. The size of landslides ranged from 16 to 118,108 m^2 and the average area was 2474 m^2. The resulting distribution of landslides caused by the eight typhoon events was shown in Figure 4, which revealed that landslide sites were concentrated in the southwestern portion of the research area.

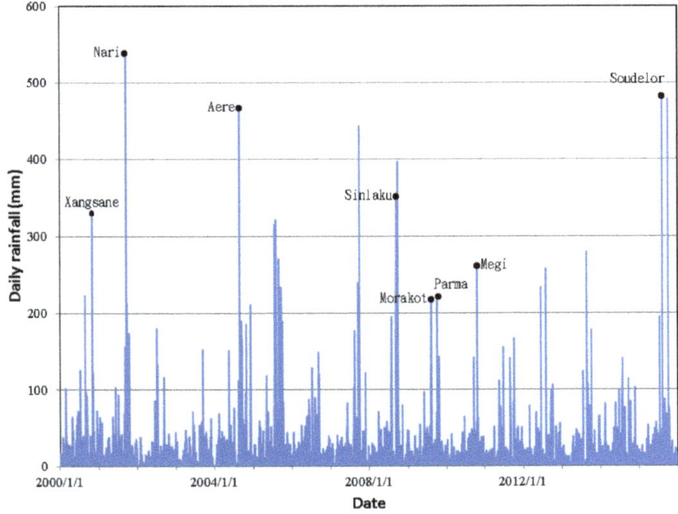

Figure 2. Average daily rainfall within the Taipei Water Source Domain, 2000–2015. The dots represent the eight typhoon events causing the corresponding landslide inventory.

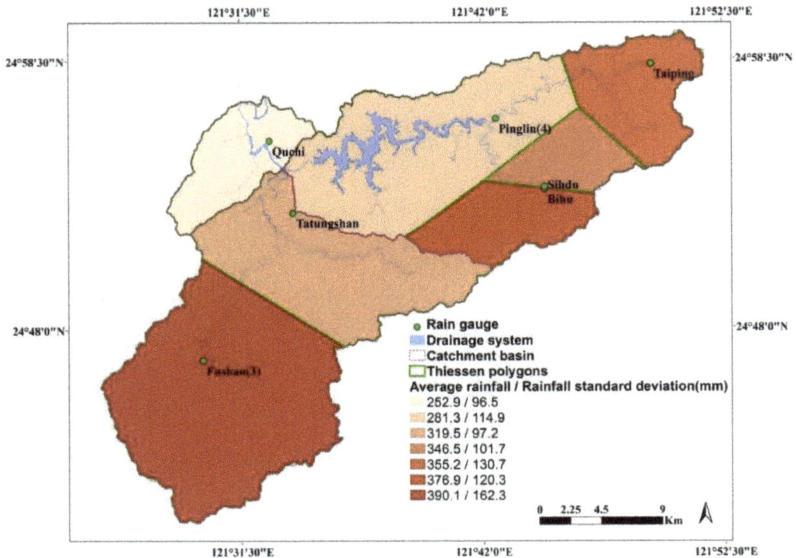

Figure 3. Average daily rainfall and standard deviation of the eight typhoon events in each control area of the rain gauge divided by the modified Thiessen polygon method.

Table 1. Landslide inventory for the eight typhoon events.

Typhoon Event	Date (MM/DD/YYYY)	Average Rainfall at the Date (mm)	Number of New Landslide Sites	Smallest Landslide Area (m^2)	Largest Landslide Area (m^2)	Total Area of Landslides (m^2)	Average Area of Landslides (m^2)
Xangsane	11/01/2000	326.67	42	326	19,619	131,148	3123
Nari	09/16/2001	538.05	92	107	68,032	261,650	2844
Aere	08/24/2004	465.57	97	140	27,270	239,856	2473
Sinlaku	09/13/2008	348.18	32	475	21,101	71,111	2222
Morakot	08/07/2009	219.83	173	16	118,108	1,016,448	5875
Parma	10/05/2009	221.79	302	47	49,369	484,785	1605
Megi	10/21/2010	262.35	47	407	27,318	118,874	2529
Soudelor	08/08/2015	478.99	589	257	48,041	1,075,263	1826

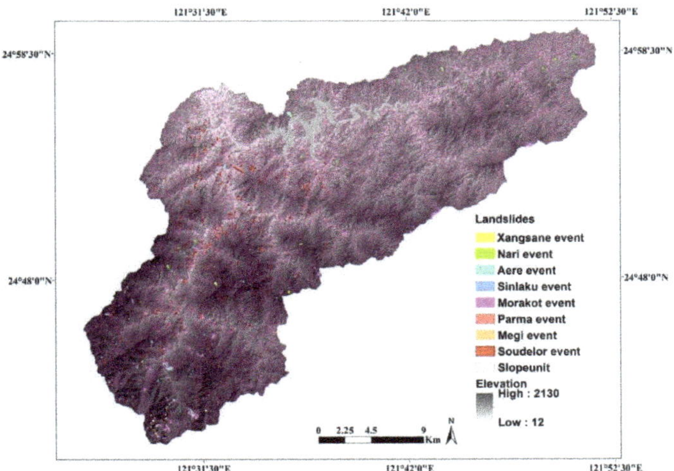

Figure 4. Distribution of slope units and landslide sites caused by the eight typhoon events in the Taipei Water Source Domain.

2.4. Analytical Units and Rain Gauge Control Areas

We employed slope units as analytical units due to their relatively well-defined topographic boundaries, as well as topographic and geological meaning. We employed the subdivision method used by Xie et al. [51] to divide the watershed into slope units. The original topography could be divided into sub-watersheds, and the combination of sub-watershed units before and after reversal yielded the slope units. We ensured that the smallest area of slope units was larger than the average area of landslides [47], which minimized the chance that any specific landslide site would be a part of different slope units, and thereby confuse the analysis results. We also divided the research area into rain gauge control areas (Figure 3) based on rain gauge locations and using the modified Thiessen polygon method. The rainfall measured by each rain gauge was taken as representative of the control area in which that gauge was located, and we expected this approach to reflect the different rainfall distribution characteristics within the research area.

3. Methods

3.1. Analysis of Discrete Rainfall Groups

The two rainfall parameters considered in this study consisted of daily rainfall (I) and effective accumulated rainfall (R_t). After selecting rain gauges near the research area with rainfall data for recent years, we obtained daily rainfall data for the 1987–2016 period from the Water Resources Agency and Central Weather Bureau. This study calculated the effective accumulated rainfall based on rainfall for that day and rainfall during the previous 7 days using the method proposed by Jan [52]; this calculation was performed using Equation (1):

$$R_t = R_0 + \sum_{i=1}^{7} \alpha^i R_i = \sum_{i=0}^{7} \alpha^i R_i \quad (1)$$

where R_0 is the rainfall amount on that day, R_1 is the rainfall amount on the day before that day, and so on, and the weighting coefficient $\alpha = 0.7$ proposed by Jan [52].

Adopting the concept proposed by Tsai [53], after using daily rainfall data to calculate effective accumulated rainfall (R_t), we obtained a group of daily rainfall and effective accumulated rainfall (I, R_t) for each day. The daily rainfall and effective accumulated rainfall were continuous variables and would not facilitate subsequent calculation of a joint cumulative distribution function, therefore we rounded off the daily rainfall and effective accumulated rainfall values to the 10th place and made them discrete variables. The group of daily rainfall and effective accumulated rainfall (I, R_t) for each day was termed as "discrete rainfall group" in this study.

We defined different rainfall events by the continuity of daily rainfall. Consecutive days of non-zero daily rainfall were considered to be the same rainfall event, and the number of the consecutive days varied from event to event. We then calculated the distance (d) from each discrete rainfall group to the origin (0, 0), and assumed that the greater the value of d, the greater the likelihood of landslides. The discrete rainfall group with the greatest d in each rainfall event was used to represent that rainfall event in subsequent analysis.

3.2. Joint Cumulative Distribution Function

The joint cumulative distribution function was obtained from the joint probability mass function of the foregoing discrete rainfall groups. The probability ($P_{I,Rt}$ (I_i, R_{tj})) of each discrete rainfall group (I_i, R_{tj}) was defined [54] as shown in Equation (2):

$$P_{I,\,R_t}\left(I_i,\,R_{tj}\right) = P\left(I = I_i \cap\, R_t = R_{tj}\right) \quad (2)$$

where i = 0, 10, 20, 30, ... ; j = 0, 10, 20, 30, ... ; the joint probability mass function has a probability value only when I and R_t are multiples of 10 and the probability values in other places are 0.

The foregoing joint probability mass function yielded a joint cumulative distribution function using:

$$F_{I,\,R_t}\left(I_i,\,R_{tj}\right) = \sum_0^i \sum_0^j P_{I,\,R_t}\left(I_i,\,R_{tj}\right). \quad (3)$$

The joint cumulative distribution function was a monotonic increasing function with a range between 0 and 1, and had the form of a three-dimensional curved surface when plotted on coordinate axes. The farther the point (I_i, R_{tj}) from the origin, the greater its probability value. The probability of a discrete rainfall group on the curved surface expressed the cumulative probability of all discrete rainfall groups, which were nearer to the origin than this discrete rainfall group (I_i, R_{tj}).

3.3. Selection of a Rainfall Probability Threshold

After establishing a joint cumulative distribution function, taking each 0.05 as an interval, we set 20 rainfall probability thresholds ranging from 0.05 to 1.00, and employed the error matrix concept to calculate the true positive rate (TPR), true negative rate (TNR), and positive predictive value (PPV) for each rainfall probability threshold at each rain gauge control area. The rainfall probability threshold was treated as the threshold of cumulative probability of the discrete rainfall groups which was used to predict whether rainfall events could trigger landslides. Here, TPR expresses the ratio of discrete rainfall groups that correctly predicted landslide occurrence to discrete rainfall groups triggering landslides actually, TNR expresses the ratio of discrete rainfall groups that correctly predicted no landslide occurrence to discrete rainfall groups triggering no landslides actually, and PPV expresses the ratio of discrete rainfall groups that correctly predicted landslide occurrence to discrete rainfall groups predicting landslides. To capture the performance of each threshold, PPV and Youden's index were used for comprehensive consideration. The higher the PPV and Youden's index values, the more accurate the rainfall probability threshold at classifying landslide occurrence and landslide occurrence for discrete rainfall groups. The TPR, TNR, PPV, and Youden's index calculations were performed employing Equations (4)–(7).

$$\text{TPR }(\%) = \frac{\text{Number of discrete rainfall groups predicting landslides when landslides actually occurred}}{\text{Number of all discrete rainfall groups triggering landslides actually}} \quad (4)$$

$$\text{TRN } (\%) = \frac{\text{Number of discrete rainfall groups predicting no landslides when no landslides occurred}}{\text{Number of all discrete rainfall groups triggering no landslides actually}} \quad (5)$$

$$\text{PPV } (\%) = \frac{\text{Number of discrete rainfall groups predicting landslides when landslides actually occurred}}{\text{Number of all discrete rainfall groups predicting landslides}} \quad (6)$$

$$\text{Youden's index} = \text{TPR} + \text{TNR} - 1 \quad (7)$$

3.4. Poisson Probability Model

A Poisson probability model relies on the past frequency of events to predict their occurrence probability in the future. The basic assumption underlying this type of model is that future events will occur with the same frequency as past events. In this model, the probability of at least one event occurring in the time interval (t) is given by Equation (8):

$$P(N(t) \geq 1) = 1 - e^{-\lambda t} \quad (8)$$

where $P(N(t) \geq 1)$ indicates the probability of at least one event occurring within a period of t years; this probability is known as the exceedance probability.

We calculated the number of discrete rainfall groups exceeding the threshold at each rain gauge in the past using the optimal rainfall probability thresholds and then divided by the years of the rainfall data to obtain the occurrence frequency (λ), which was used to calculate the exceedance probability. The exceedance probability indicated the probability of at least one rainfall event exceeding the threshold of discrete rainfall groups within any one year.

3.5. Conditional Probability

We employed the concept of conditional probability in the analysis. We first used the Poisson probability model to calculate the exceedance probability of at least one rainfall event exceeding the threshold of discrete rainfall groups within any one year at each rain gauge control area. We divided the number of landslides occurring in each slope unit by the number of rainfall events exceeding the threshold of discrete rainfall groups to estimate the probability that a landslide would occur in that slope unit when the rainfall exceeded the threshold. Lastly, we multiplied the two probabilities together to obtain the probability that a rainfall event would exceed the threshold of discrete rainfall groups and at least one landslide would also occur in each slope unit within any one year, as shown in Equation (9):

$$P(R \geq RT \cap L) = P(R \geq RT) \times P(L|R \geq RT) \quad (9)$$

where $R \geq RT$ indicates rainfall events exceed the threshold of the discrete rainfall group and L indicates the occurrence of a landslide.

4. Results and Discussion

4.1. Joint Cumulative Distribution Functions of the Rain Gauges

In this study, we collected multi-year daily rainfall data from each rain gauge and calculated the effective accumulated rainfall (R_t) by employing Equation (1), which yielded rainfall and effective accumulated rainfall for each day. We then rounded off the daily rainfall and effective accumulated rainfall values to the 10th place, which yielded discrete rainfall groups including both daily rainfall and effective accumulated rainfall. The next step was establishing frequency tables for different discrete rainfall groups, which we used to show the frequency of the discrete rainfall groups. Figure 5 shows the frequency of discrete rainfall groups at the Bihu rain gauge with daily rainfall and effective accumulated rainfall (R_t) ranging from 0 to 100 mm. The depth axis represents daily rainfall, the horizontal axis represents the effective accumulated rainfall (R_t), and the vertical axis represents the frequency of a discrete rainfall group. We then calculated the cumulative frequency of each discrete rainfall group on this basis, and this represented the frequency of all discrete rainfall groups with values lower than that of any designated discrete rainfall group. The cumulative frequency was then divided by the total frequency of all discrete rainfall groups, which yielded the cumulative probability of each discrete rainfall group.

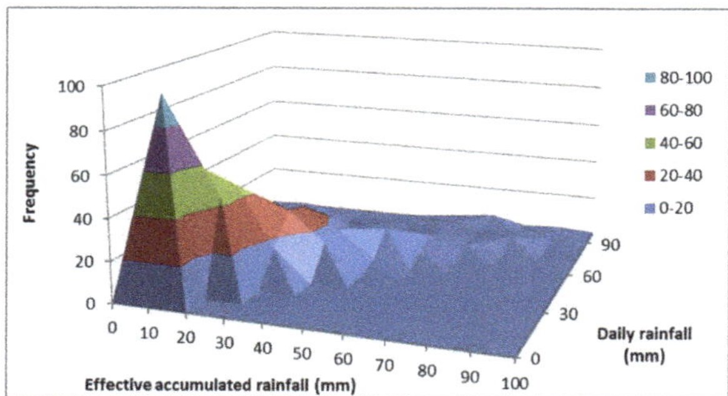

Figure 5. Frequency distribution of discrete rainfall groups at the Bihu rain gauge.

The joint cumulative distribution function of each rain gauge was then obtained from the cumulative probability of the discrete rainfall groups, and this function was used to plot a joint cumulative distribution chart. Figures 6 and 7 are joint cumulative distribution functions for the Bihu and Fushan (3) rain gauges, and daily rainfall and effective accumulated rainfall (R_t) are shown within a 0–300 mm range. The joint cumulative distribution functions have areas with gentler slopes indicating fewer and more dispersed discrete rainfall groups within a certain interval, and areas with steeper slopes indicating more and more concentrated discrete rainfall groups within a certain interval.

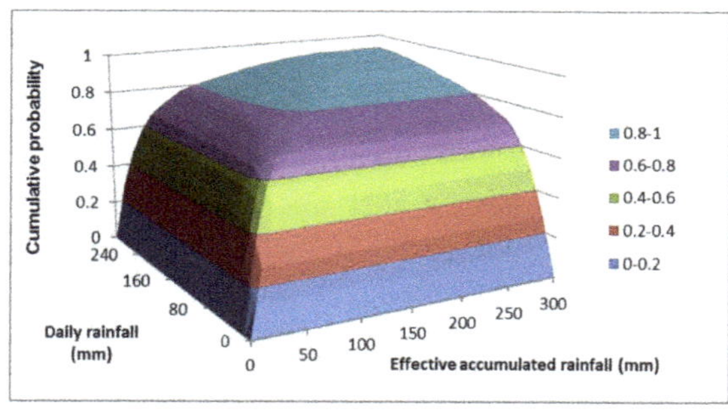

Figure 6. The joint cumulative distribution function for the Bihu rain gauge.

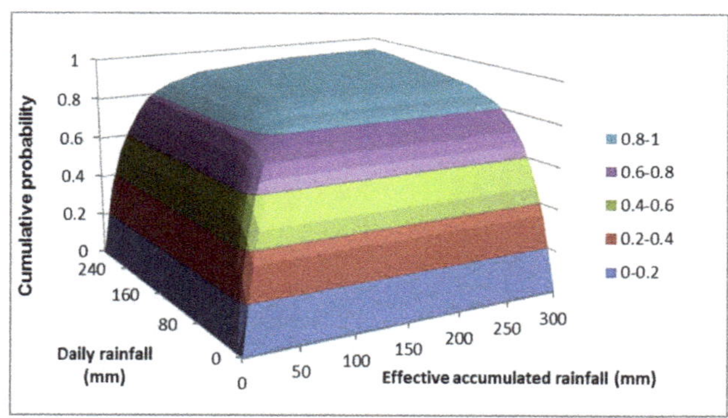

Figure 7. The joint cumulative distribution function for the Fushan (3) rain gauge.

4.2. Selection of Rainfall Probability Thresholds of the Rain Gauges

Following the analysis results of the joint cumulative distribution functions of the rain gauges, we used rainfall data from the rain gauges during the eight rainfall events triggering landslides to select rainfall probability thresholds. The rainfall probability threshold was treated as the threshold of cumulative probability of the discrete rainfall groups which was used to predict whether rainfall events could trigger landslides. Starting with a rainfall probability threshold value of 0.05, we set a rainfall probability threshold at each interval of 0.05 until a value of 1.00 was reached, and then calculated the TPR, TNR, PPV, and Youden's index of each rainfall probability threshold. Here, the number of landslide events predicted correctly divided by the number of rainfall events triggering landslides actually equaled TPR, the number of no landslide events predicted correctly divided by the number of rainfall events triggering no landslides actually equaled TNR, and the number of landslide events predicted correctly divided by the number of rainfall events predicting landslides equaled PPV. Table 2 shows the results of these calculations for the Bihu rain gauge. In the analysis results for the individual rain gauges, the rainfall probability thresholds with the highest Youden's index were within the probability interval of 0.85–0.95, and the rainfall probability thresholds with the highest PPV were at the probability of 0.95 in all cases. We consequently opted to use a rainfall probability threshold value of 0.95 for the whole area. The TPR, TNR, PPV, and Youden's index for all rain gauges when the rainfall probability threshold was 0.95 are shown in Table 3.

Table 2. True positive rate (TPR), true negative rate (TNR), positive predictive value (PPV), and Youden's index for Bihu rain gauge at different rainfall probability thresholds.

Rainfall Probability Threshold	0.05	0.10	0.15	0.20	0.25	0.30	0.35	0.40	0.45	0.50
Number of landslide events predicted correctly	8	8	8	8	8	8	8	8	8	8
Number of rainfall events triggering landslides actually	8	8	8	8	8	8	8	8	8	8
TPR	100%	100%	100%	100%	100%	100%	100%	100%	100%	100%
Number of no landslide events predicted correctly	0	0	92	92	173	237	237	292	332	361
Number of rainfall events triggering no landslides actually	608	608	608	608	608	608	608	608	608	608
TNR	0%	0%	15%	15%	28%	39%	39%	48%	55%	59%
PPV	1.3%	1.3%	1.5%	1.5%	1.8%	2.1%	2.1%	2.5%	2.8%	3.1%
Youden's index	0%	0%	15%	15%	28%	39%	39%	48%	55%	59%

Rainfall Probability Threshold	0.55	0.60	0.65	0.70	0.75	0.80	0.85	0.90	0.95	1.00
Number of landslide events predicted correctly	8	8	8	8	8	8	8	8	7	0
Number of rainfall events triggering landslides actually	8	8	8	8	8	8	8	8	8	8
TPR	100%	100%	100%	100%	100%	100%	100%	100%	88%	0%
Number of no landslide events predicted correctly	407	431	469	494	526	549	575	588	602	608
Number of rainfall events triggering no landslides actually	608	608	608	608	608	608	608	608	608	608
TNR	67%	71%	77%	81%	87%	90%	95%	97%	99%	100%
PPV	3.8%	4.3%	5.4%	6.6%	8.9%	11.9%	19.5%	28.6%	53.8%	-
Youden's index	67%	71%	77%	81%	87%	90%	95%	97%	87%	0%

Table 3. TPR, TNR, PPV, and Youden's index for all rain gauges at a rainfall probability threshold of 0.95.

Rainfall Gauge	Bihu	Fushan (3)	Tatungshan	Pinglin (4)	Sihdu	Taiping	Quchi
Number of landslide events predicted correctly	7	4	8	5	8	7	7
Number of rainfall events triggering landslides actually	8	8	8	8	8	8	8
TPR	88%	50%	100%	63%	100%	88%	88%
Number of no landslide events predicted correctly	602	586	652	598	563	549	610
Number of rainfall events triggering no landslides actually	608	599	662	605	583	565	629
TNR	99%	98%	98%	99%	97%	97%	97%
PPV	53.8%	23.5%	44.4%	41.7%	28.6%	30.4%	26.9%
Youden's index	87%	48%	98%	61%	97%	85%	84%

4.3. Landslide Probability Analysis Employing a Rainfall Probability Threshold and a Long-Term Landslide Inventory

After determining a rainfall probability threshold for the rain gauges, we calculated the number of discrete rainfall groups exceeding this threshold at each rain gauge during the 1987–2016 period. We then divided these values by the years of statistics at each gauge, which yielded the λ values in Equation (8). Substituting $t = 1$ year into Equation (8) allowed us to calculate the probability of at least one rainfall event exceeding the threshold of discrete rainfall group within any one year (i.e., $P(R$

$\geq RT$) in Equation (9)) under the assumption that future rainfall conditions will be the same as past conditions. Figure 8 shows the exceedance probability value calculated for each rain gauge overlaid on each rain gauge control area. The Quchi rain gauge control area had the highest probability of 0.76502 that at least one rainfall event will exceed the threshold of the discrete rainfall group within any one year, whereas the Bihu rain gauge control area had the lowest probability of 0.43886.

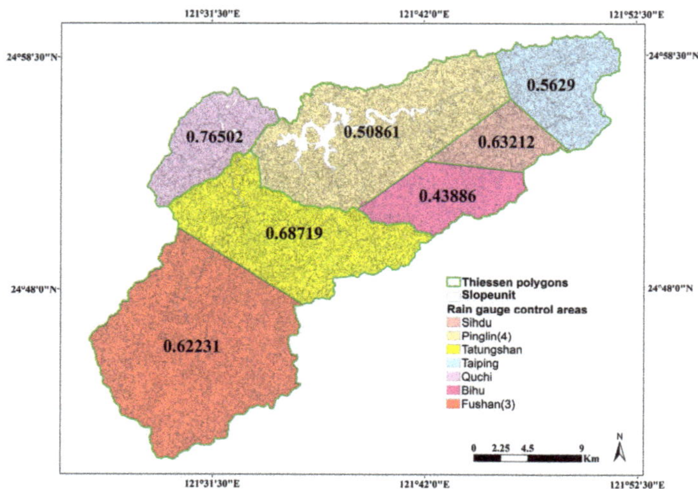

Figure 8. The exceedance probability that at least one rainfall event will exceed the threshold of discrete rainfall group within any one year in each rain gauge control area.

In this study, we also divided the number of landslides occurring in each slope unit during the 2000–2015 period by the number of rainfall events exceeding the threshold of discrete rainfall group at the rain gauges to which the slope units were assigned during the same period to estimate the landslide probability in the slope units when the rainfall exceeded the threshold, which is $P(L|R \geq RT)$ in Equation (9). The resulting probability distribution is shown in Figure 9. Figure 9 shows that the different slope units within a single rain gauge control area have different landslide probabilities, and these differences should be attributed to different geomorphological conditions in the slope units.

Lastly, employing Equation (9), we multiplied the probability $P(R \geq RT)$ that at least one rainfall event will exceed the threshold of discrete rainfall group within any one year in each rain gauge control area by the landslide probability $P(L|R \geq RT)$ in each slope unit when rainfall exceeds the threshold, which yielded the probability that at least one rainfall event exceeds the threshold of discrete rainfall group at the same time that one landslide will occur in each slope unit during the future one-year period (Figure 10). The two probability maps shown in Figures 9 and 10 were validated by the landslide inventory data respectively. The landslides were mainly distributed in the slope units where the landslide probability values were greater than 0.01. The top 2% of slope units ranked with landslide probabilities included 50.40% of slope units where landslides occurred while the top 6% of slope units ranked with landslide probabilities included 100.00% of slope units where landslides occurred in Figure 10. The results indicated these maps had reasonable landslide probability distributions. Figure 10 reveals that the Fushan (3) rain gauge control area, which is located in the southwest part of the research area, contained relatively many slope units with high landslide probability, and the highest probability value was 0.26. Apart from having fragile lithology consisting of hard sandstone and shale as well as slate, this area has a higher elevation and steeper slopes than other control areas, which suggests that elevation and slope have a definite correlation with landslide occurrence.

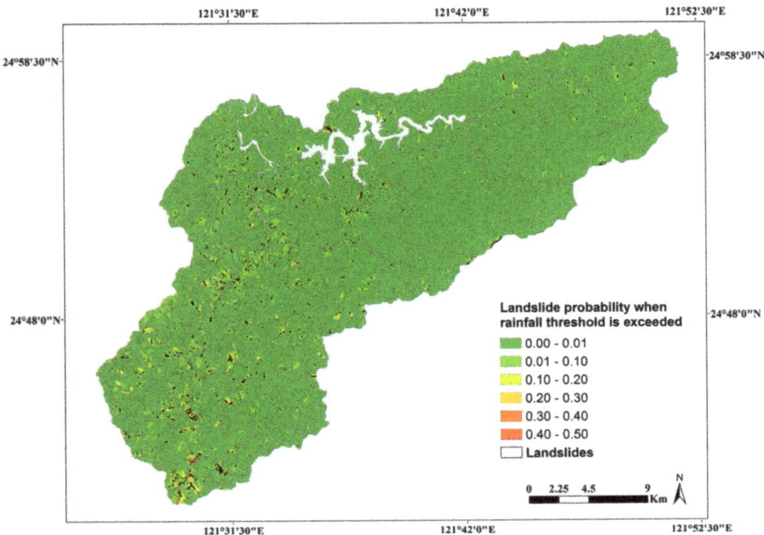

Figure 9. The landslide probability of slope units in the Taipei Water Source Domain when rainfall exceeds the threshold of discrete rainfall group.

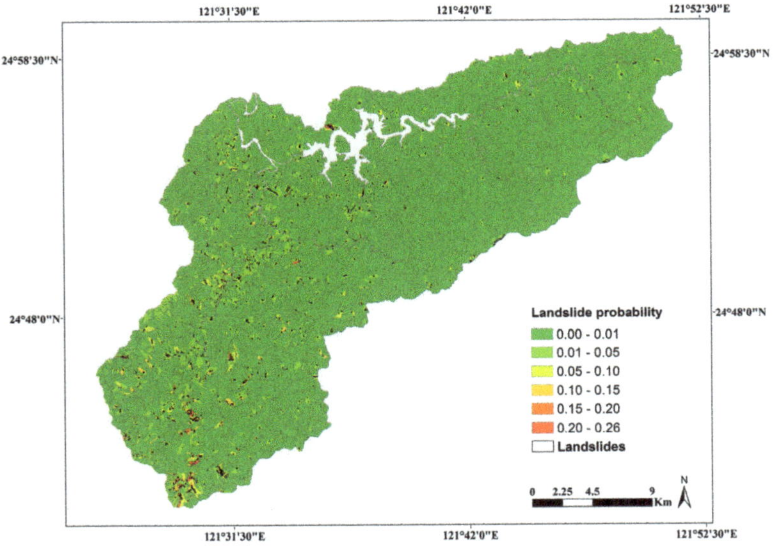

Figure 10. The probability that at least one rainfall event exceeds the threshold of discrete rainfall group and one landslide will also occur during the future one-year period within the Taipei Water Source Domain.

4.4. Discussion

In comparison with a landslide probability model based solely on the use of landslide inventories, our landslide probability model based on the use of landslide inventories and rainfall factors reflect different basic assumed conditions. The assumption of the landslide probability model incorporating rainfall factors is that the frequency of future rainfall events exceeding the threshold and the frequency of landslides occurring when the threshold has been exceeded are the same as in the past. In contrast, the assumption of a landslide probability model based solely on the use of landslide inventories is that the frequency of future landslides occurring is the same as in the past. As a consequence, landslide probability models incorporating rainfall factors possess the following advantages: (1) This model can reflect the differences in landslide probability between the rain gauge control areas that have different rainfall conditions. (2) When rainfall data were added in the analysis, the probability model we obtained yielded more reliable results because the rainfall data were collected from a longer period (29–45 years) than the landslide inventory (16 years). (3) If we know how the probability of at least one rainfall event exceeding the threshold will change in the future, the incorporation of rainfall factors in the landslide probability model will allow the effect of possible rainfall changes on the landslide probability to be assessed.

However, several aspects connected to the application of this landslide probability model still require further investigation: (1) The method of analyzing landslide probability proposed in this study requires a long-term landslide inventory and rainfall data, therefore attention must be paid to the completeness of rainfall data for the research area and handling methods when data are incomplete. (2) Whereas the rainfall factors used in this study reflect daily rainfall and effective accumulated rainfall, the use of different rainfall factors will yield different analysis results, which may be explored further in future research. (3) Apart from the modified Thiessen polygon method, the division of rain gauge control areas can be performed using other methods, such as the height–balance polygon method. A finer division method should yield more precise results of a landslide probability distribution, therefore future research can also compare the applicability of different methods of division into rain gauge control areas. (4) We obtained a long-term landslide inventory consisting of only eight events, therefore all events collected were used in the process of building the model. The landslide inventory covering the period of other events may be collected to verify the predictive ability of this landslide probability model.

5. Conclusions

In this study, we employed joint cumulative distribution functions to calculate the TPR, TNR, PPV, and Youden's index for different rainfall probability thresholds, selected a threshold of 0.95 as suitable for the research area, and used this rainfall probability threshold to calculate the Poisson probability of at least one rainfall event exceeding the threshold of discrete rainfall groups at each rain gauge within the future one-year period. We then combined this probability with the landslide probability in individual slope units when rainfall exceeded the threshold value, which allowed us to estimate the probability that a landslide will occur in individual slope units during the future one-year period. Many of the slope units with a high landslide probability are located in the Fushan (3) rain gauge control area, and the highest probability is 0.26. Apart from fragile lithology, this area is characterized by high elevations and steep slopes, which indicates that the elevation and slope have a significant influence on the occurrence of landslides. This finding suggests that this area should be a focal area for landslide prevention and mitigation efforts.

The landslide probability model established based on the use of a long-term landslide inventory and rainfall factor had a finer spatial resolution and data for a longer period, which yielded more reliable results and enabled the effect of possible rainfall changes on the landslide probability to be assessed. The effects of the completeness of rainfall data for the research area, the use of different rainfall factors, as well as the different methods of division into rain gauge control areas on the landslide probability analysis results can be significant and still require further investigation.

Author Contributions: Conceptualization, C.-Y.W.; data curation, Y.-C.Y.; formal analysis, Y.-C.Y.; methodology, C.-Y.W. and Y.-C.Y.; supervision, C.-Y.W.; visualization, Y.-C.Y.; writing—original draft, C.-Y.W. and Y.-C.Y.; writing—review & editing, C.-Y.W.; funding acquisition, C.-Y.W. All authors have read and agreed to the published version of the manuscript.

Funding: This research was funded by the Ministry of Science and Technology, Taiwan (MOST 107-2311-B-005-001).

Conflicts of Interest: The authors declare no conflict of interest.

References

1. Varnes, D.J. *IAEG Landslide Hazard Zonation: A Review of Principles and Practice (No.3)*; United Nations: New York, NY, USA, 1984.
2. Carrara, A. Multivariate models for landslide hazard evaluation. *J. Int. Assoc. Math. Geol.* **1983**, *15*, 403–426. [CrossRef]
3. Chung, C.J.F.; Fabbri, A.G. The representation of geoscience information for data integration. *Nonrenew. Resour.* **1993**, *2*, 122–139. [CrossRef]
4. Baeza, C.; Corominas, J. Assessment of shallow landslide susceptibility by means of multivariate statistical techniques. *Earth Surf. Process. Landf.* **2001**, *26*, 1251–1263. [CrossRef]
5. Guzzetti, F.; Galli, M.; Reichenbach, P.; Ardizzone, F.; Cardinali, M. Landslide hazard assessment in the Collazzone area, Umbria, Central Italy. *Nat. Hazards Earth Syst. Sci.* **2006**, *6*, 115–131. [CrossRef]
6. Zêzere, J.L.; Oliveira, S.C.; Garcia, R.A.C.; Reis, E. Landslide risk analysis in the area North of Lisbon (Portugal): Evaluation of direct and indirect costs resulting from a motorway disruption by slope movements. *Landslides* **2007**, *4*, 123–136. [CrossRef]
7. Carrara, A.; Crosta, G.; Frattini, P. Comparing models of debris-flow susceptibility in the alpine environment. *Geomorphology* **2008**, *94*, 353–378. [CrossRef]
8. Lee, C.T.; Huang, C.C.; Lee, J.F.; Pan, K.L.; Lin, M.L.; Dong, J.J. Statistical approach to storm event-induced landslides susceptibility. *Nat. Hazards Earth Syst. Sci.* **2008**, *8*, 941–960. [CrossRef]
9. Rossi, M.; Guzzetti, F.; Reichenbach, P.; Mondini, A.C.; Peruccacci, S. Optimal landslide susceptibility zonation based on multiple forecasts. *Geomorphology* **2010**, *114*, 129–142. [CrossRef]
10. Nefeslioglu, H.A.; Gokceoglu, C. Probabilistic risk assessment in medium scale for rainfall-induced earthflows: Catakli catchment area (Cayeli, Rize, Turkey). *Math. Probl. Eng.* **2011**, *2011*, 1–21. [CrossRef]
11. Chan, H.C.; Chen, P.A.; Lee, J.T. Rainfall-induced landslide susceptibility using a rainfall–runoff model and logistic regression. *Water* **2018**, *10*, 1354. [CrossRef]
12. Roccati, A.; Faccini, F.; Luino, F.; Ciampalini, A.; Turconi, L. Heavy rainfall triggering shallow landslides: A susceptibility assessment by a GIS-approach in a Ligurian Apennine catchment (Italy). *Water* **2019**, *11*, 605. [CrossRef]
13. Corominas, J.; van Westen, C.; Frattini, P.; Cascini, L.; Malet, J.P.; Fotopoulou, S.; Catani, F.; Van Den Eeckhaut, M.; Mavrouli, O.; Agliardi, F.; et al. Recommendations for the quantitative analysis of landslide risk. *Bull. Eng. Geol. Environ.* **2014**, *73*, 209–263. [CrossRef]
14. Guzzetti, F.; Reichenbach, P.; Cardinali, M.; Galli, M.; Ardizzone, F. Probabilistic landslide hazard assessment at the basin scale. *Geomorphology* **2005**, *72*, 272–299. [CrossRef]
15. Jaiswal, P.; van Westen, C.J.; Jetten, V. Quantitative landslide hazard assessment along a transportation corridor in southern India. *Eng. Geol.* **2010**, *116*, 236–250. [CrossRef]
16. Das, I.; Stein, A.; Kerle, N.; Dadhwal, V.K. Probabilistic landslide hazard assessment using homogeneous susceptible units (HSU) along a national highway corridor in the northern Himalayas, India. *Landslides* **2011**, *8*, 293–308. [CrossRef]
17. Ghosh, S.; van Westen, C.J.; Carranza, E.J.M.; Jetten, V.G.; Cardinali, M.; Rossi, M.; Guzzetti, F. Generating event-based landslide maps in a data-scarce Himalayan environment for estimating temporal and magnitude probabilities. *Eng. Geol.* **2012**, *128*, 49–62. [CrossRef]
18. Segoni, S.; Tofani, V.; Rosi, A.; Catani, F.; Casagli, N. Combination of Rainfall Thresholds and Susceptibility Maps for Dynamic Landslide Hazard Assessment at Regional Scale. *Front. Earth Sci.* **2018**, *6*, 85. [CrossRef]
19. Salvatici, T.; Tofani, V.; Rossi, G.; D'Ambrosio, M.; Stefanelli, C.T.; Masi, E.B.; Rosi, A.; Pazzi, V.; Vannocci, P.; Petrolo, M.; et al. Application of a physically based model to forecast shallow landslides at a regional scale. *Nat. Hazards Earth Syst. Sci.* **2018**, *18*, 1919–1935. [CrossRef]
20. Rosi, A.; Canavesi, V.; Segoni, S.; Nery, T.D.; Catani, F.; Casagli, N. Landslides in the Mountain Region of Rio de Janeiro: A Proposal for the Semi-Automated Definition of Multiple Rainfall Thresholds. *Geosciences* **2019**, *9*, 203. [CrossRef]
21. Segoni, S.; Piciullo, L.; Gariano, S.L. A review of the recent literature on rainfall thresholds for landslide occurrence. *Landslides* **2018**, *15*, 1483–1501. [CrossRef]

22. Vasu, N.N.; Lee, S.R.; Pradhan, A.M.S.; Kim, Y.T.; Kang, S.H.; Lee, D.H. A new approach to temporal modelling for landslide hazard assessment using an extreme rainfall induced-landslide index. *Eng. Geol.* **2016**, *215*, 36–49. [CrossRef]
23. Lee, C.T. Multi-Stage Statistical Landslide Hazard Analysis: Earthquake-Induced Landslides. In *Landslide Science for a Safer Geoenvironment, Proceedings of the International Consortium on Landslides (ICL) Third Landslide Forum, Beijing, China, 2–6 June 2014*; Springer: Cham, Switzerland, 2014.
24. Lee, C.T.; Chung, C.C. Common Patterns Among Different Landslide Susceptibility Models of the Same Region. In *Advancing Culture of Living with Landslides, Proceedings of the 4th World Landslide Forum, Ljubljana, Slovenia, 29 May–2 June 2017*; Springer: Cham, Switzerland, 2017.
25. Önöz, B.; Bayazit, M. Effect of the occurrence process of the peaks over threshold on the flood estimates. *J. Hydrol.* **2001**, *244*, 86–96. [CrossRef]
26. Jaiswal, P.; van Westen, C.J. Estimating temporal probability for landslide initiation along transportation routes based on rainfall thresholds. *Geomorphology* **2009**, *112*, 96–105. [CrossRef]
27. Wu, C.Y.; Chen, S.C. Integrating spatial, temporal, and size probabilities for the annual landslide hazard maps in the Shihmen watershed, Taiwan. *Nat. Hazards Earth Syst. Sci.* **2013**, *13*, 2353–2367. [CrossRef]
28. Althuwaynee, O.F.; Pradhan, B.; Ahmad, N. Estimation of rainfall threshold and its use in landslide hazard mapping of Kuala Lumpur metropolitan and surrounding areas. *Landslides* **2015**, *12*, 861–875. [CrossRef]
29. Afungang, R.N.; Bateira, C.V. Temporal probability analysis of landslides triggered by intense rainfall in the Bamenda Mountain Region, Cameroon. *Environ. Earth Sci.* **2016**, *75*, 1032. [CrossRef]
30. Sangelantoni, L.; Gioia, E.; Marincioni, F. Impact of climate change on landslides frequency: The Esino river basin case study (Central Italy). *Nat. Hazards* **2018**, *93*, 849–884. [CrossRef]
31. Endo, T. *Probable Distribution of the Amount of Rainfall Causing Landslides*; Annual Report 1968; Hokkaido Branch: Sapporo, Japan, 1969; pp. 122–136.
32. Onodera, T.; Yoshinaka, R.; Kazama, H. Slope failures caused by heavy rainfall in Japan. *J. Jpn. Soc. Eng. Geol.* **1974**, *15*, 191–200. [CrossRef]
33. Campbell, R.H. *Soil Slip, Debris Flows, and Rainstorms in the Santa Monica Mountains and Vicinity, Southern California*; US Government Printing Office: Washington, DC, USA, 1975; Volume 851, p. 51.
34. Aleotti, P. A warning system for rainfall-induced shallow failures. *Eng. Geol.* **2004**, *73*, 247–265. [CrossRef]
35. Caine, N. The rainfall intensity-duration control of shallow landslides and debris flows. *Geogr. Ann. Ser. A Phys. Geogr.* **1980**, *62*, 23–27.
36. Guzzetti, F.; Peruccacci, S.; Rossi, M.; Stark, C.P. Rainfall thresholds for the initiation of landslides in central and southern Europe. *Meteorol. Atmos. Phys.* **2007**, *98*, 239–267. [CrossRef]
37. Cannon, S.H.; Gartner, J.E.; Wilson, R.C.; Bowers, J.C.; Laber, J.L. Storm rainfall conditions for floods and debris flows from recently burned areas in southwestern Colorado and southern California. *Geomorphology* **2008**, *96*, 250–269. [CrossRef]
38. Brunetti, M.T.; Peruccacci, S.; Rossi, M.; Luciani, S.; Valigi, D.; Guzzetti, F. Rainfall thresholds for the possible occurrence of landslides in Italy. *Nat. Hazards Earth Syst. Sci.* **2010**, *10*, 447–458. [CrossRef]
39. Ciervo, F.; Rianna, G.; Mercogliano, P.; Papa, M.N. Effects of climate change on shallow landslides in a small coastal catchment in southern Italy. *Landslides* **2017**, *14*, 1043–1055. [CrossRef]
40. Tien Bui, D.; Pradhan, B.; Lofman, O.; Revhaug, I.; Dick, Ø.B. Regional prediction of landslide hazard using probability analysis of intense rainfall in the Hoa Binh province, Vietnam. *Nat. Hazards* **2013**, *66*, 707–730. [CrossRef]
41. Peruccacci, S.; Brunetti, M.T.; Luciani, S.; Vennari, C.; Guzzetti, F. Lithological and seasonal control on rainfall thresholds for the possible initiation of landslides in central Italy. *Geomorphology* **2012**, *139*, 79–90. [CrossRef]
42. Crozier, M.J. *Landslides: Causes, Consequences & Environment*; Taylor & Francis: Oxfordshire, UK, 1986.
43. Wieczorek, G.F. *Landslides: Investigation and Mitigation. Chapter 4-Landslide Triggering Mechanisms*; Special Report 247; Turner, A.K., Schuster, R.L., Eds.; Transportation Research Board: Washington, DC, USA, 1996.
44. Lee, S.; Won, J.S.; Jeon, S.W.; Park, I.; Lee, M.J. Spatial landslide hazard prediction using rainfall probability and a logistic regression model. *Math. Geosci.* **2015**, *47*, 565–589. [CrossRef]
45. Saadatkhah, N.; Kassim, A.; Lee, L.M. Hulu Kelang, Malaysia regional mapping of rainfall-induced landslides using TRIGRS model. *Arab. J. Geosci.* **2015**, *8*, 3183–3194. [CrossRef]
46. Guzzetti, F. Landslide Hazard and Risk Assessment. Ph.D. Thesis, Universitäts und Landesbibliothek Bonn, Bonn, Germany, 2006.

47. Van Den Eeckhaut, M.; Reichenbach, P.; Guzzetti, F.; Rossi, M.; Poesen, J. Combined landslide inventory and susceptibility assessment based on different mapping units: An example from the Flemish Ardennes, Belgium. *Nat. Hazards Earth Syst. Sci.* **2009**, *9*, 507–521. [CrossRef]
48. Lin, C.W.; Lin, M.L.; Chang, C.P.; Wu, M.C.; Wang, T.T.; Chen, T.C. *Geohazards Susceptibility Analysis of Damaged Areas in Typhoon Morakot (1/3)*; Central Geological Survey MOEA: Taipei, Taiwan, 2010. (In Chinese)
49. Wu, C.Y.; Tsai, C.W.; Chen, S.C. Topographic characteristic analysis of landslides in Kaoping River Watershed. *J. Chin. Soil Water Conserv.* **2016**, *47*, 156–164. (In Chinese)
50. Liu, J.K.; Weng, T.C.; Hung, C.H.; Yang, M.T. Remote Sensing Analysis of Heavy Rainfall Induced Landslide. In Proceedings of the 21st Century Civil Engineering Technology and Management Conference, Hsinchu, Taiwan, 28 December 2001; Minghsin University of Science and Technology: Xinfeng Township, Taiwan, 2001; pp. C21–C31. (In Chinese).
51. Xie, M.; Esaki, T.; Zhou, G. GIS-based probabilistic mapping of landslide hazard using a three-dimensional deterministic model. *Nat. Hazards* **2004**, *33*, 265–282. [CrossRef]
52. Jan, C.D. *Using Rainfall Factors to Determine Debris-Flow Warning Criteria*; Soil and Water Conservation Bureau, Council of Agriculture, Executive Yuan: Tainan, Taiwan, 2003. (In Chinese)
53. Tsai, M.C. Establishment of Critical Line for Early Warning of Debris Flow Based onTwo-Factors Survival Analysis—A Case Study in Sheng-Mu Area. Ph.D. Thesis, Feng Chia University, Taichung, Taiwan, 2 June 2016. (In Chinese).
54. Song, W.M. *Principle of Probability and Inferential Statistics*; McGraw-Hill: Taipei, Taiwan, 2007. (In Chinese)

 water

MDPI

Article

A New Method for Wet-Dry Front Treatment in Outburst Flood Simulation

Dingzhu Liu [1,2,3], Jinbo Tang [1,2], Hao Wang [4,*], Yang Cao [5], Nazir Ahmed Bazai [1,2,3], Huayong Chen [1,2] and Daochuan Liu [6]

1 Key Laboratory of Mountain Hazards and Earth Surface Process, Chinese Academy of Sciences, Chengdu 610041, China; liudingzhu@imde.ac.cn (D.L.); jinbotang@imde.ac.cn (J.T.); nazirbazai61@yahoo.com (N.A.B.); hychen@imde.ac.cn (H.C.)
2 Institute of Mountain Hazards and Environment, Chinese Academy of Sciences, Chengdu 610041, China
3 University of Chinese Academy of Sciences, Beijing 100049, China
4 Key Laboratory of Land Surface Pattern and Simulation, Institute of Geographic Sciences and Natural Resources Research, Chinese Academy of Sciences, Beijing 100101, China
5 The Engineering & Technical College of Chengdu University of Technology, Leshan 614000, China; cyy_1225@sina.com
6 Sichuan Highway Planning, Survey, Design and Research Institute Ltd., Chengdu 610041, China; Ldaochuan@163.com
* Correspondence: hgoodspeed2008@163.com; Tel.: +86-150-0828-3546

Abstract: When utilizing a finite volume method to predict outburst flood evolution in real geometry, the processing of wet-dry front and dry cells is an important step. In this paper, we propose a new approach to process wet-dry front and dry cells, including four steps: (1) estimating intercell properties; (2) modifying interface elevation; (3) calculating dry cell elevations by averaging intercell elevations; and (4) changing the value of the first term of slope limiter based on geometry in dry cells. The Harten, Lax, and van Leer with the contact wave restored (HLLC) scheme was implemented to calculate the flux. By combining the MUSCL (Monotone Upstream–centred Scheme for Conservation Laws)-Hancock method with the minmod slope limiter, we achieved second-order accuracy in space and time. This approach is able to keep the conservation property (C-property) and the mass conservation of complex bed geometry. The results of numerical tests in this study are consistent with experimental data, which verifies the effectiveness of the new approach. This method could be applied to acquire wetting and drying processes during flood evolution on structured meshes. Furthermore, a new settlement introduces few modification steps, so it could be easily applied to matrix calculations. The new method proposed in this study can facilitate the simulation of flood routing in real terrain.

Keywords: shallow water equations; wet-dry front; outburst flood; TVD-scheme; MUSCL-Hancock method

Citation: Liu, D.; Tang, J.; Wang, H.; Cao, Y.; Bazai, N.A.; Chen, H.; Liu, D. A New Method for Wet-Dry Front Treatment in Outburst Flood Simulation. *Water* **2021**, *13*, 221. https://doi.org/10.3390/w13020221

Received: 9 November 2020
Accepted: 14 January 2021
Published: 18 January 2021

1. Introduction

Glacier avalanche [1,2], debris flow [3–5], and landslide [6–8] in mountain areas could trigger the occurrence of river blocking [9–12]. Some of this blocking produces large-scale lakes, which leads to back flooding upstream and may inundate roads and villages. Most dammed lakes breach in a short time after their formation, causing massive water to be released catastrophically [9,13]. Yigong Lake was blocked by catastrophic landslides in 1902 and 2000 [14] and formed outburst floods with peak discharges of around 18.9×10^4 m^3/s [15] and 12.4×10^4 m^3/s [8], respectively; the Yarlung Tsangpo gorge was blocked twice in 2018, with a peak discharge of 3.2×10^4 m^3/s in the second outburst flood [3,4].

This kind of dynamic process can impose catastrophic damage to downstream people and infrastructure [16]. Outburst floods may also have significant geomorphic and geologic impacts; they have substantial erosive and transport capacity that can rapidly transform river channels and bedforms [17–19], and may even lead to climate change [20] and a global sea level decrease [21]. Outburst floods and their impacts even appear in the myths and stories of many civilizations, such as the Bible and the Koran [22].

Back analysis of outburst flood is an impressive method to determine risk, which has been used to reconstruct large-scale geomorphological dynamic processes that occurred ten thousand years ago. In general, the submerge area and related velocity determine the risk of outburst floods, and a shallow water dynamic model is a widely used and reliable method to predict it [23–26].

Shallow water equations are popular in long-wave hydrodynamic simulation [27] and are an effective way to analyze outburst flood routing. The Godunov-type finite volume method is an effective and convenient method to calculate flood evolution in complex geometry and is widely used in structured cells and unstructured cells [27]. There are two popular forms for shallow water equations: (1) not consider gravity source term in advection terms [28] and (2) consider the geometry in advection terms [29,30].

A TVD (total variation diminishing) scheme is used to limit numerical oscillations near discontinuity [31–33]. Slope limiters such as the minmod limiter, double limiter, and van-Leer limiter are popularly used to keep the solving scheme that has a TVD property [33]. By using a slope limiter, a monotone upstream-centered scheme for conservation laws (MUSCL) reconstruction in the cell center provides second-order accuracy in space [34,35]. The MUSCL method is one of the most successful high-resolution schemes for hyperbolic conservation laws and is applied widely [24,29,33].

Wet-dry front treatment is a key problem when applying shallow water equations to real geometry. Sharp slope geometry especially can over-predict flux and generate negative flow depth [27,29]. Specific treatments during calculation have been applied to limit flux and the gravity source term or to modify geometry [27,29,36], thus or avoiding extremely high flux in intercells and velocity in the cell center. In the process of variable modifications, the limiter's value of the dry cell would equal zero after modifying the local geometry [27,29,36,37].

Many traditional treatments to the wet-dry front change the elevation of the dry cell equal to the wet cell's free surface elevation as shown in Figure 1a [27,29]. If the dry cell is surrounded by four wet cells with different free surface elevations, four elevation modifications are necessary to achieve a balanced flux in the surrounded four cells (Figure 1b,c), and it is very hard to achieve a matrix calculation during simulation as well. A matrix calculation and less cell modification save time because matrix operators are faster than cell loops [38]. In order to apply shallow water equations to a river with a complex geometry and avoid more elevation modifications, we propose processing dry cells by adopting the first term of the slope limiter function in dry cells to solve the wet-dry front problem and accomplish matrix simulation in the whole calculation area. This method can avoid modifications in the dry cell's elevation and achieve a matrix calculation. This method was tested with many cases and is applicable to a complex geometry for outburst flood analysis.

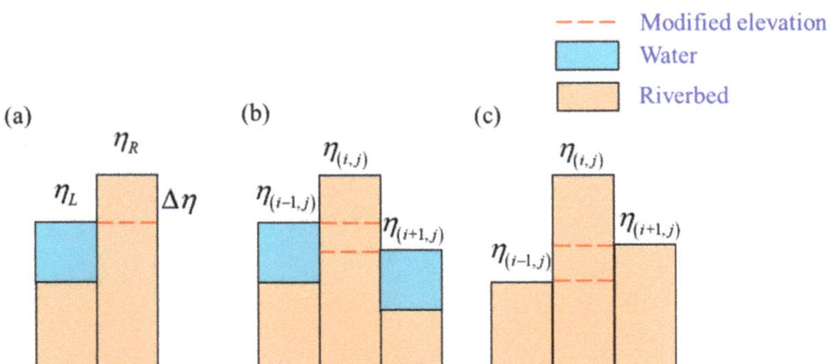

Figure 1. The traditional elevation modification of wet-dry front. (**a**) Modify elevation to the same as wet cell; (**b**,**c**) Two times elevation modification of one dry cell.

2. Governing Equations and Schemes

2.1. Governing Equations

Two-dimensional shallow water equations are integral forms of Reynolds-averaged Navier–Stokes equations. This equation presumptively neglects vertical momentum exchange and sets the pressure distribution as hydrostatic [39]:

$$U_{,t} + F_{,x} + G_{,y} = S, \tag{1}$$

where t represents time direction, x and y are two Cartesian coordinates, U is a variable with vector form, F and G are fluxes vectors at two directions, and S is a vector represents source term. The equation is a conserved equation. For general use, the conserved equation is written as:

$$\begin{bmatrix} \eta \\ hu \\ hv \end{bmatrix}_{,t} + \begin{bmatrix} hu \\ hu^2 + g(\eta^2 - 2\eta Z)/2 \\ huv \end{bmatrix}_{,x} + \begin{bmatrix} hu \\ huv \\ hv^2 + g(\eta^2 - 2\eta Z)/2 \end{bmatrix}_{,y} = \begin{bmatrix} 0 \\ -\tau_{bx}/\rho - g\eta Z_{,x} \\ -\tau_{by}/\rho - g\eta Z_{,y} \end{bmatrix}, \tag{2}$$

$$\tau_{bx} = \rho g n^2 u \sqrt{u^2 + v^2} h^{-1/3}, \tag{3}$$

$$\tau_{by} = \rho g n^2 v \sqrt{u^2 + v^2} h^{-1/3}, \tag{4}$$

where $\eta = Z + h$ is the elevation of the flood free surface, where the specific treatment to initial shallow water equations adds geometry information to the advections [29], Z is the elevation of the river bed, h is the flow depth, u is the flow velocity in the x direction, v is the flow velocity in the y direction, τ_{bx} and τ_{by} are the bottom shear stress in the x and y directions, g is gravity acceleration, and n is the Manning coefficient.

2.2. Finite Volume Method

The finite volume method has been used in many areas to solve partial equations [40]. The method is implemented by integrating partial equations over the space area for an arbitrary grid. In this study, shallow water equations are hyperbolic equations, which can be integrated as follows:

$$\frac{\partial}{\partial t}\int_{\varepsilon} U d\Omega + \int_{\varepsilon} (\frac{\partial F}{\partial x} + \frac{\partial G}{\partial y}) d\Omega = \int_{\varepsilon} S d\Omega. \tag{5}$$

By using Green's formula, Equation (6) can be described as:

$$\frac{\partial}{\partial t}\int_{\varepsilon} U d\Omega + \int_{L} (F+G) dL = \int_{\varepsilon} S d\Omega, \tag{6}$$

where L is the mesh boundary of the integral line, and ε is the integral area, which is a rectangular grid here. By using the integral form equation at mesh (i, j), the second term becomes:

$$\int_{L} F_i dL + \int_{L} G_j dL = (F_{i+1/2} - F_{i-1/2})\Delta y + (G_{j+1/2} - G_{j-1/2})\Delta x, \tag{7}$$

$$U_{i,j}^{n+1} = U_{i,j}^{n} - \frac{\Delta t}{\Delta x}\left(F_{i+1/2,j} - F_{i-1/2,j}\right) - \frac{\Delta t}{\Delta y}\left(G_{i,j+1/2} - G_{i,j-1/2}\right) + \Delta t S_i, \tag{8}$$

where n is the time, and $i + 1/2$ and $j + 1/2$ are the predicted flux at the interface, predicted by two Riemann states.

2.3. HLLC Riemann Solver for Fluxes Prediction

In order to solve the Riemann problem approximately, Harten Lax and van Leer proposed the famous HLL Riemann solver in 1983, which is widely used by researchers to solve shallow water equations today. The scheme requires estimations for the fastest signal velocities from the discontinuity at the interface, resulting in a two-wave model including shock waves, rarefaction waves, and discontinuity. Toro modified the scheme to a three-wave model [33], and the solver was suited to calculate cases involving a wet-dry front, so the HLLC (Harten, Lax and van Leer) approximate Riemann solver by Toro is used in this paper.

2.4. Slope Limiter

The face value of variables required for the MUSCL-Hancock reconstruction step and for the time updating step is:

$$U_{i+1/2} = U_i + r\nabla U_i, \tag{9}$$

where r is the distance vector, and ∇U_i is the gradient vector of variable in space. In order to avoid numerical oscillations, we adopt a single slope limiter in this study. The formula becomes:

$$U_{i+1/2} = U_i + \varphi(r) r \nabla U_i, \tag{10}$$

where $\varphi(r)$ is a limiter function. We adopted the Minmod limiter in case tests. Special gradients of variables were predicted by:

$$r_{i,j} = \begin{bmatrix} \frac{\eta_{i+Fn,j+Gn} - \eta_{i,j}}{\eta_{i,j} - \eta_{i-Fn,j-Gn}} \\ \frac{hu_{i+Fn,j+Gn} - hu_{i,j}}{hu_{i,j} - hu_{i-Fn,j-Gn}} \\ \frac{hv_{i+Fn,j+Gn} - hv_{i,j}}{hv_{i,j} - hv_{i-Fn,j-Gn}} \end{bmatrix}, \tag{11}$$

where $r_{i,j}$ is slope in mesh (i, j), which includes two directions' values. If intercell interpolation is in the x direction, $F_n = 1$ and $G_n = 0$; if intercell interpolation is in the y direction, $F_n = 0$ and $G_n = 1$.

2.5. MUSCL-Hancock Method

In the MUSCL-Hancock reconstruction step, the calculation is limited in a single cell. Thus, it does not use the HLLC Riemann solver to predict the flux at the intercell. The corrected value in the cell center is $U_i^{n+1/2}$, and the flux is calculated based on cell face reconstruction, which is predicted by the cell slope limiter:

$$UM_{i+1/2}^n = U_i^n + \frac{1}{2}\varphi(r)\left(U_i^n - U_{i-1}^n\right), \tag{12}$$

where $UM_{i+1/2}^n$ is the reconstructed cell boundary vector. The predicted cell center value is calculated by:

$$U_i^{t+1/2} = U_i^t + k_x\left(F\left(UM_{i+1/2}^n\right) - F_{i+1/2}\left(UM_{i-1/2}^n\right)\right) + k_y\left(G\left(UM_{j+1/2}^n\right) - G\left(UM_{j-1/2}^n\right)\right) + \frac{\Delta t}{2}S_i \tag{13}$$

$$k_x = \frac{\Delta t}{2\Delta x};\ k_y = \frac{\Delta t}{2\Delta y}.$$

As for the Riemann flux calculation, we use results from the MUSCL-Hancock step to reconstruct the value around the interface. The slope limiter is the same as the MUSCL-Hancock reconstruction step. The formula is:

$$U_{i+1/2}^L = U_i^{n+1/2} + \frac{1}{2}\varphi(r)\left(U_i^n - U_{i-1}^n\right). \tag{14}$$

Riemann states in another direction to use the same method.

2.6. Stability Criteria

The numerical scheme is explicit. The stability is defined by the Courant–Friedrichs–Lewy (CFL) criterion. Since this is a two-dimensional calculation case, the time step is limited by local real-time results:

$$\Delta t = \min\left(\frac{C\Delta x}{|u_i| + \sqrt{gh_i}}, \frac{C\Delta y}{|v_i| + \sqrt{gh_i}}, \Delta T\right), \tag{15}$$

where C is the Courant number, ranging between 0 and 1. In some cases, a stable ΔT could give a more stable result. If the export results include a specific time point, ΔT should be modified to a smaller time step to match the predicted time point.

3. Intercell Bed Elevation and Dry Cell

Since the flux calculation should follow the real physics law in the real world, the interface property determines the flux calculation during flow routing in real river geometry. We classified the interface property into four types based on flow depth and surface elevation (as shown in Figure 2): (1) Two cells' flow depth is higher than 0, which would generate flux in these specific two cells. (2) Two cells between the interface are dry cells such that both flow depths are equal to zero. (3) One is a wet cell and another is a dry cell, but the elevation of the wet cell is higher than the dry cell. (4) One is a wet cell and another is a dry cell, but the dry cell is higher than the wet cell.

Based on the physical property, the interface in the first and third type should consider mass and momentum exchanges between the two cells during calculation. It is not necessary to consider this effect for the cell interface in Type B and Type D.

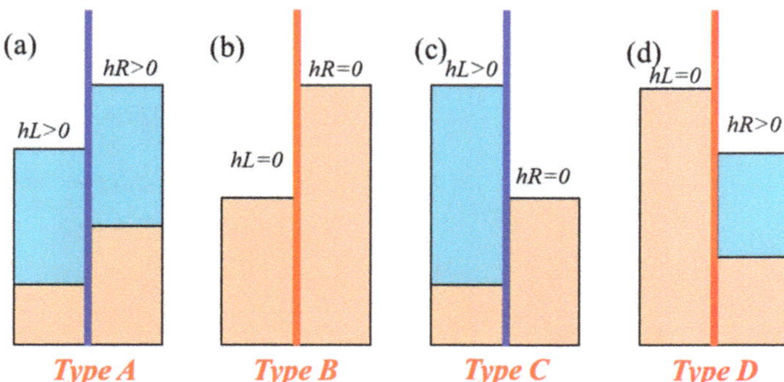

Figure 2. Classification of interface property. (**a**) Type A: wet cells at the left and right side, $hL > 0$, $hR > 0$, hL and hR are flow depth in the left and right side of intercell respectively; (**b**) Type B: dry cells at the left and right side of the intercell face; (**c**) Type C: wet and dry cells are connected through the intercell face, the free surface elevation of the wet cell is higher than the dry cell; (**d**) Type D: wet and dry cells are connected between the intercell face, and the free surface elevation of the wet cell is lower than the dry cell.

Local modification of Z at the intercell is adopted. The modification is used based on the physical property of the real condition (as shown in Figure 3); e.g., (1) the reflection boundary would stop the flow from moving forward; (2) the dry cell has no flux. The intercell property in Types A, B and C do not need modifications, and the intercell bed elevation is:

$$Z_{i+1/2} = (Z_i + Z_{i+1})/2, \tag{16}$$

where $Z_{i+1/2}$ is the elevation at the intercell; Z_i and Z_{i+1} are cell center elevations at the ith and (i + 1)th cell. Type D of the intercell face's elevation is modified as:

$$Z_{i+1/2} = min(\eta_i, \eta_{i+1}). \tag{17}$$

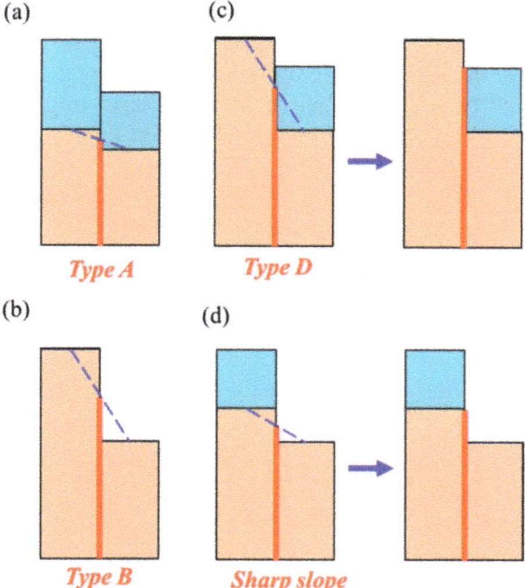

Figure 3. Modification of the intercell elevation. (**a**,**b**) The intercell does not need modification, which is related to Type A and Type B; (**c**) the intercell elevation is modified to the wet cell's elevation, which is related to Type D; (**d**) the sharp slope cell is modified to the dry cell's bed elevation.

In the Type C intercell property, a sharp slope would produce an overpredicted flux in the intercell. Based on the intercell property, the intercell bed elevation was modified as:

$$Z_{i+1/2} = max(Z_i, Z_{i+1}). \tag{18}$$

Momentum needs to be modified while the intercell property is Type D. The velocity component that is perpendicular and the limiter of the three variables of the shallow water equations should be set to zero. For rectangular cell simulation, the calculation area could be treated as a matrix. Many simulations are based on circulation to calculate the whole simulated area, and they include a step that checks for cells that do not need flux calculations. We want to skip this step due to the running circulation cost time. The specific form of the shallow water equation includes η, and the unbalanced flux would be predicted during our simulation which formed by a complex real geometry if the matrix is used directly, for example (Figure 4):

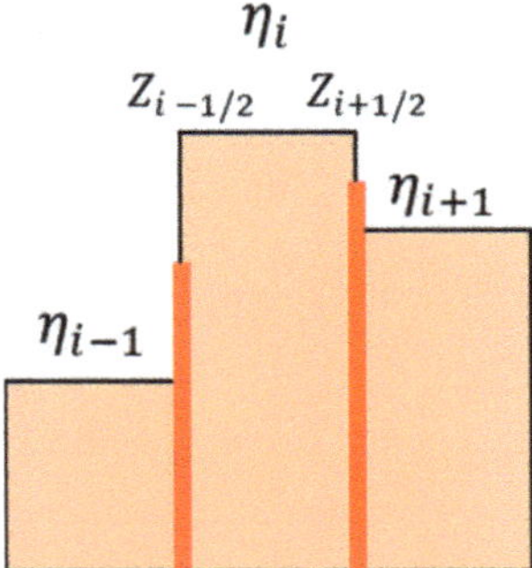

Figure 4. The calculated parameters of a shallow water equation with no special treatment.

If the dry cell's slope limiter function, Equation (11), is zero, the calculated flux would be unbalanced:

$$\left(g(\eta^2 - 2\eta Z)/2\right)_{,x} = \frac{g\left[(\eta_i{}^2 - 2\eta_i Z_{i-1/2}) - (\eta_i{}^2 - 2\eta_i Z_{i+1/2})\right]}{2\Delta \mathrm{x}} \neq g\eta_i Z_{,x}. \tag{19}$$

In order to achieve a matrix calculation and an automatic flux balance during simulation, we adopted the "zero" slope-limiter function and modified the first term based on the geometry. The elevation of dry cell was modified to:

$$\eta_i = \frac{(Z_{i+1/2} + Z_{i-1/2})}{2}, \tag{20}$$

and the slope of the surface elevation of the dry cell was calculated as:

$$r_{i,j(\eta_i)} = \frac{(Z_{i+1/2} - Z_{i-1/2})}{2\Delta x}, \tag{21}$$

where $r_{i,j(\eta_i)}$ is the value of the first term of the slope limiter function, and Δx is the cell length in the x direction.

If the flow depth in the dry cell is zero, $\eta_{i+1/2} = Z_{i+1/2}$ in the interface, and cell center's value is given by Equation (20). The specific treatment to the dry cell is shown below (as shown in Figure 5):

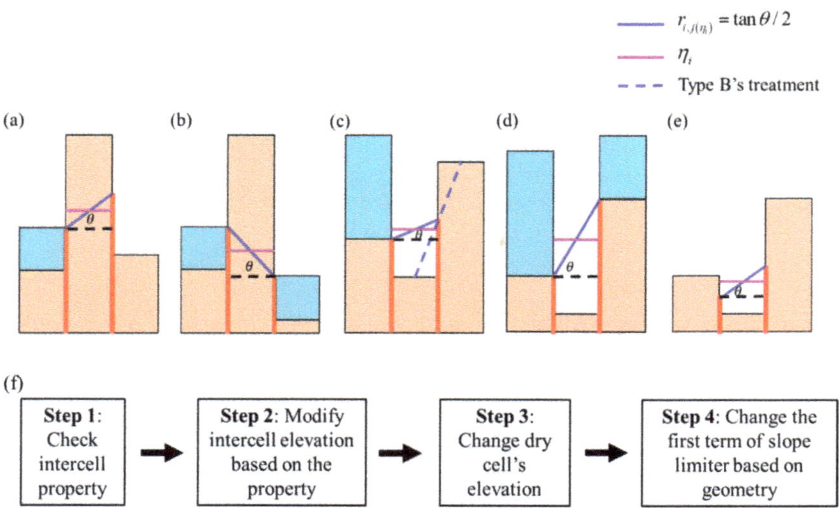

Figure 5. The dry cell's center elevation is calculated by the average of two intercell elevations. Intercell elevations are predicted from the latest two steps that are based on the intercell type and the real conditions. (**a**,**c**) One side is a wet-dry front and the one side is dry-dry; (**b**) both sides are a wet-dry front; (**d**) both sides are sharp slopes; (**e**) both sides are dry cells; (**f**) flow chart of the method.

The balance in the dry cell is automatically reached:

$$\left(g(\eta^2-2\eta Z)/2\right)_{,x}=\frac{g({Z_{i+1/2}}^2-{Z_{i-1/2}}^2)}{2\Delta x}=\frac{g(Z_{i+1/2}+Z_{i-1/2})(Z_{i+1/2}-Z_{i-1/2})}{2\Delta x}=g\eta_i Z_{,x} \tag{22}$$

In the reflection boundary, where a higher left dry cell and a lower right wet cell surround the intercell, $\eta_{i+1/2}=\eta_{i-1/2}=\eta_i$ and $\eta_{i-1/2}=Z_{i-1/2}$. The flux balance is reached automatically:

$$\left(g(\eta^2-2\eta Z)/2\right)_{,x}=\frac{g(-{\eta_{i-1/2}}^2+{\eta_{i+1/2}}^2+2\eta_{i+1/2}Z_{i+1/2}-2\eta_{i-1/2}Z_{i-1/2})}{2\Delta x}=\frac{g\eta_{i+1/2}(Z_{i-1/2}-Z_{i+1/2})}{\Delta x}=g\eta_i Z_{,x}. \tag{23}$$

If the flow velocity at all described cells is zero, the flux balance is controlled by the wet-dry boundary and the dry cells. All the steps of this method are summarized in Figure 5f.

4. Results and Discussion

4.1. Steady Condition Calculation of Flood

A test case was used to test the numerical scheme's C-property. A static lake is kept steady, and there is no disturbance. The calculation area is an 8000 m × 8000 m. In the dry bed, there are two bumps:

$$Z(x,y)=max(0,Z_{B1},Z_{B2}), \tag{24}$$

$$\begin{cases} Z_{B1}=2000-0.00032\left[(x-3000)^2+(y-5000)^2\right] \\ Z_{B2}=900-0.000144\left[(x-5000)^2+(y-3000)^2\right] \end{cases}. \tag{25}$$

The lake elevation is 1000 m, and the lower bump is submerged by the lake. The mesh size is a rectangular mesh of 1 m × 1 m. The calculation time step is 1 s. The finish time is 8000 s.

After 8000 s, the lake remained static, the results in Figure 6 show that this approach follows a C-property, the static keep balance automatically.

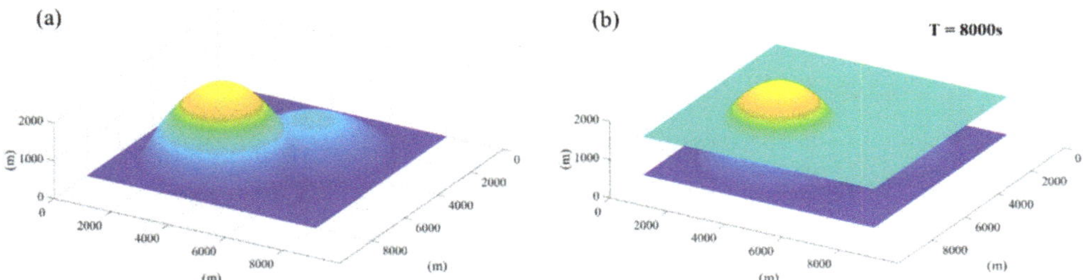

Figure 6. C-property checking for a static lake. (**a**) Lake geometry; (**b**) results after 8000 s.

4.2. Two-Dimensional Smooth River Bed Test

A two-dimensional smooth bed test was adopted here. The case has an analytical solution smooth bed. This test was adopted by many researchers to test their algorithm's wet-dry treatment and calculation accuracy [27,29,41,42]. The calculation area is a 4 m × 4 m, and the origin of the coordinates is in the center of the calculation area. The mesh size is 0.1 m × 0.1 m. The bed is a parabola rotation:

$$Z(x,y) = h_0\left(\frac{x^2+y^2}{a^2} - 1\right), \tag{26}$$

where h_0 is the initial flow depth of the origin of the coordinates, a is the distance between the origin and the elevation equal to zero, and x and y are coordinate variables. Under this condition, water flows on the smooth bed and cannot stop. The frequency of flow is $\omega = 2\pi/T = \sqrt{8gh_0}/a$, in which T is the time of one cycle. In the analytical solution for the process, the moving range is small:

$$\eta(x,y,t) = \max\left[Z(x,y), h_0\left(\frac{\sqrt{1-A^2}}{1-A\cos(\omega t)} - \frac{x^2+y^2}{a^2}\left(\frac{1-A^2}{(1-A\cos(\omega t))^2} - 1\right) - 1\right)\right], \tag{27}$$

where $A = \left(a^4 - r_0^4\right)/\left(a^4 + r_0^4\right)$, and r_0 is the farthest distance to the center. In the simulation test, we consider the same parameter treatments as Song et al. [42], a = 1 m, h_0 = 0.1 m, and r_0 = 0.8 m. We adopted a mesh size of 0.01 × 0.01 m. The initial condition is the same as the analytical solutions in T/6, T/3, T/2 and T (Figure 7).

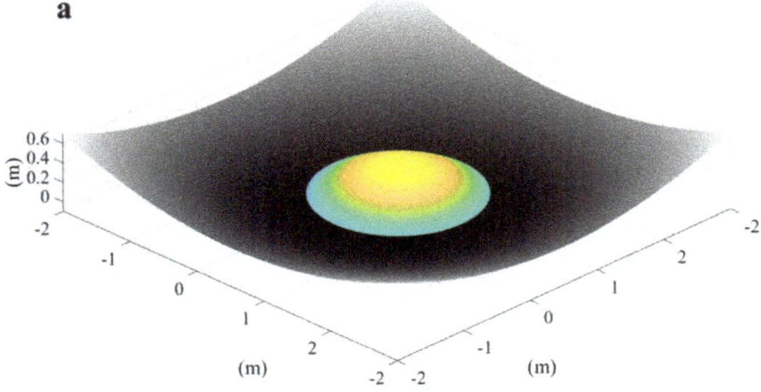

Figure 7. *Cont.*

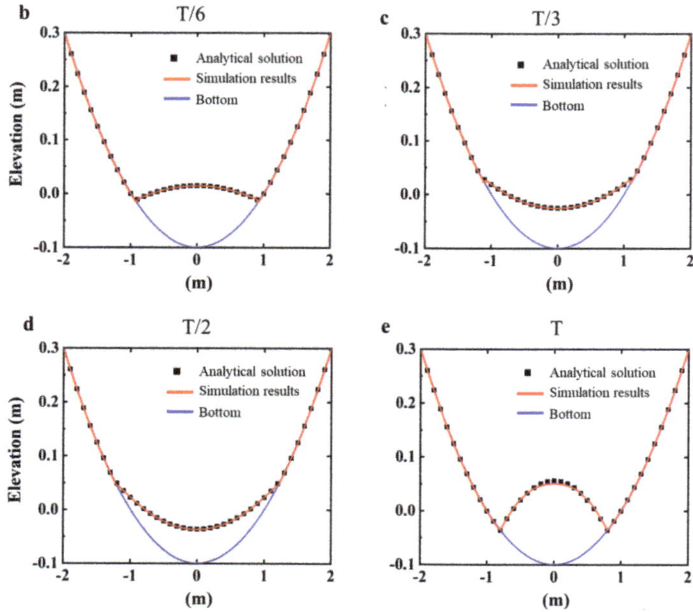

Figure 7. Simulated results compared with real analytical results. (**a**) Geometry of the calculation area and the initial condition; (**b–e**) comparison between the simulated results and the analytical solution at T/6, T/3, T/2, T.

4.3. Dam Breach over a Thump

This test case is a dam break flow over a thump. The experiment was carried at the University of Brussels, Belgium [43]. Many researchers have used this case to test their model on complex geometries [44,45].

The test simulated a sudden dam breach of flood flowing over a triangular hump. The calculation area is a 38 × 1.75 m flume. A hump was set at 15.5 m, and a barrier lake was formed upstream (as shown in Figure 8). The static lake's flow depth is 0.75 m. The peak of the triangular thump is at 28.5 m, with a height and bottom width of 0.4 and 6 m, respectively. In the tail of the obstacle, there is a 0.15 m high gate, where flow depth is also 0.15 m. Downstream, the first gate is the dry bed. Roughness of the calculation area is n = 0.0125 s × $m^{-1/3}$. Four downstream monitoring locations were set, named G1, G2, G3, and G4, and the measured data is the flow depth, located at 19.5, 25.5, 26.5, and 28.5 m respectively. The mesh size for the calculation is 0.1 m × 0.1 m.

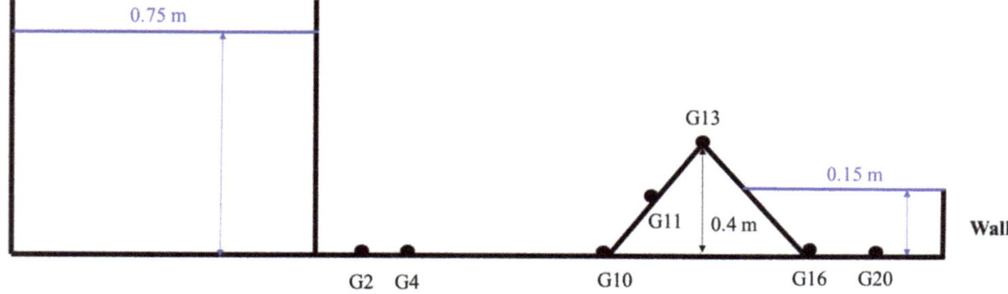

Figure 8. Flume test setup of the experiment.

Figure 9 shows four representative moments of simulation. After 1 s, the flood front arrives at the 19 m point. At 8 s, the flood flows over the obstacle, which causes backwater and imposes disturbance on the tail lake. At 16 s, a higher run-up upstream lake formed at the front of the obstacle, with waves upstream of the hump. A distinct hydraulic jump develops at the tail lake. At 40 s, the water surface before the obstacle is dominated by strong waves, while the tail lake becomes static. The flow upstream cannot flow over the obstacle.

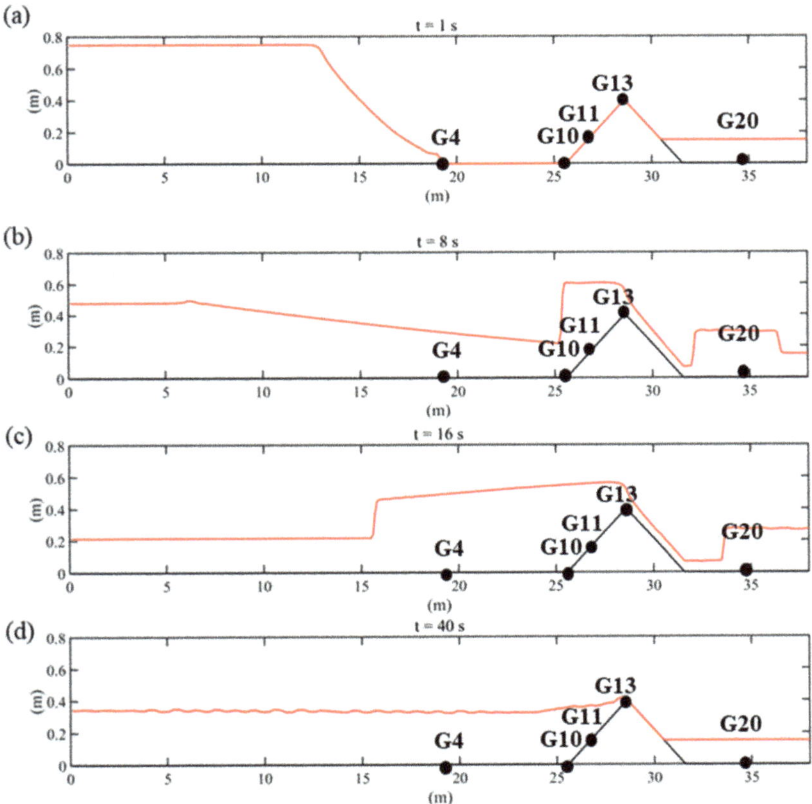

Figure 9. Free surface elevation during the flood evolution in the experiment. (**a**) At 1 s, flood flows at the dry bed; (**b**) at 8 s, the flood flows up to the obstacle and has an influence downstream; (**c**) at 16 s, all the upstream water flows to the obstacle and a run-up forms; (**d**) at 40 s, the flow downstream remains static, with waves at the upstream lake.

We extracted surface elevation data from the simulation results for comparison. Simulated results at G4 and G13 fit the monitored data very well, but the predicted water surface at G10 and G11 is slightly lower than the monitored data, G20 is slightly higher than measured data, which has been captured in many cases [44]. At the lower stage, the simulated results were similar to simulated results later. The short-term-simulated higher flow depth did not influence the real flood evolution at a later stage. Compared with the same simulated work did by Tomas and Liao [44,45], our simulated results show similar result in G10, G11, and G20. In G4 and G13, our result is closer to measured data compared with their results, which shows better results (as shown in Figure 10).

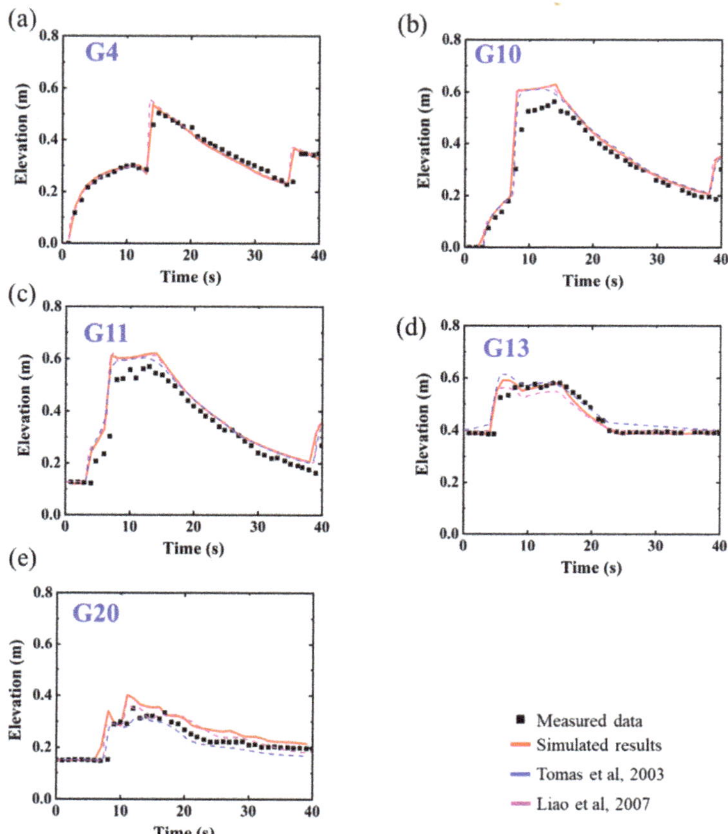

Figure 10. Monitored data compared with the simulated results at the four locations. (**a**) The simulated result is similar to the measured data at G4; (**b**) initially, the simulated results at G10 is lower but did not influence successive results; (**c**) the same higher simulated flow depth at G11 is similar to G10, a short-term lower elevation; (**d**) the simulated results fit well with the measured data at G13; (**e**) the simulated results fit well with the measured data at G20.

4.4. Dam Break Wave Propagating over Three Humps

The three humps test is a very famous test case proposed by Kawahara in 1986 [46,47]. Initially, the case was adopted to test the finite element model, which is wildly used. The calculation area in this study is a 75 × 30 m flume, which has three humps. The boundary is a fixed reflection boundary. The centers of the humps are A (30 m, 6 m), B (30 m, 24 m), and C (47.5 m, 15 m). The maximum height of the humps is 1, 1, and 3 m, respectively. In the upstream of x = 16 m, there is a lake with a depth of 1.875 m. The bed roughness is n = 0.018 sm$^{-1/3}$. The calculation geometry was calculated from the formulas below:

$$\begin{cases} a = 1 - \frac{1}{8}\sqrt{(x-30)^2 + (y-6)^2} \\ b = 1 - \frac{1}{8}\sqrt{(x-30)^2 + (y-24)^2} \\ c = 3 - \frac{3}{10}\sqrt{(x-47.5)^2 + (y-15)^2} \\ Z(x,y) = \max(0, a, b, c) \end{cases}, \tag{28}$$

where a and b are geometric functions of the two lower humps, c is the geometry function of the higher humps, and the elevation of the bed bottom is the maximum value of a, b, and c. The mesh size is 0.5 m $\times$ 0.5 m.

Figure 11 shows the simulated results of six important moments. At 2 s, the water reached two lower humps and started to flow over them. At 6 s, the flood flowed over the two lower humps and started to reached the higher hump. At 12 s, the flood bypassed the higher hump because it could not completely inundate the higher hump. At 30 s, the flood occupied the calculation area. The formed higher flow depth downstream caused backflow. At 100 s, there was still weak flow in the tank. At 300 s, the flow almost stopped and formed a static lake in the tank, and the peaks of all three humps did not submerge. The numerical model properly simulated complex wetting and drying processes and produced similar results to those of other researchers [29,48].

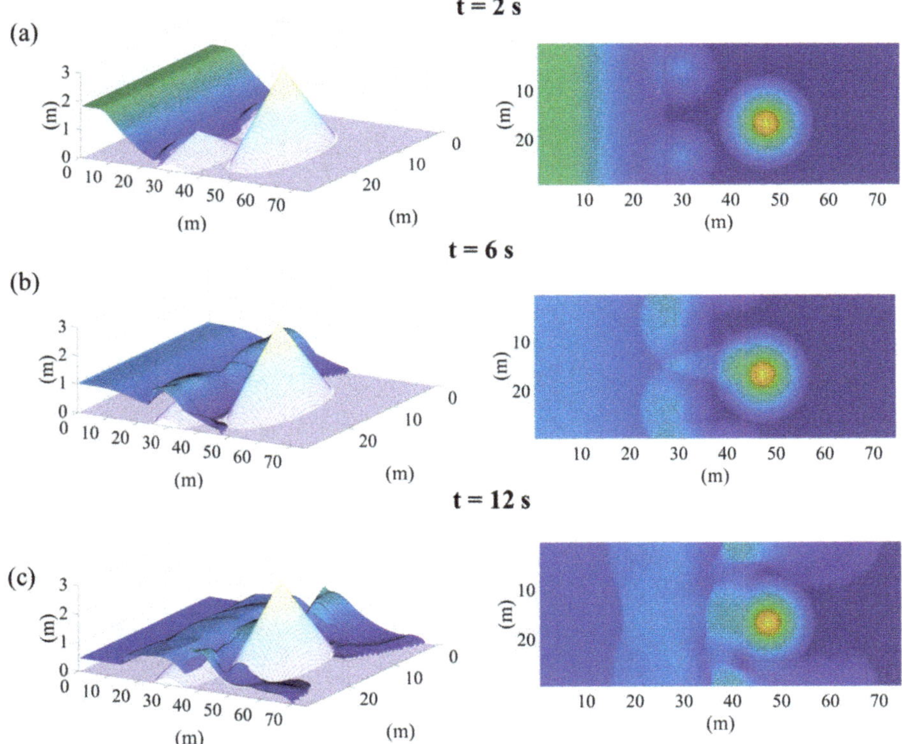

Figure 11. *Cont.*

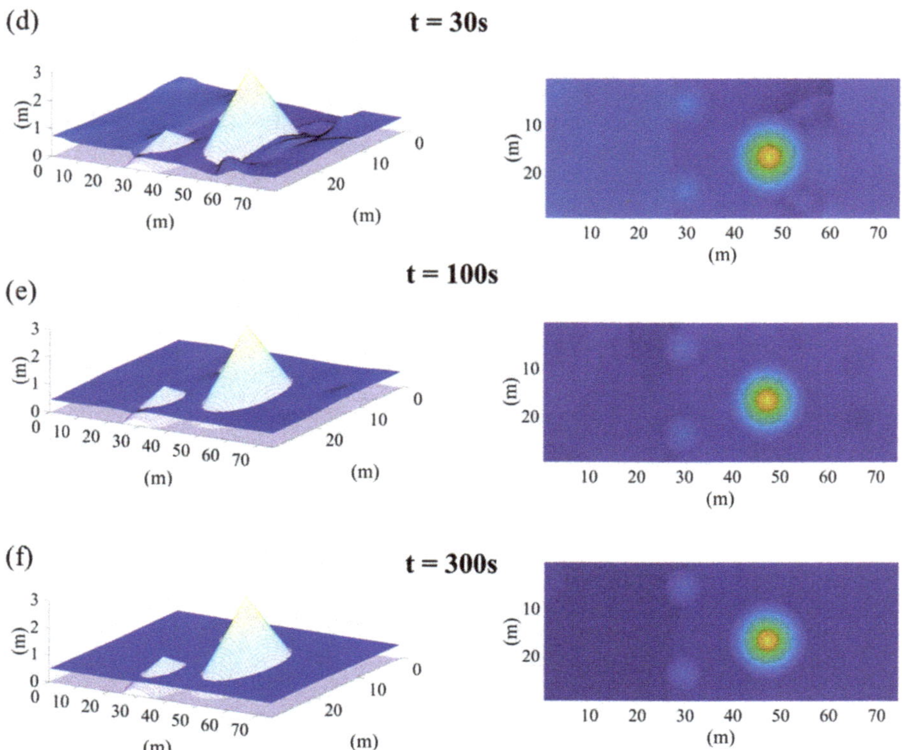

Figure 11. Simulated flood evolution on a complex three-hump condition. (**a**) The flood starts to reach the first two low humps at 2 s; (**b**) the flood flows over the two low humps at 6 s; (**c**) the flood flows downstream of the high humps at 12 s; (**d**) the flood forms a higher flow depth downstream at 30 s; (**e**) there is some weak flow in the tank at 100 s; (**f**) the tank maintains a static condition at 300 s.

5. Conclusions

We propose a new approach to process dry cells and wet-dry front cells via a Godunov-type finite volume prediction method of flood evolution. Shallow water equations automatically balance the gravity source term. The modification includes four steps: (1) identify four types of intercells based on flow depth and surface elevation difference; (2) based on the physical properties of the intercells, modify the bed elevation of the intercell, so as to avoid non-physical flux predictions and gravity balance; (3) modify the dry cell's center elevation to equal the averaged elevation of the two surrounding intercell elevations; (4) change the first term of the slope limiter at the dry cell equal to the ratio of the elevation difference between two intercell bed elevations dividing two times of mesh size. This method was applied to a second-order MUSCL-Hancock-HLLC scheme in time and space for flux and variable prediction in a real geometry. The intercell flux predicted by the reconstructed method remained balanced with the gravity source term automatically, which was proved by mathematical derivations. Four simulated cases showed that the method has a C-property in a complex geometry and achieves the same results as those of many other researchers. Results in the analytical case and the experiment monitoring cases fit each other very well. During all the processing steps, modification could be finished in one step, such that cells did not need to be checked through circulation. This new method can increase the convenience and efficiency of matrix calculations and has a potential for faster GPU (Graphics Processing Unit) simulation and parallel computing. It could be used in real world outburst flood simulation with high efficiency.

Author Contributions: D.L. (Dingzhu Liu) and H.W. design this research and draft the manuscript. J.T. gave many suggestions on testing cases. Y.C. help with coding the program. N.A.B. help with the language. H.C. and D.L. (Daochuan Liu) help with find checking data. All authors have read and agreed to the published version of the manuscript.

Funding: This research was funded by the National Natural Science Foundation of China, grant no. 41941017; the Applied Fundamental Research Program of Sichuan Province, grant no. 2019JY0387; Key Research Program of Frontier Sciences, Chinese Academy of Sciences, grant no. QYZDY-SSW-DQC006 and the National Natural Science Foundation of China.

Institutional Review Board Statement: Not applicable.

Informed Consent Statement: Not applicable.

Data Availability Statement: The data presented in this study is contained within the article.

Conflicts of Interest: The authors declare no conflict of interest.

References

1. Capps, D.M.; Clague, J.J. Evolution of glacier-dammed lakes through space and time; Brady Glacier, Alaska, USA. *Geomorphology* **2014**, *210*, 59–70. [CrossRef]
2. Cook, S.J.; Kougkoulos, I.; Edwards, L.A.; Dortch, J.M.; Hoffmann, D. Glacier change and glacial lake outburst flood risk in the Bolivian Andes. *Cryosphere* **2016**, *10*, 2399–2413. [CrossRef]
3. Chen, C.; Zhang, L.; Xiao, T.; He, J. Barrier lake bursting and flood routing in the Yarlung Tsangpo Grand Canyon in October 2018. *J. Hydrol.* **2020**, *583*. Available online: https://www.sciencedirect.com/science/article/abs/pii/S0022169420300639 (accessed on 16 January 2021). [CrossRef]
4. Hu, K.; Zhang, X.; You, Y.; Hu, X.; Liu, W.; Li, Y. Landslides and dammed lakes triggered by the 2017 Ms6.9 Milin earthquake in the Tsangpo gorge. *Landslides* **2019**, *16*, 993–1001. [CrossRef]
5. Wei, R.; Zeng, Q.; Davies, T.; Yuan, G.; Wang, K.; Xue, X.; Yin, Q. Geohazard cascade and mechanism of large debris flows in Tianmo gully, SE Tibetan Plateau and implications to hazard monitoring. *Eng. Geol.* **2018**, *233*, 172–182. [CrossRef]
6. Fan, X.; Xu, Q.; Alonsorodriguez, A.; Subramanian, S.S.; Li, W.; Zheng, G.; Dong, X.; Huang, R. Successive landsliding and damming of the Jinsha River in eastern Tibet, China: Prime investigation, early warning, and emergency response. *Landslides* **2019**, *16*, 1003–1020. [CrossRef]
7. Li, B.; Feng, Z.; Wang, G.; Wang, W. Processes and behaviors of block topple avalanches resulting from carbonate slope failures due to underground mining. *Environ. Earth. Sci.* **2016**, *75*, 694. [CrossRef]
8. Liu, W.; Carling, P.A.; Hu, K.; Wang, H.; Zhou, Z.; Zhou, L.; Liu, D.; Lai, Z.; Zhang, X. Outburst floods in China: A review. *Earth Sci. Rev.* **2019**, *197*, 102895. [CrossRef]
9. Cui, P.; Dang, C.; Zhuang, J.Q.; You, Y.; Chen, X.Q.; Scott, K.M. Landslide-dammed lake at Tangjiashan, Sichuan province, China (triggered by the Wenchuan Earthquake, May 12, 2008): Risk assessment, mitigation strategy, and lessons learned. *Environ. Earth. Sci.* **2012**, *65*, 1055–1065. [CrossRef]
10. Wang, G.Q.; Fan, L. Simulation of dam breach development for emergency treatment of the Tangjiashan Quake Lake in China. *Sci. China Ser. E Technol. Sci.* **2008**, *51*, 82–94. [CrossRef]
11. Yan, Y.; Cui, Y.F.; Xin, T.; Hu, S.; Guo, J.; Wang, Z.; Yin, S.Y.; Liao, L.F. Seismic Signal Recognition and Interpretation of the 2019 "7.23" Shuicheng Landslide by Seismogram Stations. *Landslides* **2020**, *17*, 1206–1911. [CrossRef]
12. Zhang, L.; Xiao, T.; He, J.; Chen, C. Erosion-based analysis of breaching of Baige landslide dams on the Jinsha River, China, in 2018. *Landslides* **2019**, *16*, 1965–1979. [CrossRef]
13. Costa, J.E.; Schuster, R.L. The formation and failure of natural dams. *Geol. Soc. Am. Bull.* **1988**, *100*, 1054–1068. [CrossRef]
14. Zhou, J.; Cui, P.; Hao, M. Comprehensive analyses of the initiation and entrainment processes of the 2000 Yigong catastrophic landslide in Tibet, China. *Landslides* **2016**, *13*, 39–54. [CrossRef]
15. Ma, D.T. Study on Influences of Mountain Hazards in Yigong Zangbu River Basin to Mitigation and Reconstruction of Sichuan-Tibetan Highway Line. Ph.D. Thesis, The Graduate School of Chinese Academy of Sciences, Beijing, China, 2006.
16. Cui, P.; Zhu, Y.Y.; Han, Y.S.; Chen, X.Q.; Zhuang, J.Q. The 12 May Wenchuan earthquake-induced landslide lakes: Distribution and preliminary risk evaluation. *Landslides* **2009**, *6*, 209–223. [CrossRef]
17. Carling, P.A. Freshwater megaflood sedimentation: What can we learn about generic processes? *Earth Sci. Rev.* **2013**, *125*, 87–113. [CrossRef]
18. Carling, P.A.; Fan, X. Particle comminution defines megaflood and superflood energetics. *Earth Sci. Rev.* **2020**, *204*, 103087. [CrossRef]

19. Turzewski, M.D.; Huntington, K.W.; LeVeque, R.J. The Geomorphic Impact of Outburst Floods: Integrating Observations and Numerical Simulations of the 2000 Yigong Flood, Eastern Himalaya. *J. Geophys. Res Earth.* **2019**, *124*, 1056–1079. [CrossRef]
20. Teller, J.T.; Leverington, D.W.; Mann, J.D. Freshwater outbursts to the oceans from glacial Lake Agassiz and their role in climate change during the last deglaciation. *Quat. Sci. Rev.* **2002**, *21*, 1–887. [CrossRef]
21. Garcia-Castellanos, D.; Estrada, F.; Jimenez-Munt, I.; Gorini, C.; Fernandez, M.; Verges, J.; De Vicente, R. Catastrophic flood of the Mediterranean after the Messinian salinity crisis. *Nature.* **2009**, *462*, 778–781. [CrossRef]
22. Burr, D.M.; Carling, P.A.; Baker, V.R. *Megaflooding on Earth and Mars*; Cambridge University Press: Cambridge, UK, 2009.
23. Anacona, P.I.; Mackintosh, A.; Norton, K. Reconstruction of a glacial lake outburst flood (GLOF) in the Engano Valley, Chilean Patagonia: Lessons for GLOF Risk management. *Sci. Total Environ.* **2015**, *527–528*, 1–11. [CrossRef] [PubMed]
24. Bohorquez, P.; Cañada-Pereira, P.; Jimenez-Ruiz, P.J.; del Moral-Erencia, J.D. The fascination of a shallow-water theory for the formation of megaflood-scale dunes and antidunes. *Earth Sci. Rev.* **2019**, *193*, 91–108. [CrossRef]
25. George, D.L. Adaptive finite volume methods with well-balanced Riemann solvers for modeling floods in rugged terrain: Application to the Malpasset dam-break flood (France, 1959). *Int. J. Numer. Methods Fluids* **2011**, *66*, 1000–1018. [CrossRef]
26. Swartenbroekx, C.; Zech, Y.; Soares-Frazão, S. Two-dimensional two-layer shallow water model for dam break flows with significant bed load transport. *Int. J. Numer. Methods Fluids* **2013**, *73*, 477–508. [CrossRef]
27. Hou, J.; Liang, Q.; Simons, F.; Hinkelmann, R. A 2D well-balanced shallow flow model for unstructured grids with novel slope source term treatment. *Adv. Water Resour.* **2013**, *52*, 107–131. [CrossRef]
28. Ma, D.J.; Sun, D.J.; Yin, X.Y. Solution of the 2D shallow water equations with source terms in surface elevation splitting form. *Int. J. Numer. Methods Fluids* **2007**, *55*, 431–454. [CrossRef]
29. Liang, Q.; Borthwick, A.G.L. Adaptive quadtree simulation of shallow flows with wet-dry fronts over complex topography. *Comput. Fluids* **2009**, *38*, 221–234. [CrossRef]
30. Rogers, B.; And, M.F.; Borthwick, A.G.L. Adaptive Q-tree Godunov-type scheme for shallow water equations. *Int. J. Numer. Methods Fluids* **2001**, *3*, 247–280. [CrossRef]
31. Moukalled, F.; Mangani, L.; Darwish, M. *The Finite Volume Method in Computational Fluid Dynamics. An Advanced Introduction with OpenFoam®and Matlab®*; Springer: Berlin, Germany, 2016.
32. Toro, E.F. *Shock-Capturing Methods for Free-Surface Shallow Flows*; Wiley: Hoboken, NJ, USA, 2001.
33. Toro, E.F. *Riemann Solvers and Numerical Methods for Fluid Dynamics: A Practical Introduction*; Springer: Berlin, Germany, 2013.
34. Van Leer, B. Towards the ultimate conservative difference scheme, V: Asecond-order sequel to Godunov's method. *J. Comput. Phys.* **1979**, *32*, 101–136. [CrossRef]
35. Harten, A.; Lax, P.D.; van Leer, B. On upstream differencing and Godunov-type schemes for hyperbolic conservation laws. *SIAM Rev.* **1983**, *25*, 35–61. [CrossRef]
36. Cea, L.; Puertas, J.; Vazquezcendon, M. Depth Averaged Modelling of Turbulent Shallow Water Flow with Wet-Dry Fronts. *Arch. Comput. Method E* **2007**, *14*, 303–341. [CrossRef]
37. Hu, P.; Cao, Z.; Pender, G.; Tan, G. Numerical modelling of turbidity currents in the Xiaolangdi reservoir, Yellow River, China. *J. Hydrol.* **2012**, *464*, 41–53. [CrossRef]
38. Liu, L.; Liu, L.; Yang, G.W. Cache performance optimization of irregular sparse matrix multiplication on modern multi-core CPU and GPU. *High Technol. Lett.* **2013**, 339–345. [CrossRef]
39. Wu, W. *Computational River Dynamics*; Crc Press: Boca Raton, FL, USA, 2007.
40. Cao, Z.; Pender, G.; Wallis, S.; Carling, P. Computational dam-break hydraulics over erodible sediment bed. *J. Hydraul. Eng.* **2004**, *130*, 689–703. [CrossRef]
41. Liang, Q. Flood Simulation Using a Well-Balanced Shallow Flow Model. *J. Hydraul. Eng.* **2010**, *136*, 669–675. [CrossRef]
42. Song, L.; Zhou, J.; Li, Q.Q.; Yang, X.L.; Zhang, Y.C. An unstructured finite volume model for dam-break floods with wet/dry fronts over complex topography. *Int. J. Numer. Methods Fluids* **2011**, *67*, 960–980. [CrossRef]
43. Hiver, J. Adverse-slope and slope (bump). In Proceedings of the Concerted Action on Dam Break Modelling: Objectives, Project Report, Test Cases, Civil Engineering Department, Hydraulic Division, Universitè Catholique de Lille, Lille, France, 26–28 June 2000.
44. Liao, C.; Wu, M.S.; Liang, S.J. Numerical simulation of a dam break for an actual river terrain environment. *Hydrol. Processes* **2007**, *21*, 447–460. [CrossRef]
45. Rebollo, T.C.; Nieto, E.D.; Marmol, M.G. A flux-splitting solver for shallow water equations with source terms. *Int. J. Numer. Methods Fluids* **2003**, *42*, 23–55. [CrossRef]
46. Zhou, J.G.; Causon, D.M.; Minghan, C.G. Numerical prediction of dam-break flows in general geometries with complex bed topography. *J. Hydraul. Eng.* **2004**, *130*, 332–340. [CrossRef]
47. Kawahara, M.; Umetsu, T. Finite element method for moving boundary problems in river flow. *Int. J. Numer. Methods Fluids* **1986**, *6*, 365–386. [CrossRef]
48. Brufau, P.; Vázquez-Cendón, M.E.; García-Navarro, P. A numerical model for the flooding and drying of irregular domains. *Int. J. Numer. Methods Fluids* **2002**, *39*, 247–275. [CrossRef]

Article

On Dam Failure Induced Seismic Signals Using Laboratory Tests and on Breach Morphology Due to Overtopping by Modeling

Chi-Yao Hung [1], I-Fan Tseng [1], Su-Chin Chen [1,2] and Zheng-Yi Feng [1,*]

[1] Department of Soil and Water Conservation, National Chung Hsing University, Taichung 40227, Taiwan; cyhung@nchu.edu.tw (C.-Y.H.); d0436694@gmail.com (I.-F.T.); scchen@nchu.edu.tw (S.-C.C.)

[2] Innovation and Development Centre of Sustainable Agriculture (IDCSA), National Chung Hsing University, Taichung 40227, Taiwan

* Correspondence: tonyfeng@nchu.edu.tw

Abstract: Dam models were constructed in an indoor flume to test dam breach failure processes to study the seismic signals induced. A simple dam breach model was also proposed to estimate hydrographs for dam breach floods. The test results showed that when the retrogressive erosion due to seepage of the dam continues, it will eventually reach the crest at the upstream side of the dam, and then trigger overtopping and breaching. The seismic signals corresponding to the failure events during retrogressive erosion and overtopping of the dam models were evaluated. Characteristics of the seismic signals were analyzed by Hilbert–Huang transform. Based on the characteristics of the seismic signals, we found four types of mass movement during the retrogressive erosion process, i.e., the single, intermittent, and successive slides and fall. There were precursor seismic signals found caused by cracking immediately before the sliding events of the dam. Furthermore, the dam breach modeling results coincided well with the test results and the field observations. From the test and modeling results, we confirmed that the overtopping discharge and the lateral sliding masses of the dam are also among the important factors influencing the evolution of the breach. In addition, the widening rate of the breach decreases with decreased discharge. The proposed dam breach model can be a useful tool for dam breach warning and hazard reduction.

Keywords: dam breach; seepage; overtopping; seismic signal; flume test; breach model

Citation: Hung, C.-Y.; Tseng, I.-F.; Chen, S.-C.; Feng, Z.-Y. On Dam Failure Induced Seismic Signals Using Laboratory Tests and on Breach Morphology due to Overtopping by Modeling. *Water* **2021**, *13*, 2757. https://doi.org/10.3390/w13192757

Academic Editor: António Pinheiro

Received: 22 August 2021
Accepted: 1 October 2021
Published: 5 October 2021

1. Introduction

Large landslides induced by rainfall or earthquakes may form landslide dams and inundate upstream areas. If the dam breaches, it will pose a serious threat to the area downstream. Nearly 89% of landslide dam failures are caused by overtopping [1,2]. The large-scale landslide caused by Typhoon Morakot in 2009 in Xiaolin village in southern Taiwan caused over 400 fatalities. It also formed a landslide dam [3–5]. Feng (2012) [6] indicated that the dam breached 1 h and 24 min after its formation according to the seismic signal recorded and the time–frequency spectrum. They estimated the velocity of flood propagation downstream to be 8.3 m/s. The dam breach produced large turbulent flows downstream in a short period of time, causing flooding downstream and the failure of many bridges.

In 1951, heavy rainfall caused a large-scale landslide in Tsaoling, Yunlin County, Taiwan, and a large landslide dam was formed. As a result, 137 army engineers unfortunately sacrificed their lives during the installation of an emergency spillway due to the sudden overtopping failure of the dam [7]. In 1999, a large landslide occurred in Tsaoling again due to the 1999 Chichi earthquake [8]. Five landslide-dammed lakes were subsequently formed, of which three were cleared soon after the landslide. However, the other two were not easily cleared, so were strengthened to improve the stability of the dam and emergency spillways were setup to prevent overtopping erosion [9]. Due to the establishment of emergency spillways that controlled the maximum water storage capacity of the landslide dam, the impacts of four typhoons (Typhoon Bilis in 2000, Torajiin 2001, Nari and Mindulle in 2004) were reduced [10]. Cui et al. (2013) [11] and Zhou et al. (2015) [12] reported that on 8 August 2010, the intense rainfall and cascading failure of landslide dams along two gullies induced a fatal debris flows to Zhouqu County, China, that claimed the lives of 1765 people and damaged infrastructure and many homes. Their preliminary field and experimental studies showed that landslide dam cluster modes (i.e., different dam types and their combination) in upstream gullies accounted for the amplification of the scale of Zhouqu debris flows downstream. Cai et al. (2019) [13] analyzed the cascade dam system using a dam breach analysis model (DB-IWHR) for continuous breaking failure paths. They also created a Bayesian network model to determine the failure probability of the cascade dam system. Říha et al. (2020) [14] also modeled cascade dams and indicated that the peak discharge of a dam cascade system may be underestimated by up to 10% when applying an empirical formula derived for a single dam breach. Shrestha and Nakagawa (2016) [15] studied the large-scale landslide in Nepal that resulted from heavy rainfall on 2 August 2014 and the landslide dam on the Sunkoshi River. The retained water overflowed 36 days after the landslide. However, there was no serious damage downstream and no casualties because an emergency spillway was setup before the overflow and the spillway controlled the overflowing water. From these reviews, it is clear that landslides and dam breaches can cause a large number of casualties and property destruction. Therefore, research on the failure processes of landslide dams and hazard prevention of landslide dams is very important.

Hazard prevention and monitoring of landslide dams are always compulsory. Many scholars have used seismic signals recorded from geophones and/or accelerometers for analyses for the creation of warning systems [16–20]. Because both landslide and dam breach events generate seismic signals, the signals can be faithfully recorded by seismometers, and so can be used for interpreting the processes of landslides and dam breach events. However, the seismic signals of landslide and dam breach events cannot always be successfully recorded due to limitations. Additionally, because natural phenomena cannot be repeated, a series of event data cannot be obtained for analysis. Therefore, researchers have often adopted numerical simulations, outdoor large-scale experiments, and indoor small-scale experiments to conduct research on seismic signals induced by landslide dam failure.

Yan et al. (2020) [21] reconstructed the dynamic behavior of the 2017 landslide event in Xinmo village, China, by using the seismic signal characteristics and discrete element method. They categorized the landslide processes into five stages: stationary, slipping, transition, entrainment–transportation, and deposition stages, according to the characteristics of the seismic signals and time–frequency spectra. They identified the transition stage, which is caused by ancient colluvial materials hindering sliding from upslope. However, as the sliding materials continued to accumulate and produce more downward dragging forces, another larger landslide was triggered. This can be observed from the seismic signal as the amplitude first decreases and then increases at the transition stage. Feng et al. (2017) [22] used PFC coupled with FLAC to simulate the 2009 Xiaolin Village landslide process and compared the results with the seismic signal recorded by a broadband seismometer. They found that the types of movement and terrain significantly affect the seismic signal. Although a numerical simulation can readily reconstruct the failure process and influence zones of a landslide, sometimes it is not easy to select accurate physical parameters; thus, the simulation results may be different from the actual landslide.

Some researchers also choose large-scale outdoor experiments. Yan et al. (2017) [23] monitored the seismic signals generated during an outdoor dam-breach test. According to their results based on time–frequency analysis, the low-frequency band (0–1.5 Hz) was mainly due to dam collapse events; the intermediate-frequency band (1.5–10 Hz) was due to rock slide events; the high-frequency band (10–45 Hz) was a result of water flow and sediment transport. Feng et al. (2020a) [24] conducted a large-scale dam breach test at Huisun Forest Station in Nantou, Taiwan. They discussed the characteristics of seismic signals during the dam breach processes and flooding. The flood speed of the test was also estimated from the seismic signals. They indicated that seismic signals can be applied as a basis for early warnings for floods. As we know, a large-scale outdoor test better reflects natural dam breach behavior than the numerical simulation and the small-scale indoor flume test. However, due to the limitations of outdoor large-scale tests that require a large area, long preparation time, and are relatively more expensive, they are less widely used than indoor flume tests.

Many researchers use small indoor flumes to perform tests for landslide dam breaches [25–28]. Most of the tests explored the erosion of the dam body, including the effects of different flume slopes, dam geometry, and material properties. However, the seismic signals caused by the destruction of the dam are less discussed. Hu et al. (2018) [29] used a small flume to test the seismic signals of internal dam erosions. They found a precursor seismic signal prior to the sliding of dam materials. Seismic signals caused by internal erosion due to seepage were mainly high-frequency. If the dam materials were loosely packed, there were more high-frequency seismic signals induced due to the internal erosion. However, they only discussed the seismic signals of dam failure due to seepage and did not discuss the overtopping dam breach and subsequent flooding. There are seismic precursors detected prior to landslides reported in the literatures, e.g., Poli (2017) [30] and Butler (2019) [31]. Feng et al. (2020b) [32] used an indoor flume to study the seismic signals of landslides caused by riverbank erosion. They also found precursor seismic signals before the riverbank sliding. They classified the river bank sliding into three types: single, intermittent, and successive. The three types correspond to the three different characteristics of seismic signals. They also pointed out that the higher frequency seismic signal decays faster than the lower frequency signal. However, this research did not perform tests for retrogressive erosion of a dam due to seepage and overtopping failure.

This research performed tests for retrogressive erosion of a dam due to seepage and the subsequent overtopping failure of dam models. A theoretical dam breach model was proposed and used to compare the flooding process of the test. The experimental setup was a modified version of the test used by Feng et al. (2020b) [32]. In the tests, accelerometers were installed inside the dam to monitor the seismic signals during the dam failure and to understand the correspondence between the failure processes and characteristics of the seismic signals. Hilbert–Huang transform [33] was used for time–frequency analysis for the seismic signals recorded. The setting of the test conditions in this study is not a simple overtopping failure but is similar to the progressive (retrogressive) failure of an outdoor dam breach test by Takayama et al. (2021) [34]. The progressive failure was mainly induced when the dam body was retrogressively eroded towards the upstream crest by seepage. Initially, only small slides and erosions occurred at the toe of the slope, and then as the phreatic surface of the seeping water gradually rose, an increasing number of slides and erosions occurred from the toe towards the crest. At this point, overtopping occurs and the dam starts to breach with vertical downcutting and lateral erosion. The overtopping flood gradually expands the width of the breach, and then a larger amount of floodwater is discharged downstream.

During overtopping, the breach is widened and deepened by the overflowing water. To model the dam breach process, Wu (2011) [35] listed different model approaches and pointed out that modeling can be classified as: (1) parametric breach models and (2) physically based breach models. Parametric breach models usually use statistical regression equations based on laboratory experiments or field dam failure cases. Physically based breach models were highly developed during the past decades and can simulate the dam breach process in a more complete and detailed way. However, the models require heavy numerical calculation requiring extensive calculation time. In addition, these detailed simulation models can be limited due to a lack of understanding of sediment transport under the flow conditions and require multiple runs to calibrate. Unlike the detailed simulation,

Alhasan et al. (2015) [36] proposed a conceptual model, simplified the three-dimensional problem to a one-dimensional problem, and successfully compared the results with field observations. To simulate the horizontal expansion of the dam breach, Tian et al. (2021) [37] proposed a model to combine a theory of sediment transport for vertical incision and a horizontal expansion model based on geotechnical theory. Similarly, the framework proposed in this research is based on the simplified analytical dam breach model by Capart (2013) [38]. We considered the sediment mass conservation Exner equation [39] and a simple sediment transport law [40] to describe the changing of the dam and channel bed profile. We then use this proposed dam breach model to predict the discharge, the height of the crest, and width of the breach for our test. Comparison and analyses were made between the dam breach model calculations and the test results to verify the feasibility of the model.

The major purposes of this study are to discuss (1) the seismic signal precursors prior to the sliding of the dam, (2) the types of movement of the sliding mass of the dam during the retrogressive erosion due to seepage, and (3) the dam breach model proposed and its comparison with the test results.

In this study, only the most representative test result was chosen for presentation; however, many tests were performed and similar results were obtained.

2. Materials and Methods

2.1. Test Configuration

The laboratory flume is shown in Figure 1, which is the same equipment used in Feng et al. (2020b) [32]. The size of the flume is 15 × 0.6 × 0.6 m and the slope of the channel bed is 0.1%. The pump is mainly used to pump water from the underground storage tank to the headwater. Water is introduced from the headwater into the water tank and controlled by a sluice (a valve). The inflow is regulated by a screening device before flowing into the flume channel to enable stable flow conditions. The design of the water supply setup is similar to that of Alhasan et al. (2016) [41] in that the water was stilled/regulated before entering the flume channel. Dimensions of the dam model are listed in Table 1. The slope of the dam is 1:1 (45°) before water impounding. After the construction of the dam model, it was left to sit for 1 h before the test. This study assumes that the right side of the flume is an axis of symmetry; therefore, an overflow notch of 0.05 m depth was set on the right side of the flume. The material used to construct the dam model was sieved uniform sand with a median particle diameter of D_{50} = 1.5 mm, D_{10} = 0.9 mm, and D_{90} = 12.1 mm. The unit weight of the dam material averaged 13.93 kN/m^3 and the density of solid particles ρ_s = 2583 kg/m^3. The void ratio was 0.816 and porosity 45%. The initial moisture content of the dam materials was measured as 5–8%. The initial internal friction angle of the dam material was estimated to be 38–40°. A sand bed with a length of 3.8 m and a thickness of 0.05 m was placed downstream of the dam (Figure 1). Figure 2 shows the side and front view of the dam model.

The dam model was instrumented with sensors including 4 accelerometers, 4 piezometers, and 2 moisture sensors. Figure 3 shows the configuration of the sensors in the dam model and are numbered for identification. The locations of the sensors were selected so that they are not washed out during the tests. Therefore, they were mostly installed on the left side of the dam, with the exception of PP-3 and PP-4. The piezometers were installed close to the bottom of the dam to reflect pore pressure. Moisture sensors 1 and 2 were installed higher, at 0.2 and 0.3 m above the bottom to detect when the seepage water reached those levels. The accelerometers were of the Type 731A produced by Wilcoxon Sensing Technologies with a response frequency between 0.1 and 450 Hz and sensitivity of 10 V/g. The sampling rate was set at 5.12 kHz for the accelerometers to record seismic signals. The piezometers were of the Type KPE-200KPB made by Tokyo Measuring Instruments Laboratory Co, Ltd., and were installed 3 cm above the bottom of the dam model to trace pore pressure (PP) variation during the test. The moisture sensors were of the Type EC-5 produced by METER Group, Inc., and were installed at two different levels to monitor variation in volumetric water content (VWC).

Three cameras were set up at the right side, front, and top of the dam model to record the test process. Because the shooting angle of the sideview file was skewed, for subsequent analyses we used a projection to estimate the water level changes and dam dimensions during the breach.

Brief test process: When water was released, the upstream water level gradually increased until the maximum water level was reached. The maximum water level was controlled by the sluice and maintained at 0.3 m until overtopping. The pump was turned off at 116 s for better seismic signal quality. At 130 s, a tapping was made to leave a time marker, which was used to match the time axis between seismic signals and the test videos. After water seeped into the dam and outflowed at the downstream toe, retrogressive erosion and landslides then started. When the retrogressive erosion reached the upstream crest of the dam, overtopping occurred. A breach was then formed, down cut, and widened. The detailed test process and results are discussed in Section 3.

Table 1. The dimensions of the dam model.

Slope (°)	Height (m)	Crest Width (m)	Bottom Length (m)	Channel Width (m)	Maximum Water Level (m)
45	0.4	0.35	1.15	0.6	0.3

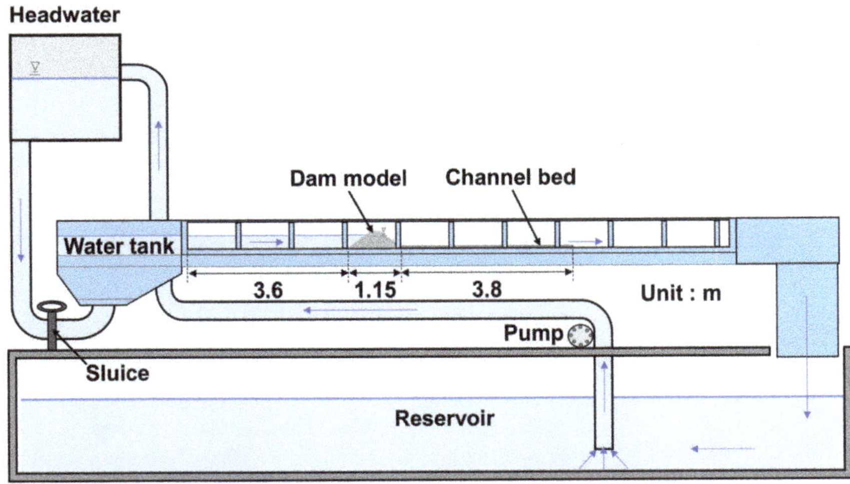

Figure 1. Layout of the test flume and the dam model.

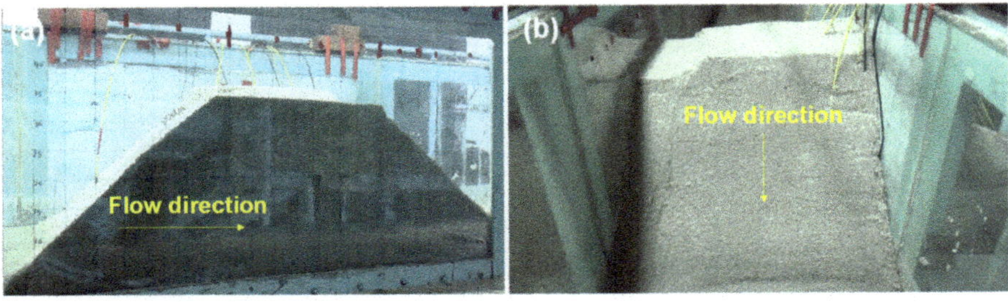

Figure 2. (**a**) The side view of the dam model and (**b**) front view of the dam model.

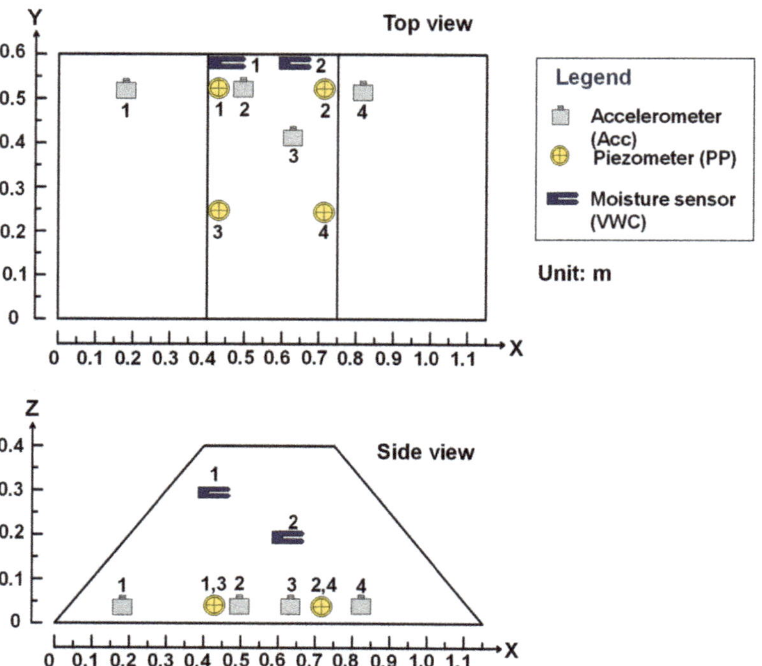

Figure 3. Sensor configuration in the dam model (the numbering of the sensors is marked near each sensor).

2.2. Seismic Signal Processing by Hilbert–Huang Transform (HHT)

The seismic signals recorded during the tests were processed using Hilbert–Huang transform (HHT, Huang et al. 1998) [33]. HHT includes empirical mode decomposition (EMD) to calculate the intrinsic mode functions (IMFs) and a Hilbert transform (HT) to obtain the corresponding time–frequency spectra from the IMFs. HHT can process unsteady and nonlinear signals and analyze the relationship between time, seismic frequency, and energy distribution of signals. The HHT analyses in this study were performed by Visual Signal Ver. 1.6 software (AnCad, Inc. (2018) [42]). The characteristics of the seismic signals due to various sliding events can be identified more easily with the help of time–frequency spectra. The HHT was also applied in Feng et al. (2020b) [32] and Feng et al. (2020a) [24] to successfully interpret flood and landslide events.

2.3. Dam Breach Model—Overtopping

As we described in the previous section, the test can be separated into two stages. In the first stage, the seepage water flows out of the downstream surface of the dam, causing retrogressive erosion. Due to the retrogressive erosion, the shape of the dam deforms from a trapezoid to triangle and the breach process leads to the second stage. In the second stage, the water overtops the crest and begins the overtopping process. In the second stage, the outflow from the crest dominates the breach process and reduces the water level in the lake.

To simulate the overtopping incision process and compare it with the test results, we propose a simplified dam breach model based on Capart (2013) [38]. In this model, we neglect the discharge due to seepage and simulate the breaching as a continuous process. As illustrated in Figure 4, we assumed a triangular-shaped dam with upstream slope R_D, downstream slope S_D, and initial elevation of the dam z_D; $z_C(t)$ is the crest level and

$z_L(t)$ is the lake level. The initial crest level $z_C(t=0)$ is equal to z_D. The initial lake level $z_L(t=0)$ before overtopping is assumed to be maintained at a steady level.

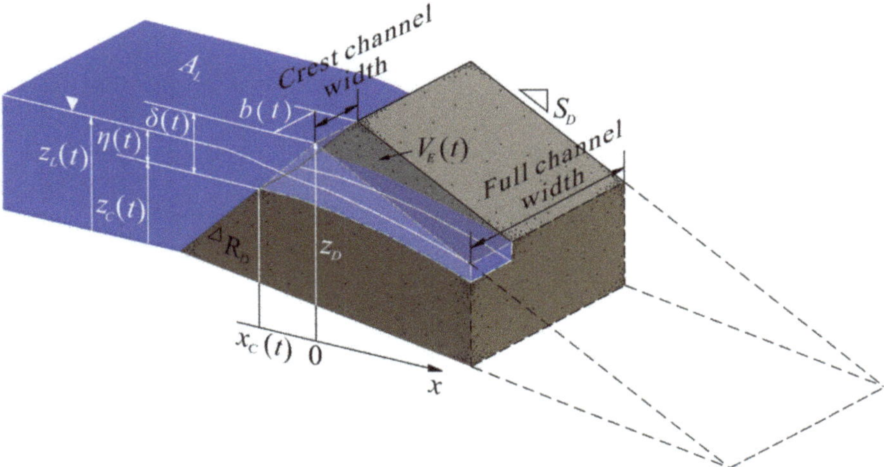

Figure 4. Schematic and parameters of the proposed dam breach model.

The Exner equation governs the breach process based on the sediment mass balance (Paola and Voller 2005) [43].

$$b\frac{\partial z}{\partial t} + \frac{\partial J}{\partial x} = 0 \tag{1}$$

where b is the channel width, z is the bottom elevation of the breach channel, x is the streamwise direction, and J is the sediment transport rate. In this equation, Paola and Voller (2005) [43] simply demonstrated the mass of bedload balancing in a controlled volume; when sediment influx $J(x)$ is larger than outflux $J(x+\Delta x)$ the elevation of the sediment in the control volume increases. For the sediment transport rate, Visser (1995) [44], Alhasan et al. (2016) [41], and Haddadchi et al. (2013) [45] collected different empirical sediment transport formulas used in sand–dike breach erosion. The formulas were verified with experimental results or field cases. However, the formulas contained too many detailed variables and coefficients (e.g., internal friction coefficient, bed shear velocity), which increases the complexity of the model [46]. In the 1950s, Lane (1955) [40] presented a qualitative law of sediment transport rate, which demonstrated J can be generally scaled by the water flux and the channel gradient as:

$$J = KQS = -KQ\frac{\partial z}{\partial x} \tag{2}$$

where K is a dimensionless transport coefficient; Q is the local discharge (the discharge through the breach); S is the channel gradient, which can be written as the derivative of channel elevation in the streamwise direction. By substituting (2) into (1), a variable rate diffusion equation can be obtained. Next, we assumed the outflow channel at the toe of the dam converges to the initial shape of the dam, and the level of the crest level evolves along with the water level of the lake. A zero-sediment flux is assumed at the crest ($J(x_C(t),t) = 0$) as the second upstream boundary condition [47,48]. By defining operational time $d\tau(t) = KQ(t)/b$, the varying rate diffusion equation can be reduced to a standard diffusion equation:

$$\frac{\partial z}{\partial \tau} - \frac{\partial^2 z}{\partial x^2} = 0 \tag{3}$$

For this dam breach problem (triangle dam with moving upstream boundary condition), a similar structure of the solution can be found in Capart et al. (2007) [47], Voller et al. (2004) [49], and Lai and Capart (2007, 2009) [50,51], and the detailed derivation can be found in Capart (2013) [38]. By using the boundary conditions, the profile of the dam was solved as:

$$z(x,\tau) = z_D - S_D x - \frac{(R_D + S_D)\lambda_s}{ierfc(-0.5\lambda_s)}\sqrt{\tau}\, ierfc\left(-\frac{1}{2}x/\sqrt{\tau}\right) \tag{4}$$

where S_D and R_D represent the downstream slope and upstream slope of the dam, and $ierfc(\xi)$ is a special function introduced by Carslaw and Jaeger (1959) [52]:

$$ierfc(\xi) = \int_{\xi}^{\infty} erfc(x)dx = \frac{1}{\sqrt{\pi}}exp\left(-\xi^2\right) - \xi erfc(\xi) \tag{5}$$

where $erfc(\xi)$ is the complementary error function and λ_s is a constant related to the shape of the dam. By taking the upstream sediment flux boundary condition at the dam crest position, λ_s can be solved numerically in Equation (6):

$$\frac{0.5\lambda_s\, erfc(-0.5\lambda_s)}{ierfc(-0.5\lambda_s)} - \frac{S_D}{S_D + R_D} = 0 \tag{6}$$

Following the profile of the dam, focusing on the dam crest, the drop of crest $\delta(\tau)$ can also be written as:

$$\delta(\tau) = z_D - z(x_C(\tau),\tau) = R_D\lambda_s\sqrt{\tau} \tag{7}$$

We replaced the operational time τ with the real time, and the time evolution of the breach drop can be given by the ODE:

$$\frac{d\delta(t)}{dt} = \frac{1}{2}\frac{KQ}{b}\frac{R_D^2\lambda_s^2}{\delta(t)} \tag{8}$$

V_E indicates the erosion volume at the crest during the overtopping process. Here, we find that V_E can be scaled as the drop of the crest and the width of the width $\delta^2 b$. To include the widening effect and simplify the governing equation, we rewrite the equation as:

$$\frac{dV_E(t)}{dt} = \frac{d}{dt}\left(\delta^2 b\right) = K_T Q = (K_V + K_L)Q \tag{9}$$

where K_T is the scaled coefficient of the total sediment transport coefficient; K_L and K_V represent the coefficients of the erosion rate of the dam in lateral and vertical directions, respectively. To separate the process of vertical incision and lateral erosion, we applied chain rules to simplify the partial differential equation (PDE) (Equation (9)) into two ordinary differential equations (ODEs):

$$\frac{d\delta(t)}{dt} = \frac{K_V Q(t)}{2b(t)\delta(t)}\ \frac{db(t)}{dt} = \frac{K_L Q(t)}{\delta(t)^2} \tag{10}$$

To simulate the lake drainage, the level-pool routing equation [53] is adopted:

$$A_L\frac{dz_L}{dt} = -Q \tag{11}$$

where A_L is the lake area, which we assume constant during breaching. For the outflow discharge Q, the broad-crested discharge equation is used:

$$Q = \sqrt{\frac{8g}{27}}b\eta(t)^{3/2} \tag{12}$$

where $\eta(t) = z_L(t) - z_C(t)$ represents flow depth at the crest. By substituting the broad-crested-weir discharge equation (Equation (12)) into the level-pooling routing equation, Equation (11) can be rewritten as:

$$\frac{d\eta(t)}{dt} = \frac{d\delta(t)}{dt} - \frac{1}{A_L}\sqrt{\frac{8g}{27}}b\eta(t)^{3/2} \tag{13}$$

Now, the three ODEs (the two ODEs from Equation (10) and the ODE of Equation (13)) with the three variables, $\eta(t)$, $\delta(t)$, and $b(t)$, were successfully derived. The forward Euler method, a first-order numerical procedure for solving ODEs, was applied to calculate the solution of the dam breach model with the following initial conditions:

$$\delta(0) = 0,\ \eta(0) = 0,\ b(0) = b_0 \tag{14}$$

The water level of the lake is assumed to be the same height of the crest as the initial condition at $t = 0$, and b_0 is the initial channel width.

The proposed dam breach model is then successfully derived and can be solved numerically. This model is simple because we only have to solve the simpler ODEs instead of the PDEs. The model results were compared with the field observations from the literature and test result in Section 3.

3. Results and Discussions

3.1. Measurements of the Test

In the discussion of the test results hereafter, the term "tank" was used instead of "lake". They both represent the water level at the upstream side of the dam in this study. The test processes included tank level raising, water seeping into the dam, sliding of the dam due to retrogressive erosion, overtopping, breach downcutting, and widening and lateral sliding of the dam. Table 2 lists the timing, test stages, selected important events, and variation of the seismic signals of the test processes. We defined T_0 = 76 s as the time when the upstream water reached the toe of the dam. Sliding Events 1–6 occurred between 180.6 and 231.9 s, of which Events 1–3 occurred during the rising of the tank level and Events 4–6 occurred when the tank level was maintained at 0.3 m. Events 1–6 are discussed in detail later in Section 3.2. There were also many slides of different magnitudes that occurred during 260–592 s, but they were not discussed in this study. T_6 = 592 s is defined as the timing of overtopping and thus the breach downcutting and widening processes started as Event 7. There are six lateral slides of the dam detected during Event 7. We monitored the test until 930 s.

Table 2. Test processes and the selected important events.

Time, s	Event	Variation of the Seismic Signals
0–116	The flow pump was opened and the sluice was opened.	Frequency of the flow pump: 106 Hz, 227 Hz
76	T_0 = 76 s when water reached the upstream toe of the dam model; the water continued to fill the tank up to 0.3 cm.	-
116	The flow pump was closed; the sluice remained opened.	Frequency of the environment: 101 Hz;
130	A tapping was made to leave a time marker.	Amplitudes of the signals at 130 s were very high.
180.6–183.5	T_1 =180.6 s; Event 1 and 2 precursor signals; one smaller crack and one larger crack developed prior to a slide (see Supplementary Materials Video S1)	Two precursor seismic signals occurred prior to a slide; the 1st with lower energy; the 2nd with higher energy.
184–185.6	T_2 = 184 s; a single slide (Event 3) occurred (see Supplementary Materials Video S2)	Seismic signals significantly increased
189–207	T_3 = 189 s; an intermittent slide (Event 4) occurred (see Supplementary Materials Video S3); retrogressive erosion	Seismic signals significantly increased nearly periodically in about 2 to 5 s.
221–251	T_4 = 221 s; a successive slide (Event 5) occurred (see Supplementary Materials Video S4); retrogressive erosion	Seismic signals significantly increased aperiodically.
309.5–310	T_5 = 309.5 s; a fall (or collapse) (Event 6) occurred (see Supplementary Materials Video S5)	Seismic signals suddenly increased and the duration was very short.
412–510	Many single slides with different magnitudes occurred; retrogressive erosion	Seismic signals significantly increased with different corresponding amplitudes
592–930	T_6 = 592 s; Overtopping process (Event 7) (see Supplementary Materials Videos S6 and S7); the sluice was then closed; six lateral slides occurred.	Seismic signals significantly increased when the six lateral slides occurred. Strong spectral traces appear around 220–420 Hz in the time–frequency spectrum due to the overtopping water flow.
930	The end of the test	

Figure 5 shows the tank water level, the signals measured by the accelerometers, piezometers, and water content sensors. The tank level gradually increased from T_0 = 76 s (Figure 5a). When the water reached the dam slope, some shallow surface sands started to slide down after 150 s as the water level rose. The slope of the upstream face of the dam was slightly reduced to about 40°, but no serious deep sliding occurred. During the test, the upstream face of the dam was still stable. The pore pressure of PP-1 started to increase at about T = 100 s (Figure 5d). The volumetric water content measured by VWC-2 began to increase at T = 176 s (Figure 5c) because the seepage water had reached the level of VWC-2 at 0.2 m. The elevation (0.3 m) of VWC-1 was higher than that of VWC-2 so that there was about 10 s lag at T = 166 s for VWC-1 to detect the seepage water. When overtopping occurred at T_6 = 592 s, the pore pressures and volumetric water contents began to gradually decrease simultaneously. Therefore, we confirmed that the piezometers and the moisture sensors can accurately reflect the rising and falling of the water inside the dam.

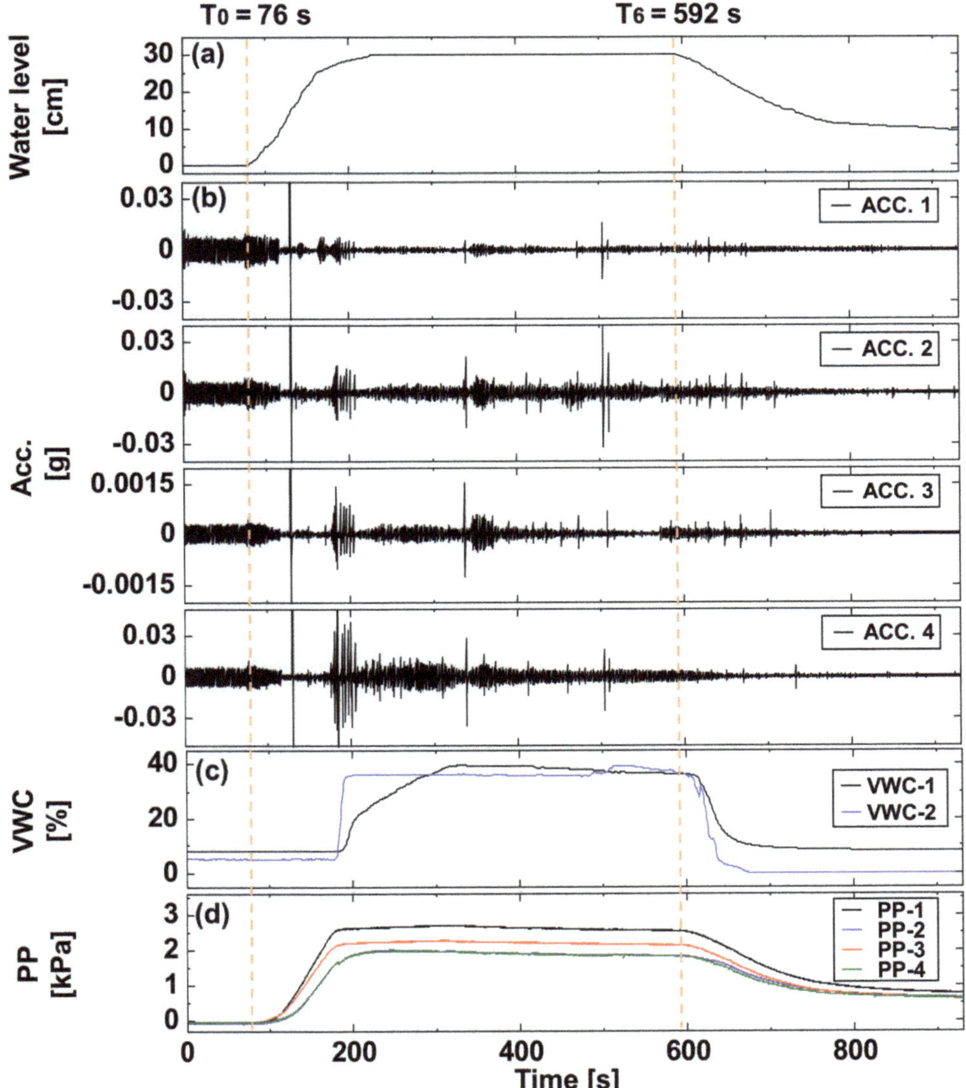

Figure 5. Test measurements: (**a**) water level of the tank, (**b**) seismic signal of Acc. 2, (**c**) volumetric water content, (**d**) pore pressure.

Referring to Figure 6, six side-view images were captured to show the overall test processes: (a) at 72 s (T_0), the water reached the toe of the dam. The tank level was still zero. (b) At 150 s, the tank level had increased and water seeped into the dam. Once the water seeped out of the downstream surface of the dam, retrogressive erosion commenced and induced slides that gradually eroded the dam towards the upstream crest until overtopping. (c) At 300 s, the tank level was controlled at 0.3 m. Retrogressive erosion continued with many slides at the downstream slope of the dam. (d) At 592 s (T_6), overtopping occurred and the dam became a triangular shape. (e) At 650s, the tank level was reduced to 0.24 m and the breach was downcut and widened due to overtopping flow. Lateral slides also occurred. (f) At 930 s, the test was ended. Overtopping flow stopped. The tank level was the same as the crest level.

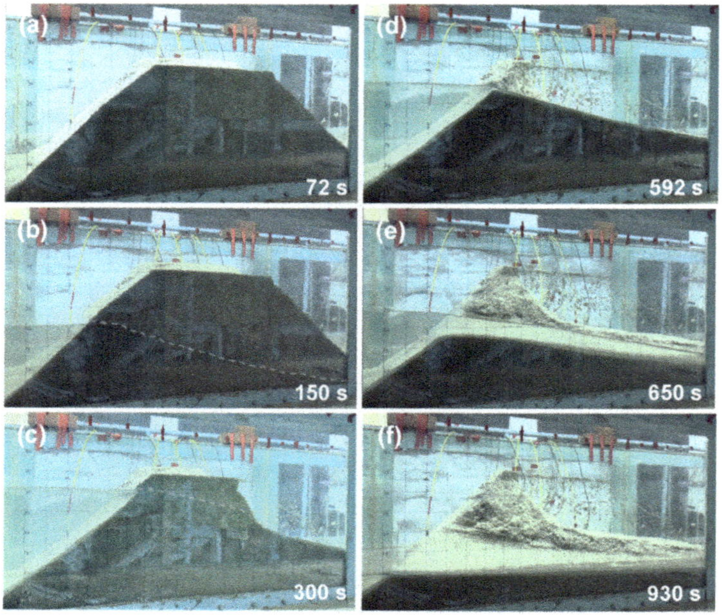

Figure 6. Side-view images at different test stages.

3.2. Seismic Signals Due to Retrogressive Erosion before Overtopping

While the dam experienced retrogressive erosion, many slides occurred with different types of movements. Based on the seismic signals and with the help of the test videos (Supplementary Materials Videos S1–S5), we identified four types of movements from the test result. In addition, precursor seismic signals were found prior to a slide. These can be useful in categorizing landslide types and prewarning based on the seismic signals in hazard prevention work. There were many slides that occurred during retrogressive erosion; however, we only chose six typical events, Events 1–6 for illustration.

Figure 7 shows two seismic signals (Events 1 and 2) induced by two cracks, Cracks 1 and 2, the corresponding time–frequency spectrum, and the top view of Cracks 1 and 2. The occurrence time of the two signals was 180.6 s (T_1) to 181.3 s and 182.8 to 183.5 s, respectively. The two signals are recognized as precursor signals prior to a slide, because a large slide, Event 3, occurred immediately after these two signals. The major seismic frequency of Event 1 and 2 was analyzed as 374 Hz, as shown in Figure 7b, and heavier energy traces in dark red can be observed near 374 Hz. This very high frequency is likely due to the particulate sands sliding on or colliding with each other during the crack development. The duration of Event 1 is about 0.5 s corresponding to the shorter crack length of Crack 1, while the duration of Event 2 is longer at 0.7 s corresponding to the longer Crack 2. From these results, we know that seismic signals induced by precursor events can be properly recorded and very useful for application to landslide prewarning.

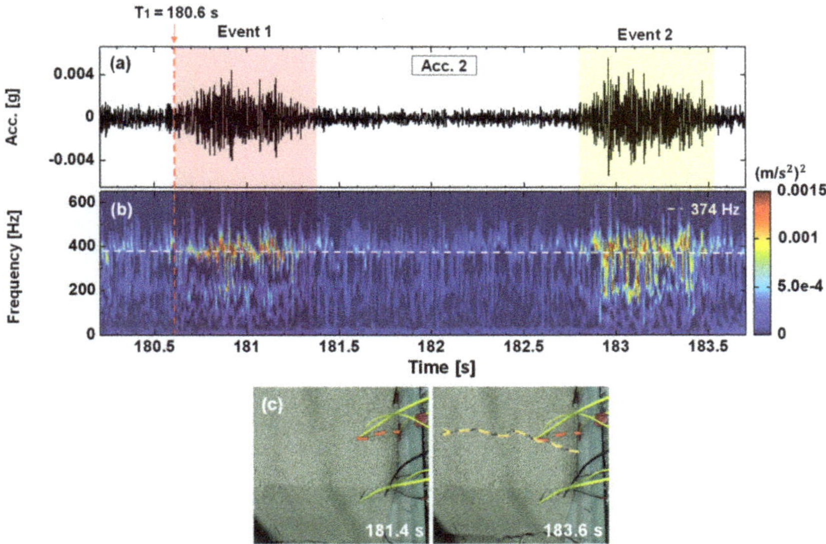

Figure 7. Events 1 and 2, the precursors prior to the Event 3 slide: (**a**) seismic signal, (**b**) time–frequency spectrum, (**c**) top view of the Crack 1 in red dashed line and Crack 2 in orange dashed line.

From the results of the retrogressive erosion process, we categorized four types of mass movements: (1) single slide, (2) intermittent slide, (3) successive slide, and (4) fall. Events 3–6 correspond to these four types of movements, respectively.

(1) A single slide—Event 3: A single slide is defined as a rapid one-time sliding that displaces a large volume of mass [54]. Strong seismic energy is often released. Event 3 was a single slide that occurred immediately after the precursor Events 1 and 2 during 184 s (T_2) to 185.6 s. The seismic signal, time–frequency spectrum, and the front and top view of Event 3 are shown in Figure 8. The amplitudes and spectral traces were larger and stronger than other events in this test due to a larger displaced mass and displacement. The duration of Event 3 was also longer than other events. It is noted that the average frequency of the seismic signal of Event 3 was 375 Hz.

(2) An intermittent slide—Event 4: An intermittent slide is defined as the same mass sliding down multiple times on the same rupture surface [54]. An intermittent slide usually generates nearly periodic seismic signals. Event 4, an intermittent slide, occurred during 189 s (T_3) ~ 207 s, which was about 3 s after Event 3. The displaced mass of the first two movements was the same mass in Event 3 and moved short distances on the same rupture surface, while another deeper rupture surface developed and part of the displaced mass of the third–fifth movements slid on it (see Supplementary Materials Video S3 for Event 4). The corresponding seismic signal, spectrum, and images of Event 4 are presented in Figure 9. For every 2–5 s there was a movement in this event. The nearly periodic seismic signals and spectral traces can easily be identified.

(3) A successive slide—Event 5: A successive slide is defined as the many irregularly and randomly occurrences of sliding on multiple rupture surfaces and the displaced masses are from many different locations [54]. A successive slide usually generates aperiodic or random seismic signals. Event 5, a successive slide, occurred during 221 s (T_4) ~ 251 s. In Figure 10, the signal shown was induced by random sliding of many different masses from place to place. The seismic signal was much more irregular without a pattern and the intervals between slides were random.

(4) A fall—Event 6: A fall is that mass falls down through the air and falls almost vertically. Figure 11 shows the fall, Event 6, which occurred at 309.5 s (T_5). The duration of Event 6 was short. The amplitudes of the signal were small and the energy traces in the spectrum were also weak owing to the fact that only a small mass fell in Event 6.

The movement types of single, intermittent, and successive slides identified in this study also occurred in the riverbank erosion tests performed by Feng et al. (2020b) [32] and the landslide tests by Feng and Chen (2021) [54]. Their failure types and failure mechanism are consistent with the results found in this study. Besides these three types of movement, we presented the fourth type, fall, which should also be helpful when correlating seismic signals to landslide events. All four types of movements are very common in retrogressive erosion due to seepage out of a dam and can be easily and correspondingly identified from induced seismic signals.

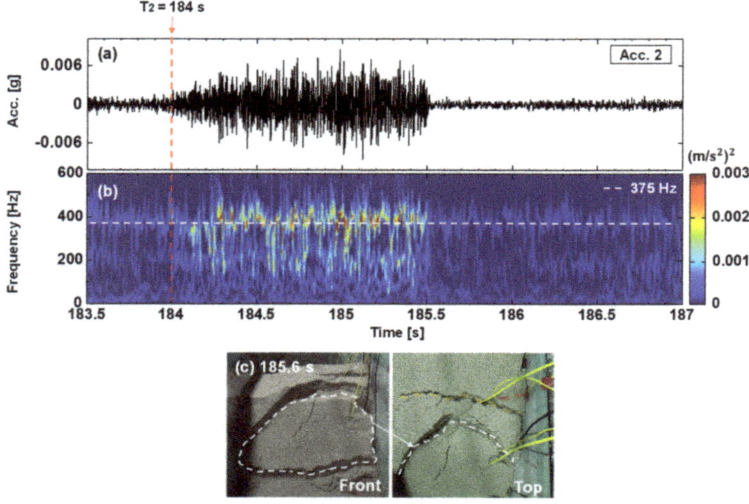

Figure 8. Event 3, the single slide: (**a**) seismic signal, (**b**) time–frequency spectrum, (**c**) front and top view of the single slide.

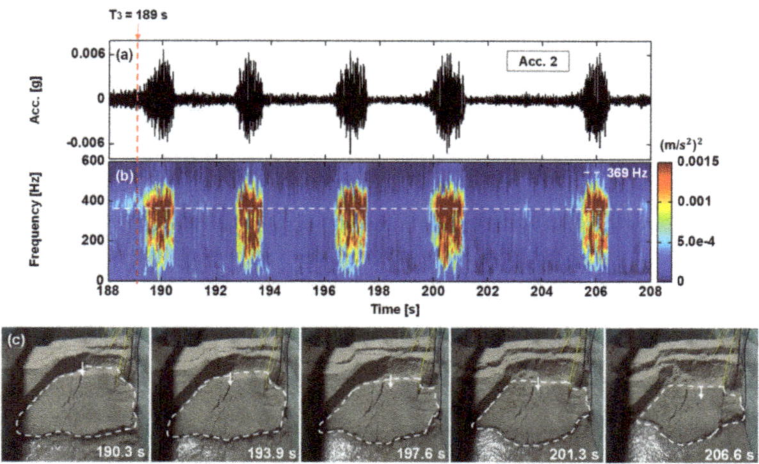

Figure 9. Event 4, the intermittent slide: (**a**) seismic signal, (**b**) time–frequency spectrum, (**c**) series of front views the intermittent slide.

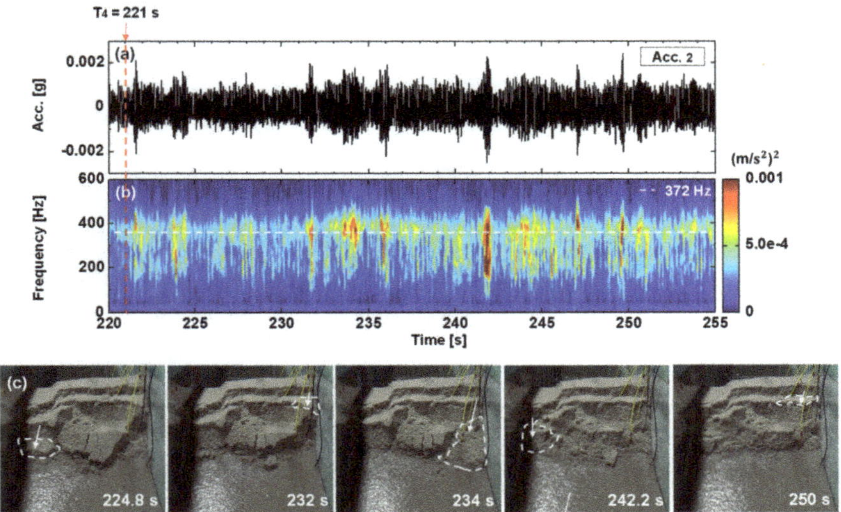

Figure 10. Event 5, the successive slide: (**a**) seismic signal, (**b**) time–frequency spectrum, (**c**) series of front views of the successive slide.

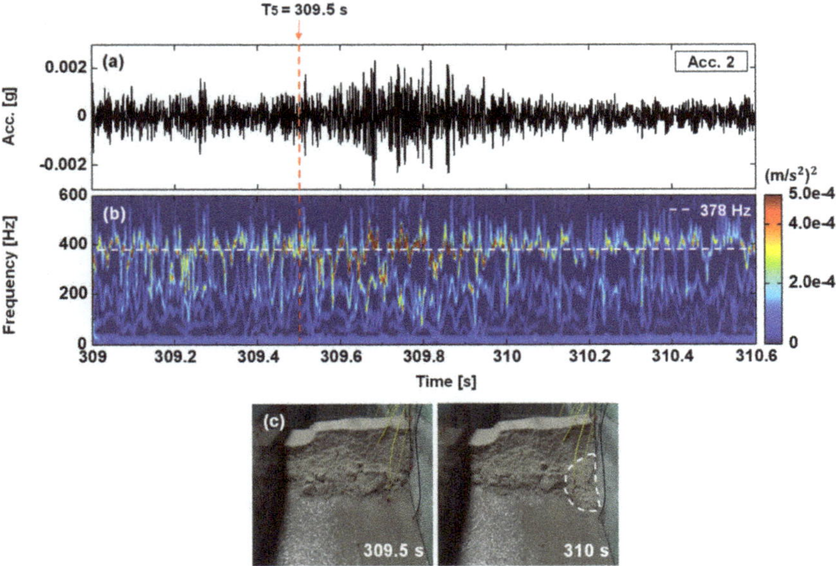

Figure 11. Event 6—the fall (**a**) seismic signal, (**b**) time–frequency spectrum, (**c**) front view images of the fall.

3.3. Comparison of the Dam Breach Model Result with the Breach Events from Literature

The model-predicted discharge was first verified with field measurements from literature and the test results. We reviewed four field dam breach cases: (1) Tangjiashan landslide dam breach in China [55], (2) Lake Ha! Ha! breakout flood in Canada [56], and (3) two lahar dam breaches from Mapanuepe Lake in the Philippines [57]. The essential information is shown in Table 3.

Table 3. Information on dam breach events from literature [38].

Dam/Lake	Date	A_L (10^6 m^2)	b (m)	Q_p (10^3 m^3/s)	Source
Mapanuepe	25–27 August 1991 12–16 October 1991	6.7 6.7	70 70	0.65 0.39	Umbal and Rodolfo (1996) [57]
Ha!Ha!	19–22 July 1996	6	90	0.85	Capart et al. (2007) [56]
Tangjiashan	10–11 June 2008	6.4	110	6.5	Liu et al. (2010) [55]

To compare the model-predicted results with the field measurements, it is convenient to present the data in dimensionless form. Since it is difficult to identify the point in time when the overtopping reached peak discharge (T_P) from the field, we adopted the approximate equation from Capart (2013) [38] as Equation (8):

$$T_P = \frac{3A_L}{(gb^2Q_P)^{1/3}} \tag{15}$$

where Q_P is the field observed peak discharge, A_L is the dammed lake area, and b is the breach channel width. The dimensionless hydrographs recorded are shown in Figure 12. The hydrographs from these different field measurements are reasonably close.

For the breach model, there are three critical parameters to control the shape and the magnitude of the hydrograph: λ, K_V, and K_L. To eliminate the complexity of the model, we used dimensionless hydrographs to compare with that of the test result. In that case, the hydrograph is only controlled by the ratio between the coefficients of lateral erosion (K_L) and the vertical erosion (K_V). We varied the ratio (K_L/K_V) between 0 and 1. When the ratio was zero, only vertical erosion was considered. When equal erodibility in lateral and vertical directions is considered equal, i.e., their ratio is 1 (unity). The model-predicted results are shown in Figure 12 with different grayscale lines for the different coefficient ratios of erosion K_L/K_V. The field observation cases of Table 3 are plotted with lines and symbols for comparison. Two hydrographs are shown for the Tangjiashan event based on the data shown by Liu et al. (2010) [55], in which they used two approaches to estimate the hydrographs. The hydrography estimated from the test is represented by the blue solid curve.

In Figure 12, the predicted hydrograph becomes narrower when the ratio of K_L/K_V is gradually increased. The narrower curve represents the stronger lateral erosion strength. By comparing the field measurement data, most of the hydrographs showed agreement with the model prediction when the K_L/K_V ratio ranged from 0.5 to 1. Therefore, we can say that the lateral erosion of the dam should be considered in modeling; the lateral expansion could affect the shape of the hydrograph.

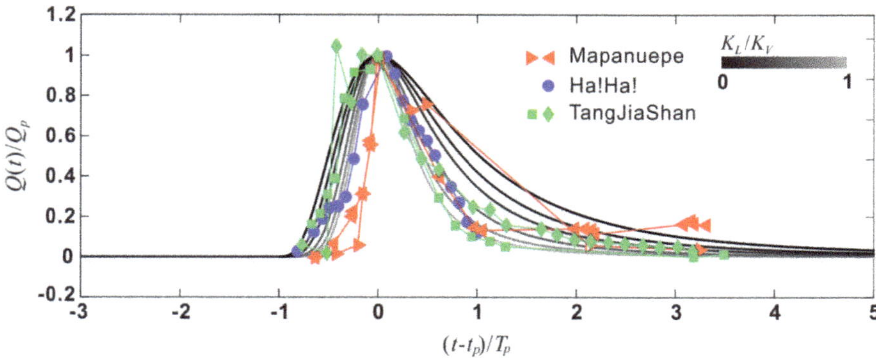

Figure 12. Dimensionless discharge hydrographs of the field-observation cases (the lines with symbols) and model-predicted hydrographs with different coefficient ratio of K_L/K_V (grayscale lines).

3.4. Comparision of the Dam Breach Model Result with the Test Result after Overtopping (Event 7)

To further verify the performance of the proposed dam breach model, the model result was compared with the test results. Table 4 lists the parameters used in the breach model to simulate the breach morphology of the test. To calibrate the coefficients in the model, we simply assigned the initial values of the channel width, tank level, and crest height, and then used the model to compute the values in the final stage. According to the differences between the model predictions and the test measurements, we adjusted the coefficients that showed the minimum difference. Although the calibration of the coefficients is simple in this study, the predictability of the model on breach evolution is still satisfactory.

Table 4. Parameters for the proposed dam breach model for the test dam.

Parameters/Unit	Value	Source
z_{D0} [m]	0.3	Measured from the initial dimensions of the test dam
z_{C0} [m]	0.3	
z_{L0} [m]	0.3	
b_0 [m]	0.015	
A_L [m^2]	2.22	
dt [s]	0.005	Numerical calculation time step
K_V	1.875	Calibrated from the test result
K_L/K_V	0.1375	

The modeled result is presented in Figure 13 alongside Event 7 of the test result. The videos of Event 7 can be seen in Supplementary Materials Videos S6 and S7. Figure 13a,b displays the seismic signal and its time–frequency spectrum, respectively, after the overtopping (T6 = 592 s) for the test result. Figure 13c shows the spectral magnitude cross-sectional profile of 366 Hz from the spectrum in Figure 13b. The dominant seismic frequency of the signals during 592~930 s was 366 Hz. There were six noticeable amplitude and magnitude peaks, and they are marked by S1–S6 at 607, 616, 632, 651, 670, and 677 s in Figure 13a–c. They were induced by the six lateral slides of the dam during the breach. Note that the magnitude peaks in Figure 13c are more easily identified than the amplitude peaks in Figure 13a,b when corresponded to the six slides. Figure 14 shows the before-and-after photos of the six lateral slides of the dam.

In the early stages of the breach before 651 s, we can observe that the magnitudes of the lateral slides S1 to S4 gradually increased (Figure 13c) with increasing discharge. This indicates that the magnitudes of these lateral slides of the dam became increasingly larger (Figure 14a–d). As a result, vertical downcutting of the breach was more significant than lateral erosion in the early stage of the test such that the toe of the dam in the lateral direction was lowered and induced the subsequent larger lateral slides.

Figure 13d shows changes in lake level $z_L(t)$ and crest level $z_C(t)$. Figure 13e shows the variation of the width of the breach $b(t)$. From the test results in Figure 13d,e, we know that the tank level did not greatly decrease until 630 s, and the lowering of the tank level became prominent after 630 s. However, the lowering rate of the crest level and widening rate of the width were relatively high before 630 s, indicating that vertical downcutting and lateral erosion were quite strong.

The materials of the lateral slides (S1–S6) of the dam had an influence on the width of the breach. The width of the breach was reduced by 4.1, 2, 2.9, 0.8, 0.45, and 0.1 cm, correspondingly. When the slides occurred, the sliding materials partially blocked the breach, causing a reduction in the width of the beach. The breach was gradually opened wider again while the slid materials were taken by the overtopping flow downstream. Based on the test result (Figure 13d,e), the width evolution gradually becomes stable after the final sliding S6 at 677 s. However, the vertical downcutting of the crest was still active and caused the crest level z_C to decrease until 750 s. After 750 s, the crest level became stable.

Generally, if the volume of a lateral slide is larger, the reduction in the width of the breach should be more observable. However, as a result of the test, the reduction of the width of the breach did not necessarily become larger when the volume of the lateral slides became larger. For example, when slide S1 occurred (Figure 14a), although the volume of the slide was not large, the width of the breach was significantly reduced. This is because the overtopping flow was small at that time and the transport of the slid materials was small, resulting in more slid material accumulating in the breach. When slide S3 occurred (Figure 14c), the slid volume was large, but the width change of the breach was not obvious, due to the large discharge and because the transport of the slid materials was significant such that the accumulation of slid materials in the breach was greatly reduced. Therefore, there is no obvious correlation between the change in the width of the breach and the volume of the slid materials in the test. This is mainly because larger discharges will transport more slid materials. Therefore, it may not be accurate to use the width of the breach to estimate the overtopping discharge when many lateral slides of the dam occur. In addition, we can find that after about 645 s, the lowering rate of tank level and crest level gradually decreased. The rate of change in the width of the breach from 651 to 677 s gradually decreased. Additionally, as the volume of the lateral slides gradually decreased, the influence on the width of the breach was also less; e.g., slide S5 caused marginal changes in breach width. After slide S6, the width of the breach was stable and no longer changed, and furthermore, sliding events no longer occurred.

We then compared the morphological evolutions of the breach between the model and test results. From Figure 13d, the lake level and the crest level results of the model and test results are in generally good agreement, with the exception of a slight difference at the early stage before 645 s. The difference is due to the test starting from retrogressive erosion due to seepage and then overtopping. The test dam's strength was weaker than that of the breach model. In other words, many parts of the materials of the test dam model were formed from the very loose slid materials of the retrogressive slides. This situation is not considered in the model. In the model, we only considered the overtopping process. Therefore, the downcutting erosion rate in the test was higher during the early stage than those set in the model.

In Figure 13e, the evolution of the breach channel width of the model results shows a smoothly increasing curve. In contrast, from the test result, the evolution of the breach width was not smooth and the breach width was reduced when the four lateral slides S1–S4 occurred. These discrepancies were caused by the masses of lateral sliding materials not being considered in the model; i.e., the lateral sediment influx was not considered in the model equations. However, the model result fits the trends of the test result qualitatively well with the exception of the breach width reduction kinks in the curves. Additionally, the small pulses from the lateral slides did not affect the continuous evolution of the crest and the tank level. Therefore, to estimate the dam breach discharge, we could use the lateral erosion (or width expansion) to represent the lateral slides, especially for the loose materials.

In this study, due to the complexity of the test measurement, we did not measure the discharge directly but estimated the discharge during the breach by two different approaches. In the first approach, we considered the mass balance in the lake and assumed the lake area was maintained at a constant size during the breach and thus the discharge can be simply estimated using Equation (16) by differentiating the lake level with respect to time and multiplying by the lake area.

$$Q_{Test}(t) = -A_L \frac{dz_L(t)}{dt} \tag{16}$$

The estimated discharge hydrograph of the test is shown with the thin red curve in Figure 13f. The curve is obviously zigzag in shape. This is because the discharge is estimated by the differential of the lake level. Small changes in the tank level cause large changes in the discharge; for example, the significant change in the hydrograph during t = 700~716 s, as shown in Figure 13f.

Therefore, we developed the second approach to estimate the discharge hydrograph for the test. Similar to the model derivation in Section 3, we applied the broad-crested-weir discharge equation again to estimate the discharge hydrograph of the test using Equation (17):

$$Q_{Test}(t) = c_D\sqrt{\frac{8g}{27}}b(t)\eta(t)^{3/2} \tag{17}$$

where c_D is a discharge coefficient to be determined. In previous research, Pařílková et al. (2012) [58] and Imanian et al. (2021) [59] both pointed out the choice of c_D could be highly affected by the roughness of the crest and different hydraulic head ratios. However, it is difficult to measure the roughness of the crest in the dam breach experiments, and the coefficients in our dam breach tests were not comparable with those dams without breach in the literature [58,59]. To calibrate the coefficients, we considered the tank level becoming stable in the late stage of the test, and thus used the estimated hydrograph of the late stage derived from Equation (16) to calibrate c_D; c_D was set as 0.75, with $b(t)$ and $\eta(t)$ from the test measurement. The resultant hydrograph is shown in Figure 13f as the thin blue curve. To discuss in further detail, we also considered the moving average of the early stage of the test and estimate c_D as 0.5. The result is shown as the dashed blue curve. However, we found that the peak discharge of the hydrograph obtained by the second approach (Equation (17)) with c_D = 0.75 was close to the peak discharge obtained by the first approach (Equation (16)), i.e., the thin blue curve and thin red curve in Figure 13f, respectively. In contrast, the peak discharge of the hydrograph using c_D = 0.5 was smaller than that obtained by Equation (16) (dashed blue curve and thin red curve, respectively). The hydrograph estimated with a lower coefficient may underestimate the real peak discharge due to the moving averaging process. Overall, the two hydrographs are smoother and more stable than those obtained by Equation (16). To more accurately estimate the discharge, the second approach that applied the broad-crested-weir discharge equation with c_D = 0.75 is preferred.

In Figure 13f, the three hydrographs estimated for the test using the two approaches show an increase in the discharge, maintain at a fairly "stable" discharge, and then a decrease with time, which resembles a trapezoidal shape. In contrast, the modeled discharge hydrograph (thick blue line in Figure 13f) shows a higher peak discharge at about 645 s, and this curve is similar to a bell shape without a stable discharge. The rise and decline of the hydrographs match the general trends of the estimated hydrographs from the test result.

From the perspective of disaster prevention, we hope to determine the timing of the peak discharge as preparation for early warning. Therefore, it is important to know when the peak discharge of a dam breach occurs. The timing of the peak discharge of this model is at around 645 s, which approximately matches the result of the test result. The dam breach model we proposed is simple and its estimation of hydrographs, lake and crest levels, and widths of the breach are equivalent to the test results. Therefore, it has the potential to be extended and applied to dam-breach assessment and early warning in actual situations.

From the perspective of dam breach warning and hazard reduction, the timing and magnitude of peak discharge arrival are both very important factors. From the test results, we found that when the timing of peak discharge was approaching, the lateral slides occurred more frequently and the slid masses increased. If we apply this finding to the real world, the frequency of lateral slides and their magnitudes can be indicated by seismometers and cameras to give early warning on the timing of peak discharge of a dam breach. In addition, the peak discharge and hydrograph can be calculated through the proposed dam breach model with variations of the breach width and flow depth at the crest. This model shows consistency in comparison with the field observation data and test results, and the model can be a useful tool in dam breach warning and hazard reduction.

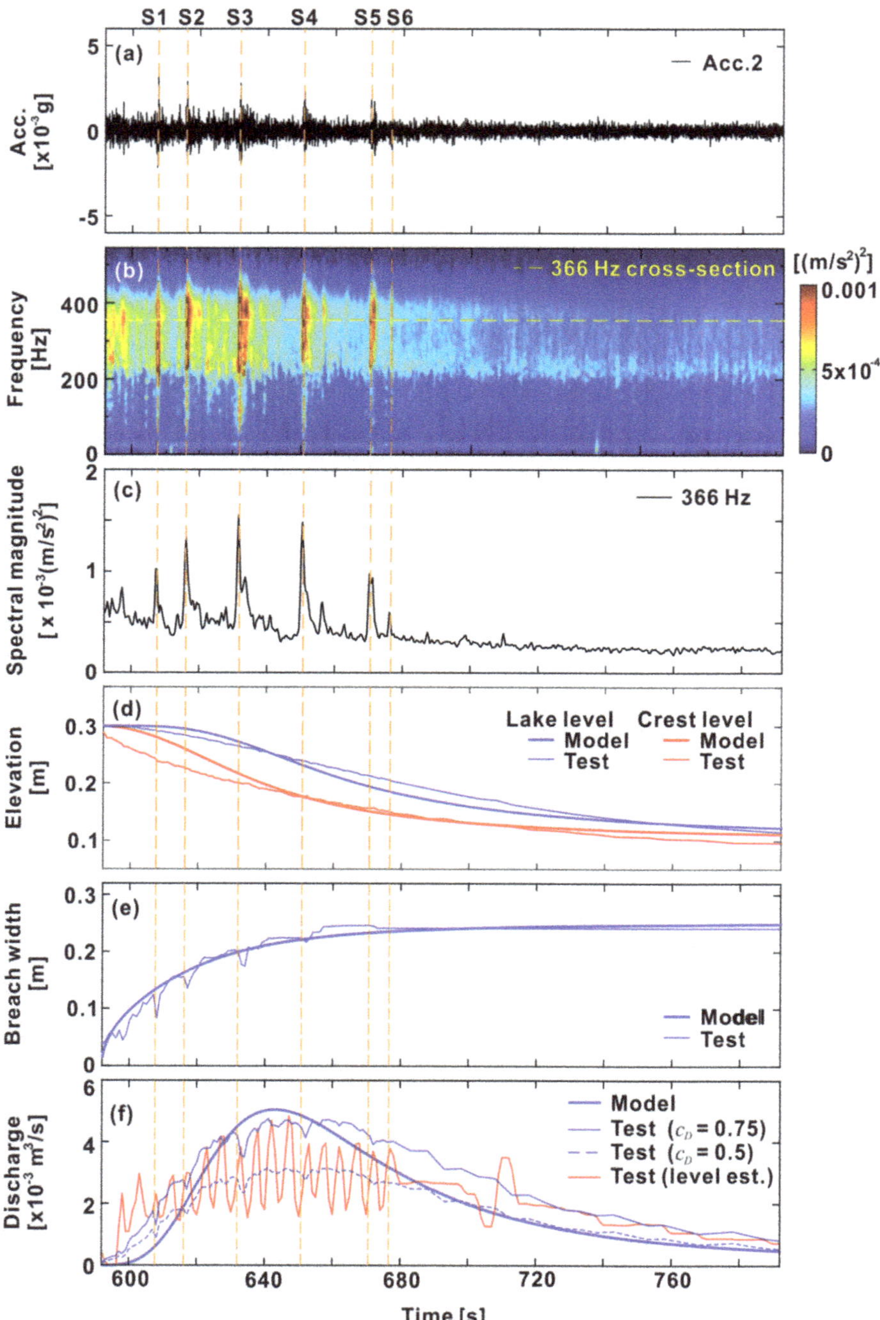

Figure 13. Test and model results of breaching: (**a**) seismic signal of Acc. 2, (**b**) time–frequency spectrum of the seismic signal of Acc. 2, (**c**) spectral magnitude profile of 366 Hz cross-section shown in Figure 13b, (**d**) lake (tank) and crest levels of the model and test results, (**e**) breach width (crest channel width), (**f**) discharge hydrographs of the test and model.

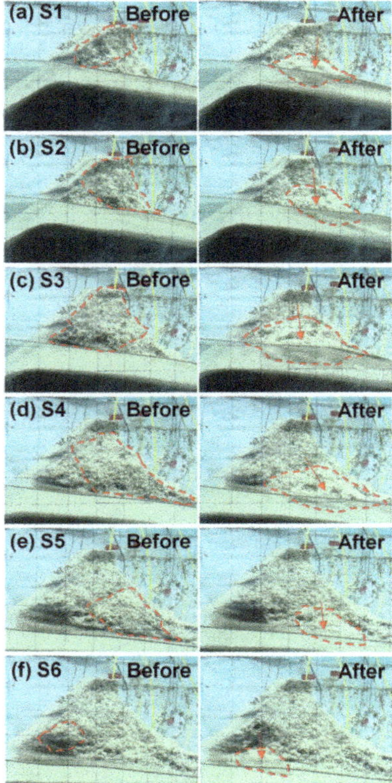

Figure 14. Before-and-after photos of the six lateral slides (S1–S6) of the dam at 607, 616, 632, 651, 670, and 677 s. The red dashed lines indicate the slid areas.

4. Conclusions

This study constructed test dam models in an indoor flume to examine dam failure processes with seismic signal monitoring. A simple dam breach model was proposed and used to compare the flood process of the test. We explored the seismic signals corresponding to the sliding events during retrogressive erosion due to seepage and breaching. The monitored seismic signals corresponded clearly to the sliding events. As retrogressive erosion continued, the erosion eventually reached the crest at the upstream side of the dam, and then triggering overtopping and breaching. We verified satisfactorily this simple dam breach model by using our test result together with field observations from the literature.

Precursor seismic signals generated by cracking prior to the sliding events of the dam model were detected. Further research is strongly recommended on how to extend this observation for dam safety monitoring and prewarning to the real world. Based on the characteristics of the seismic signals, we found four types of mass movements during the retrogressive erosion process, i.e., single, intermittent, and successive slides, and fall. This result is also very useful when categorizing landslide types using seismic signals.

Overtopping discharge and lateral sliding masses of the dam are also among the important factors influencing the evolution of the breach. The masses of the lateral slides will suddenly reduce the width of the breach, but the overtopping flow will transport the masses downstream, making the width of the breach wider again. Two approaches that apply the lake level data or the broad-crested-weir discharge equation were used to estimate the hydrograph for the test. The hydrographs, obtained by the approach using the broad-crested-weir discharge equation, showed smoother and more stable hydrograph curves than those obtained by the other approach using lake level data.

The proposed simple dam breach model satisfactorily simulated the hydrograph of dam breaching and successfully assessed the vertical and lateral variations of the breach. The model can be a useful tool to help explain the dam breach process and dam breach prewarning. However, in future study, the mass input from the lateral slides can be further considered in the model.

Supplementary Materials: The following are available online at https://zenodo.org/record/5220558#.YR5PwdMzZTY, Video S1: Event 1 and 2, Video S2: Event 3, Video S3: Event 4, Video S4: Event 5, Video S5: Event 6, Video S6: Event 7 side view, Video S7: Event 7 front view.

Author Contributions: Conceptualization, Z.-Y.F., C.-Y.H. and S.-C.C.; methodology, Z.-Y.F. and C.-Y.H.; validation, Z.-Y.F., C.-Y.H. and I.-F.T.; formal analysis, Z.-Y.F. and C.-Y.H.; investigation, Z.-Y.F., C.-Y.H. and I.-F.T.; resources, S.-C.C.; data curation, C.-Y.H. and I.-F.T.; writing—original draft preparation, Z.-Y.F. and C.-Y.H.; writing—review and editing, Z.-Y.F., C.-Y.H., I.-F.T. and S.-C.C.; visualization, C.-Y.H. and I.-F.T. All authors have read and agreed to the published version of the manuscript.

Funding: This research was funded by Ministry of Science and Technology, Taiwan, R.O.C., grant number 109-2625-M-005-009-MY2.

Institutional Review Board Statement: Not applicable.

Informed Consent Statement: Not applicable.

Data Availability Statement: The data presented in this study are available on request from the corresponding author.

Acknowledgments: The authors acknowledge Hallam Atherton for reviewing the manuscript style.

Conflicts of Interest: The authors declare no conflict of interest.

List of symbols

z	elevation of the breach channel (m)
x	streamwise direction (m)
b	breach channel width (m)
b_0	initial breach channel width (m)
t	time (s)
J	sediment flux (m^3/s)
z_L	water level (elevation) of the lake (m)
z_D	dam initial elevation (before overtopping) (m)
z_C	dam breach crest elevation (m)
x_C	streamwise position of dam breach crest (m)
A_L	lake area (m^2)
V_E	erosion volume of breach crest (m^3)
δ	drop of the dam elevation to the breach crest elevation:$z_D - z_C$ (m)
η	flow depth of the breach crest: $z_L - z_C$ (m)
S_D	downstream slope of the dam (initial) (-)
R_D	upstream slope of the dam (initial) (-)
K	dimensionless sediment transport coefficient (-)
S	breach channel local gradient (-)
K_T	scaled dimensionless sediment transport coefficient (-)
K_V	scaled dimensionless sediment transport coefficient in vertical direction (-)
K_L	scaled dimensionless sediment transport coefficient in lateral direction (-)
Q	local discharge (m^3/s)
Q_p	peak discharge (m^3/s)
Q_{Test}	estimated test discharge (m^3/s)
c_D	dimensionless discharge coefficient (-)
τ	defined operation time (accumulated discharge) (m^2)
λ_s	dimensionless scaling constant, related to the shape of the dam (-)

References

1. Costa, J.E.; Schuster, R.L. The formation and failure of natural dams. *Geol. Soc. Am. Bull.* **1988**, *100*, 1054–1068. [CrossRef]
2. Fan, X.M.; Dufresne, A.; Subramanian, S.S.; Strom, A.; Hermanns, R.; Stefanelli, C.T.; Hewitt, K.; Yunus, A.P.; Dunning, S.; Capra, L.; et al. The formation and impact of landslide dams—State of the art. *Earth Sci. Rev.* **2020**, *203*, 103116. [CrossRef]
3. Chen, S.-C.; Hsu, C.-L.; Wu, T.; Chou, H.-T.; Cui, P. Landslide dams induced by typhoon Morakot and risk assessment. In Proceedings of the 5th International Conference on Debris-Flow Hazards Mitigation: Mechanics, Prediction and Assessment, Padua, Italy, 14–17 June 2011; pp. 653–660.
4. Tsou, C.-Y.; Feng, Z.-Y.; Chigira, M. Catastrophic landslide induced by typhoon Morakot, Shiaolin, Taiwan. *Geomorphology* **2011**, *127*, 166–178. [CrossRef]
5. Wu, C.-H.; Chen, S.-C.; Feng, Z.-Y. Formation, failure, and consequences of the Xiaolin landslide dam, triggered by extreme rainfall from Typhoon Morakot. Taiwan. *Landslides* **2014**, *11*, 357–367. [CrossRef]
6. Feng, Z.-Y. The seismic signatures of the surge wave from the 2009 Xiaolin landslide—Dam breach in Taiwan. *Hydrol. Process.* **2011**, *26*, 1342–1351. [CrossRef]
7. Chen, R.-F.; Chang, K.-J.; Angelier, J.; Chan, Y.-C.; Deffontaines, B.; Lee, C.-T.; Lin, M.-L. Topographical changes revealed by high-resolution airborne LiDAR data: The 1999 Tsaoling landslide induced by the Chi–Chi earthquake. *Eng. Geol.* **2006**, *88*, 160–172. [CrossRef]
8. Chigira, M.; Wang, W.-N.; Furuya, T.; Kamai, T. Geological causes and geomorphological precursors of the Tsaoling landslide triggered by the 1999 Chi-Chi earthquake, Taiwan. *Eng. Geol.* **2003**, *68*, 259–273. [CrossRef]
9. Li, M.-H.; Hsu, M.-H.; Hsieh, L.-S.; Teng, W.-H. Inundation Potentials Analysis for Tsao-Ling Landslide Lake Formed by Chi-Chi Earthquake in Taiwan. *Nat. Hazards* **2002**, *25*, 289–303. [CrossRef]
10. Hsu, Y.-S.; Hsu, Y.-H. Impact of earthquake-induced dammed lakes on channel evolution and bed mobility: Case study of the Tsaoling landslide dammed lake. *J. Hydrol.* **2009**, *374*, 43–55. [CrossRef]
11. Cui, P.; Zhou, G.G.D.; Zhu, X.H.; Zhang, J.Q. Scale amplification of natural debris flows caused by cascading landslide dam failures. *Geomorphology* **2013**, *182*, 173–189. [CrossRef]
12. Zhou, G.G.D.; Cui, P.; Zhu, X.; Tang, J.; Chen, H.; Sun, Q. A preliminary study of the failure mechanisms of cascading landslide dams. *Int. J. Sediment Res.* **2015**, *30*, 223–234. [CrossRef]
13. Cai, W.; Zhu, X.; Peng, A.; Wang, X.; Fan, Z. Flood Risk Analysis for Cascade Dam Systems: A Case Study in the Dadu River Basin in China. *Water* **2019**, *11*, 1365. [CrossRef]
14. Říha, J.; Kotaška, S.; Petrula, L. Dam Break Modeling in a Cascade of Small Earthen Dams: Case Study of the Čižina River in the Czech Republic. *Water* **2020**, *12*, 2309. [CrossRef]
15. Shrestha, B.B.; Nakagawa, H. Hazard assessment of the formation and failure of the Sunkoshi landslide dam in Nepal. *Nat. Hazards* **2016**, *82*, 2029–2049. [CrossRef]
16. Suriñach, E.; Vilajosana, I.; Khazaradze, G.; Biescas, B.; Furdada, G.; Vilaplana, J.M. Seismic detection and characterization of landslides and other mass movements. *Nat. Hazards Earth Syst. Sci.* **2005**, *5*, 791–798. [CrossRef]
17. Suwa, H.; Mizuno, T.; Ishii, T. Prediction of a landslide and analysis of slide motion with reference to the 2004 Ohto slide in Nara, Japan. *Geomorphology* **2010**, *124*, 157–163. [CrossRef]
18. Yamada, M.; Matsushi, Y.; Chigira, M.; Mori, J. Seismic recordings of landslides caused by Typhoon Talas (2011), Japan. *Geophys. Res. Lett.* **2012**, *39*, 13301. [CrossRef]
19. Provost, F.; Malet, J.-P.; Hibert, C.; Helmstetter, A.; Radiguet, M.; Amitrano, D.; Langet, N.; Larose, E.; Abancó, C.; Hürlimann, M.; et al. Towards a standard typology of endogenous landslide seismic sources. *Earth Surf. Dyn.* **2018**, *6*, 1059–1088. [CrossRef]
20. Yan, Y.; Cui, Y.F.; Liu, D.Z.; Tang, H.; Li, Y.J.; Tian, X.; Zhang, L.; Hu, S. Seismic signal characteristics and interpretation of the 2020 "6.17" Danba landslide dam failure hazard chain process. *Landslides* **2021**, *18*, 1–18. [CrossRef]
21. Yan, Y.; Cui, Y.; Guo, J.; Hu, S.; Wang, Z.; Yin, S. Landslide reconstruction using seismic signal characteristics and numerical simulations: Case study of the 2017 "6.24" Xinmo landslide. *Eng. Geol.* **2020**, *270*, 105582. [CrossRef]
22. Feng, Z.-Y.; Lo, C.-M.; Lin, Q.-F. The characteristics of the seismic signals induced by landslides using a coupling of discrete element and finite difference methods. *Landslides* **2017**, *14*, 661–674. [CrossRef]
23. Yan, Y.; Cui, P.; Chen, S.-C.; Chen, X.-Q.; Chen, H.-Y.; Chien, Y.-L. Characteristics and interpretation of the seismic signal of a field-scale landslide dam failure experiment. *J. Mt. Sci.* **2017**, *14*, 219–236. [CrossRef]
24. Feng, Z.-Y.; Huang, H.-Y.; Chen, S.-C. Analysis of the characteristics of seismic and acoustic signals produced by a dam failure and slope erosion test. *Landslides* **2020**, *17*, 1605–1618. [CrossRef]
25. Jiang, X.G.; Wei, Y.W.; Wu, L.; Lei, Y. Experimental investigation of failure modes and breaching characteristics of natural dams. *Geomat. Nat. Hazards Risk* **2018**, *9*, 33–48. [CrossRef]
26. Jiang, X.; Huang, J.; Wei, Y.; Niu, Z.; Chen, F.; Zou, Z.; Zhu, Z. The influence of materials on the breaching process of natural dams. *Landslides* **2018**, *15*, 243–255.
27. Nian, T.-K.; Wu, H.; Li, D.-Y.; Zhao, W.; Takara, K.; Zheng, D.-F. Experimental investigation on the formation process of landslide dams and a criterion of river blockage. *Landslides* **2020**, *17*, 2547–2562. [CrossRef]
28. Zhu, X.; Liu, B.; Peng, J.; Zhang, Z.; Zhuang, J.; Huang, W.; Leng, Y.; Duan, Z. Experimental study on the longitudinal evolution of the overtopping breaching of noncohesive landslide dams. *Eng. Geol.* **2021**, *288*, 106137. [CrossRef]

29. Hu, W.; Scaringi, G.; Xu, Q.; Huang, R.Q. Acoustic Emissions and Microseismicity in Granular Slopes Prior to Failure and Flow-Like Motion: The Potential for Early Warning. *Geophys. Res. Lett.* **2018**, *45*, 10406–10415. [CrossRef]
30. Poli, P. Creep and slip: Seismic precursors to the Nuugaatsiaq landslide (Greenland). *Geophys. Res. Lett.* **2017**, *44*, 8832–8836. [CrossRef]
31. Butler, R. Seismic precursors to a 2017 Nuugaatsiaq, Greenland, earthquake–landslide–tsunami event. *Nat. Hazards* **2019**, *96*, 961–973. [CrossRef]
32. Feng, Z.-Y.; Hsu, C.-M.; Chen, S.-H. Discussion on the Characteristics of Seismic Signals Due to Riverbank Landslides from Laboratory Tests. *Water* **2019**, *12*, 83. [CrossRef]
33. Huang, N.E.; Shen, Z.; Long, S.R.; Wu, M.L.C.; Shih, H.H.; Zheng, Q.N.; Yen, N.C.; Tung, C.C.; Liu, H.H. The empirical mode decomposition and the Hilbert spectrum for nonlinear and non-stationary time series analysis. *Proc. R. Soc. A Math. Phys. Eng. Sci.* **1998**, *454*, 903–995. [CrossRef]
34. Takayama, S.; Miyata, S.; Fujimoto, M.; Satofuka, Y. Numerical simulation method for predicting a flood hydrograph due to progressive failure of a landslide dam. *Landslides* **2021**. [CrossRef]
35. Wu, W. Simplified physically based model of earthen embankment breaching. *J. Hydraul. Eng.* **2011**, *137*, 1549–1564. [CrossRef]
36. Alhasan, Z.; Jandora, J.; Říha, J. Study of dam-break due to overtopping of four small dams in the Czech Republic. *Acta Univ. Agric. Silvic. Mendel. Brun.* **2015**, *63*, 717–729. [CrossRef]
37. Tian, S.; Dai, X.; Wang, G.; Lu, Y.; Chen, J. Formation and evolution characteristics of dam breach and tailings flow from dam failure: An experimental study. *Nat. Hazards* **2021**, *107*, 1–18. [CrossRef]
38. Capart, H. Analytical solutions for gradual dam breaching and downstream river flooding. *Water Resour. Res.* **2013**, *49*, 1968–1987. [CrossRef]
39. Capart, H.; Hsu, J.P.C.; Lai, S.Y.J.; Hsieh, M.-L. Formation and decay of a tributary-dammed lake, Laonong River, Taiwan. *Water Resour. Res.* **2010**, *46*, W11522. [CrossRef]
40. Lane, E.W. The importance of fluvial morphology in hydraulic engineering. *Proc. Am. Soc. Civ. Eng.* **1955**, *81*, 1–17.
41. Alhasan, Z.; Jandora, J.; Říha, J. Comparison of specific sediment transport rates obtained from empirical formulae and dam breaching experiments. *Environ. Fluid Mech.* **2016**, *16*, 997–1019. [CrossRef]
42. AnCAD, Inc. Visual Signal Reference Guide, Version1.6. Available online: http://www.ancad.com.tw/VisualSignal/doc/1.6/RefGuide.html (accessed on 15 July 2021). (In Chinese).
43. Paola, C.; Voller, V.R. A generalized Exner equation for sediment mass balance. *J. Geophys. Res. Earth Surf.* **2005**, *110*, F04014. [CrossRef]
44. Visser, P.J. Application of sediment transport formulae to sand-dike breach erosion. *Oceanogr. Lit. Rev.* **1996**, *43*, 954.
45. Haddadchi, A.; Omid, M.H.; Dehghani, A.A. Bedload equation analysis using bed load-material grain size. *J. Hydrol. Hydromech.* **2013**, *61*, 241–249. [CrossRef]
46. Cao, Z.; Yue, Z.; Pender, G. Landslide dam failure and flood hydraulics. Part II: Coupled mathematical modelling. *Nat. Hazards* **2011**, *59*, 1021–1045. [CrossRef]
47. Capart, H.; Bellal, M.; Young, D.-L. Self-similar evolution of semi-infinite alluvial channels with moving boundaries. *J. Sediment. Res.* **2007**, *77*, 13–22. [CrossRef]
48. Hsu, J.P.C.; Capart, H. Onset and growth of tributary-dammed lakes. *Water Resour. Res.* **2008**, *44*, W11201. [CrossRef]
49. Voller, V.R.; Swenson, J.B.; Paola, C. An analytical solution for a Stefan problem with variable latent heat. *Int. J. Heat Mass Transf.* **2004**, *47*, 5387–5390. [CrossRef]
50. Lai, S.Y.J.; Capart, H. Two-diffusion description of hyperpycnal deltas. *J. Geophys. Res. Space Phys.* **2007**, *112*, F03005. [CrossRef]
51. Lai, S.Y.J.; Capart, H. Reservoir infill by hyperpycnal deltas over bedrock. *Geophys. Res. Lett.* **2009**, *36*, L08402. [CrossRef]
52. Carslaw, H.S.; Jaeger, J.C. *Conduction of Heat in Solids*; Oxford University Press: Oxford, UK, 1959.
53. Henderson, F.M. *Open Channel Flow*; Macmillan: New York, NY, USA, 1966.
54. Feng, Z.-Y.; Chen, S.-H. Discussions on landslide types and seismic signals produced by the soil rupture due to seepage and retrogressive erosion. *Landslides* **2021**, *18*, 2265–2279. [CrossRef]
55. Liu, N.; Chen, Z.; Zhang, J.; Lin, W.; Chen, W.; Xu, W. Draining the Tangjiashan barrier lake. *J. Hydraul. Eng.* **2010**, *136*, 914–923. [CrossRef]
56. Capart, H.; Spinewine, B.; Young, D.; Zech, Y.; Brooks, G.R.; Leclerc, M.; Secretan, Y. The 1996 Lake Ha! Ha! breakout flood, Québec: Test data for geomorphic flood routing methods. *J. Hydraul. Res.* **2007**, *45* (Suppl. 1), 97–109. [CrossRef]
57. Umbal, J.V.; Rodolfo, K.S. The 1991 Lahars of Southwestern Mount Pinatubo and Evolution of the Lahar-Dammed Mapanuepe Lake. In *Fire and Mud: Eruptions and Lahars of Mount Pinatubo, Philippines*; Newhall, C.G., Punongbayan, R.S., Eds.; University of Washington Press: Washington, DC, USA, 1996; pp. 951–970.
58. Pařílková, J.; Říha, J.; Zachoval, Z. The influence of roughness on the discharge coefficient of a broad-crested weir. *J. Hydrol. Hydromech.* **2012**, *60*, 101–114. [CrossRef]
59. Imanian, H.; Mohammadian, A.; Hoshyar, P. Experimental and numerical study of flow over a broad-crested weir under different hydraulic head ratios. *Flow Meas. Instrum.* **2021**, *80*, 102004. [CrossRef]

MDPI
St. Alban-Anlage 66
4052 Basel
Switzerland
Tel. +41 61 683 77 34
Fax +41 61 302 89 18
www.mdpi.com

Water Editorial Office
E-mail: water@mdpi.com
www.mdpi.com/journal/water